THE IMPACTS OF CLIMATE CHANGE ON
HUMAN HEALTH
IN THE UNITED STATES

A Scientific Assessment

THE IMPACTS OF CLIMATE CHANGE ON
HUMAN HEALTH
IN THE UNITED STATES

A Scientific Assessment

U.S. Global Change Research Program

Skyhorse Publishing

To read the full report, go to: health2016.globalchange.gov

Recommended Citation: USGCRP, 2016: *The Impacts of Climate Change on Human Health in the United States: A Scientific Assessment.* Crimmins, A., J. Balbus, J.L. Gamble, C.B. Beard, J.E. Bell, D. Dodgen, R.J. Eisen, N. Fann, M.D. Hawkins, S.C. Herring, L. Jantarasami, D.M. Mills, S. Saha, M.C. Sarofim, J. Trtanj, and L. Ziska, Eds. U.S. Global Change Research Program, Washington, DC, 312 pp. http://dx.doi.org/10.7930/J0R49NQX

First published 2016 by the US Global Change Research Program.
First Skyhorse edition 2018.

Skyhorse Publishing books may be purchased in bulk at special discounts for sales promotion, corporate gifts, fund-raising, or educational purposes. Special editions can also be created to specifications. For details, contact the Special Sales Department, Skyhorse Publishing, 307 West 36th Street, 11th Floor, New York, NY 10018 or info@skyhorsepublishing.com.

Skyhorse® and Skyhorse Publishing® are registered trademarks of Skyhorse Publishing, Inc.®, a Delaware corporation.

Visit our website at www.skyhorsepublishing.com.

10 9 8 7 6 5 4 3 2 1

Library of Congress Cataloging-in-Publication Data is available on file.

Cover design by the US Global Change Research Program.

ISBN: 978-1-5107-2609-3
Ebook ISBN: 978-1-5107-2621-5

Printed in China

April 2016

Dear Colleagues:

On behalf of the National Science and Technology Council and the U.S. Global Change Research Program, I am pleased to share this report, *The Impacts of Climate Change on Human Health in the United States: A Scientific Assessment*. It advances scientific understanding of the impacts of climate change on public health, highlights social and environmental disparities that make some communities particularly vulnerable to climate change, and confirms that climate change is a significant threat to the health of all Americans.

This report was developed by over 100 experts from across the Nation representing eight Federal agencies. I want to thank in particular the efforts of the U.S. Environmental Protection Agency (EPA), the U.S. Department of Health and Human Services (HHS), and the National Oceanic and Atmospheric Administration (NOAA) for leading in the development of this report. It was called for under the President's Climate Action Plan and is a major contribution to the sustained National Climate Assessment process. The report was informed by input gathered in listening sessions and scientific and technical information contributed through open solicitations. It underwent rigorous reviews by the public and by scientific experts inside and outside of the government, including a special committee of the National Academies of Sciences, Engineering, and Medicine.

I applaud the authors, reviewers, and staff who have developed this scientific assessment. Their dedication over the past three years has been remarkable and their work has advanced our knowledge of how human health is impacted by climate change now and in the future.

Combating the health threats from climate change is a top priority for President Obama and a key driver of his Climate Action Plan. I strongly and respectfully urge decision makers across the Nation to use the scientific information contained within to take action and protect the health of current and future generations.

John P. Holdren

Dr. John P. Holdren
Assistant to the President for Science and Technology
Director, Office of Science and Technology Policy
Executive Office of the President

THE IMPACTS OF CLIMATE CHANGE ON HUMAN HEALTH IN THE UNITED STATES
A Scientific Assessment

About the USGCRP Climate and Health Assessment

The U.S. Global Change Research Program (USGCRP) Climate and Health Assessment has been developed to enhance understanding and inform decisions about the growing threat of climate change to the health and well-being of residents of the United States. This scientific assessment is part of the ongoing efforts of USGCRP's sustained National Climate Assessment (NCA) process and was called for under the President's Climate Action Plan.[1] USGCRP agencies identified human health impacts as a high-priority topic for scientific assessment.

This assessment was developed by a team of more than 100 experts from 8 U.S. Federal agencies (including employees, contractors, and affiliates) to inform public health officials, urban and disaster response planners, decision makers, and other stakeholders within and outside of government who are interested in better understanding the risks climate change presents to human health.

The USGCRP Climate and Health Assessment draws from a large body of scientific peer-reviewed research and other publicly available sources; all sources meet the standards of the Information Quality Act (IQA). The report was extensively reviewed by the public and experts, including a committee of the National Academies of Sciences, Engineering, and Medicine,[2] the 13 Federal agencies of the U.S. Global Change Research Program, and the Federal Committee on Environment, Natural Resources, and Sustainability (CENRS).

About the National Climate Assessment

The Third National Climate Assessment (2014 NCA)[3] assessed the science of climate change and its impacts across the United States, now and throughout this century. The report documents climate change related impacts and responses for various sectors and regions, with the goal of better informing public and private decision making at all levels. The 2014 NCA included a chapter on human health impacts,[4] which formed the foundation for the development of this assessment.

TABLE OF CONTENTS

CHAPTERS

ABOUT THIS REPORT

Climate change threatens human health and well-being in the United States. The U.S. Global Change Research Program (USGCRP) Climate and Health Assessment has been developed to enhance understanding and inform decisions about this growing threat. This scientific assessment, called for under the President's Climate Action Plan,[1] is a major report of the sustained National Climate Assessment (NCA) process. The report responds to the 1990 Congressional mandate[5] to assist the Nation in understanding, assessing, predicting, and responding to human-induced and natural processes of global change. The agencies of the USGCRP identified human health impacts as a high-priority topic for scientific assessment.

The purpose of this assessment is to provide a comprehensive, evidence-based, and, where possible, quantitative estimation of observed and projected climate change related health impacts in the United States. The USGCRP Climate and Health Assessment has been developed to inform public health officials, urban and disaster response planners, decision makers, and other stakeholders within and outside of government who are interested in better understanding the risks climate change presents to human health.

The authors of this assessment have compiled and assessed current research on human health impacts of climate change and summarized the current state of the science for a number of key topics. This assessment provides a comprehensive update to the most recent detailed technical assessment for the health impacts of climate change, the 2008 Synthesis and Assessment Product 4.6 (SAP 4.6), *Analyses of the Effects of Global Change on Human Health and Welfare and Human Systems*.[6] It also updates and builds upon the health chapter of the 2014 NCA.[4] While Chapter 1: Introduction: Climate Change and Human Health includes a brief overview of observed and projected climate change impacts in the United States, a detailed assessment of climate science is outside the scope of this report. This report relies on the 2014 NCA[3] and other peer-reviewed scientific assessments of climate change and climate scenarios as the basis for describing health impacts.

Each chapter of this assessment summarizes scientific literature on specific health outcomes or climate change related exposures that are important to health. The chapters emphasize research published between 2007 and 2015 that quantifies either observed or future health impacts associated with climate change, identifies risk factors for health impacts, and recognizes populations that are at greater risk. In addition, four chapters (Temperature-Related Death and Illness, Air Quality Impacts, Vector-Borne Disease, and Water-Related Illness) highlight recent modeling analyses that project national-scale impacts in these areas.

The geographic focus of this assessment is the United States. Studies at the regional level within the United States, analyses or observations in other countries where the findings have implications for potential U.S. impacts, and studies of global linkages and implications are also considered where relevant. For example, global studies are considered for certain topics where there is a lack of consistent, long-term historical monitoring in the United States. In some instances it is more appropriate to consider regional studies, such as where risk and impacts vary across the Nation.

While climate change is observed and measured on long-term time scales (30 years or more), decision frameworks for public health officials and regional planners are often based on much shorter time scales, determined by epidemiological, political, or budgeting factors. This assessment focuses on observed and current impacts as well as impacts projected in 2030, 2050, and 2100.

The focus of this assessment is on the *health impacts* of climate change. The assessment provides timely and relevant information, but makes no policy recommendations. It is beyond the scope of this report to assess the peer-reviewed literature on climate change mitigation, adaptation, or economic valuation or on health co-benefits that may be associated with climate mitigation, adaptation, and resilience strategies. The report does assess scientific literature describing the role of adaptive capacity in creating, moderating, or exacerbating vulnerability to health impacts where appropriate. The report also cites analyses that include modeling parameters that make certain assumptions about emissions pathways or adaptive capacity in order to project climate impacts on human health. This scientific assessment of impacts helps build the integrated knowledge base needed to understand, predict, and respond to these changes, and it may help inform mitigation or adaptation decisions and other strategies in the public health arena.

Climate and health impacts do not occur in isolation, and an individual or community could face multiple threats at the same time, at different stages in one's life, or accumulating over the course of one's life. Though important to consider as part of a comprehensive assessment of changes in risks, many types of cumulative, compounding, or secondary impacts are beyond the scope of this report. Though this assessment does not focus on health research needs or gaps, brief insights gained on research needs while conducting this assessment can be found at the end of each chapter to help inform research decisions.

The first chapter of this assessment provides background information on observations and projections of climate change in the United States and the ways in which climate change, acting in combination with other factors and stressors, influences human health. It also provides an overview of the approaches and methods used in the quantitative projections of health impacts of climate change conducted for this assessment. The next seven chapters focus on specific climate-related health impacts and exposures: Temperature-Related Death and Illness; Air Quality Impacts; Extreme Events; Vector-Borne Diseases; Water-Related Illness; Food Safety, Nutrition, and Distribution; and Mental Health and Well-Being. A final chapter on Populations of Concern identifies factors that create or exacerbate the vulnerability of certain population groups to health impacts from climate change. That chapter also integrates information from the topical health impact chapters to identify specific groups of people in the United States who may face greater health risks associated with climate change.

The Sustained National Climate Assessment

The Climate and Health Assessment has been developed as part of the U.S. Global Change Research Program's (USGCRP's) sustained National Climate Assessment (NCA) process. This process facilitates continuous and transparent participation of scientists and stakeholders across regions and sectors, enabling new information and insights to be synthesized as they emerge. The Climate and Health Assessment provides a more comprehensive assessment of the impacts of climate change on human health, a topic identified as a priority for assessment by USGCRP and its Interagency Crosscutting Group on Climate Change and Human Health (CCHHG) and featured in the President's Climate Action Plan.[1]

Report Sources

The assessment draws from a large body of scientific, peer-reviewed research and other publicly available resources. Author teams carefully reviewed these sources to ensure a reliable assessment of the state of scientific understanding. Each source of information was determined to meet the four parts of the Information Quality Act (IQA): utility, transparency and traceability, objectivity, and integrity and security (see Appendix 2: Process for Literature Review). More information on the process each chapter author team used to review, assess, and determine whether a literature source should be cited can be found in the Supporting Evidence section of each chapter. Report authors made use of the findings of the 2014 NCA, peer-reviewed literature and scientific assessments, and government statistics (such as population census reports). Authors also updated the literature search[7] conducted by the National Institute of Environmental Health Sciences (NIEHS) as technical input to the Human Health chapter of the 2014 NCA.

Overarching Perspectives

Five overarching perspectives, derived from decades of observations, analysis, and experience, have helped to shape this report: 1) climate change is happening in the context of other ongoing changes across the United States and around the globe; 2) there are complex linkages and important non-climate stressors that affect individual and community health; 3) many of the health threats described in this report do not occur in isolation but may be cumulative, compounding, or secondary; 4) climate change impacts can either be amplified or reduced by individual, community, and societal decisions; and 5) climate change related impacts, vulnerabilities, and opportunities in the United States are linked to impacts and changes outside the United States, and vice versa. These overarching perspectives are briefly discussed below.

Global Change Context

This assessment follows the model of the 2014 NCA, which recognized that climate change is one of a number of global changes affecting society, the environment, the economy, and public health.[3] While changes in demographics, socio-economic factors, and trends in health status are discussed in Chapter 1: Introduction: Climate Change and Human Health, discussion of other global changes, such as land-use change, air and water pollution, and rising consumption of resources by a growing and wealthier global population, are limited in this assessment.

Complex Linkages and the Role of Non-Climate Stressors

Many factors may exacerbate or moderate the impact of climate change on human health. For example, a population's vulnerability 1) may be affected by direct climate changes or by non-climate factors (such as changes in population, economic development, education, infrastructure, behavior, technology, and ecosystems); 2) may differ across regions and in urban, rural, coastal, and other communities; and 3) may be influenced by individual vulnerability factors such as age, socioeconomic status, and existing physical and/or mental illness or disability. These considerations are summarized in Chapter 1: Introduction: Climate Change and Human Health and Chapter 9: Populations of Concern. There are limited studies that quantify how climate impacts interact with the factors listed above or how these interactions can lead to many other compounding, secondary, or indirect health effects. However, where possible, this assessment identifies key environmental, institutional, social, and behavioral influences on health impacts.

Cumulative, Compounding, or Secondary Impacts

Climate and health impacts do not occur in isolation and an individual or community could face multiple threats at the same time, at different stages in one's life, or accumulating over the course of one's life. Some of these impacts, such as the combination of high ozone levels on hot days (see Ch. 3: Air Quality Impacts) or cascading effects during extreme events (see Ch. 4: Extreme Events), have clear links to one another. In other cases, people may be threatened simultaneously by seemingly unconnected risks, such as increased exposure to Lyme disease and extreme heat. These impacts can also be compounded by secondary or tertiary impacts, such as climate change impacts on access to or disruption of healthcare services, damages to infrastructure, or effects on the economy.

Societal Choices and Adaptive Behavior

Environmental, cultural, and socioeconomic systems are tightly coupled, and as a result, climate change impacts can either be amplified or reduced by cultural and socioeconomic decisions.[3] Adaptive capacity ranges from an individual's ability to acclimatize to different meteorological conditions to a community's ability to prepare for and recover from damage, injuries, and lives lost due to extreme weather events. Awareness and communication of health threats to the public health community, practitioners, and the public is an important factor in the incidence, diagnosis, and treatment of climate-related health outcomes. Recognition of these interactions, together with recognition of multiple sources of vulnerability, helps identify what information decision makers need as they manage risks.

International Context

Climate change is a global phenomenon; the causes and the impacts involve energy-use, economic, and risk-management decisions across the globe.[3] Impacts, vulnerabilities, and opportunities in the United States are related in complex and interactive ways with changes outside the United States, and vice versa. The health of Americans is affected by climate changes and health impacts experienced in other parts of the world.

GUIDE TO THE REPORT

The following describes the format of the report and the structure of each chapter.

Executive Summary

The Executive Summary describes the impacts of climate change on the health of the American public. It summarizes the overall findings and represents each chapter with a brief overview, the Key Findings, and a figure from the chapter.

Chapters

Key Findings and Traceable Accounts

Topical chapters include Key Findings, which are based on the authors' consensus expert judgment of the synthesis of the assessed literature. The Key Findings include confidence and likelihood language as appropriate (see "Documenting Uncertainty" below and Appendix 4: Documenting Uncertainty).

Each Key Finding is accompanied by a Traceable Account which documents the process and rationale the authors used in reaching these conclusions and provides additional information on sources of uncertainty. The Traceable Accounts can be found in the Supporting Evidence section of each chapter.

Chapter Text

Each chapter assesses the state of the science in terms of observed and projected impacts of climate change on human health in the United States, describes the link between climate change and health outcomes, and summarizes the authors' assessment of risks to public health. Both positive and negative impacts on health are reported as supported by the scientific literature. Where appropriate and supported by the literature, authors include descriptions of critical non-climate stressors and other environmental and institutional context; social, behavioral, and adaptive factors that could increase or moderate impacts; and underlying trends in health that affect vulnerability (see "Populations of Concern" below). While the report is designed to inform decisions about climate change, it does not include an assessment of literature on climate change mitigation, adaptation, or economic valuation, nor does it include policy recommendations.

Exposure Pathway Diagram

Each topical chapter includes an exposure pathway diagram (see Figure 1). These conceptual diagrams illustrate a key example by which climate change affects health within the area of interest of that chapter. These diagrams are not meant to be comprehensive representations of all the factors that affect human health. Rather, they summarize the key connections between climate drivers and health outcomes while recognizing that these pathways exist within the context of other factors that positively or negatively influence health outcomes.

The exposure pathway diagram in Chapter 1: Introduction: Climate Change and Human Health is a high-level overview of the main routes by which climate change affects health, summarizing the linkages described in the following chapters. Because the exposure pathway diagrams rely on examples from a specific health topic area, a diagram is not included in Chapter 9: Populations of Concern, as that chapter describes crosscutting issues relevant to all health topics.

Research Highlights

Four chapters include research highlights: Temperature-Related Death and Illness, Air Quality Impacts, Vector-Borne Disease, and Water-Related Illness. Six research highlight sections across these four chapters describe the findings of recently published quantitative analyses of projected impacts conducted for inclusion in this report. Each analysis is summarized with a brief description of the study's 1) Importance, 2) Objectives, 3) Methods, 4) Results, and 5) Conclusions. The analyses are all published in external peer-reviewed sources, and the full description of modeling methods and findings can be found in those citations. While authors of these analyses were provided with modeling guidance and conferred on opportunities for consistency in approach, no comprehensive set of assumptions, timeframes, or scenarios were applied across modeling analyses. Therefore, these six studies do not represent an integrated modeling assessment. The findings of these analyses are considered as part of the overall assessment of the full body of literature when developing the chapter Key Findings. For more information on modeling methods see Appendix 1: Technical Support Document.

Understanding the Exposure Pathway Diagrams

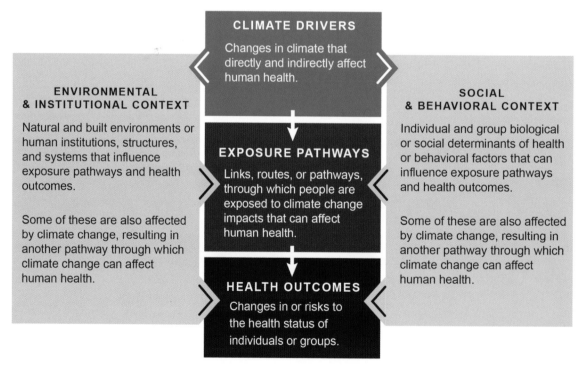

Figure 1: The center boxes include selected examples of climate drivers, the primary pathways by which humans are exposed to health threats from those drivers, and the key health outcomes that may result from exposure. The left gray box indicates examples of the larger environmental and institutional context that can affect a person's or community's vulnerability to health impacts of climate change. The right gray box indicates the social and behavioral context that also affects a person's vulnerability to health impacts of climate change. This path includes factors such as race, gender, and age, as well as socioeconomic factors like income and education or behavioral factors like individual decision making. The examples listed in these two gray boxes can increase or reduce vulnerability by influencing the exposure pathway (changes in exposure) or health outcomes (changes in sensitivity or adaptive capacity). The diagram shows that climate change can affect health outcomes directly and by influencing the environmental, institutional, social, and behavioral contexts of health.

Populations of Concern

One of the main goals of this assessment was to identify populations that are particularly vulnerable to specific health impacts associated with climate change. Each chapter includes discussion of this topic in addition to the full chapter devoted to populations of concern. In these discussions, the authors identify segments of the general population that the peer-reviewed literature has identified as being at increased risk for health-related climate impacts, now or in the future.

Emerging Issues

The Emerging Issues sections briefly describe emerging areas of research including areas of potential future concern; health impacts not currently prevalent or severe in the United States but with potential to become a health concern; or areas where the links between climate change and a human health outcome are in early stages of study and for which a more comprehensive synthesis is outside the scope of this report.

Research Needs

While the goal of this assessment is to highlight the current state of the science on climate impacts on health, research needs identified through the development of this assessment are briefly summarized in each chapter. These research needs could inform research beyond the current state of the science or outside the scope of this report.

Supporting Evidence

The Traceable Accounts supporting each Key Finding are provided at the end of each chapter in the Supporting Evidence section.

Documenting Uncertainty: Confidence and Likelihood

Two kinds of language are used when describing the uncertainty associated with specific statements in this report: confidence language and likelihood language (see table below and Appendix 4: Documenting Uncertainty). Confidence in the validity of a finding is based on the type, amount, quality, strength, and consistency of evidence and the degree of expert agreement on the finding. Confidence is expressed qualitatively and ranges from low confidence (inconclusive evidence or disagreement among experts) to very high confidence (strong evidence and high consensus).

Likelihood language describes the likelihood of occurrence based on measures of uncertainty expressed probabilistically (in other words, based on statistical analysis of observations or model results or based on expert judgment). Likelihood, or the probability of an impact, is a term that allows a quantita-

tive estimate of uncertainty to be associated with projections. Thus, likelihood statements have a specific probability associated with them, ranging from very unlikely (less than or equal to a 1 in 10 chance of the outcome occurring) to very likely (greater than or equal to a 9 in 10 chance).

Likelihood and Confidence Evaluation

All Key Findings include a description of confidence. Where it is considered scientifically justified to report the likelihood of particular impacts within the range of possible outcomes, Key Findings also include a likelihood designation. Confidence and likelihood levels are based on the expert assessment and consensus of the chapter author teams. The author teams determined the appropriate level of confidence or likelihood by assessing the available literature, determining the quality and quantity of available evidence, and evaluating the level of agreement across different studies. For specific descriptions of the process by which each chapter author team came to consensus on the Key Findings and assessment of confidence and likelihood, see the Traceable Account section for each chapter. More information is also available in Appendix 1: Technical Support Document and Appendix 4: Documenting Uncertainty.

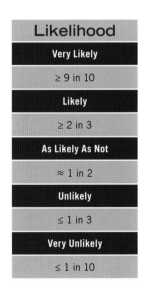

Confidence Level	Likelihood
Very High	**Very Likely**
Strong evidence (established theory, multiple sources, consistent results, well documented and accepted methods, etc.), high consensus	≥ 9 in 10
High	**Likely**
Moderate evidence (several sources, some consistency, methods vary and/or documentation limited, etc.), medium consensus	≥ 2 in 3
	As Likely As Not
Medium	≈ 1 in 2
Suggestive evidence (a few sources, limited consistency, models incomplete, methods emerging, etc.), competing schools of thought	**Unlikely**
	≤ 1 in 3
Low	**Very Unlikely**
Inconclusive evidence (limited sources, extrapolations, inconsistent findings, poor documentation and/or methods not tested, etc.), disagreement or lack of opinions among experts	≤ 1 in 10

LIST OF CONTRIBUTORS

Report Steering Committee

Lead Coordinator
Allison Crimmins, U.S. Environmental Protection Agency

Committee Members
John Balbus, National Institutes of Health
Charles B. Beard, Centers for Disease Control and Prevention
Rona Birnbaum, U.S. Environmental Protection Agency
Neal Fann, U.S. Environmental Protection Agency
Janet L. Gamble, U.S. Environmental Protection Agency
Jada Garofalo, Centers for Disease Control and Prevention
Vito Ilacqua, U.S. Environmental Protection Agency
Lesley Jantarasami, U.S. Environmental Protection Agency
George Luber, Centers for Disease Control and Prevention
Shubhayu Saha, Centers for Disease Control and Prevention
Paul Schramm, Centers for Disease Control and Prevention
Mark M. Shimamoto, U.S. Global Change Research Program, National Coordination Office
Kimberly Thigpen Tart, National Institutes of Health
Juli Trtanj, National Oceanic and Atmospheric Administration

Chapter Authors

Carl Adrianopoli, U.S. Department of Health and Human Services
Allan Auclair, U.S. Department of Agriculture
John Balbus, National Institutes of Health
Christopher M. Barker, University of California, Davis
Charles B. Beard, Centers for Disease Control and Prevention
Jesse E. Bell, Cooperative Institute for Climate and Satellites–North Carolina
Kaitlin Benedict, Centers for Disease Control and Prevention
Martha Berger, U.S. Environmental Protection Agency
Karen Bouye, Centers for Disease Control and Prevention
Terry Brennan, Camroden Associates, Inc.
Joan Brunkard, Centers for Disease Control and Prevention
Vince Campbell, Centers for Disease Control and Prevention
Karletta Chief, The University of Arizona
Tracy Collier, National Oceanic and Atmospheric Administration and University Corporation for Atmospheric Research
Kathryn Conlon, Centers for Disease Control and Prevention
Allison Crimmins, U.S. Environmental Protection Agency
Stacey DeGrasse, U.S. Food and Drug Administration
Daniel Dodgen, U.S. Department of Health and Human Services, Office of the Assistant Secretary for Preparedness and Response

Patrick Dolwick, U.S. Environmental Protection Agency
Darrin Donato, U.S. Department of Health and Human Services, Office of the Assistant Secretary for Preparedness and Response
David R. Easterling, National Oceanic and Atmospheric Administration
Kristie L. Ebi, University of Washington
Rebecca J. Eisen, Centers for Disease Control and Prevention
Vanessa Escobar, National Aeronautics and Space Administration
Neal Fann, U.S. Environmental Protection Agency
Barry Flanagan, Centers for Disease Control and Prevention
Janet L. Gamble, U.S. Environmental Protection Agency
Jada F. Garofalo, Centers for Disease Control and Prevention
Cristina Gonzalez-Maddux, formerly Institute for Tribal Environmental Professionals
Micah Hahn, Centers for Disease Control and Prevention
Elaine Hallisey, Centers for Disease Control and Prevention
Michelle D. Hawkins, National Oceanic and Atmospheric Administration
Mary Hayden, National Center for Atmospheric Research
Stephanie C. Herring, National Oceanic and Atmospheric Administration
Jeremy Hess, University of Washington
Radley Horton, Columbia University
Sonja Hutchins, Centers for Disease Control and Prevention
Vito Ilacqua, U.S. Environmental Protection Agency
John Jacobs, National Oceanic and Atmospheric Administration
Lesley Jantarasami, U.S. Environmental Protection Agency
Ali S. Khan, University of Nebraska Medical Center
Patrick Kinney, Columbia University
Laura Kolb, U.S. Environmental Protection Agency
Nancy Kelly, U.S. Department of Health and Human Services, Substance Abuse and Mental Health Services Administration
Samar Khoury, Association of Schools and Programs of Public Health
Max Kiefer, Centers for Disease Control and Prevention, National Institute for Occupational Safety and Health
Jessica Kolling, Centers for Disease Control and Prevention
Kenneth E. Kunkel, Cooperative Institute for Climate and Satellite–North Carolina,
Annette La Greca, University of Miami
Erin Lipp, The University of Georgia
Irakli Loladze, Bryan College of Health Sciences
Jeffrey Luvall, National Aeronautics and Space Administration
Kathy Lynn, University of Oregon
Arie Manangan, Centers for Disease Control and Prevention
Marian McDonald, Centers for Disease Control and Prevention

Sandra McLellan, University of Wisconsin-Milwaukee

David M. Mills, Abt Associates

Andrew J. Monaghan, National Center for Atmospheric Research

Stephanie Moore, National Oceanic and Atmospheric Administration and University Corporation for Atmospheric Research

Rachel Morello-Frosch, University of California, Berkeley

Joshua Morganstein, Uniformed Services University of the Health Sciences

Christopher G. Nolte, U.S. Environmental Protection Agency

Nicholas H. Ogden, Public Health Agency of Canada

Hans Paerl, The University of North Carolina at Chapel Hill

Adalberto A. Pérez de León, U.S. Department of Agriculture

Carlos Perez Garcia-Pando, Columbia University

Dale Quattrochi, National Aeronautics and Space Administration

John Ravenscroft, U.S. Environmental Protection Agency

Margaret H. Redsteer, U.S. Geological Survey

Joseph Reser, Griffith University

Jennifer Runkle, Cooperative Institute for Climate and Satellites– North Carolina

Josef Ruzek, U.S. Department of Veterans Affairs

Shubhayu Saha, Centers for Disease Control and Prevention

Marcus C. Sarofim, U.S. Environmental Protection Agency

Paul J. Schramm, Centers for Disease Control and Prevention

Carl J. Schreck III, Cooperative Institute for Climate and Satellites– North Carolina

Shulamit Schweitzer, U.S. Department of Health and Human Services, Office of the Assistant Secretary for Preparedness and Response

Mario Sengco, U.S. Environmental Protection Agency

Mark M. Shimamoto, U.S. Global Change Research Program, National Coordination Office

Allan Showler, U.S. Department of Agriculture

Tanya L. Spero, U.S. Environmental Protection Agency

Joel Schwartz, Harvard University

Perry Sheffield, Icahn School of Medicine at Mount Sinai, New York

Alexis St. Juliana, Abt Associates

Kimberly Thigpen Tart, National Institutes of Health

Jeanette Thurston, U.S. Department of Agriculture

Juli Trtanj, National Oceanic and Atmospheric Administration

Robert Ursano, Uniformed Services University of the Health Sciences

Isabel Walls, U.S. Department of Agriculture

Joanna Watson, Centers for Disease Control and Prevention, National Institute for Occupational Safety and Health

Kyle Powys Whyte, Michigan State University

Amy F. Wolkin, Centers for Disease Control and Prevention

Lewis Ziska, U.S. Department of Agriculture

Chapter Coordinators

Allison Crimmins, U.S. Environmental Protection Agency

Jada F. Garofalo, Centers for Disease Control and Prevention

Lesley Jantarasami, U.S. Environmental Protection Agency

Andrea Maguire, U.S. Environmental Protection Agency

Daniel Malashock, U.S. Department of Health and Human Services, Public Health Service

Jennifer Runkle, Cooperative Institute for Climate and Satellites– North Carolina

Marcus C. Sarofim, U.S. Environmental Protection Agency

Mark M. Shimamoto, U.S. Global Change Research Program, National Coordination Office

United States Global Change Research Program

Michael Kuperberg, Executive Director, USGCRP, White House Office of Science and Technology Policy (OSTP)

Ben DeAngelo, Deputy Executive Director, USGCRP, White House OSTP

Subcommittee on Global Change Research Leadership and Executive Committee

Chair

Thomas Karl, U.S. Department of Commerce

Vice Chairs

Michael Freilich, National Aeronautics and Space Administration

Gerald Geernaert, U.S. Department of Energy

Richard Spinrad, U.S. Department of Commerce

Roger Wakimoto, National Science Foundation

Jeffrey Arnold, U.S. Army Corps of Engineers (Adjunct)

Principals

John Balbus, U.S. Department of Health and Human Services

William Breed, U.S. Agency for International Development (Acting)

Joel Clement, U.S. Department of the Interior

Pierre Comizzoli, Smithsonian Institution

Wayne Higgins, U.S. Department of Commerce

Scott Harper, U.S. Department of Defense (Acting)

William Hohenstein, U.S. Department of Agriculture

Jack Kaye, National Aeronautics and Space Administration

Dorothy Koch, U.S. Department of Energy

C. Andrew Miller, U.S. Environmental Protection Agency

Craig Robinson, National Science Foundation

Arthur Rypinski, U.S. Department of Transportation (Acting)

Trigg Talley, U.S. Department of State

Executive Office of the President Liaisons

Tamara Dickinson, Principal Assistant Director for Environment and Energy, White House OSTP

Afua Bruce, Executive Director, National Science and Technology Council, White House OSTP (from June 2015)

Jayne Morrow, Executive Director, National Science and Technology Council, White House OSTP (through June 2015)

Richard Duke, White House Council on Environmental Quality

Kimberly Miller, White House Office of Management and Budget

Fabien Laurier, Director (Acting), National Climate Assessment, White House OSTP (from December 2013)

USGCRP Climate and Health Assessment Staff

USGCRP National Coordination Office

Michael Kuperberg, Executive Director, USGCRP, White House OSTP

Ben DeAngelo, Deputy Executive Director, USGCRP, White House OSTP

Katharine Jacobs, Director, National Climate Assessment, White House OSTP (through December 2013)

Thomas Armstrong, Executive Director, USGCRP NCO, White House OSTP (through December 2014)

Christopher P. Weaver, Executive Director (Acting, through August 2015), formerly Deputy Director, USGCRP NCO, White House OSTP

Glynis C. Lough, Chief of Staff, National Climate Assessment

Bradley Akamine, Chief Digital Officer

Mark M. Shimamoto, Health Program Lead

Ilya Fischhoff, Senior Scientist, National Climate Assessment

Emily Therese Cloyd, Engagement and Outreach Lead

Steve Aulenbach, GCIS Content Curator (through September 2015)

Samantha Brooks, SGCR Executive Secretary (through July 2015)

Tess Carter, Student Assistant, National Climate Assessment

Brian Duggan, GCIS Lead System Engineer (through September 2015)

Bryce Golden-Chen, Coordinator, National Climate Assessment (through September 2015)

Justin Goldstein, Advance Science Climate Data and Observing Systems Coordinator

Alexa Jay, Science Writer (from December 2015)

Amanda Jensen, Student Assistant, The George Washington University (January-May 2015)

Amanda McQueen, SGCR Executive Secretary (from July 2015)

Alena Marovitz, Student Assistant, Amherst College (June-August 2015)

Tanya Maslak, Chief of Operations (through May 2015)

Julie Morris, Associate Director of Implementation and Strategic Planning

Brent Newman, GCIS Data Coordinator (from January 2015)

Katie Reeves, Engagement Support Associate (from December 2015)

Catherine Wolner, Science Writer (through June 2015)

Robert Wolfe, Technical Lead for the Global Change Information System (GCIS), NASA (through March 2016)

NOAA Technical Support Unit, National Centers for Environmental Information

David R. Easterling, NCA Technical Support Unit Director, NOAA National Centers for Environmental Information (NCEI)

Paula Ann Hennon, NCA Technical Support Unit Deputy Director, Cooperative Institute for Climate and Satellites–North Carolina (CICS-NC) (through December 2015)

Kenneth E. Kunkel, Lead Scientist, CICS-NC

Sara W. Veasey, Creative Director, NOAA NCEI

Andrew Buddenberg, Software Engineer/Scientific Programmer, CICS-NC

Sarah Champion, Data Architect, CICS-NC

Daniel Glick, Editor, CICS-NC

Jessicca Griffin, Lead Graphic Designer, CICS-NC

Angel Li, Web Developer, CICS-NC

Liz Love-Brotak, Graphic Designer, NOAA NCEI

Tom Maycock, Project Manager/Editor, CICS-NC

Deborah Misch, Graphic Designer, LMI Consulting

Susan Osborne, Copy Editor, LMI Consulting

Deborah B. Riddle, Graphic Designer, NOAA NCEI

Jennifer Runkle, Editor, CICS-NC

April Sides, Web Developer, CICS-NC

Mara Sprain, Copy Editor, LAC Group

Laura E. Stevens, Research Scientist, CICS-NC

Brooke C. Stewart, Science Editor, CICS-NC

Liqiang Sun, Research Scientist/Modeling Support, CICS-NC

Devin Thomas, Metadata Specialist, ERT Inc.

Kristy Thomas, Metadata Specialist, ERT Inc.

Teresa Young, Print Specialist, ERT Inc.

UNC Asheville's National Environmental Modeling and Analysis Center (NEMAC)

Karin Rogers, Director of Operations/Research Scientist

Greg Dobson, Director of Geospatial Technology/Research Scientist

Caroline Dougherty, Principal Designer

John Frimmel, Applied Research Software Developer

Ian Johnson, Geospatial and Science Communications Associate

USGCRP Interagency Crosscutting Group on Climate Change and Human Health (CCHHG)

Co-Chairs
John Balbus, National Institutes of Health
George Luber, Centers for Disease Control and Prevention
Juli Trtanj, National Oceanic and Atmospheric Administration

Coordinator
Mark M. Shimamoto, U.S. Global Change Research Program, National Coordination Office

National Aeronautics and Space Administration
Sue Estes, Universities Space Research Association
John Haynes, Science Mission Directorate

U.S. Department of Agriculture
Isabel Walls, National Institute of Food and Agriculture

U.S. Department of Commerce
Michelle Hawkins, National Oceanic and Atmospheric Administration
Hunter Jones, National Oceanic and Atmospheric Administration
Juli Trtanj, National Oceanic and Atmospheric Administration

U.S. Department of Defense
Jean-Paul Chretien, Armed Forces Health Surveillance Center
James Persson, U.S. Army Research Institute of Environmental Medicine

U.S. Department of Health and Human Services
John Balbus, National Institutes of Health
Charles B. Beard, Centers for Disease Control and Prevention
Ross Bowling, Office of the Assistant Secretary for Administration
Kathleen Danskin, Office of the Assistant Secretary for Preparedness and Response
Stacey Degrasse, Food and Drug Administration
Renee Dickman, Office of the Assistant Secretary for Planning and Evaluation
Caroline Dilworth, National Institutes of Health
Jada F. Garafalo, Centers for Disease Control and Prevention
Christine Jessup, National Institutes of Health
Maya Levine, Office of Global Affairs
George Luber, Centers for Disease Control and Prevention
Joshua Rosenthal, National Institutes of Health
Shubhayu Saha, Centers for Disease Control and Prevention
Bono Sen, National Institutes of Health
Paul J. Schramm, Centers for Disease Control and Prevention
Joanna Watson, Centers for Disease Control and Prevention - NIOSH
Kimberly Thigpen Tart, National Institutes of Health

U.S. Department of Homeland Security
Jeffrey Stiefel, Office of Health Affairs

U.S. Department of Housing and Urban Development
J. Kofi Berko, Jr., Office of Lead Hazard Control & Healthy Homes

U.S. Department of the Interior
Patricia Bright, U.S. Geological Survey
Joseph Bunnell, U.S. Geological Survey

U.S. Department of State
Joshua Glasser, Bureau of Oceans and International Environmental and Scientific Affairs

U.S. Environmental Protection Agency
Martha Berger, Office of Children's Health Protection
Rona Birnbaum, Office of Air and Radiation
Bryan Bloomer, Office of Research and Development
Allison Crimmins, Office of Air and Radiation
Amanda Curry Brown, Office of Air and Radiation
Janet L. Gamble, Office of Research and Development
Vito Ilacqua, Office of Research and Development
Michael Kolian, Office of Air and Radiation
Marian Rutigliano, Office of Research and Development

White House National Security Council
David V. Adams

Review Editors

Rupa Basu, California Office of Environmental Health Hazard Assessment
Paul English, Public Health Institute, Oakland, CA
Kim Knowlton, Columbia University Mailman School of Public Health
Patricia Romero-Lankao, National Center for Atmospheric Research
Bart Ostro, University of California, Davis
Jan Semenza, European Centre for Disease Prevention and Control
Fran Sussman, ICF International
Felicia Wu, Michigan State University

Acknowledgements

The authors acknowledge RTI International, ICF International, Abt Associates, and Abt Environmental Research (formerly Stratus Consulting) for their support in the development of this report.

References:

1. Executive Office of the President, 2013: The President's Climate Action Plan. Washington, D.C. https://http://www.whitehouse.gov/sites/default/files/image/president27sclimate-actionplan.pdf

2. National Academies of Sciences Engineering and Medicine, 2015: Review of the Draft Interagency Report on *the Impacts of Climate Change on Human Health in the United States*. National Academies Press, Washington, D.C. http://www.nap.edu/catalog/21787/review-of-the-draft-interagency-report-on-the-impacts-of-climate-change-on-human-health-in-the-united-states

3. 2014: *Climate Change Impacts in the United States: The Third National Climate Assessment*. Melillo, J.M., T.C. Richmond, and G.W. Yohe, Eds. U.S. Global Change Research Program, Washington, D.C., 842 pp. http://dx.doi.org/10.7930/J0Z31WJ2

4. Luber, G., K. Knowlton, J. Balbus, H. Frumkin, M. Hayden, J. Hess, M. McGeehin, N. Sheats, L. Backer, C.B. Beard, K.L. Ebi, E. Maibach, R.S. Ostfeld, C. Wiedinmyer, E. Zielinski-Gutiérrez, and L. Ziska, 2014: Ch. 9: Human health. *Climate Change Impacts in the United States: The Third National Climate Assessment*. Melillo, J.M., T.C. Richmond, and G.W. Yohe, Eds. U.S. Global Change Research Program, Washington, D.C., 220-256. http://dx.doi.org/10.7930/J0PN93H5

5. GCRA, 1990: Global Change Research Act of 1990, Pub. L. No. 101-606, 104 Stat. 3096-3104. http://www.gpo.gov/fdsys/pkg/STATUTE-104/pdf/STATUTE-104-Pg3096.pdf

6. CCSP, 2008: Analyses of the Effects of Global Change on Human Health and Welfare and Human Systems. A Report by the U.S. Climate Change Science Program and the Subcommittee on Global Change Research. 205 pp. Gamble, J. L., (Ed.), Ebi, K.L., F.G. Sussman, T.J. Wilbanks, (Authors). U.S. Environmental Protection Agency, Washington, D.C. http://downloads.globalchange.gov/sap/sap4-6/sap4-6-final-report-all.pdf

7. USGCRP, 2012: National Climate Assessment Health Sector Literature Review and Bibliography. Technical Input for the Interagency Climate Change and Human Health Group. National Institute of Environmental Health Sciences. http://www.globalchange.gov/what-we-do/assessment/nca-activities/available-technical-inputs

EXECUTIVE SUMMARY

Climate change threatens human health and well-being in the United States. The U.S. Global Change Research Program (USGCRP) Climate and Health Assessment has been developed to enhance understanding and inform decisions about this growing threat. This scientific assessment, called for under the President's Climate Action Plan, is a major report of the sustained National Climate Assessment (NCA) process. The report responds to the 1990 Congressional mandate to assist the Nation in understanding, assessing, predicting, and responding to human-induced and natural processes of global change. The agencies of the USGCRP identified human health impacts as a high-priority topic for scientific assessment.

The purpose of this assessment is to provide a comprehensive, evidence-based, and, where possible, quantitative estimation of observed and projected climate change related health impacts in the United States. The USGCRP Climate and Health Assessment has been developed to inform public health officials, urban and disaster response planners, decision makers, and other stakeholders within and outside of government who are interested in better understanding the risks climate change presents to human health.

Lead Authors

Allison Crimmins
U.S. Environmental Protection Agency

John Balbus
National Institutes of Health

Janet L. Gamble
U.S. Environmental Protection Agency

Charles B. Beard
Centers for Disease Control and Prevention

Jesse E. Bell
Cooperative Institute for Climate and Satellites–North Carolina

Daniel Dodgen
U.S. Department of Health and Human Services, Office of the Assistant Secretary for Preparedness and Response

Rebecca J. Eisen
Centers for Disease Control and Prevention

Neal Fann
U.S. Environmental Protection Agency

Michelle D. Hawkins
National Oceanic and Atmospheric Administration

Stephanie C. Herring
National Oceanic and Atmospheric Administration

Lesley Jantarasami
U.S. Environmental Protection Agency

David M. Mills
Abt Associates

Shubhayu Saha
Centers for Disease Control and Prevention

Marcus C. Sarofim
U.S. Environmental Protection Agency

Juli Trtanj
National Oceanic and Atmospheric Administration

Lewis Ziska
U.S. Department of Agriculture

Recommended Citation: Crimmins, A., J. Balbus, J.L. Gamble, C.B. Beard, J.E. Bell, D. Dodgen, R.J. Eisen, N. Fann, M.D. Hawkins, S.C. Herring, L. Jantarasami, D.M. Mills, S. Saha, M.C. Sarofim, J. Trtanj, and L. Ziska, 2016: Executive Summary. *The Impacts of Climate Change on Human Health in the United States: A Scientific Assessment.* U.S. Global Change Research Program, Washington, DC, page 1–24. http://dx.doi.org/10.7930/J00P0WXS

On the web: health2016.globalchange.gov

Executive Summary of
THE IMPACTS OF CLIMATE CHANGE ON HUMAN HEALTH IN THE UNITED STATES

Climate change is a significant threat to the health of the American people. The impacts of human-induced climate change are increasing nationwide. Rising greenhouse gas concentrations result in increases in temperature, changes in precipitation, increases in the frequency and intensity of some extreme weather events, and rising sea levels. These climate change impacts endanger our health by affecting our food and water sources, the air we breathe, the weather we experience, and our interactions with the built and natural environments. As the climate continues to change, the risks to human health continue to grow.

Current and future climate impacts expose more people in more places to public health threats. Already in the United States, we have observed climate-related increases in our exposure to elevated temperatures; more frequent, severe, or longer-lasting extreme events; degraded air quality; diseases transmitted through food, water, and disease vectors (such as ticks and mosquitoes); and stresses to our mental health and well-being.

Almost all of these threats are expected to worsen with continued climate change. Some of these health threats will occur over longer time periods, or at unprecedented times of the year; some people will be exposed to threats not previously experienced in their locations. Overall, instances of potentially beneficial health impacts of climate change are limited in number and pertain to specific regions or populations. For example, the reduction in cold-related deaths is projected to be smaller than the increase in heat-related deaths in most regions.

Every American is vulnerable to the health impacts associated with climate change. Increased exposure to multiple health threats, together with changes in sensitivity and the ability to adapt to those threats, increases a person's vulnerability to climate-related health effects. The impacts of climate change on human health interact with underlying health, demographic, and socioeconomic factors. Through the combined influence of these factors, climate change exacerbates some existing health threats and creates new public health challenges. While all Americans are at risk, some populations are disproportionately vulnerable, including those with low income, some communities of color, immigrant groups (including those with limited English proficiency), Indigenous peoples, children and pregnant women, older adults, vulnerable occupational groups, persons with disabilities, and persons with preexisting or chronic medical conditions.

Changes in aquatic habitats and species may affect subsistence fishing among Indigenous populations.

In recent years, scientific understanding of how climate change increases risks to human health has advanced significantly. Even so, the ability to evaluate, monitor, and project health effects varies across climate impacts. For instance, information on health outcomes differs in terms of whether complete, long-term datasets exist that allow quantification of observed changes, and whether existing models can project impacts at the timescales and geographic scales of interest. Differences also exist in the metrics available for observing or projecting different health impacts. For some health impacts, the available metrics only describe changes in risk of exposure, while for others, metrics describe changes in actual health outcomes (such as the number of new cases of a disease or an increase in deaths).

While all Americans are at risk, some populations are disproportionately vulnerable, including children and pregnant women.

This assessment strengthens and expands our understanding of climate-related health impacts by providing a more definitive description of climate-related health burdens in the United States. It builds on the 2014 National Climate Assessment[1] and reviews and synthesizes key contributions to the published literature. Acknowledging the rising demand for data that can be used to characterize how climate change affects health, this report assesses recent analyses that quantify observed and projected health impacts. Each chapter characterizes the strength of the scientific evidence for a given climate–health exposure pathway or "link" in the

Every American is vulnerable to the health impacts associated with climate change

causal chain between a climate change impact and its associated health outcome. This assessment's findings represent an improvement in scientific confidence in the link between climate change and a broad range of threats to public health, while recognizing populations of concern and identifying emerging issues. These considerations provide the context for understanding Americans' changing health risks and allow us to identify, project, and respond to future climate change health threats. The overall findings underscore the significance of the growing risk climate change poses to human health in the United States.

Los Angeles, California, May 22, 2012. Unless offset by additional emissions reductions of ozone precursors, climate-driven increases in ozone will cause premature deaths, hospital visits, lost school days, and acute respiratory symptoms.

1 CLIMATE CHANGE AND HUMAN HEALTH

The influences of weather and climate on human health are significant and varied. Exposure to health hazards related to climate change affects different people and different communities to different degrees. While often assessed individually, exposure to multiple climate change threats can occur simultaneously, resulting in compounding or cascading health impacts.

With climate change, the frequency, severity, duration, and location of weather and climate phenomena—like rising temperatures, heavy rains and droughts, and some other kinds of severe weather—are changing. This means that areas already experiencing health-threatening weather and climate phenomena, such as severe heat or hurricanes, are likely to experience worsening impacts, such as higher temperatures and increased storm intensity, rainfall rates, and storm surge.

It also means that some locations will experience new climate-related health threats. For example, areas previously unaffected by toxic algal blooms or waterborne diseases because of cooler water temperatures may face these hazards in the future as increasing water temperatures allow the organisms that cause these health risks to thrive. Even areas that currently experience these health threats may see a shift in the timing of the seasons that pose the greatest risk to human health.

Climate change can therefore affect human health in two main ways: first, by changing the severity or frequency of health problems that are already affected by climate or weather factors; and second, by creating unprecedented or unanticipated health problems or health threats in places where they have not previously occurred.

Climate Change and Health

CLIMATE DRIVERS
- Increased temperatures
- Precipitation extremes
- Extreme weather events
- Sea level rise

ENVIRONMENTAL & INSTITUTIONAL CONTEXT
- Land-use change
- Ecosystem change
- Infrastructure condition
- Geography
- Agricultural production & livestock use

EXPOSURE PATHWAYS
- Extreme heat
- Poor air quality
- Reduced food & water quality
- Changes in infectious agents
- Population displacement

SOCIAL & BEHAVIORAL CONTEXT
- Age & gender
- Race & ethnicity
- Poverty
- Housing & infrastructure
- Education
- Discrimination
- Access to care & community health infrastructure
- Preexisting health conditions

HEALTH OUTCOMES
- Heat-related illness
- Cardiopulmonary illness
- Food-, water-, & vector-borne disease
- Mental health consequences & stress

Conceptual diagram illustrating the exposure pathways by which climate change affects human health. Here, the center boxes list some selected examples of the kinds of changes in climate drivers, exposure, and health outcomes explored in this report. Exposure pathways exist within the context of other factors that positively or negatively influence health outcomes (gray side boxes). Some of the key factors that influence vulnerability for individuals are shown in the right box, and include social determinants of health and behavioral choices. Some key factors that influence vulnerability at larger scales, such as natural and built environments, governance and management, and institutions, are shown in the left box. All of these influencing factors can affect an individual's or a community's vulnerability through changes in exposure, sensitivity, and adaptive capacity and may also be affected by climate change.

Examples of Climate Impacts on Human Health

	Climate Driver	Exposure	Health Outcome	Impact
Extreme Heat	More frequent, severe, prolonged heat events	Elevated temperatures	Heat-related death and illness	Rising temperatures will lead to an increase in heat-related deaths and illnesses.
Outdoor Air Quality	Increasing temperatures and changing precipitation patterns	Worsened air quality (ozone, particulate matter, and higher pollen counts)	Premature death, acute and chronic cardiovascular and respiratory illnesses	Rising temperatures and wildfires and decreasing precipitation will lead to increases in ozone and particulate matter, elevating the risks of cardiovascular and respiratory illnesses and death.
Flooding	Rising sea level and more frequent or intense extreme precipitation, hurricanes, and storm surge events	Contaminated water, debris, and disruptions to essential infrastructure	Drowning, injuries, mental health consequences, gastrointestinal and other illness	Increased coastal and inland flooding exposes populations to a range of negative health impacts before, during, and after events.
Vector-Borne Infection (Lyme Disease)	Changes in temperature extremes and seasonal weather patterns	Earlier and geographically expanded tick activity	Lyme disease	Ticks will show earlier seasonal activity and a generally northward range expansion, increasing risk of human exposure to Lyme disease-causing bacteria.
Water-Related Infection (*Vibrio vulnificus*)	Rising sea surface temperature, changes in precipitation and runoff affecting coastal salinity	Recreational water or shellfish contaminated with *Vibrio vulnificus*	*Vibrio vulnificus* induced diarrhea & intestinal illness, wound and bloodstream infections, death	Increases in water temperatures will alter timing and location of *Vibrio vulnificus* growth, increasing exposure and risk of water-borne illness.
Food-Related Infection (*Salmonella*)	Increases in temperature, humidity, and season length	Increased growth of pathogens, seasonal shifts in incidence of *Salmonella* exposure	*Salmonella* infection, gastrointestinal outbreaks	Rising temperatures increase *Salmonella* prevalence in food; longer seasons and warming winters increase risk of exposure and infection.
Mental Health and Well-Being	Climate change impacts, especially extreme weather	Level of exposure to traumatic events, like disasters	Distress, grief, behavioral health disorders, social impacts, resilience	Changes in exposure to climate- or weather-related disasters cause or exacerbate stress and mental health consequences, with greater risk for certain populations.

The diagram shows specific examples of how climate change can affect human health, now and in the future. These effects could occur at local, regional, or national scales. The examples listed in the first column are those described in each underlying chapter's exposure pathway diagram. Moving from left to right along one health impact row, the three middle columns show how climate drivers affect an individual's or a community's exposure to a health threat and the resulting change in health outcome. The overall climate impact is summarized in the final gray column. For a more comprehensive look at how climate change affects health, and to see the environmental, institutional, social, and behavioral factors that play an interactive role in determining health outcomes, see the exposure pathway diagrams in chapters 2–8 in the full report.

2 TEMPERATURE-RELATED DEATH AND ILLNESS

Increasing concentrations of greenhouse gases lead to an increase of both average and extreme temperatures. This is expected to lead to an increase in deaths and illness from heat and a potential decrease in deaths from cold, particularly for a number of communities especially vulnerable to these changes, such as children, the elderly, and economically disadvantaged groups.

Days that are hotter than the average seasonal temperature in the summer or colder than the average seasonal temperature in the winter cause increased levels of illness and death by compromising the body's ability to regulate its temperature or by inducing direct or indirect health complications. Loss of internal temperature control can result in a cascade of illnesses, including heat cramps, heat exhaustion, heatstroke, and hyperthermia in the presence of extreme heat, and hypothermia and frostbite in the presence of extreme cold.

Temperature extremes can also worsen chronic conditions such as cardiovascular disease, respiratory disease, cerebrovascular disease, and diabetes-related conditions. Prolonged exposure to high temperatures is associated with increased hospital admissions for cardiovascular, kidney, and respiratory disorders.

Climate change will increase the frequency and severity of future extreme heat events while also resulting in generally warmer summers and milder winters, with implications for human health.

Future Increases in Temperature-Related Deaths

Key Finding 1: Based on present-day sensitivity to heat, an increase of thousands to tens of thousands of premature heat-related deaths in the summer *[Very Likely, High Confidence]* and a decrease of premature cold-related deaths in the winter *[Very Likely, Medium Confidence]* are projected each year as a result of climate change by the end of the century. Future adaptation will very likely reduce these impacts (see the Changing Tolerance to Extreme Heat Finding). The reduction in cold-related deaths is projected to be smaller than the increase in heat-related deaths in most regions *[Likely, Medium Confidence]*.

Even Small Differences from Seasonal Average Temperatures Result in Illness and Death

Key Finding 2: Days that are hotter than usual in the summer or colder than usual in the winter are both associated with increased illness and death *[Very High Confidence]*. Mortality effects are observed even for small differences from seasonal average temperatures *[High Confidence]*. Because small temperature differences occur much more frequently than large temperature differences, not accounting for the effect of these small differences would lead to underestimating the future impact of climate change *[Likely, High Confidence]*.

Projected Changes in Deaths in U.S. Cities by Season

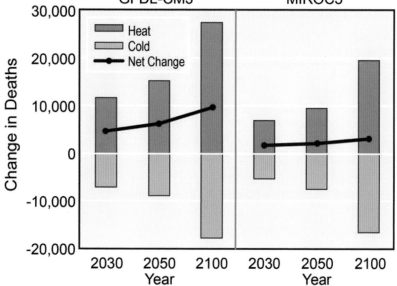

This figure shows the projected increase in deaths due to warming in the summer months (hot season, April–September), the projected decrease in deaths due to warming in the winter months (cold season, October–March), and the projected net change in deaths compared to a 1990 baseline period for the 209 U.S. cities examined, using the GFDL–CM3 and MIROC5 climate models (see Ch. 2: Temperature-Related Deaths and Illness). (Figure source: adapted from Schwartz et al. 2015)[2]

Changing Tolerance to Extreme Heat

Key Finding 3: An increase in population tolerance to extreme heat has been observed over time *[Very High Confidence]*. Changes in this tolerance have been associated with increased use of air conditioning, improved social responses, and/or physiological acclimatization, among other factors *[Medium Confidence]*. Expected future increases in this tolerance will reduce the projected increase in deaths from heat *[Very Likely, Very High Confidence]*.

Some Populations at Greater Risk

Key Finding 4: Older adults and children have a higher risk of dying or becoming ill due to extreme heat *[Very High Confidence]*. People working outdoors, the socially isolated and economically disadvantaged, those with chronic illnesses, as well as some communities of color, are also especially vulnerable to death or illness *[Very High Confidence]*.

Outdoor workers spend a great deal of time exposed to temperature extremes, often while performing vigorous activities.

3 AIR QUALITY IMPACTS

Changes in the climate affect the air we breathe, both indoors and outdoors. The changing climate has modified weather patterns, which in turn have influenced the levels and location of outdoor air pollutants such as ground-level ozone (O_3) and fine particulate matter. Increasing carbon dioxide (CO_2) levels also promote the growth of plants that release airborne allergens (aeroallergens). Finally, these changes to outdoor air quality and aeroallergens also affect indoor air quality as both pollutants and aeroallergens infiltrate homes, schools, and other buildings. Poor air quality, whether outdoors or indoors, can negatively affect the human respiratory and cardiovascular systems. Higher pollen concentrations and longer pollen seasons can increase allergic sensitization and asthma episodes and thereby limit productivity at work and school.

Ragweed pollen frequently triggers hay fever and asthma episodes during the fall.

Projected Change in Temperature, Ozone, and Ozone-Related Premature Deaths in 2030

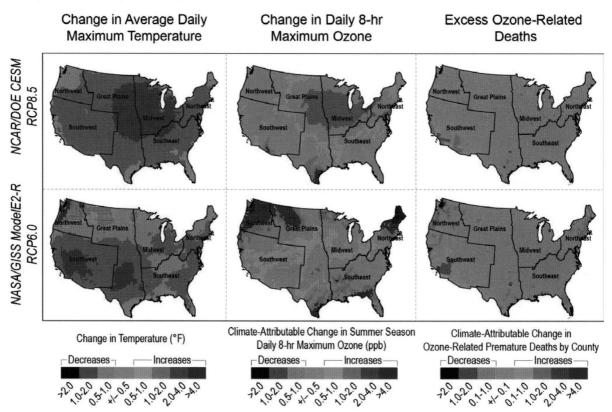

The air quality response to climate change can vary substantially by region across scenarios. Two downscaled global climate model projections using two greenhouse gas concentration pathways estimate increases in average daily maximum temperatures of 1.8°F to 7.2°F (1°C to 4°C) and increases of 1 to 5 parts per billion (ppb) in daily 8-hour maximum ozone in the year 2030 relative to the year 2000 throughout the continental United States. Unless reductions in ozone precursor emissions offset the influence of climate change, this "climate penalty" of increased ozone concentrations due to climate change would result in tens to thousands of additional ozone-related premature deaths per year, shown here as incidences per year by county (see Ch. 3: Air Quality Impacts). (Figure source: adapted from Fann et al. 2015)[3]

Exacerbated Ozone Health Impacts

Key Finding 1: Climate change will make it harder for any given regulatory approach to reduce ground-level ozone pollution in the future as meteorological conditions become increasingly conducive to forming ozone over most of the United States *[Likely, High Confidence]*. Unless offset by additional emissions reductions of ozone precursors, these climate-driven increases in ozone will cause premature deaths, hospital visits, lost school days, and acute respiratory symptoms *[Likely, High Confidence]*.

Increased Health Impacts from Wildfires

Key Finding 2: Wildfires emit fine particles and ozone precursors that in turn increase the risk of premature death and adverse chronic and acute cardiovascular and respiratory health outcomes *[Likely, High Confidence]*. Climate change is projected to increase the number and severity of naturally occurring wildfires in parts of the United States, increasing emissions of particulate matter and ozone precursors and resulting in additional adverse health outcomes *[Likely, High Confidence]*.

Worsened Allergy and Asthma Conditions

Key Finding 3: Changes in climate, specifically rising temperatures, altered precipitation patterns, and increasing concentrations of atmospheric carbon dioxide, are expected to contribute to increases in the levels of some airborne allergens and associated increases in asthma episodes and other allergic illnesses *[High Confidence]*.

(Top) Dampness and mold in U.S. homes are linked to approximately 4.6 million cases of worsened asthma. (Left) Wildfires are a major source of airborne particulate matter, especially in the western United States during summer. Climate change has already led to an increased frequency of large wildfires, as well as longer durations of individual wildfires and longer wildfire seasons in the western United States. (Right) Nearly 6.8 million children in the United States are affected by asthma, making it a major chronic disease of childhood.

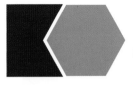

4 IMPACTS OF EXTREME EVENTS ON HUMAN HEALTH

Climate change projections show that there will be continuing increases in the occurrence and severity of some extreme events by the end of the century, while for other extremes the links to climate change are more uncertain. Some regions of the United States have already experienced costly impacts—in terms of both lives lost and economic damages—from observed changes in the frequency, intensity, or duration of certain extreme events.

While it is intuitive that extremes can have health impacts such as death or injury during an event (for example, drowning during floods), health impacts can also occur before or after an extreme event, as individuals may be involved in activities that put their health at risk, such as disaster

preparation and post-event cleanup. Health risks may also arise long after the event, or in places outside the area where the event took place, as a result of damage to property, destruction of assets, loss of infrastructure and public services, social and economic impacts, environmental degradation, and other factors.

Extreme events also pose unique health risks if multiple events occur simultaneously or in succession in a given location. The severity and extent of health effects associated with extreme events depend on the physical impacts of the extreme events themselves as well as the unique human, societal, and environmental circumstances at the time and place where events occur.

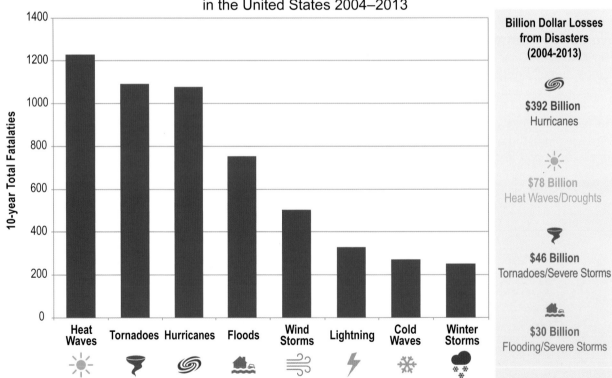

Estimated Deaths and Billion Dollar Losses from Extreme Events in the United States 2004–2013

Billion Dollar Losses from Disasters (2004-2013)

$392 Billion
Hurricanes

$78 Billion
Heat Waves/Droughts

$46 Billion
Tornadoes/Severe Storms

$30 Billion
Flooding/Severe Storms

This figure provides 10-year estimates of fatalities related to extreme events from 2004 to 2013,[4] as well as estimated economic damages from 58 weather and climate disaster events with losses exceeding $1 billion (see Smith and Katz 2013 to understand how total losses were calculated).[5] These statistics are indicative of the human and economic costs of extreme weather events over this time period. Climate change will alter the frequency, intensity, and geographic distribution of some of these extremes,[1] which has consequences for exposure to health risks from extreme events. Trends and future projections for some extremes, including tornadoes, lightning, and wind storms are still uncertain (see Ch. 4: Extreme Events).

Increased Exposure to Extreme Events

Key Finding 1: Health impacts associated with climate-related changes in exposure to extreme events include death, injury, or illness; exacerbation of underlying medical conditions; and adverse effects on mental health *[High Confidence]*. Climate change will increase exposure risk in some regions of the United States due to projected increases in the frequency and/or intensity of drought, wildfires, and flooding related to extreme precipitation and hurricanes *[Medium Confidence]*.

(Top) A truck gets stuck in the storm surge covering Highway 90 in Gulfport, Mississippi, during Hurricane Isaac. (Bottom) Power lines damaged in Plaquemines Parish, Louisiana, by Hurricane Isaac. September 3, 2012.

Family farmer in drought-stressed peanut field, Unadilla, Georgia. July 24, 2012.

Disruption of Essential Infrastructure

Key Finding 2: Many types of extreme events related to climate change cause disruption of infrastructure, including power, water, transportation, and communication systems, that are essential to maintaining access to health care and emergency response services and safeguarding human health *[High Confidence]*.

Vulnerability to Coastal Flooding

Key Finding 3: Coastal populations with greater vulnerability to health impacts from coastal flooding include persons with disabilities or other access and functional needs, certain populations of color, older adults, pregnant women and children, low-income populations, and some occupational groups *[High Confidence]*. Climate change will increase exposure risk to coastal flooding due to increases in extreme precipitation and in hurricane intensity and rainfall rates, as well as sea level rise and the resulting increases in storm surge *[High Confidence]*.

5 VECTOR-BORNE DISEASES

Vector-borne diseases are illnesses that are transmitted by vectors, which include mosquitoes, ticks, and fleas. These vectors can carry infective pathogens such as viruses, bacteria, and protozoa, which can be transferred from one host (carrier) to another. The seasonality, distribution, and prevalence of vector-borne diseases are influenced significantly by climate factors, primarily high and low temperature extremes and precipitation patterns.

Climate change is likely to have both short- and long-term effects on vector-borne disease transmission and infection patterns, affecting both seasonal risk and broad geographic changes in disease occurrence over decades. While climate variability and climate change both alter the transmission of vector-borne diseases, they will likely interact with many other factors, including how pathogens adapt and change, the availability of hosts, changing ecosystems and land use, demographics, human behavior, and adaptive capacity. These complex interactions make it difficult to predict the effects of climate change on vector-borne diseases.

In the eastern United States, Lyme disease is transmitted to humans primarily by blacklegged (deer) ticks.

Changes in Lyme Disease Case Report Distribution

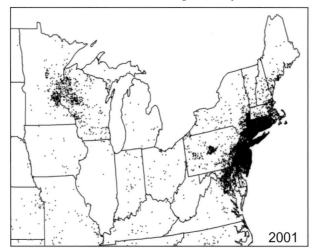

2001

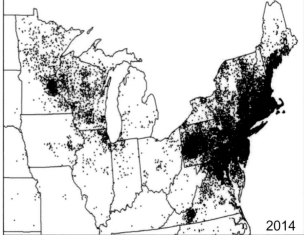

2014

Maps show the reported cases of Lyme disease in 2001 and 2014 for the areas of the country where Lyme disease is most common (the Northeast and Upper Midwest). Both the distribution and the numbers of cases have increased (see Ch. 5: Vector-Borne Diseases). (Figure source: adapted from CDC 2015)[6]

Changing Distributions of Vectors and Vector-Borne Diseases

Key Finding 1: Climate change is expected to alter the geographic and seasonal distributions of existing vectors and vector-borne diseases *[Likely, High Confidence]*.

Earlier Tick Activity and Northward Range Expansion

Key Finding 2: Ticks capable of carrying the bacteria that cause Lyme disease and other pathogens will show earlier seasonal activity and a generally northward expansion in response to increasing temperatures associated with climate change *[Likely, High Confidence]*. Longer seasonal activity and expanding geographic range of these ticks will increase the risk of human exposure to ticks *[Likely, Medium Confidence]*.

Changing Mosquito-Borne Disease Dynamics

Key Finding 3: Rising temperatures, changing precipitation patterns, and a higher frequency of some extreme weather events associated with climate change will influence the distribution, abundance, and prevalence of infection in the mosquitoes that transmit West Nile virus and other pathogens by altering habitat availability and mosquito and viral reproduction rates *[Very Likely, High Confidence]*. Alterations in the distribution, abundance, and infection rate of mosquitoes will influence human exposure to bites from infected mosquitoes, which is expected to alter risk for human disease *[Very Likely, Medium Confidence]*.

Emergence of New Vector-Borne Pathogens

Key Finding 4: Vector-borne pathogens are expected to emerge or reemerge due to the interactions of climate factors with many other drivers, such as changing land-use patterns *[Likely, High Confidence]*. The impacts to human disease, however, will be limited by the adaptive capacity of human populations, such as vector control practices or personal protective measures *[Likely, High Confidence]*.

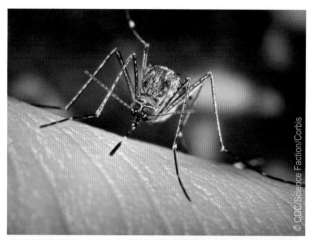

Birds such as the house finch are the natural host of West Nile virus. Humans can be infected from a bite of a mosquito that has previously bitten an infected bird.

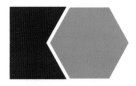

6 CLIMATE IMPACTS ON WATER-RELATED ILLNESSES

Across most of the United States, climate change is expected to affect fresh and marine water resources in ways that will increase people's exposure to water-related contaminants that cause illness. Water-related illnesses include waterborne diseases caused by pathogens, such as bacteria, viruses, and protozoa. Water-related illnesses are also caused by toxins produced by certain harmful algae and cyanobacteria and by chemicals introduced into the environment by human activities. Exposure occurs through ingestion, inhalation, or direct contact with contaminated drinking or recreational water and through consumption of contaminated fish and shellfish. Factors related to climate change—including temperature, precipitation and related runoff, hurricanes, and storm surge—affect the growth, survival, spread, and virulence or toxicity of agents (causes) of water-related illness. Whether or not illness results from exposure to contaminated water, fish, or shellfish is dependent on a complex set of factors, including human behavior and social determinants of health that may affect a person's exposure, sensitivity, and adaptive capacity. Water resource, public health, and environmental agencies in the United States provide many public health safeguards to reduce risk of exposure and illness even if water becomes contaminated. These include water quality monitoring, drinking water treatment standards and practices, beach closures, and issuing advisories for boiling drinking water and harvesting shellfish.

Links between Climate Change, Water Quantity and Quality, and Human Exposure to Water-Related Illness

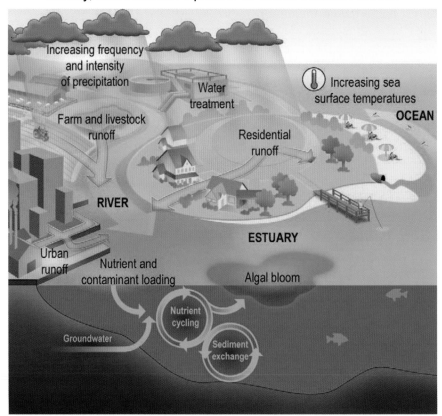

Precipitation and temperature changes affect fresh and marine water quantity and quality primarily through urban, rural, and agriculture runoff. This runoff in turn affects human exposure to water-related illnesses primarily through contamination of drinking water, recreational water, and fish or shellfish (see Ch. 6: Water-Related Illness).

U.S. Global Change Research Program   14   Impacts of Climate Change on Human Health in the United States

Red tide bloom, Hood Canal, Puget Sound, Washington State.

Seasonal and Geographic Changes in Waterborne Illness Risk

Key Finding 1: Increases in water temperatures associated with climate change will alter the seasonal windows of growth and the geographic range of suitable habitat for freshwater toxin-producing harmful algae *[Very Likely, High Confidence]*, certain naturally occurring *Vibrio* bacteria *[Very Likely, Medium Confidence]*, and marine toxin-producing harmful algae *[Likely, Medium Confidence]*. These changes will increase the risk of exposure to waterborne pathogens and algal toxins that can cause a variety of illnesses *[Medium Confidence]*.

Runoff from Extreme Precipitation Increases Exposure Risk

Key Finding 2: Runoff from more frequent and intense extreme precipitation events will increasingly compromise recreational waters, shellfish harvesting waters, and sources of drinking water through increased introduction of pathogens and prevalence of toxic algal blooms *[High Confidence]*. As a result, the risk of human exposure to agents of water-related illness will increase *[Medium Confidence]*.

Water Infrastructure Failure

Key Finding 3: Increases in some extreme weather events and storm surges will increase the risk that infrastructure for drinking water, wastewater, and stormwater will fail due to either damage or exceedance of system capacity, especially in areas with aging infrastructure *[High Confidence]*. As a result, the risk of exposure to water-related pathogens, chemicals, and algal toxins will increase in recreational and shellfish harvesting waters, and in drinking water where treatment barriers break down *[Medium Confidence]*.

Young women walk through floodwater in the historic district of Charleston, South Carolina, as Hurricane Joaquin passes offshore. October 4, 2015.

7 FOOD SAFETY, NUTRITION, AND DISTRIBUTION

A safe and nutritious food supply is a vital component of food security. The impacts of climate change on food production, prices, and trade for the United States and globally have been widely examined, including in the recent report "Climate Change, Global Food Security, and the U.S. Food System."[7] An overall finding of that report was that "climate change is very likely to affect global, regional, and local food security by disrupting food availability, decreasing access to food, and making utilization more difficult."

This chapter focuses on some of the less reported aspects of food security, specifically the impacts of climate change on food safety, nutrition, and distribution. There are two overarching means by which increasing carbon dioxide (CO_2) and climate change alter safety, nutrition, and distribution of food. The first is associated with rising global temperatures and the subsequent changes in weather patterns and extreme climate events. Current and anticipated changes in climate and the physical environment have consequences for contamination, spoilage, and the disruption of food distribution. The second pathway is through the direct CO_2 "fertilization" effect on plant photosynthesis. Higher concentrations of CO_2 stimulate growth and carbohydrate production in some plants, but can lower the levels of protein and essential minerals in a number of widely consumed crops, including wheat, rice, and potatoes, with potentially negative implications for human nutrition.

Farm to Table
The Potential Interactions of Rising CO_2 and Climate Change on Food Safety

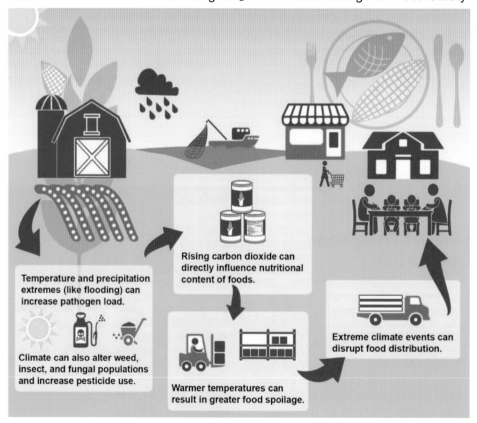

The food system involves a network of interactions with our physical and biological environments as food moves from production to consumption, or from "farm to table." Rising CO_2 and climate change will affect the quality and distribution of food, with subsequent effects on food safety and nutrition (see Ch. 7: Food Safety).

Increased Risk of Foodborne Illness

Key Finding 1: Climate change, including rising temperatures and changes in weather extremes, is expected to increase the exposure of food to certain pathogens and toxins *[Likely, High Confidence]*. This will increase the risk of negative health impacts *[Likely, Medium Confidence]*, but actual incidence of foodborne illness will depend on the efficacy of practices that safeguard food in the United States *[High Confidence]*.

Chemical Contaminants in the Food Chain

Key Finding 2: Climate change will increase human exposure to chemical contaminants in food through several pathways *[Likely, Medium Confidence]*. Elevated sea surface temperatures will lead to greater accumulation of mercury in seafood *[Likely, Medium Confidence]*, while increases in extreme weather events will introduce contaminants into the food chain *[Likely, Medium Confidence]*. Rising carbon dioxide concentrations and climate change will alter incidence and distribution of pests, parasites, and microbes *[Very Likely, High Confidence]*, leading to increases in the use of pesticides and veterinary drugs *[Likely, Medium Confidence]*.

Rising Carbon Dioxide Lowers Nutritional Value of Food

Key Finding 3: The nutritional value of agriculturally important food crops, such as wheat and rice, will decrease as rising levels of atmospheric carbon dioxide continue to reduce the concentrations of protein and essential minerals in most plant species *[Very Likely, High Confidence]*.

Extreme Weather Limits Access to Safe Foods

Key Finding 4: Increases in the frequency or intensity of some extreme weather events associated with climate change will increase disruptions of food distribution by damaging existing infrastructure or slowing food shipments *[Likely, High Confidence]*. These impediments lead to increased risk for food damage, spoilage, or contamination, which will limit availability of and access to safe and nutritious food depending on the extent of disruption and the resilience of food distribution infrastructure *[Medium Confidence]*.

(Left) The risk of foodborne illness is higher when food is prepared outdoors. (Right) Crop dusting of a corn field in Iowa.

8 MENTAL HEALTH AND WELL-BEING

The effects of global climate change on mental health and well-being are integral parts of the overall climate-related human health impacts. Mental health consequences of climate change range from minimal stress and distress symptoms to clinical disorders, such as anxiety, depression, post-traumatic stress, and suicidality. Other consequences include effects on the everyday life, perceptions, and experiences of individuals and communities attempting to understand and respond appropriately to climate change and its implications. The mental health and well-being consequences of climate change related impacts rarely occur in isolation, but often interact with other social and environmental stressors. The interactive and cumulative nature of climate change effects on health, mental health, and well-being are critical factors in understanding the overall consequences of climate change on human health.

© Aurora/Aurora Photo/Corbis

Children are at particular risk for distress, anxiety, and other adverse mental health effects in the aftermath of an extreme event.

The Impact of Climate Change on Physical, Mental, and Community Health

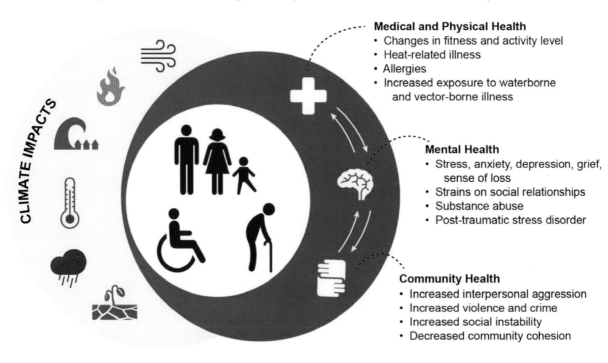

CLIMATE IMPACTS

Medical and Physical Health
- Changes in fitness and activity level
- Heat-related illness
- Allergies
- Increased exposure to waterborne and vector-borne illness

Mental Health
- Stress, anxiety, depression, grief, sense of loss
- Strains on social relationships
- Substance abuse
- Post-traumatic stress disorder

Community Health
- Increased interpersonal aggression
- Increased violence and crime
- Increased social instability
- Decreased community cohesion

At the center of the diagram are human figures representing adults, children, older adults, and people with disabilities. The left circle depicts climate impacts including air quality, wildfire, sea level rise and storm surge, heat, storms, and drought. The right circle shows the three interconnected health domains that will be affected by climate impacts: Medical and Physical Health, Mental Health, and Community Health (see Ch. 8: Mental Health). (Figure source: adapted from Clayton et al. 2014)[7]

Exposure to Disasters Results in Mental Health Consequences

Key Finding 1: Many people exposed to climate-related or weather-related disasters experience stress and serious mental health consequences. Depending on the type of the disaster, these consequences include post-traumatic stress disorder (PTSD), depression, and general anxiety, which often occur at the same time *[Very High Confidence]*. The majority of affected people recover over time, although a significant proportion of exposed individuals develop chronic psychological dysfunction *[High Confidence]*.

Specific Groups of People Are at Higher Risk

Key Finding 2: Specific groups of people are at higher risk for distress and other adverse mental health consequences from exposure to climate-related or weather-related disasters. These groups include children, the elderly, women (especially pregnant and post-partum women), people with preexisting mental illness, the economically disadvantaged, the homeless, and first responders *[High Confidence]*. Communities that rely on the natural environment for sustenance and livelihood, as well as populations living in areas most susceptible to specific climate change events, are at increased risk for adverse mental health outcomes *[High Confidence]*.

Residents and volunteers in the Rockaways section of Queens in New York City filter through clothes and food supplies from donors following Superstorm Sandy. November 3, 2012.

(Top) Rescue worker receives hug from Galveston, TX, resident after Hurricane Ike, September 2008. (Bottom) People experience the threat of climate change through frequent media coverage.

Climate Change Threats Result in Mental Health Consequences and Social Impacts

Key Finding 3: Many people will experience adverse mental health outcomes and social impacts from the threat of climate change, the perceived direct experience of climate change, and changes to one's local environment *[High Confidence]*. Media and popular culture representations of climate change influence stress responses and mental health and well-being *[Medium Confidence]*.

Extreme Heat Increases Risks for People with Mental Illness

Key Finding 4: People with mental illness are at higher risk for poor physical and mental health due to extreme heat *[High Confidence]*. Increases in extreme heat will increase the risk of disease and death for people with mental illness, including elderly populations and those taking prescription medications that impair the body's ability to regulate temperature *[High Confidence]*.

9 POPULATIONS OF CONCERN

Climate change is already causing, and is expected to continue to cause, a range of health impacts that vary across different population groups in the United States. The vulnerability of any given group is a function of its sensitivity to climate change related health risks, its exposure to those risks, and its capacity for responding to or coping with climate variability and change. Vulnerable groups of people, described here as populations of concern, include those with low income, some communities of color, immigrant groups (including those with limited English proficiency), Indigenous peoples, children and pregnant women, older adults, vulnerable occupational groups, persons with disabilities, and persons with preexisting or chronic medical conditions. Characterizations of vulnerability should consider how populations of concern experience disproportionate, multiple, and complex risks to their health and well-being in response to climate change.

Vulnerability Varies Over Time and Is Place-Specific

Key Finding 1: Across the United States, people and communities differ in their exposure, their inherent sensitivity, and their adaptive capacity to respond to and cope with climate change related health threats *[Very High Confidence]*. Vulnerability to climate change varies across time and location, across communities, and among individuals within communities *[Very High Confidence]*.

Health Impacts Vary with Age and Life Stage

Key Finding 2: People experience different inherent sensitivities to the impacts of climate change at different ages and life stages *[High Confidence]*. For example, the very young and the very old are particularly sensitive to climate-related health impacts.

Determinants of Vulnerability

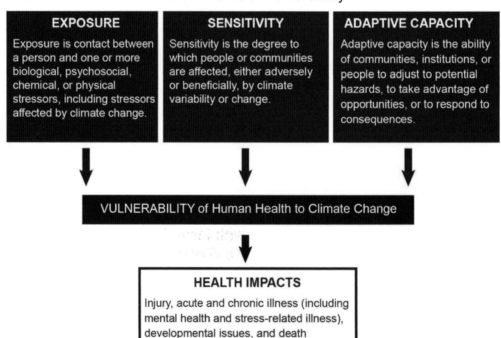

EXPOSURE	SENSITIVITY	ADAPTIVE CAPACITY
Exposure is contact between a person and one or more biological, psychosocial, chemical, or physical stressors, including stressors affected by climate change.	Sensitivity is the degree to which people or communities are affected, either adversely or beneficially, by climate variability or change.	Adaptive capacity is the ability of communities, institutions, or people to adjust to potential hazards, to take advantage of opportunities, or to respond to consequences.

VULNERABILITY of Human Health to Climate Change

HEALTH IMPACTS
Injury, acute and chronic illness (including mental health and stress-related illness), developmental issues, and death

Defining the determinants of vulnerability to health impacts associated with climate change, including exposure, sensitivity, and adaptive capacity (see Ch. 9: Populations of Concern). (Figure source: adapted from Turner et al. 2003)[8]

Mapping Tools and Vulnerability Indices Identify Climate Health Risks

Key Finding 4: The use of geographic data and tools allows for more sophisticated mapping of risk factors and social vulnerabilities to identify and protect specific locations and groups of people *[High Confidence].*

Social Determinants of Health Interact with Climate Factors to Affect Health Risk

Key Finding 3: Climate change threatens the health of people and communities by affecting exposure, sensitivity, and adaptive capacity *[High Confidence].* Social determinants of health, such as those related to socioeconomic factors and health disparities, may amplify, moderate, or otherwise influence climate-related health effects, particularly when these factors occur simultaneously or close in time or space *[High Confidence].*

(Left) Persons with disabilities often rely on medical equipment (such as portable oxygen) that requires an uninterrupted source of electricity. (Right) Climate-related exposures may lead to adverse pregnancy and newborn health outcomes.

Because of existing vulnerabilities, Indigenous people, especially those who are dependent on the environment for sustenance or who live in geographically isolated or impoverished communities, are likely to experience greater exposure and lower resilience to climate-related health effects.

References

1. Melillo, J.M., T.C. Richmond, and G.W. Yohe, Eds., 2014: *Climate Change Impacts in the United States: The Third National Climate Assessment*. U.S. Global Change Research Program, Washington, D.C., 842 pp. http://dx.doi.org/10.7930/J0Z31WJ2

2. Schwartz, J.D., M. Lee, P.L. Kinney, S. Yang, D. Mills, M. Sarofim, R. Jones, R. Streeter, A. St. Juliana, J. Peers, and R.M. Horton, 2015: Projections of temperature-attributable premature deaths in 209 U.S. cities using a cluster-based Poisson approach. *Environmental Health*, **14**. http://dx.doi.org/10.1186/s12940-015-0071-2

3. Fann, N., C.G. Nolte, P. Dolwick, T.L. Spero, A. Curry Brown, S. Phillips, and S. Anenberg, 2015: The geographic distribution and economic value of climate change-related ozone health impacts in the United States in 2030. *Journal of the Air & Waste Management Association*, **65**, 570-580. http://dx.doi.org/10.1080/10962247.2014.996270

4. NOAA, 2015: Natural Hazard Statistics: Weather Fatalities. National Oceanic and Atmospheric Administration, National Weather Service, Office of Climate, Water, and Weather Services. http://www.nws.noaa.gov/om/hazstats.shtml

5. Smith, A.B. and R.W. Katz, 2013: US billion-dollar weather and climate disasters: Data sources, trends, accuracy and biases. *Natural Hazards*, **67**, 387-410. http://dx.doi.org/10.1007/s11069-013-0566-5

6. CDC, 2015: Lyme Disease: Data and Statistics: Maps-Reported Cases of Lyme Disease – United States, 2001-2014. Centers for Disease Control and Prevention. http://www.cdc.gov/lyme/stats/

7. Brown, M.E., J.M. Antle, P. Backlund, E.R. Carr, W.E. Easterling, M.K. Walsh, C. Ammann, W. Attavanich, C.B. Barrett, M.F. Bellemare, V. Dancheck, C. Funk, K. Grace, J.S.I. Ingram, H. Jiang, H. Maletta, T. Mata, A. Murray, M. Ngugi, D. Ojima, B. O'Neill, and C. Tebaldi, 2015: Climate Change, Global Food Security, and the U.S. Food System. 146 pp. U.S. Global Change Research Program. http://www.usda.gov/oce/climate_change/FoodSecurity2015Assessment/FullAssessment.pdf

8. Clayton, S., C.M. Manning, and C. Hodge, 2014: Beyond Storms & Droughts: The Psychological Impacts of Climate Change. 51 pp. American Psychological Association and ecoAmerica, Washington, D.C. http://ecoamerica.org/wp-content/uploads/2014/06/eA_Beyond_Storms_and_Droughts_Psych_Impacts_of_Climate_Change.pdf

9. Turner, B.L., R.E. Kasperson, P.A. Matson, J.J. McCarthy, R.W. Corell, L. Christensen, N. Eckley, J.X. Kasperson, A. Luers, M.L. Martello, C. Polsky, A. Pulsipher, and A. Schiller, 2003: A framework for vulnerability analysis in sustainability science. *Proceedings of the National Academy of Sciences*, **100**, 8074-8079. http://dx.doi.org/10.1073/pnas.1231335100

1 INTRODUCTION: CLIMATE CHANGE AND HUMAN HEALTH

Lead Authors

John Balbus
National Institutes of Health

Allison Crimmins*
U.S. Environmental Protection Agency

Janet L. Gamble
U.S. Environmental Protection Agency

Contributing Authors

David R. Easterling
National Oceanic and Atmospheric Administration

Kenneth E. Kunkel
Cooperative Institute for Climate and Satellites–North Carolina

Shubhayu Saha
Centers for Disease Control and Prevention

Marcus C. Sarofim
U.S. Environmental Protection Agency

Recommended Citation: Balbus, J., A. Crimmins, J.L. Gamble, D.R. Easterling, K.E. Kunkel, S. Saha, and M.C. Sarofim, 2016: Ch. 1: Introduction: Climate Change and Human Health. *The Impacts of Climate Change on Human Health in the United States: A Scientific Assessment.* U.S. Global Change Research Program, Washington, DC, 25–42. http://dx.doi.org/10.7930/J0VX0DFW

On the web: health2016.globalchange.gov

*Chapter Coordinator

1 INTRODUCTION: CLIMATE CHANGE AND HUMAN HEALTH

Human health has always been influenced by climate and weather. Changes in climate and climate variability, particularly changes in weather extremes, affect the environment that provides us with clean air, food, water, shelter, and security. Climate change, together with other natural and human-made health stressors, threatens human health and well-being in numerous ways. Some of these health impacts are already being experienced in the United States.

Given that the impacts of climate change are projected to increase over the next century, certain existing health threats will intensify and new health threats may emerge. Connecting our understanding of how climate is changing with an understanding of how those changes may affect human health can inform decisions about mitigating (reducing) the amount of future climate change, suggest priorities for protecting public health, and help identify research needs.

1.1 Our Changing Climate

Observed Climate Change

The fact that the Earth has warmed over the last century is unequivocal. Multiple observations of air and ocean temperatures, sea level, and snow and ice have shown these changes to be unprecedented over decades to millennia. Human influence has been the dominant cause of this observed warming.[1] The 2014 U.S. National Climate Assessment (2014 NCA) found that rising temperatures, the resulting increases in the frequency or intensity of some extreme weather events, rising sea levels, and melting snow and ice are already disrupting people's lives and damaging some sectors of the U.S. economy.[2]

The concepts of climate and weather are often confused. *Weather* is the state of the atmosphere at any given time and place. Weather patterns vary greatly from year to year and from region to region. Familiar aspects of weather include temperature, precipitation, clouds, and wind that people experience throughout the course of a day. Severe weather conditions include hurricanes, tornadoes, blizzards, and droughts. *Climate* is the average weather conditions that persist over

multiple decades or longer. While the weather can change in minutes or hours, identifying a change in climate has required observations over a time period of decades to centuries or longer. Climate change encompasses both increases and decreases in temperature as well as shifts in precipitation, changing risks of certain types of severe weather events, and changes to other features of the climate system.

Observed changes in climate and weather differ at local and regional scales (Figure 1). Some climate and weather changes already observed in the United States include:[2, 3]

- U.S. average temperature has increased by 1.3°F to 1.9°F since record keeping began in 1895; most of this increase has occurred since about 1970. The first decade of the 2000s (2000–2009) was the warmest on record throughout the United States.

Major U.S. Climate Trends

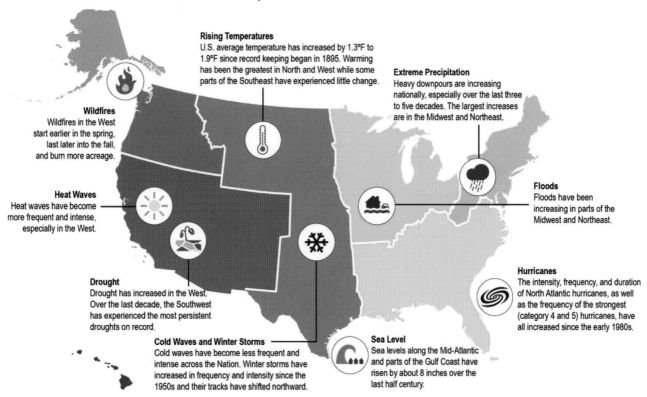

Rising Temperatures
U.S. average temperature has increased by 1.3°F to 1.9°F since record keeping began in 1895. Warming has been the greatest in North and West while some parts of the Southeast have experienced little change.

Extreme Precipitation
Heavy downpours are increasing nationally, especially over the last three to five decades. The largest increases are in the Midwest and Northeast.

Wildfires
Wildfires in the West start earlier in the spring, last later into the fall, and burn more acreage.

Heat Waves
Heat waves have become more frequent and intense, especially in the West.

Floods
Floods have been increasing in parts of the Midwest and Northeast.

Drought
Drought has increased in the West. Over the last decade, the Southwest has experienced the most persistent droughts on record.

Hurricanes
The intensity, frequency, and duration of North Atlantic hurricanes, as well as the frequency of the strongest (category 4 and 5) hurricanes, have all increased since the early 1980s.

Cold Waves and Winter Storms
Cold waves have become less frequent and intense across the Nation. Winter storms have increased in frequency and intensity since the 1950s and their tracks have shifted northward.

Sea Level
Sea levels along the Mid-Atlantic and parts of the Gulf Coast have risen by about 8 inches over the last half century.

Figure 1: Major U.S. national and regional climate trends. Shaded areas are the U.S. regions defined in the 2014 NCA.[2, 4]

- Average U.S. precipitation has increased since 1900, but some areas have experienced increases greater than the national average, and some areas have experienced decreases.

- Heavy downpours are increasing nationally, especially over the last three to five decades. The largest increases are in the Midwest and Northeast, where floods have also been increasing. Figure 2 shows how the annual number of heavy downpours, defined as extreme two-day precipitation events, for the contiguous United States has increased, particularly between the 1950s and the 2000s.

- Drought has increased in the West. Over the last decade, the Southwest has experienced the most persistent droughts since record keeping began in 1895.[4] Changes in precipitation and runoff, combined with changes in consumption and withdrawal, have reduced surface and groundwater supplies in many areas.

- There have been changes in some other types of extreme weather events over the last several decades. Heat waves have become more frequent and intense, especially in the West. Cold waves have become less frequent and intense across the nation.

Change in Number of Extreme Precipitation Events

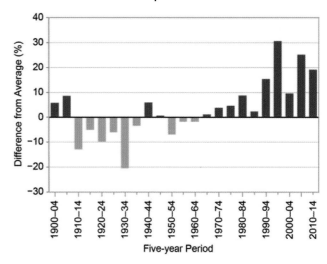

Figure 2: Time series of 5-year averages of the number of extreme 2-day duration precipitation events, averaged over the United States from 1900 to 2014. The number is expressed as the percent difference from the average for the entire period. This is based on 726 stations that have precipitation data for at least 90% of the days in the period. An event is considered extreme if the precipitation amount exceeds a threshold for a once-per-year recurrence. (Figure source: adapted from Melillo et al. 2014)[2]

- The intensity, frequency, and duration of North Atlantic hurricanes, as well as the frequency of the strongest (category 4 and 5) hurricanes, have all increased since the early 1980s. The relative contributions of human and natural causes to these increases remain uncertain.

Projected Climate Change

Projections of future climate conditions are based on results from climate models—sophisticated computer programs that simulate the behavior of the Earth's climate system. These climate models are used to project how the climate system is expected to change under different possible scenarios. These scenarios describe future changes in atmospheric greenhouse gas concentrations, land use, other human influences on climate, and natural factors. The most recent set of coordinated climate model simulations use a set of scenarios called Representative Concentration Pathways (RCPs), which describe four possible trajectories in greenhouse gas concentrations.[1] Actual future greenhouse gas concentrations, and the resulting amount of future climate change, will still largely be determined by choices society makes about emissions.[2] The RCPs, and the temperature increases associated with these scenarios, are described in more detail in Appendix 1: Technical Support Document and in the 2014 NCA.[3,5,6]

Some of the projected changes in climate in the United States as described in the 2014 NCA are listed below:[2,3]

- Temperatures in the United States are expected to continue to rise. This temperature rise has not been, and will not be, uniform across the country or over time (Figure 3, top panels).

- Increases are also projected for extreme temperature conditions. The temperature of both the hottest day and coldest night of the year are projected to increase (Figure 4, top panels).

- More winter and spring precipitation is projected for the northern United States, and less for the Southwest, over this century (Figure 3, bottom panels).

- Increases in the frequency and intensity of extreme precipitation events are projected for all U.S. areas (Figure 4, bottom panels).

- Short-term (seasonal or shorter) droughts are expected to intensify in most U.S. regions. Longer-term droughts are expected to intensify in large areas of the Southwest, the southern Great Plains, and the Southeast. Trends in reduced surface and groundwater supplies in many areas are expected to continue, increasing the likelihood of water shortages for many uses.

Projected Changes in Temperature and Precipitation by Mid-Century

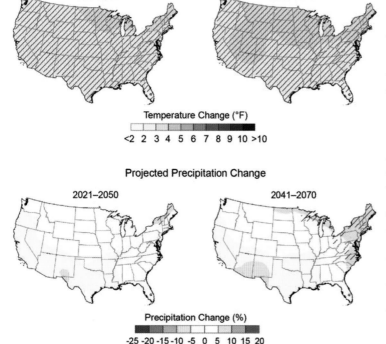

2021–2050   **2041–2070**

Temperature Change (°F)

<2 2 3 4 5 6 7 8 9 10 >10

Projected Precipitation Change

2021–2050   **2041–2070**

Precipitation Change (%)

-25 -20 -15 -10 -5 0 5 10 15 20

Figure 3: Projected changes in annual average temperature (top) and precipitation (bottom) for 2021–2050 (left) and 2041–2070 (right) with respect to the average for 1971–2000 for the RCP6.0 scenario. The RCP6.0 pathway projects an average global temperature increase of 5.2°F in 2100 over the 1901–1960 global average temperature (the RCPs are described in more detail in Appendix 1: Technical Support Document). Temperature increases in the United States for this scenario (top panels) are in the 2°F to 3°F range for 2021 to 2050 and 2°F to 4°F for 2041 to 2070. This means that the increase in temperature projected in the United States over the next 50 years under this scenario would be larger than the 1°F to 2°F increase in temperature that has already been observed over the previous century. Precipitation is projected to decrease in the Southwest and increase in the Northeast (bottom panels). These projected changes are statistically significant (95% confidence) in small portions of the Northeast, as indicated by the hatching. (Figure source: adapted from Sun et al. 2015)[54]

Projected Changes in the Hottest/Coldest and Wettest/Driest Day of the Year

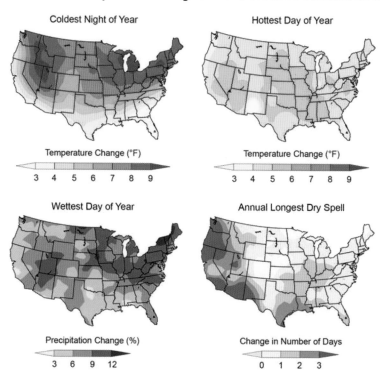

Figure 4: Projected changes in several climate variables for 2046–2065 with respect to the 1981–2000 average for the RCP6.0 scenario. These include the coldest night of the year (top left) and the hottest day of the year (top right). By the middle of this century, the coldest night of the year is projected to warm by 6°F to 10°F over most of the country, with slightly smaller changes in the south. The warmest day of the year is projected to be 4°F to 6°F warmer in most areas. Also shown are projections of the wettest day of the year (bottom left) and the annual longest consecutive dry day spell (bottom right). Extreme precipitation is projected to increase, with an average change of 5% to 15% in the precipitation falling on the wettest day of the year. The length of the annual longest dry spell is projected to increase in most areas, but these changes are small: less than two days in most areas. (Figure source: adapted from Sun et al. 2015)[54]

- Heat waves are projected to become more intense, and cold waves less intense, everywhere in the United States.

- Hurricane-associated storm intensity and rainfall rates are projected to increase as the climate continues to warm.

1.2 How Does Climate Change Affect Health?

The influences of weather and climate on human health are significant and varied. They range from the clear threats of temperature extremes and severe storms to connections that may seem less obvious. For example, weather and climate affect the survival, distribution, and behavior of mosquitoes, ticks, and rodents that carry diseases like West Nile virus or Lyme disease. Climate and weather can also affect water and food quality in particular areas, with implications for human health. In addition, the effects of global climate change on mental health and well-being are integral parts of the overall climate-related human health impact.

A useful approach to understand how climate change affects health is to consider specific exposure pathways and how they can lead to human disease. The concept of exposure pathways is adapted from its use in chemical risk assessment, and in this context describes the main routes by which climate change affects health (see Figure 5). Exposure pathways differ over time and in different locations, and climate change related exposures can affect different people and different communities to different degrees. While often assessed individually, exposure to multiple climate change threats can occur simultaneously,

resulting in compounding or cascading health impacts. Climate change threats may also accumulate over time, leading to longer-term changes in resilience and health.

Whether or not a person is exposed to a health threat or suffers illness or other adverse health outcomes from that exposure depends on a complex set of vulnerability factors. Vulnerability is the tendency or predisposition to be adversely affected by climate-related health effects, and encompasses three elements: exposure, sensitivity or susceptibility to harm, and the capacity to adapt or to cope (see also Figure 1 in Ch. 9: Populations of Concern). Because multiple disciplines use these terms differently and multiple definitions exist in the literature, the distinctions between them are not always clear.[7] All three of these elements can change over time and are place- and system-specific.[8] In the context of this report, we define the three elements of vulnerability as follows (definitions adapted from IPCC 2014 and NRC 2012)[9, 10]

- ***Exposure*** is contact between a person and one or more biological, psychosocial, chemical, or physical stressors, including stressors affected by climate change. Contact may occur in a single instance or repeatedly over time, and may occur in one location or over a wider geographic area.

- ***Sensitivity*** is the degree to which people or communities are affected, either adversely or beneficially, by climate variability or change.

• ***Adaptive capacity*** is the ability of communities, institutions, or people to adjust to potential hazards, to take advantage of opportunities, or to respond to consequences. A related term, *resilience*, is the ability to prepare and plan for, absorb, recover from, and more successfully adapt to adverse events.

Vulnerability, and the three components of vulnerability, are factors that operate at multiple levels, from the individual and community to the country level, and affect all people to some degree.[8] For an individual, these factors include human behavioral choices and the degree to which that person is vulnerable based on his or her level of exposure, sensitivity, and adaptive capacity. Vulnerability is also influenced by social determinants of health (see Ch. 9 Populations of Concern), including those that affect a person's adaptive capacity, such as social capital

and social cohesion (for example, the strength of interpersonal networks and social patterns in a community).

At a larger community or societal scale, health outcomes are strongly influenced by adaptive capacity factors, including those related to the natural and built environments (for example, the state of infrastructure), governance and management (health-protective surveillance programs, regulations and enforcement, or community health programs), and institutions (organizations operating at all levels to form a national public health system).[11, 12] For example, water resource, public health, and environmental agencies in the United States provide many public health safeguards, such as monitoring water quality and issuing advisories to reduce risk of exposure and illness if water becomes contaminated. Some aspects of climate change health impacts in the United States may therefore be

Climate Change and Health

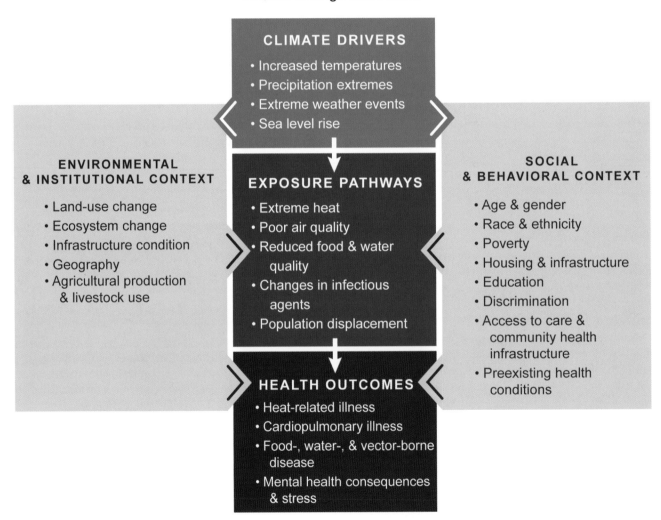

Figure 5: Conceptual diagram illustrating the exposure pathways by which climate change affects human health. Exposure pathways exist within the context of other factors that positively or negatively influence health outcomes (gray side boxes). Key factors that influence vulnerability for individuals are shown in the right box, and include social determinants of health and behavioral choices. Key factors that influence vulnerability at larger scales, such as natural and built environments, governance and management, and institutions, are shown in the left box. All of these influencing factors can affect an individual's or a community's vulnerability through changes in exposure, sensitivity, and adaptive capacity and may also be affected by climate change.

mediated by factors like strong social capital, fully functional governance/management, and institutions that maintain the Nation's generally high level of adaptive capacity. On the other hand, the evidence base regarding the effectiveness of public health interventions in a climate change context is still relatively weak.[13] Current levels of adaptive capacity may not be sufficient to address multiple impacts that occur simultaneously or in close succession, or impacts of climate change that result in unprecedented damages.[2, 12]

The three components of vulnerability (exposure, sensitivity, and adaptive capacity) are associated with social and demographic factors, including level of wealth and education, as well as other characteristics of people and places, such as the condition of infrastructure and extent of ecosystem degradation. For example, poverty can leave people more exposed to climate and weather threats, increase sensitivity because of associations with higher rates of illness and nutritional deficits, and limit people's adaptive capacity. As another example, people living in a city with degraded coastal ecosystems and inadequate water and wastewater infrastructure may be at greater risk of health consequences from severe storms. Figure 5 demonstrates the interactions among climate drivers, health impacts, and other factors that influence people's vulnerability to health impacts.

We are already experiencing changes in the frequency, severity, and even the location of some weather and climate

> *Current levels of adaptive capacity may not be sufficient to address multiple impacts that occur simultaneously or in close succession, or impacts of climate change that result in unprecedented damages.*

phenomena, including extreme temperatures, heavy rains and droughts, and some other kinds of severe weather, and these changes are projected to continue. This means that areas already experiencing health-threatening weather and climate phenomena, such as severe heat or hurricanes, are likely to experience worsening impacts, such as even higher temperatures and increased storm intensity, rainfall rates, and storm surge. It also means that some areas will experience new climate-related health threats. For example, areas previously unaffected by toxic algal blooms or waterborne diseases because of cooler water temperatures may face these hazards in the future as increasing water temperatures allow the organisms that cause these health risks to thrive. Even areas that currently experience these health threats may see a shift in the timing of the seasons that pose the greatest risk to human health.

Climate change can therefore affect human health in two main ways: first, by changing the severity or frequency of health problems that are already affected by climate or weather factors; and second, by creating unprecedented or unanticipated health problems or health threats in places where they have not previously occurred.

1.3 Our Changing Health

In order to understand how climate change creates or exacerbates health problems, assessments of climate change health impacts must start with what is known about the current state and observed trends in a wide array of health conditions. In addition, because preexisting health conditions, socioeconomic status, and life stage all contribute to *vulnerability* to climate-related and weather-related health effects, assessments of climate change health impacts should be informed by projected changes in these factors. In cases where people's health or socioeconomic status is getting worse, climate change may accentuate the health burdens associated with those worsening trends. Conversely, in cases where people's health or socioeconomic status is improving, the effect of climate change may be to slow or reduce that improvement. Where the state of scientific understanding allows, the inclusion of projected trends in health and socioeconomic conditions into models of climate change impacts on health can provide useful insights into these interactions between non-climate factors and climate change effects.

Demographic and Socioeconomic Trends

The United States is in the midst of several significant demographic changes: the population is aging, growing in number, becoming more ethnically diverse, and demonstrating greater disparities between the wealthy and the poor. Immigration is

Storm-damaged home after Hurricane Sandy.

Terminology

Incidence: A measure of the frequency with which an event, such as a new case of illness, occurs in a population over a period of time.

Morbidity: A disease or condition that reduces health and the quality of life. The morbidity rate is a measure of the frequency of disease among a defined population during a specified time period.

Mortality: Death as a health outcome. The mortality rate is the number of deaths in a defined population during a specified time period.

Premature (early) mortality or death: Deaths that occur earlier than a specified age, often the average life expectancy at birth.

Prevalence: A measure of the number or proportion of people with a specific disease or condition at a specific point in time.

Surveillance: The collection, analysis, interpretation, and dissemination of health data.

having a major influence on both the size and age distribution of the population.[14] Each of these demographic trends has implications for climate change related human health impacts (see Ch. 9: Populations of Concern). Some of these trends and projections are summarized below:

Trends in population growth

• The total U.S. population has more than doubled since 1950, from 151,325,798 persons in 1950 to 308,745,538 in 2010.[15]

• The Census Bureau projects that the U.S. population will grow to almost 400 million by 2050 (from estimates of about 320 million in 2014).[16]

Trends in the elderly population

• The nation's older adult population (ages 65 and older) will nearly double in number from 2015 through 2050, from approximately 48 million to 88 million.[17] Of those 88 million older adults, a little under 19 million will be 85 years of age and older.[18]

Trends in racial and ethnic diversity

• As the United States becomes more diverse, the aggregate minority population is projected to become the majority by 2042.[17] The non-Hispanic or non-Latino White population will increase, but more slowly than other racial groups. Non-Hispanic Whites are projected to become a minority by 2050.[19]

• Projections for 2050 suggest that nearly 19% of the population will be immigrants, compared with 12% in 2005.[19]

• The Hispanic population is projected to nearly double from 12.5% of the U.S. population in 2000 to 24.6% in 2050.[20]

Trends in economic disparity

• Income inequality rose and then stabilized during the last 30 years, and is projected to resume rising over the next 20 years, though at a somewhat slower overall rate that declines to near zero by 2035.[21] For example, the Gini coefficient, a measure of income inequality, is estimated to have risen by 18% between 1984 and 2000, and is projected to rise by an additional 17% for all workers between 2009 and 2035.[21]

• America's communities of color have disproportionately higher poverty rates and lower income levels. While racial disparities in household wealth were higher in the late 1980s than now, trends in more recent years have been toward greater inequality. The ratio of the median net household worth of White, non-Hispanic versus non-White or Hispanic households increased from 6.0 to 7.8 between 2007 and 2013.[22] In 2009, 25.8% of non-Hispanic Blacks and 25.3% of Hispanics had incomes below the poverty level as compared to 9.4% of non-Hispanic Whites and 12.5% of Asian Americans.[23] In 2014, the median income level for a non-Hispanic Black household was approximately $35,000, $25,000 lower than a non-Hispanic White household.[24]

Population growth and migration in the United States may place more people at risk of the health impacts of climate change, especially as more people are located in and around vulnerable areas, such as coastal, low-lying, or flood-prone zones;[25] densely populated urban areas;[26] and drought-stricken or wildfire-prone regions. Increases in racial and ethnic diversity and in the number of persons living near the poverty line may increase the risk of health impacts from climate change. Economic disparity can make it difficult for some populations to respond to dangerous weather conditions, especially when evacuation is necessary or when the aftermath requires rebuilding of homes and businesses not covered by home or property insurance.

Trends in Health Status

As a nation, trends in the population's health are mixed. Some major indicators of health, such as life expectancy, are consistently improving, while others, such as rate and number of diabetes deaths, are getting worse. Changes in these metrics may differ across populations and over time. For example, though rates of obesity have increased in both children and adults over the last 30 years or more, rates over just the last decade have remained steady for adults but increased among children.[27]

Climate change impacts to human health will act on top of these underlying trends. Some of these underlying health conditions can increase sensitivity to climate change effects such as heat waves and worsening air quality (see Ch. 2: Temperature-Related Death and Illness; Ch. 3: Air Quality Impacts; Ch. 9: Populations of Concern). Understanding the trends in these conditions is therefore important in considering how many people are likely to experience illness when exposed to these climate change effects. Potential climate change related health impacts may reduce the improvements that would otherwise be expected in some indicators of health status and accentuate trends towards poorer health in other health indicators.[1, 28]

Examples of health indicators that have been improving between 2000 and 2013 include the following:

• Life expectancy at birth increased from 76.8 to 78.8 years.[29]

• Death rates per 100,000 people from heart disease and cancer decreased from 257.6 to 169.8 and from 199.6 to 163.2, respectively.[29]

• The percent of people over age 18 who say they smoke decreased from 23.2% to 17.8%.[29]

At the same time, some health trends related to the prevalence of chronic diseases, self-reported ill health, and disease risk factors have been getting worse. For example:

Diabetes increases sensitivity to heat stress.

• The percentage of adult (18 years and older) Americans describing their health as "poor or fair" increased from 8.9% in 2000 to 10.3% in 2012.[29]

• Prevalence of physician-diagnosed diabetes among adults aged 20 and over increased from 5.2% in 1988–1994 to 8.4% in 2009-2012.[29]

• The prevalence of obesity among adults (aged 20–74) increased by almost three-fold from 1960–1962 (13.4% of adults classified as obese) to 2009–2010 (36.1% of adults classified as obese).[30]

• In the past 30 years, obesity has more than doubled in children and quadrupled in adolescents in the United States. The percentage of children aged 6–11 who were obese increased from 7% in 1980 to nearly 18% in 2012. Similarly, the percentage of adolescents aged 12–19 years who were obese increased from 5% to nearly 21% over the same period. In 2012, approximately one-third of American children and adolescents were overweight or obese.[31]

Table 1 shows some examples of underlying health conditions that are associated with increased vulnerability to health effects from climate change related exposures (see Ch. 9: Populations of Concern for more details) and provides information on current status and future trends.

Health status is often associated with demographics and socio-economic status. Changes in the overall size of the population, racial and ethnic composition, and age distribution affect the health status of the population. Poverty, educational attainment, access to care, and discrimination all contribute to disparities in the incidence and prevalence of a variety of medical conditions (see Ch. 9: Populations of Concern). Some examples of these interactions include:

Older Adults. In 2013, the percentage of adults age 75 and older described as persons in fair or poor health totaled 27.6%, as compared to 6.2% for adults age 18 to 44.[29] Among adults age 65 and older, the number in nursing homes or other residential care facilities totaled 1.8 million in 2012, with more than 1 million utilizing home health care.[32]

Children. Approximately 9.0% of children in the United States have asthma. Between 2011 and 2013, rates for Black (15.3%) and Hispanic (8.6%) children were higher than the rate for White (7.8%) children.[29] Rates of asthma were also higher in poor children who live below 100% of the poverty level (12.4%).[29]

Table 1: Current estimates and future trends in chronic health conditions that interact with the health risks associated with climate change.

Health Conditions	Current Estimates	Future Trends	Possible Influences of Climate Change
Alzheimer's Disease	Approximately 5 million Americans over 65 had Alzheimer's disease in 2013.[33]	Prevalence of Alzheimer's is expected to triple to 13.8 million by 2050.[33]	Persons with cognitive impairments are vulnerable to extreme weather events that require evacuation or other emergency responses.
Asthma	Average asthma prevalence in the U.S. was higher in children (9% in 2014)[29] than in adults (7% in 2013).[34] Since the 1980s, asthma prevalence increased, but rates of asthma deaths and hospital admissions declined.[35, 36]	Stable incidence and increasing prevalence of asthma is projected in the U.S. in coming decades.	Asthma is exacerbated by changes in pollen season and allergenicity and in exposures to air pollutants affected by changes in temperature, humidity, and wind.[28]
Chronic Obstructive Pulmonary Disease (COPD)	In 2012, approximately 6.3% of adults had COPD. Deaths from chronic lung diseases increased by 50% from 1980 to 2010.[37, 38]	Chronic respiratory diseases are the third leading cause of death and are expected to become some of the most costly illnesses in coming decades.[37]	COPD patients are more sensitive than the general population to changes in ambient air quality associated with climate change.
Diabetes	In 2012, approximately 9% of the total U.S. population had diabetes. Approximately 18,400 people younger than age 20 were newly diagnosed with type 1 diabetes in 2008–2009; an additional 5,000 were diagnosed with type 2.[39]	New diabetes cases are projected to increase from about 8 cases per 1,000 in 2008 to about 15 per 1,000 in 2050. If recent increases continue, prevalence is projected to increase to 33% of Americans by 2050.[40]	Diabetes increases sensitivity to heat stress; medication and dietary needs may increase vulnerability during and after extreme weather events.
Cardiovascular Disease	Cardiovascular disease (CVD) is the leading cause of death in the U.S.[41]	By 2030, approximately 41% of the U.S. population is projected to have some form of CVD.[42]	Cardiovascular disease increases sensitivity to heat stress.
Mental Illness	Depression is one of the most common types of mental illness, with approximately 7% of adults reporting a major episode in the past year. Lifetime prevalence is approximately twice as high for women as for men.[43] Lifetime prevalence is more than 15% for anxiety disorders and nearly 4% for bipolar disorder.[44]	By 2050, the total number of U.S. adults with depressive disorder is projected to increase by 35%, from 33.9 million to 45.8 million, with those over age 65 having a 117% increase.[43]	Mental illness may impair responses to extreme events; certain medications increase sensitivity to heat stress.
Obesity	In 2009–2010, approximately 35% of American adults were obese.[31] In 2012, approximately 32% of youth (aged 2–19) were overweight or obese.[45, 46]	By 2030, 51% of the U.S. population is expected to be obese. Projections suggest a 33% increase in obesity and a 130% increase in severe obesity.[47]	Obesity increases sensitivity to high ambient temperatures.
Disability	Approximately 18.7% of the U.S. population has a disability. In 2010, the percent of American adults with a disability was approximately 16.6% for those age 21–64 and 49.8% for persons 65 and older.[48]	The number of older adults with activity limitations is expected to grow from 22 million in 2005 to 38 million in 2030.[49]	Persons with disabilities may find it hard to respond when evacuation is required and when there is no available means of transportation or easy exit from residences

Non-Hispanic Blacks. In 2014, the percentage of non-Hispanic Blacks of all ages who were described as persons in fair or poor health totaled 14.3% as compared to 8.7% for Whites. Health risk factors for this population include high rates of smoking, obesity, and hypertension in adults, as well as high infant death rates.[29]

Hispanics. The percentage of Hispanics of all ages who were described as persons in fair or poor health totaled 12.7% in 2014. Health disparities for Hispanics include moderately higher rates of smoking in adults, low birth weights, and infant deaths.[29]

The impacts of climate change may worsen these health disparities by exacerbating some of the underlying conditions they create. For example, disparities in life expectancy may be exacerbated by the effects of climate change related heat and air pollution on minority populations that have higher rates of hypertension, smoking, and diabetes. Conversely, public health measures that reduce disparities and overall rates of illness in populations would lessen vulnerability to worsening of health status from climate change effects.

1.4 Quantifying Health Impacts

For some changes in exposures to health risks related to climate change, the future rate of a health impact associated with any given environmental exposure can be estimated by multiplying three values: 1) the baseline rate of the health impact, 2) the expected change in exposure, and 3) the exposure–response function. An exposure–response function is an estimate of how the risk of a health impact changes with changes in exposures, and is related to sensitivity, one of the three components of vulnerability. For example, an exposure–response function for extreme heat might be used to quantify the increase in heat-related deaths in a region (the change in health impact) for every 1°F increase in daily ambient temperature (the change in exposure).

Asthma affects approximately 9% of children in the United States.

Where there is a lack of data or these relationships are poorly understood, health impacts are harder to project. For more information on exposure–response (also called dose–response or concentration–response) functions, see the Exposure–Response section in Appendix 1: Technical Support Document.

Information on trends in underlying health or background rates of health impacts is summarized in Section 1.3, "Our Changing Health." Data on the incidence and prevalence of health conditions are obtained through a complicated system of state- and city-level surveillance programs, national health surveys, and national collection of data on hospitalizations, emergency room visits, and deaths. For example, data on the incidence of a number of infectious diseases are captured through the National Notifiable Diseases Surveillance System.[50]

| Future Rate of Health Impact | = | Baseline Health Status | × | Expected Change in Exposure | × | Exposure Response Function |

The ability to quantify many types of health impacts is dependent on the availability of data on the baseline incidence or prevalence of the health impact, the ability to characterize the future changes in the types of exposures relevant to that health impact, and how well the relationship between these exposures and health impacts is understood. Health impacts with many intervening factors, like infectious diseases, may require different and more complex modeling approaches. Where our understanding of these relationships is strong, some health impacts, even those occurring in unprecedented places or times of the year, may in fact be predictable.

This system relies first on the mandatory reporting of specific diseases by health care providers to state, local, territorial, and tribal health departments. These reporting jurisdictions then have the option of voluntarily providing the Centers for Disease Control and Prevention (CDC) with data on a set of nationally notifiable diseases. Because of challenges with getting health care providers to confirm and report specific diagnoses of reportable diseases in their patients, and the lack of requirements for reporting a consistent set of diseases and forwarding data to CDC, incidence of infectious disease is generally believed to be underreported, and actual rates are uncertain.[51]

Characterizing certain types of climate change related exposures can be a challenge. Exposures can consist of temperature changes and other weather conditions, inhaling air pollutants and pollens, consuming unsafe food supplies or contaminated water, or experiencing trauma or other mental health consequences from weather disasters. For some health impacts, the ability to understand the relationships between climate-related exposures and health impacts is limited by these difficulties in characterizing exposures or in obtaining accurate data on the occurrence of illnesses. For these health impacts, scientists may not have the capability to project changes in a health outcome (like incidence of diseases), and can only estimate how risks of exposure will change. For example, modeling capabilities allow projections of the impact of rising water temperatures on the concentration of *Vibrio* bacteria, which provides an understanding of geographic changes

in exposure but does not capture how people may be exposed and how many will actually become sick (see Ch. 6: Water-Related Illness). Nonetheless, the ability to project changes in exposure or in intermediate determinants of health impacts may improve understanding of the change in health *risks*, even if modeling quantitative changes in health *impacts* is not possible. For example, seasonal temperature and precipitation projections may be combined to assess future changes in ambient pollen concentrations (the exposure that creates risk), even though the potential associated increase in respiratory and allergic diseases (the health impacts) cannot be directly modeled (see Ch. 3: Air Quality Impacts).

Modeling Approaches Used in this Report

Four chapters within this assessment—Ch. 2: Temperature-Related Death and Illness, Ch. 3: Air Quality Impacts, Ch. 5:

Sources of Uncertainty

CLIMATE DRIVERS

Changes in climate that directly or indirectly affect human health.

PROJECTING CLIMATE CHANGE IMPACTS

- Future concentrations of GHGs (greenhouse gases)
- Future warming that will occur from a given increase in GHG concentration

EXPOSURE PATHWAYS

Links, routes, or pathways, through which people are exposed to climate change impacts that can affect human health.

UNDERSTANDING CHANGES IN VULNERABLITY

- Underlying health context, including demographic and socioeconomic trends and health status
- Interaction of changes in exposure, sensitivity, and adaptive capacity at individual, community, and institutional scales

HEALTH IMPACTS

Changes in or risks to the health status of individuals or groups.

ESTIMATING EXPOSURE–RESPONSE RELATIONSHIPS

- Change in health effects caused by different levels of exposure (linear or non-linear)
- Role of factors that modify the relationship between exposure and health outcomes

HEALTH OUTCOMES

Overall change in public health burden inclusive of intervention, adaptation, and mitigation.

PUBLIC HEALTH SURVEILLANCE & MONITORING

- Source, access to, and quality of socioeconomic, geographic, demographic, and health data
- Spatial and temporal variability in disease patterns or trends across populations

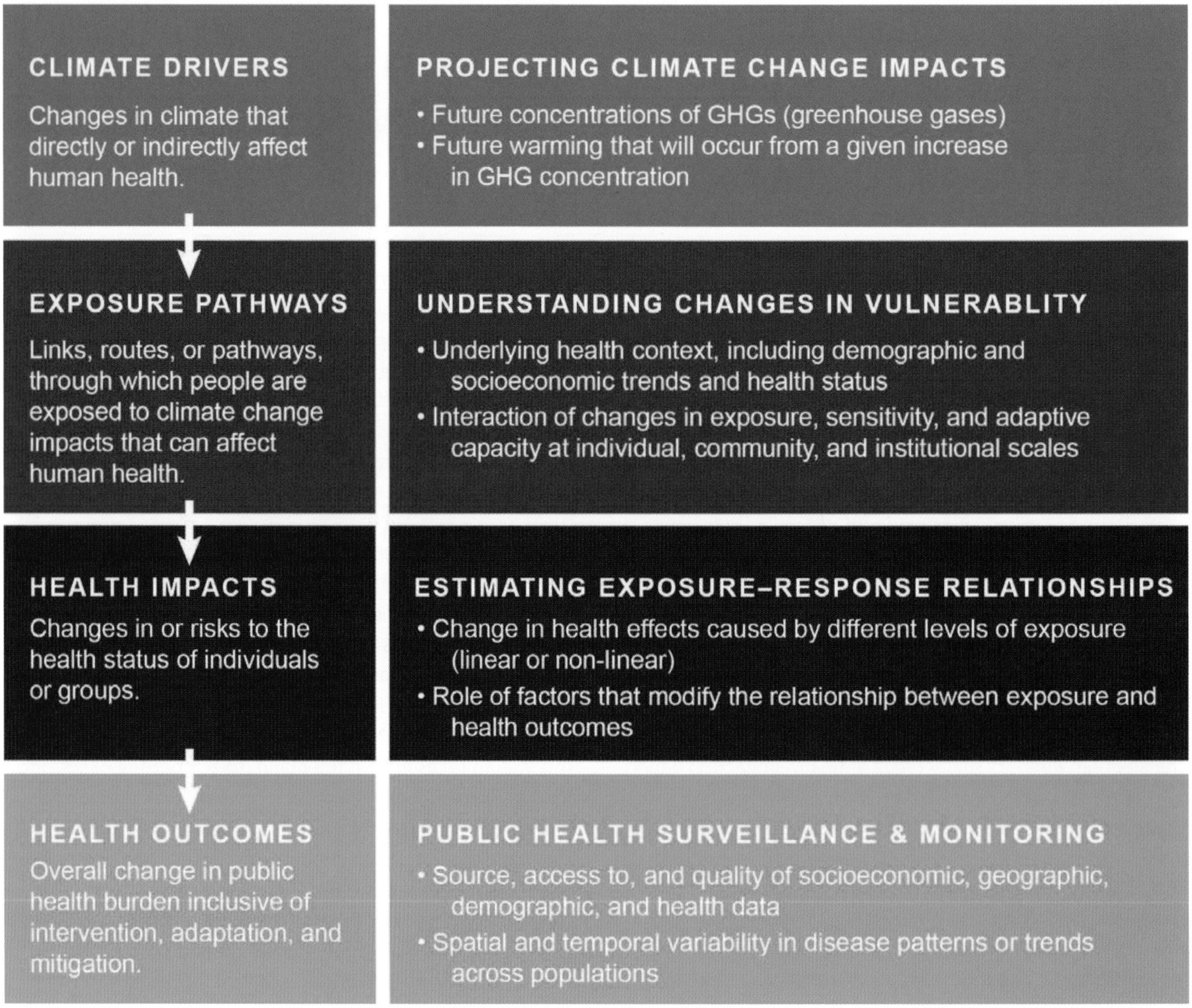

Figure 6: Examples of sources of uncertainty in projecting impacts of climate change on human health. The left column illustrates the exposure pathway through which climate change can affect human health. The right column lists examples of key sources of uncertainty surrounding effects of climate change at each stage along the exposure pathway.

Vector-Borne Diseases, and Ch. 6: Water-Related Illness—include new peer-reviewed, quantitative analyses based on modeling. The analyses highlighted in these chapters mainly relied on climate model output from the Coupled Model Intercomparison Project Phase 5 (CMIP5). Due to limited data availability and computational resources, the studies highlighted in the four chapters analyzed only a subset of the full CMIP5 dataset, with most of the studies including at least one analysis based on RCP6.0, an upper midrange greenhouse gas concentration pathway, to facilitate comparisons across chapters. For example, the air quality analysis examined results from two different RCPs, with a different climate model used for each, while the waterborne analyses examined results from 21 of the CMIP5 models for a single RCP. See the Guide to the Report and Appendix 1: Technical Support Document for more on modeling and scenarios.

Adverse health effects attributed to climate change can have many economic and social consequences, including direct medical costs, work loss, increased care giving, and other limitations on everyday activities

Adverse health effects attributed to climate change can have many economic and social consequences, including direct medical costs, work loss, increased care giving, and other limitations on everyday activities. Though economic impacts are a crucial component to understanding risk from climate change, and may have important direct and secondary impacts on human health and well-being by reducing resources available for other preventative health measures, economic valuation of the health impacts was not reported in this assessment.

Uncertainty in Health Impact Assessments

Figure 6 illustrates different sources of uncertainty along the exposure pathway.

Two of the key uncertainties in projecting future global temperatures are 1) uncertainty about future concentrations of greenhouse gases, and 2) uncertainty about how much warming will occur for a given increase in greenhouse gas concentrations. The Intergovernmental Panel on Climate Change's Fifth Assessment Report found that the most likely response of the climate system to a doubling of carbon dioxide concentrations lies between a 1.5°C and 4.5°C (2.7°F to 8.1°F) increase in global average temperature.[1] Future concentrations depend on both future emissions and how long these emissions remain in the atmosphere (which can vary depending on how natural systems process those emissions). To capture these uncertainties, climate modelers often use multiple models, analyze multiple scenarios, and conduct sensitivity analyses to assess the significance of these uncertainties.

Uncertainty in current and future estimates of health or socioeconomic status is related to several factors. In general, estimates are more uncertain for less-prevalent health conditions

(such as rare cancers versus cardiovascular disease), smaller subpopulations (such as Hispanic subpopulations versus White adults), smaller geographic areas (census tracts versus state or national scale), and time periods further into the future (decades versus seasons or years). Most current estimates of disease prevalence or socioeconomic status have uncertainty expressed as standard errors or confidence intervals that are derived from sampling methods and sample sizes. When modeling health impacts using data on health prevalence or socioeconomic status, these measures of uncertainty are typically included in the analysis to help establish a range of plausible results. Expert judgment is typically used to assess the overall effects of uncertainty from estimates of health or socioeconomic status when assessing the scientific literature.

The factors related to uncertainty in exposure–response functions are similar to those for the projections of health or socioeconomic status. Estimates are more uncertain for smaller subpopulations, less-prevalent health conditions, and smaller geographic areas. Because these estimates are based on observations of real populations, their validity when applied to populations in the future is more uncertain the further into the future the application occurs. Uncertainty in the estimates of the exposure–outcome relationship also comes from factors related to the scientific quality of relevant studies, including appropriateness of methods, source of data, and size of study populations. Expert judgment is used to evaluate the validity of an individual study as well as the collected group of relevant studies in assessing uncertainty in estimates of exposure–outcome relationships.

Approach to Reporting Uncertainty in Key Findings

Despite the sources of uncertainty described above, the current state of the science allows an examination of the likely direction of and trends in the health impacts of climate change. Over the past ten years, the models used for climate and health assessments have become more useful and more accurate (for example, Melillo et al. 2014).[6, 52, 53] This assessment builds on that improved capability. A more detailed discussion of the approaches to addressing uncertainty from the various sources can be found in the Guide to the Report and Appendix 1: Technical Support Document.

Two kinds of language are used when describing the uncertainty associated with specific statements in this report: confidence language and likelihood language. Confidence in the validity of a finding is expressed qualitatively and is based on the type, amount, quality, strength, and consistency of evidence and the degree of expert agreement on the finding. Likelihood, or the projected probability of an impact occurring, is based on quantitative estimates or measures of uncertainty expressed probabilistically (in other words, based on statistical analysis of observations or model results, or on expert judgment). Whether a Key Finding has a confidence level associated with it or, where findings can be quantified, both a confidence and likelihood level associated with it, involves the expert assessment and consensus of the chapter author teams.

Confidence Level

Very High
Strong evidence (established theory, multiple sources, consistent results, well documented and accepted methods, etc.), high consensus
High
Moderate evidence (several sources, some consistency, methods vary and/or documentation limited, etc.), medium consensus
Medium
Suggestive evidence (a few sources, limited consistency, models incomplete, methods emerging, etc.), competing schools of thought
Low
Inconclusive evidence (limited sources, extrapolations, inconsistent findings, poor documentation and/or methods not tested, etc.), disagreement or lack of opinions among experts

Likelihood

Very Likely
$\geq$ 9 in 10
Likely
$\geq$ 2 in 3
As Likely As Not
$\approx$ 1 in 2
Unlikely
$\leq$ 1 in 3
Very Unlikely
$\leq$ 1 in 10

PHOTO CREDITS

Pg. 25–Los Angeles, California skyline: © Lisa Romerein/Corbis

Pg. 26– Doctor showing girl how to use stethoscope: © John Fedele/Blend Images/Corbis

Pg. 31–Collapsed house after Hurricane Sandy: © iStockPhoto.com/Aneese

Pg. 33– Woman checking blood sugar levels: © Monkey Business Images/Corbis

Pg. 35–Girl suffering from asthma: © Stephen Welstead/LWA/Corbis

References

1. IPCC, 2013: Climate Change 2013: The Physical Science Basis. Contribution of Working Group I to the Fifth Assessment Report of the Intergovernmental Panel on Climate Change. Stocker, T.F., D. Qin, G.-K. Plattner, M. Tignor, S.K. Allen, J. Boschung, A. Nauels, Y. Xia, V. Bex, and P.M. Midgley (Eds.), 1535 pp. Cambridge University Press, Cambridge, UK and New York, NY. http://www.climatechange2013.org

2. Melillo, J.M., T.C. Richmond, and G.W. Yohe, Eds., 2014: *Climate Change Impacts in the United States: The Third National Climate Assessment*. U.S. Global Change Research Program, Washington, D.C., 842 pp. http://dx.doi.org/10.7930/J0Z31WJ2

3. Walsh, J., D. Wuebbles, K. Hayhoe, J. Kossin, K. Kunkel, G. Stephens, P. Thorne, R. Vose, M. Wehner, J. Willis, D. Anderson, S. Doney, R. Feely, P. Hennon, V. Kharin, T. Knutson, F. Landerer, T. Lenton, J. Kennedy, and R. Somerville, 2014: Ch. 2: Our changing climate. *Climate Change Impacts in the United States: The Third National Climate Assessment*. Melillo, J.M., T.C. Richmond, and G.W. Yohe, Eds. U.S. Global Change Research Program, Washington, D.C., 19-67. http://dx.doi.org/10.7930/J0KW5CXT

4. EPA, 2014: Climate Change Indicators in the United States, 2014. 3rd edition. EPA 430-R-14-04, 107 pp. U.S. Environmental Protection Agency, Washington, D.C. http://www.epa.gov/climatechange/pdfs/climateindicators-full-2014.pdf

5. Walsh, J., D. Wuebbles, K. Hayhoe, J. Kossin, K. Kunkel, G. Stephens, P. Thorne, R. Vose, M. Wehner, J. Willis, D. Anderson, V. Kharin, T. Knutson, F. Landerer, T. Lenton, J. Kennedy, and R. Somerville, 2014: Appendix 3: Climate science supplement. *Climate Change Impacts in the United States: The Third National Climate Assessment*. Melillo, J.M., T.C. Richmond, and G.W. Yohe, Eds. U.S. Global Change Research Program, Washington, D.C., 735-789. http://dx.doi.org/10.7930/J0KS6PHH

6. Melillo, J.M., T.C. Richmond, and G.W. Yohe, 2014: Appendix 5: Scenarios and models. *Climate Change Impacts in the United States: The Third National Climate Assessment*. U.S. Global Change Research Program, Washington, D.C., 821-825. http://dx.doi.org/10.7930/J0B85625

7. Gallopin, G.C., 2006: Linkages between vulnerability, resilience, and adaptive capacity. *Global Environmental Change,* **16,** 293-303. http://dx.doi.org/10.1016/j.gloenvcha.2006.02.004

8. Smit, B. and J. Wandel, 2006: Adaptation, adaptive capacity and vulnerability. *Global Environmental Change,* **16,** 282-292. http://dx.doi.org/10.1016/j.gloenvcha.2006.03.008

9. IPCC, 2014: Climate Change 2014: Impacts, Adaptation, and Vulnerability. Part A: Global and Sectoral Aspects. Contribution of Working Group II to the Fifth Assessment Report of the Intergovernmental Panel on Climate Change. Field, C.B., V.R. Barros, D.J. Dokken, K.J. Mach, M.D. Mastrandrea, T.E. Bilir, M. Chatterjee, K.L. Ebi, Y.O. Estrada, R.C. Genova, B. Girma, E.S. Kissel, A.N. Levy, S. MacCracken, P.R. Mastrandrea, and L.L. White (Eds.), 1132 pp. Cambridge University Press, Cambridge, UK and New York, NY. http://www.ipcc.ch/report/ar5/wg2/

10. NRC, 2012: *Disaster Resilience: A National Imperative*. National Academies Press, Washington, D.C., 244 pp.

11. Ebi, K.L. and J.C. Semenza, 2008: Community-based adaptation to the health impacts of climate change. *American Journal of Preventive Medicine,* **35,** 501-507. http://dx.doi.org/10.1016/j.amepre.2008.08.018

12. Hess, J.J., J.Z. McDowell, and G. Luber, 2012: Integrating climate change adaptation into public health practice: Using adaptive management to increase adaptive capacity and build resilience. *Environmental Health Perspectives,* **120,** 171-179. http://dx.doi.org/10.1289/ehp.1103515

13. Bouzid, M., L. Hooper, and P.R. Hunter, 2013: The effectiveness of public health interventions to reduce the health impact of climate change: A systematic review of systematic reviews. *PLoS ONE,* **84,** e62041. http://dx.doi.org/10.1371/journal.pone.0062041

14. Shrestha, L.B. and E.J. Heisler, 2011: The Changing Demographic Profile of the United States. CRS Report No. RL32701, 32 pp. Congressional Research Service. http://fas.org/sgp/crs/misc/RL32701.pdf

15. U.S. Census Bureau, 2010: U.S. Census 2010: Resident Population Data. U.S. Department of Commerce. http://www.census.gov/2010census/data/apportionment-pop-text.php

16. U.S. Census Bureau, 2014: 2014 National Population Projections: Summary Tables. Table 1. Projections of the Population and Components of Change for the United States: 2015 to 2060 (NP2014-T1). U.S. Department of Commerce, Washington, D.C. http://www.census.gov/population/projections/data/national/2014/summarytables.html

17. Vincent, G.K. and V.A. Velkof, 2010: The Next Four Decades: The Older Population in the United States: 2010 to 2050. Current Population Reports #P25-1138, 16 pp. U.S. Department of Commerce, Economics and Statistics Administration, U.S. Census Bureau, Washington, D.C. http://www.census.gov/prod/2010pubs/p25-1138.pdf

18. U.S. Census Bureau, 2014: 2014 National Population Projections: Summary Tables. Table 9. Projections of the Population by Sex and Age for the United States: 2015 to 2060 U.S. Department of Commerce, Washington, D.C. http://www.census.gov/population/projections/data/national/2014/summarytables.html

19. Passel, J.S. and D. Cohen, 2008: U.S. Population Projections: 2005-2050. 49 pp. Pew Research Center, Washington, D.C. http://www.pewhispanic.org/files/reports/85.pdf

20. NRC, 2006: *Hispanics and the Future of America*. Tienda, M. and F. Mitchell, Eds. National Research Council. The National Academies Press, Washington, D.C., 502 pp. http://www.nap.edu/catalog/11539/hispanics-and-the-future-of-america

21. Schwabish, J.A., 2013: Modeling Individual Earnings in CBO's Long-term Microsimulation Model. CBO Working Paper Series 2013-04, 28 pp. Congressional Budget Office, Washington, D.C. http://www.cbo.gov/sites/default/files/cbofiles/attachments/44306_CBOLT.pdf

22. Board of Governors of the Federal Reserve System, 2014: 2013 Survey of Consumer Finances: SCF Chartbook. http://www.federalreserve.gov/econresdata/scf/files/BulletinCharts.pdf

23. Gabe, T., 2010: Poverty in the United States, 2009. CRS 7-5700. Congressional Research Service, Washington, D.C. http://digitalcommons.ilr.cornell.edu/key_workplace/764/

24. DeNavas-Walt, C. and B.D. Proctor, 2015: Income and Poverty in the United States: 2014. Current Population Reports P60-252. U.S. Census Bureau, Washington, D.C. https://http://www.census.gov/content/dam/Census/library/publications/2015/demo/p60-252.pdf

25. Moser, S.C., M.A. Davidson, P. Kirshen, P. Mulvaney, J.F. Murley, J.E. Neumann, L. Petes, and D. Reed, 2014: Ch. 25: Coastal zone development and ecosystems. *Climate Change Impacts in the United States: The Third National Climate Assessment*. Melillo, J.M., T.C. Richmond, and G.W. Yohe, Eds. U.S. Global Change Research Program, Washington, D.C., 579-618. http://dx.doi.org/10.7930/J0MS3QNW

26. Cutter, S.L., W. Solecki, N. Bragado, J. Carmin, M. Fragkias, M. Ruth, and T. Wilbanks, 2014: Ch. 11: Urban systems, infrastructure, and vulnerability. *Climate Change Impacts in the United States: The Third National Climate Assessment*. Melillo, J.M., T.C. Richmond, and G.W. Yohe, Eds. U.S. Global Change Research Program, Washington, D.C., 282-296. http://dx.doi.org/10.7930/J0F769GR

27. Johnson, N.B., L.D. Hayes, K. Brown, E.C. Hoo, and K.A. Ethier, 2014: CDC National Health Report: Leading Causes of Morbidity and Mortality and Associated Behavioral Risk and Protective Factors - United States, 2005-2013. *MMWR. Morbidity and Mortality Weekly Report*, **63**, 3-27. http://www.cdc.gov/mmwr/preview/mmwrhtml/su6304a2.htm

28. Luber, G., K. Knowlton, J. Balbus, H. Frumkin, M. Hayden, J. Hess, M. McGeehin, N. Sheats, L. Backer, C.B. Beard, K.L. Ebi, E. Maibach, R.S. Ostfeld, C. Wiedinmyer, E. Zielinski-Gutiérrez, and L. Ziska, 2014: Ch. 9: Human health. *Climate Change Impacts in the United States: The Third National Climate Assessment*. Melillo, J.M., T.C. Richmond, and G.W. Yohe, Eds. U.S. Global Change Research Program, Washington, D.C., 220-256. http://dx.doi.org/10.7930/J0PN93H5

29. NCHS, 2015: Health, United States, 2014: With Special Feature on Adults Aged 55-64. 473 pp. National Center for Health Statistics, Centers for Disease Control and Prevention, Hyattsville, MD. http://www.cdc.gov/nchs/data/hus/hus14.pdf

30. Fryar, C.D., M.D. Carroll, and C.L. Ogden, 2012: Prevalence of Overweight, Obesity, and Extreme Obesity Among Adults: United States, Trends 1960-1962 through 2009-2010. CDC National Center for Health Statistics. http://www.cdc.gov/nchs/data/hestat/obesity_adult_09_10/obesity_adult_09_10.pdf

31. Ogden, C.L., M.D. Carroll, B.K. Kit, and K.M. Flegal, 2014: Prevalence of childhood and adult obesity in the United States, 2011-2012. *JAMA - Journal of the American Medical Association*, **311**, 806-814. http://dx.doi.org/10.1001/jama.2014.732

32. Harris-Kojetin, L., M. Sengupta, E. Park-Lee, and R. Valverde, 2013: Long-Term Care Services in the United States: 2013 Overview. Vital and Health Statistics 3(37), 107 pp. National Center for Health Statistics, Hyattsville, MD. http://www.cdc.gov/nchs/data/nsltcp/long_term_care_services_2013.pdf

33. Hebert, L.E., J. Weuve, P.A. Scherr, and D.A. Evans, 2013: Alzheimer disease in the United States (2010–2050) estimated using the 2010 census. *Neurology*, **80**, 1778-1783. http://dx.doi.org/10.1212/WNL.0b013e31828726f5

34. CDC, 2015: Asthma: Data, Statistics, and Surveillance: Asthma Surveillance Data. Centers for Disease Control and Prevention, Atlanta, GA. http://www.cdc.gov/asthma/asthmadata.htm

35. Akinbami, L.J., J.E. Moorman, C. Bailey, H.S. Zahran, M. King, C.A. Johnson, and X. Liu, 2012: Trends in Asthma Prevalence, Health Care Use, and Mortality in the United States, 2001–2010. NCHS Data Brief No. 94, May 2012, 8 pp. National Center for Health Statistics, Hyattsville, MD. http://www.cdc.gov/nchs/data/databriefs/db94.pdf

36. Moorman, J.E., L.J. Akinbami, C.M. Bailey, H.S. Zahran, M.E. King, C.A. Johnson, and X. Liu, 2012: National Surveillance of Asthma: United States, 2001-2010. Vital and Health Statistics 3(35). National Center for Health Statistics. http://www.cdc.gov/nchs/data/series/sr_03/sr03_035.pdf

37. CDC, 2015: Chronic Obstructive Pulmonary Disease (COPD). Centers for Disease Control and Prevention, Atlanta, GA. http://www.cdc.gov/copd/

38. Kosacz, N.M., A. Punturieri, T.L. Croxton, M.N. Ndenecho, J.P. Kiley, G.G. Weinmann, A.G. Wheaton, E.S. Ford, L.R. Presley-Cantrell, J.B. Croft, and W.H. Giles, 2012: Chronic obstructive pulmonary disease among adults - United States, 2011. *MMWR: Morbidity and Mortality Weekly Report,* **61,** 938-943. http://www.cdc.gov/mmwr/preview/mmwrhtml/mm6146a2.htm

39. CDC, 2014: National Diabetes Statistics Report: Estimates of Diabetes and Its Burden in the United States, 2014. 12 pp. U.S. Department of Health and Human Services, Centers for Disease Control and Prevention, Atlanta, GA. http://www.cdc.gov/diabetes/pubs/statsreport14/national-diabetes-report-web.pdf

40. Boyle, J.P., T.J. Thompson, E.W. Gregg, L.E. Barker, and D.F. Williamson, 2010: Projection of the year 2050 burden of diabetes in the US adult population: Dynamic modeling of incidence, mortality, and prediabetes prevalence. *Population Health Metrics,* **8,** 29. http://dx.doi.org/10.1186/1478-7954-8-29

41. Heidenreich, P.A., J.G. Trogdon, O.A. Khavjou, J. Butler, K. Dracup, M.D. Ezekowitz, E.A. Finkelstein, Y. Hong, S.C. Johnston, A. Khera, D.M. Lloyd-Jones, S.A. Nelson, G. Nichol, D. Orenstein, P.W.F. Wilson, and Y.J. Woo, 2011: Forecasting the future of cardiovascular disease in the United States: A policy statement from the American Heart Association. *Circulation,* **123,** 933-944. http://dx.doi.org/10.1161/CIR.0b013e31820a55f5

42. Roger, V.L., A.S. Go, D.M. Lloyd-Jones, E.J. Benjamin, J.D. Berry, W.B. Borden, D.M. Bravata, S. Dai, E.S. Ford, C.S. Fox, H.J. Fullerton, C. Gillespie, S.M. Hailpern, J.A. Heit, V.J. Howard, B.M. Kissela, S.J. Kittner, D.T. Lackland, J.H. Lichtman, L.D. Lisabeth, D.M. Makuc, G.M. Marcus, A. Marelli, D.B. Matchar, C.S. Moy, D. Mozaffarian, M.E. Mussolino, G. Nichol, N.P. Paynter, E.Z. Soliman, P.D. Sorlie, N. Sotoodehnia, T.N. Turan, S.S. Virani, N.D. Wong, D. Woo, and M.B. Turner, 2012: Heart disease and stroke statistics—2012 update: A report from the American Heart Association. *Circulation,* **125,** e2-e220. http://dx.doi.org/10.1161/CIR.0b013e31823ac046

43. Heo, M., C.F. Murphy, K.R. Fontaine, M.L. Bruce, and G.S. Alexopoulos, 2008: Population projection of US adults with lifetime experience of depressive disorder by age and sex from year 2005 to 2050. *International Journal of Geriatric Psychiatry,* **23,** 1266-1270. http://dx.doi.org/10.1002/gps.2061

44. CDC, 2013: Mental Health: Burden of Mental Illness. Centers for Disease Control and Prevention, Atlanta, GA. http://www.cdc.gov/mentalhealth/basics/burden.htm

45. Flegal, K.M., M.D. Carroll, B.K. Kit, and C.L. Ogden, 2012: Prevalence of obesity and trends in the distribution of body mass index among US adults, 1999-2010. *JAMA-Journal of the American Medical Association,* **307,** 491-497. http://dx.doi.org/10.1001/jama.2012.39

46. Ogden, C.L., M.D. Carroll, B.K. Kit, and K.M. Flegal, 2012: Prevalence of obesity and trends in body mass index among US children and adolescents, 1999-2010. *JAMA - Journal of the American Medical Association,* **307,** 483-490. http://dx.doi.org/10.1001/jama.2012.40

47. Finkelstein, E.A., O.A. Khavjou, H. Thompson, J.G. Trogdon, L. Pan, B. Sherry, and W. Dietz, 2012: Obesity and severe obesity forecasts through 2030. *American Journal of Preventive Medicine,* **42,** 563-570. http://dx.doi.org/10.1016/j.amepre.2011.10.026

48. Brault, M.W., 2012: Americans With Disabilities: 2010. Current Population Reports #P70-131, 23 pp. U.S. Census Bureau, Washington, D.C. http://www.census.gov/prod/2012pubs/p70-131.pdf

49. Waidmann, T.A. and K. Liu, 2000: Disability trends among elderly persons and implications for the future. *The Journals of Gerontology Series B: Psychological Sciences and Social Sciences,* **55,** S298-S307. http://dx.doi.org/10.1093/geronb/55.5.S298

50. CDC, 2014: National Notifiable Disease Surveillance System (NNDSS). Centers for Disease Control and Prevention, Atlanta, GA. http://wwwn.cdc.gov/nndss/script/DataCollection.aspx

51. CDC, 2015: Lyme Disease Surveillance and Available Data. Centers for Disease Control and Prevention, Atlanta, GA. http://www.cdc.gov/lyme/stats/survfaq.html

52. Post, E.S., A. Grambsch, C. Weaver, P. Morefield, J. Huang, L.-Y. Leung, C.G. Nolte, P. Adams, X.-Z. Liang, J.-H. Zhu, and H. Mahone, 2012: Variation in estimated ozone-related health impacts of climate change due to modeling choices and assumptions. *Environmental Health Perspectives,* **120,** 1559-1564. http://dx.doi.org/10.1289/ehp.1104271

53. Tamerius, J.D., E.K. Wise, C.K. Uejio, A.L. McCoy, and A.C. Comrie, 2007: Climate and human health: Synthesizing environmental complexity and uncertainty. *Stochastic Environmental Research and Risk Assessment,* **21,** 601-613. http://dx.doi.org/10.1007/s00477-007-0142-1

54. Sun, L., K.E. Kunkel, L.E. Stevens, A. Buddenberg, J.G. Dobson, and D.R. Easterling, 2015: Regional Surface Climate Conditions in CMIP3 and CMIP5 for the United States: Differences, Similarities, and Implications for the U.S. National Climate Assessment. NOAA Technical Report NESDIS 144, 111 pp. National Oceanic and Atmospheric Administration, National Environmental Satellite, Data, and Information Service. http://www.nesdis.noaa.gov/technical_reports/NOAA_NESDIS_Technical_Report_144.pdf

2 TEMPERATURE-RELATED DEATH AND ILLNESS

Lead Authors

Marcus C. Sarofim*
U.S. Environmental Protection Agency

Shubhayu Saha
Centers for Disease Control and Prevention

Michelle D. Hawkins
National Oceanic and Atmospheric Administration

David M. Mills
Abt Associates

Contributing Authors

Jeremy Hess
University of Washington

Radley Horton
Columbia University

Patrick Kinney
Columbia University

Joel Schwartz
Harvard University

Alexis St. Juliana
Abt Associates

Recommended Citation: Sarofim, M.C., S. Saha, M.D. Hawkins, D.M. Mills, J. Hess, R. Horton, P. Kinney, J. Schwartz, and A. St. Juliana, 2016: Ch. 2: Temperature-Related Death and Illness. *The Impacts of Climate Change on Human Health in the United States: A Scientific Assessment.* U.S. Global Change Research Program, Washington, DC, 43–68. http://dx.doi.org/10.7930/J0MG7MDX

On the web: health2016.globalchange.gov   *Chapter Coordinator*

2 TEMPERATURE-RELATED DEATH AND ILLNESS

Key Findings

Future Increases in Temperature-Related Deaths

Key Finding 1: Based on present-day sensitivity to heat, an increase of thousands to tens of thousands of premature heat-related deaths in the summer *[Very Likely, High Confidence]* and a decrease of premature cold-related deaths in the winter *[Very Likely, Medium Confidence]* are projected each year as a result of climate change by the end of the century. Future adaptation will very likely reduce these impacts (see Changing Tolerance to Extreme Heat Finding). The reduction in cold-related deaths is projected to be smaller than the increase in heat-related deaths in most regions *[Likely, Medium Confidence]*.

Even Small Differences from Seasonal Average Temperatures Result in Illness and Death

Key Finding 2: Days that are hotter than usual in the summer or colder than usual in the winter are both associated with increased illness and death *[Very High Confidence]*. Mortality effects are observed even for small differences from seasonal average temperatures *[High Confidence]*. Because small temperature differences occur much more frequently than large temperature differences, not accounting for the effect of these small differences would lead to underestimating the future impact of climate change *[Likely, High Confidence]*.

Changing Tolerance to Extreme Heat

Key Finding 3: An increase in population tolerance to extreme heat has been observed over time *[Very High Confidence]*. Changes in this tolerance have been associated with increased use of air conditioning, improved social responses, and/or physiological acclimatization, among other factors *[Medium Confidence]*. Expected future increases in this tolerance will reduce the projected increase in deaths from heat *[Very Likely, Very High Confidence]*.

Some Populations at Greater Risk

Key Finding 4: Older adults and children have a higher risk of dying or becoming ill due to extreme heat *[Very High Confidence]*. People working outdoors, the socially isolated and economically disadvantaged, those with chronic illnesses, as well as some communities of color, are also especially vulnerable to death or illness *[Very High Confidence]*.

2.1 Introduction

The Earth is warming due to elevated concentrations of greenhouse gases, and will continue to warm in the future. U.S. average temperatures have increased by 1.3°F to 1.9°F since record keeping began in 1895, heat waves have become more frequent and intense, and cold waves have become less frequent across the nation (see Ch. 1: Introduction). Annual average U.S. temperatures are projected to increase by 3°F to 10°F by the end of this century, depending on future emissions of greenhouse gases and other factors.[1] These temperature changes will have direct effects on human health.

Days that are hotter than the average seasonal temperature in the summer or colder than the average seasonal temperature in the winter cause increased levels of illness and death by compromising the body's ability to regulate its temperature or by inducing direct or indirect health complications. Figure 1 provides a conceptual model of the various climate drivers, social factors, and environmental and institutional factors that can interact to result in changes in illness and deaths as a result of extreme heat. Increasing concentrations of greenhouse gases lead to an increase of both average and extreme temperatures, leading to an increase in deaths and illness from heat and a potential decrease in deaths from cold. Challenges involved in determining the temperature–death relationship include a lack of consistent diagnoses on death certificates and the fact that the health implications of extreme temperatures are not absolute, differing from location to location and changing over time. Both of these issues can be partially addressed through the use of statistical methods. Climate model projections of future temperatures can be combined with the estimated relationships between temperatures and health in order to assess how deaths and illnesses resulting from temperature could change in the future. The impact of a warming climate on deaths and illnesses will not be realized equally as a number of populations, such as children, the elderly, and economically disadvantaged groups, are especially vulnerable to temperature.

Climate Change and Health—Extreme Heat

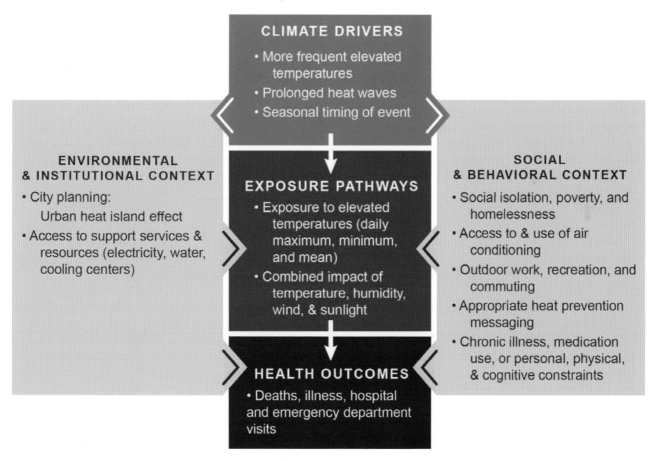

Figure 1: This conceptual diagram illustrates the key pathways by which climate change influences human health during an extreme heat event, and potential resulting health outcomes (center boxes). These exposure pathways exist within the context of other factors that positively or negatively influence health outcomes (gray side boxes). Key factors that influence vulnerability for individuals are shown in the right box, and include social determinants of health and behavioral choices. Key factors that influence vulnerability at larger scales, such as natural and built environments, governance and management, and institutions, are shown in the left box. All of these influencing factors can affect an individual's or a community's vulnerability through changes in exposure, sensitivity, and adaptive capacity and may also be affected by climate change. See Chapter 1: Introduction for more information.

2.2 Contribution of Extreme Temperatures to Death and Illness

Temperature extremes most directly affect health by compromising the body's ability to regulate its internal temperature. Loss of internal temperature control can result in a cascade of illnesses, including heat cramps, heat exhaustion, heatstroke, and hyperthermia in the presence of extreme heat, and hypothermia and frostbite in the presence of extreme cold. Temperature extremes can also worsen chronic conditions such as cardiovascular disease, respiratory disease, cerebrovascular disease, and diabetes-related conditions. Prolonged exposure to high temperatures is associated with increased hospital admissions for cardiovascular, kidney, and respiratory disorders. Exposures to high minimum temperatures may also reduce the ability of the human body to recover from high daily maximum temperatures.

2.3 Defining Temperature Exposures

Extreme temperatures are typically defined by some measure, for example, an ambient temperature, heat index (a combination of temperature and humidity), or wind chill (a combination of temperature and wind speed), exceeding predefined thresholds over a number of days.[2, 3, 4, 5, 6, 7, 8] Extremes can be defined by average, minimum, or maximum daily temperatures, by nighttime temperatures, or by daytime temperatures. However, there is no standard method for defining a heat wave or cold wave. There are dramatic differences in the observed relationships between temperature, death, and illness across different regions and seasons; these relationships vary based on average temperatures in those locations and the timing of the heat or cold event. For example, a 95°F day in Vermont will have different implications for health than a 95°F day in Texas, and similarly, a 95°F day in May will have different implications than one in August[9, 10, 11, 12] (this is further discussed in "Evidence of Adaptation to Temperature Extremes" on page 49). Therefore, in some cases, temperature extremes are defined by comparison to some local average (for example, the top 1% of warmest days recorded in a particular location) rather than to some absolute temperature (such as 95°F). While temperature extremes are generally determined based on weather station records, the exposure of individuals will depend on their location: urban heat islands, microclimates, and differences between indoor and outdoor temperatures can all lead to differences between weather station data and actual exposure. The indoor environment is particularly important as most people spend the majority of their time inside.

One exception to using relative measures of temperature is that there are some critical physical and weather condition thresholds that are absolute. For example, one combined measure of humidity and temperature is known as the wet bulb temperature. As the wet bulb temperature reaches or exceeds the threshold of 35°C (95°F), the human body can no longer cool through perspiration, and recent evidence suggests that there is a physical heat tolerance limit in humans to sustained temperatures above 35°C that is similar across diverse climates.[13] The combined effects of temperature and humidity have been incorporated in tools such as heat index tables, which reflect how combinations of heat and relative humidity "feel." The heat index in these tools is often presented with notes about the potential nature and type of health risks different combinations of temperature and humidity may pose, along with confounding conditions such as exposure to direct sunlight or strong winds.

Variations in heat wave definitions make it challenging to compare results across studies or determine the most appropriate public health warning systems.[8, 14] This is important as the associations between deaths and illnesses and extreme heat conditions vary depending on the methods used for defining the extreme conditions.[2, 15, 16]

2.4 Measuring the Health Impact of Temperature

Two broad approaches are used to study the relationship between temperatures and illness and death: direct attribution and statistical methods.[17, 18]

Direct Attribution Studies

With direct attribution, researchers link health outcomes to temperatures based on assigned diagnosis codes in medical records such as hospital admissions and death certificates. For example, the International Classification of Diseases (ICD-10) contains specific codes for attributing deaths to exposure to excessive natural heat (X30) and excessive natural cold (X31).[19] However, medical records will not include information on the weather conditions at the time of the event or preceding the event. It is generally accepted that direct attribution underestimates the number of people who die from temperature extremes. Reasons for this include difficulties in diagnosing heat-related and cold-related deaths, lack of consistent diagnostic criteria, and difficulty in identifying, or lack of reporting, heat or cold as a factor that worsened a preexisting medical condition.[9, 17] Heat-related deaths are often not reported as such if another cause of death exists and there is no well-publicized heat wave. An additional challenging factor in deaths classified as X31 (cold) deaths is that a number of these deaths result from situations involving substance use/abuse and/or contact with water, both of which can contribute to hypothermia.[20, 21]

> *Temperature extremes most directly affect health by compromising the body's ability to regulate its internal temperature.*

Statistical Studies

Statistical studies measure the impact of temperature on death and illness using methods that relate the number of cases (for example, total daily deaths in a city) to observed weather conditions and other socio-demographic factors. These statistical methods determine whether the temperature conditions were associated with increased deaths or illness above longer-term average levels. These associations establish the relationship between temperature and premature deaths and illness. In some cases, particularly with extreme temperature conditions, the increase in premature deaths and illness can be quite dramatic and the health impact may be referred to in terms of excess deaths or illnesses. Methods for evaluating the impact of temperature in these models vary.

Many studies include all the days in the study period, which makes it possible to capture changes in deaths resulting from small variations of temperatures from their seasonal averages. Other methods restrict the analysis to days that exceed some threshold for extreme heat or cold conditions.[22] Some studies incorporate methods that determine different health relationships for wind, air pressure, and cloud cover as well as the more common temperature and humidity measures.[15] Another approach is to identify a heat event and compare observed illness and deaths during the event with a carefully chosen comparison period.[23, 24, 25] Many of these methods also incorporate socio-demographic factors (for example, age, race, and poverty) that may affect the temperature–death relationship.

Comparing Results of Direct Attribution and Statistical Studies

Comparing death estimates across studies is therefore complicated by the use of different criteria for temperature extremes, different analytical methods, varying time periods, and different affected populations. Further, it is widely accepted that characteristics of extreme temperature events such as duration, intensity, and timing in season directly affect actual death totals.[2, 12] Estimates of the average number of deaths attributable to heat and cold considering all temperatures, rather than just those associated with extreme events, provide an alternative for considering the mortality impact of climate change.[26, 27] Statistical studies can also offer insights into what aspects of a temperature extreme are most important. For example, there are indications that the relationship between high nighttime temperatures and mortality is more pronounced than the relationship for daytime temperatures.[12, 16]

These two methods (direct attribution and statistical approaches) yield very different results for several reasons. First, statistical approaches generally suggest that the actual number of deaths associated with temperature is far greater than those recorded as temperature-related in medical records. Medical records often do not capture the role of heat in

exacerbating the cause of death, only recording the ultimate cause, such as a stroke or a heart attack (see, for example, Figure 2, where the excess deaths during the 1995 Chicago heat wave clearly exceeded the number of deaths recorded as heat-related on death certificates). Statistical methods focus on determining how temperature contributes to premature deaths and illness and therefore are not susceptible to this kind of undercount, though they face potential biases due to time-varying factors like seasonality. Both methods depend on temperatures measured at weather stations, though the actual temperature exposure of individuals may differ. In short, while the focus on temperature is consistent in both methods, the methods potentially evaluate very different combinations of deaths and weather conditions.

2.5 Observed Impact of Temperature on Deaths

A number of extreme temperature events in the United States have led to dramatic increases in deaths, including events in Kansas City and St. Louis in 1980, Philadelphia in 1993, Chicago in 1995, and California in 2006. (See Figure 2 for more on the July 1995 heat wave in Chicago).[28, 29, 30, 31, 32]

Recent U.S. studies in specific communities and for specific extreme temperature events continue to conclude that extreme temperatures, particularly extreme heat, result in premature deaths.[7, 30, 36, 37] This finding is further reinforced by a growing suite of regional- and national-scale studies documenting an increase in deaths following extreme temperature conditions, using both direct attribution[17] and statistical approaches.[9, 10, 12, 15, 38] The connection between heat events and deaths is also evident internationally. The European heat wave of 2003 is an especially notable example, as it is estimated to have been responsible for between 30,000 and 70,000 premature deaths.[39] However, statistical approaches find that elevated death rates are seen even for less extreme temperatures. These approaches find an optimal temperature, and show that there are more deaths at any temperatures that are higher or lower than that optimal temperature.[11, 40] Even though the increase in deaths per degree are smaller near the optimum than at more extreme temperatures, because the percentage of days that do not qualify as extreme are large,[41] it can be important to address the changes in deaths that occur for these smaller temperature differences.

A recent analysis of U.S. deaths from temperature extremes based on death records found an average of approximately 1,300 deaths per year from 2006 to 2010 coded as resulting from extreme cold exposures, and 670 deaths per year coded as resulting from exposure to extreme heat.[17] These results, and those from all similar studies that rely solely on coding within medical records to determine cause of deaths, will underestimate the actual number of deaths due to extreme temperatures.[17, 42] For example, some statistical approaches estimate that more than 1,300 deaths per year in the United

Heat-Related Deaths in Chicago in the Summer of 1995

Figure 2 illustrates an example of excess deaths following an extreme heat event. In this case, excess deaths are determined by calculating the difference between daily observed deaths in Chicago during the worst of the heat wave (starting on July 11) and longer-term daily averages for this time of year. The period of extreme heat extended from June 21 through August 10, 1995. Research into the event suggests it was the combination of high humidity, high daily maximum temperatures, and high daily minimum temperatures that made this event truly exceptional.[33] This event is estimated to have resulted in nearly 700 excess deaths in Chicago, based on a statistical approach.[35] By comparison, a direct attribution approach based on death certificates found only 465 deaths were attributed to extreme heat during this time period.[29] This kind of underestimate resulting from relying on death certificates is common. It is reasonable to expect that deaths may be even less likely to be attributed to extreme heat during a heat wave that, unlike the Chicago event, does not receive a great deal of public attention.

Heat-Related Deaths During the 1995 Chicago Heat Wave

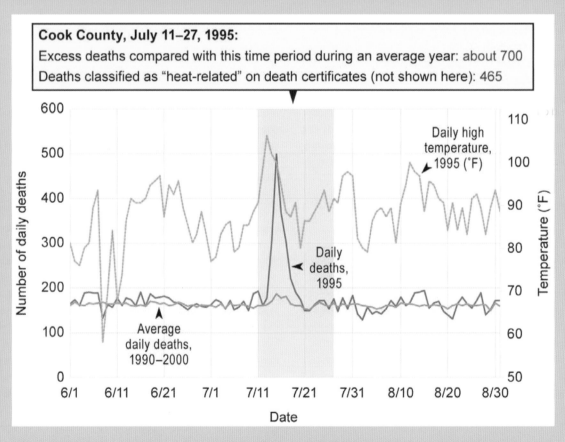

Figure 2: This figure shows the relationship between high temperatures and deaths observed during the 1995 Chicago heat wave. The large spike in deaths in mid-July of 1995 (red line) is much higher than the average number of deaths during that time of year (orange line), as well as the death rate before and after the heat wave. This increase in the rate of deaths occurred during and after the heat wave, as shown here by temperatures exceeding 100°F during the day (green line). Humidity and high nighttime temperatures were also key contributing factors to this increase in deaths.[33] The number of excess deaths has been estimated to be about 700 based on statistical methods, but only 465 deaths in Cook County were classified as "heat-related" on death certificates during this same period,[29] demonstrating the tendency of direct attribution to undercount total heat-related deaths. (Figure source: EPA 2014)[34]

States are due to extreme heat.[15, 43] Different approaches to attributing cause of death lead to differences in the relative number of deaths attributed to heat and cold.[44] Studies based on statistical approaches have found that, despite a larger number of deaths being coded as related to extreme cold rather than extreme heat, and a larger mortality rate in winter overall, the relationship between mortality and an additional day of extreme heat is generally much larger than the relationship between mortality and an additional day of extreme cold.[12]

Confounding Factors and Effect Modifiers

While the direct attribution approach underestimates the number of deaths resulting from extreme temperature events, there are a few ways in which the statistical approach may lead to an overestimation. However, any overestimation due to these potential confounding factors and effect modifiers is thought to be much smaller than the direct attribution underestimation.[12]

The first potential overestimation results from the connection between elevated temperatures and other variables that correlate with temperature, such as poor air quality. This connection involves a combination of factors, including stagnant air masses and changes in the atmospheric chemistry that affect the concentrations of air pollutants such as ozone or particulate matter (see Ch. 3: Air Quality Impacts). If some portion of the deaths during extreme heat events are actually a result of the higher levels of atmospheric pollution that are correlated with these events, then including those deaths in a statistical analysis to determine the relationship of increased heat on human health would result in double counting deaths.[10, 45, 46, 47] However, this issue is often addressed by including air pollution and other correlated variables in statistical modeling.[26]

A second consideration when using statistical approaches to determine the relationship between temperature and deaths is whether some of the individuals who died during the temperature event were already near death, and therefore the temperature event could be considered to have "displaced" the death by a matter of days rather than having killed a person not otherwise expected to die. This effect is referred to as mortality displacement. There is still no consensus regarding the influence of mortality displacement on premature death estimates, but this effect generally accounts for a smaller portion of premature deaths as events become more extreme.[7, 12, 48, 49, 50]

The relationship between mortality and an additional day of extreme heat is generally much larger than the relationship between mortality and an additional day of extreme cold.

Evidence of Adaptation to Temperature Extremes

The impact on human health of a given temperature event (for example, a 95°F day) can depend on where and when it occurs. The evidence also shows larger changes in deaths and hospitalizations in response to elevated temperatures in cities where temperatures are typically cooler as compared with warmer cities.[9, 11, 40, 51, 52] This suggests that people can adapt, at least partially, to the average temperature that they are used to experiencing. Some of this effect can be explained by differences in infrastructure. For example, locations with higher average temperature, such as the Southeast, will generally have greater prevalence and use of air conditioning. However, there is also evidence that there is a physiological acclimatization (the ability to gradually adapt to heat), with changes in sweat volume and timing, blood flow and heat transfer to the skin, and kidney function and water conservation occurring over the course of weeks to months of exposure to a hot climate.[53] For example, as a result of this type of adaptation, heat events later in the summer have less of an impact on deaths than those earlier in the summer, all else being equal,[15] although some of this effect is also due to the deaths of some of the most vulnerable earlier in the season. However, children and older adults remain vulnerable given their reduced ability to regulate their internal temperature and limited acclimatization capacities.[53]

An increased tolerance to extreme temperatures has also been observed over multiyear and multidecadal periods.[9, 10, 54, 55, 56] This improvement is likely due to some combination of physiological acclimatization, increased prevalence and use of air conditioning,[10] and general improvements in public health over time,[9, 54] but the relative importance of each is not yet clear.[56]

> *The impact on human health of a given temperature event (for example, a 95° day) can depend on where and when it occurs.*

Recent changes in urban planning and development programs reflect an adaptive trend implemented partially in response to the anticipated temperature health risks of climate change. For example, because urban areas tend to be warmer than surrounding rural areas (the "urban heat island" effect), there is an increased emphasis on incorporating green space and other technologies, such as cool roofs, in new development or redevelopment projects.[57] Similarly, programs that provide advice and services in preparation for or response to extreme temperatures continue to increase in number and expand the scope of their activity (see for example guidance documents on responses to extreme temperature developed by the Centers for Disease Control and Prevention and the Environmental Protection Agency).[58, 59] Continued changes in personal behavior as a result of these efforts, for example, seeking access to air-conditioned areas during extreme heat events or limiting outside activity, may continue to change future exposure to extreme temperatures and other climate-sensitive health stressors, such as outdoor air pollutants and vectors for disease such as ticks or mosquitoes.

Observed Trends in Heat Deaths

As discussed in Chapter 1, U.S. average temperature has increased by 1.3°F to 1.9°F since 1895, with much of that increase occurring since 1970, though this temperature increase has not been uniform geographically and some regions, such as the Southeast, have seen little increase in temperature and extreme heat over time.[1, 15] This warming is attributable to elevated concentrations of greenhouse gases and it has been estimated that three-quarters of moderately hot extremes are already a result of this historical warming.[60] As discussed in the previous section, there have also been changes in the tolerance of populations within the United States to extreme temperatures. Changes in mortality due to high temperatures are therefore a result of the combination of higher temperatures and higher heat tolerance. Use of the direct attribution approach, based on diagnosis codes in medical records, to examine national trends in heat mortality over time is challenging because of changes in classification methods over time.[34]

Certain occupational groups such as agricultural workers, construction workers, and electricity and pipeline utility workers are at increased risk for heat- and cold-related illness, especially where jobs involve heavy exertion.

The few studies using statistical methods that have presented total mortality estimates over time suggest that, over the last several decades, reductions in mortality due to increases in tolerance have outweighed increases in mortality due to increased temperatures.[15, 61]

2.6 Observed Impact of Temperature on Illness

Temperature extremes are linked to a range of illnesses reported at emergency rooms and hospitals. However, estimates for the national burden of illness associated with extreme temperatures are limited.

Using a direct attribution approach, an analysis of a nationally representative database from the Healthcare Utilization Project (HCUP) produced an annual average estimate of 65,299 emergency visits for acute heat illness during the summer months (May through September)—an average rate of 21.5 visits for every 100,000 people each year.[62] This result was based only on recorded diagnosis codes for hyperthermia and probably underestimates the true number of heat-related healthcare visits, as a wider range of health outcomes is potentially affected by extreme heat. For example, hyperthermia is not the only complication from extreme heat, and not every individual that suffers from a heat illness visits an emergency department. In a national study of Medicare patients from 2004 to 2005, an annual average of 5,004 hyperthermia cases and 4,381 hypothermia cases were reported for inpatient and outpatient visits.[63] None of these studies link health episodes to observed temperature data, thus limiting the opportunity to attribute these adverse outcomes to specific heat events or conditions.

High ambient heat has been associated with adverse impacts for a wide range of illnesses.[25] Examples of illnesses associated with extreme heat include cardiovascular, respiratory, and renal illnesses; diabetes; hyperthermia; mental health issues; and preterm births. Children spend more time outdoors and have insufficient ability for physiologic adaptation, and thus may be particularly vulnerable during heat waves.[64] Respiratory illness among the elderly population was most commonly reported during extreme heat.[65]

Statistical studies examine the association between extreme heat and illness using data from various healthcare access points (such as hospital admissions, emergency department visits, and ambulance dispatches). The majority of these studies examine the association of extreme heat with cardiovascular and respiratory illnesses. For these particular health outcomes, the evidence is mixed, as many studies observed elevated risks of illness during periods of extreme heat but others found no evidence of elevated levels of illness.[24, 51, 66, 67, 68, 69, 70] The evidence on some of the other health outcomes is more robust. Across emergency department visits and hospital admissions, high temperature have been associated with renal diseases, electrolyte imbalance, and hyperthermia.[24, 67,]

[71, 72] These health risks vary not only across types of illness but also for the same illness across different healthcare settings. In general, evidence for associations with morbidity outcomes, other than cardiovascular impacts, is strong.

While there is still uncertainty about how levels of heat-related illness are expected to change with projected increases in summer temperature from climate change,[41] advances have been made in surveillance of heat-related illness. For example, monitoring of emergency ambulance calls during heat waves can be used to establish real-time surveillance systems to identify extreme heat events.[73] The increase in emergency visits for a wide range of illnesses during the 2006 heat wave in California points to the potential for using this type of information in real-time health surveillance systems.[24]

2.7 Projected Deaths and Illness from Temperature Exposure

Climate change will increase the frequency and severity of future extreme heat events while also resulting in generally warmer summers and milder winters,[1] with implications for human health. Absent further adaptation, these changes are expected to lead to an increase in illness and death from increases in heat, and reductions in illness and death resulting from decreases in cold, due to changes in outcomes such as heat stroke, cardiovascular disease, respiratory disease, cerebrovascular disease, and kidney disorders.[41, 74]

A warmer future is projected to lead to increases in future mortality on the order of thousands to tens of thousands of additional premature deaths per year across the United States by the end of this century.[22, 38, 54, 75, 76, 77, 78, 79] Studies differ in which regions of the United States are examined and in how they account for factors such as adaptation, mortality displacement, demographic changes, definitions of heat waves and extreme cold, and air quality factors, and some studies examine only extreme events while others take into account the health effects of smaller deviations from average seasonal temperatures. Despite these differences there is reasonable agreement on the magnitude of the projected changes. Additionally, studies have projected an increase in premature deaths due to increases in temperature for Chicago, IL,[39, 80] Dallas, TX,[18] the Northeast corridor cities of Boston, MA, New York, NY, and Philadelphia, PA,[18, 26, 81, 82] Washington State,[83, 84] California,[85] or a group of cities including Portland, OR; Minneapolis and St. Paul, MN; Chicago, IL; Detroit, MI; Toledo, Cleveland, Columbus, and Cincinnati, OH; Pittsburgh and Philadelphia, PA; and Washington, DC.[86] However, these regional projections use a variety of modeling strategies and therefore show more variability in mortality estimates than studies that are national in scope.

Less is known about how non-fatal illnesses will change in response to projected increases in heat. However, hospital admissions related to respiratory, hormonal, urinary, genital, and renal problems are generally projected to increase.[72, 87] Kidney stone prevalence has been linked to high temperatures, possibly due to dehydration leading to concentration of the salts that form kidney stones. In the United States, an increased rate of kidney stones is observed in southern regions of the country, especially the Southeast. An expansion of the regions where the risk of kidney stones is higher is therefore plausible in a warmer future.[88, 89, 90]

The decrease in deaths and illness due to reductions in winter cold have not been as well studied as the health impacts of increased heat, but the reduction in premature deaths from cold are expected to be smaller than the increase in deaths from heat in the United States.[22, 26, 38, 41, 75, 77] While this is true nationally (with the exception of Barreca 2012),[75] it may not hold for all regions within the country.[27] Similarly, international studies have generally projected a net increase in deaths from a warming climate, though in some regions, decreases in cold mortality may outweigh increases in heat mortality.[91] The projected net increase in deaths is based in part on historical studies that show that an additional extreme hot day leads to more deaths than an additional extreme cold day, and in part on the fact that the decrease in extreme cold deaths is limited as the total number of cold deaths approaches zero in a given location.

It is important to distinguish between generally higher wintertime mortality rates that are not strongly associated with daily temperatures—such as respiratory infections and some cardiovascular disease [12, 92]—from mortality that is more directly related to the magnitude of the cold temperatures. Some recent studies have suggested that factors leading to higher wintertime mortality rates may not be sensitive to climate warming, and that deaths due to these factors are expected to occur with or without climate change. Considering this, some estimates of wintertime mortality may overstate the benefit of climate change in reducing wintertime deaths.[49, 93, 94]

The U.S. population has become less sensitive to heat over time. Factors that have contributed to this change include infrastructure improvements, including increased access and use of air conditioning in homes and businesses, and improved societal responses, including increased access to public health programs and healthcare.[15, 54, 61, 95, 96, 97] Projecting these trends into the future is challenging, but this trend of increasing tolerance is projected to continue, with future changes in adaptive capacity expected to reduce the future increase in mortality.[56] However, there are limits to adaptation, whether physiological[53] or sociotechnical (for example, air conditioning, awareness programs, or cooling centers). While historically adaptation has outpaced warming, most studies project a future increase in mortality even when including assumptions regarding adaptation.[18, 22, 81, 85, 91] Additionally, the occurrence of events such as power outages simultaneous with a heat wave may reduce some of these adaptive benefits. Such simultaneous events can

Research Highlight: Modeling the Effect of Warming on U.S. Deaths

Importance: A warming climate is expected to result in more days that are warmer than today's usual temperature in the summer, leading to an increase in heat-related deaths. A warming climate is also expected to result in fewer days that are colder than today's usual temperatures in the winter, leading to a decrease in cold-related deaths. Understanding these changes is an important factor in understanding the human health response to climate change.

Objective: A quantitative projection of future deaths from heat and cold for 209 U.S. cities with a total population of over 160 million inhabitants.

Method: A relationship between average daily temperature and deaths by city and month was developed using historical data on deaths and temperatures from 1996–2006, generating results for both same-day temperature and the average of the previous five-day temperatures to account for delayed responses to temperature. Cities, which are defined using county borders, were allocated to nine different clusters based on similarity of climates. Temperature–death relationships were refined for cities within a given cluster based on the other cities in that cluster. Projections of temperature in future time periods were based on the RCP6.0 scenario from two climate models: the Geophysical Fluid Dynamic Laboratory–Coupled Physical Model 3 (GFDL–CM3) and the Model for Interdisciplinary Research on Climate (MIROC5). These projections were adjusted to match the historical data from the same weather stations that were used in the statistical analysis. Further details can be found in Schwartz et al. 2015.[27]

Projected Changes in Temperature-Related Death Rates

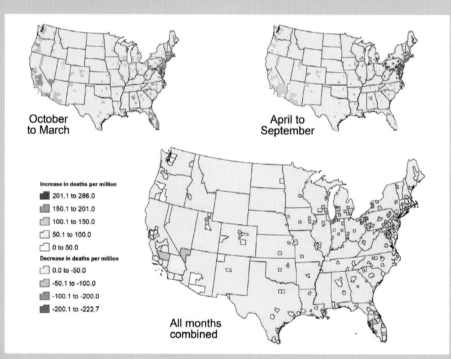

Increase in deaths per million
- 201.1 to 286.0
- 150.1 to 201.0
- 100.1 to 150.0
- 50.1 to 100.0
- 0 to 50.0

Decrease in deaths per million
- 0.0 to -50.0
- -50.1 to -100.0
- -100.1 to -200.0
- -200.1 to -222.7

Figure 3: This figure shows the projected decrease in death rates due to warming in colder months (October–March, top left), the projected increase in death rates due to warming in the warmer months (April–September, top right), and the projected net change in death rates (combined map, bottom), comparing results for 2100 to those for a 1990 baseline period in 209 U.S. cities. These results are from one of the two climate models (GFDL–CM3 scenario RCP6.0) studied in Schwartz et al. (2015). In the study, mortality data for a city is based on county-level records, so the borders presented reflect counties corresponding to the study cities. Geographic variation in the death rates are due to a combination of differences in the amount of projected warming and variation in the relationship between deaths and temperatures derived from the historical health and temperature data. These results are based on holding the 2010 population constant in the analyses, with no explicit assumptions or adjustment for potential future adaptation. Therefore, these results reflect only the effect of the anticipated change in climate over time. (Figure source: Schwartz et al. 2015)[27]

Research Highlight: Modeling the Effect of Warming on U.S. Deaths, continued

Results: The modeling done for this study projects that future warming, without any adjustments for future adaptation, will lead to an increase in deaths during hotter months, defined as April–September, and a decrease in deaths during colder months, defined as October–March. Overall, this leads to a total net increase of about 2,000 to 10,000 deaths per year in the 209 cities by the end of the century compared to a 1990 baseline (Figure 4). Net effects vary from city to city, and a small number of cities are projected to experience a decrease in deaths (Figures 3 and 4).

Conclusions: This study is an improvement on previous studies because it examines a greater proportion of the U.S. population, uses more recent data on deaths, takes advantage of similar relationships between deaths and temperature between nearby cities to generate more statistically robust results, and addresses the difference in these relationships by month of the year. The results are consistent with most of the previous studies in projecting that climate change will lead to an increase in heat deaths on the order of thousands to tens of thousands of annual deaths by the end of the century compared to the 1990 baseline, and that the increase in deaths from heat will be larger than the reduction in deaths from cold. In contrast to some previous similar studies,[22] some individual cities show a net reduction in future deaths due to future warming, mainly in locations where the population is already well-adapted to heat but poorly prepared for cold (like Florida). Barreca 2012[75] also shows net mortality benefits in some counties, though with a different spatial pattern due to humidity effects. Some other studies also have different spatial patterns, projecting high excess mortality in Southern states despite a lower risk per degree change, due to larger increases in frequency and duration of heat waves in that region.[79] Like most previous studies, this analysis does not account for the effects of further adaptation on future mortality. Results are based on the temperature–death relationships observed for the period from 1996 to 2006, which reflect historical adaptation to extreme temperatures. However, future adaptation would, all else equal, mean that these results may overestimate the potential impact of climate change on changes in both heat- and cold-related deaths.

This study increases the confidence in the key finding that the number of heat deaths will increase in the future compared to a future with no climate change, and that the increase in heat deaths will be larger than the reduction in cold deaths.

Projected Changes in Deaths in U.S. Cities by Season

Figure 4: This figure shows the projected increase in deaths due to warming in the summer months (hot season, April–September), the projected decrease in deaths due to warming in the winter months (cold season, October–March), and the projected net change in deaths for the 209 U.S. cities examined. These results compare projected deaths for future reporting years to results for the year 1990 while holding the population constant at 2010 levels and without any quantitative adjustment for potential future adaptation, so that temperature–death relationships observed in the last decade of the available data (1997–2006) are assumed to remain unchanged in projections over the 21st century.

With these assumptions, the figure shows an increasing health benefit in terms of reduced deaths during the cold season (October–March) over the 21st century from warming temperatures, while deaths during the hot season (April–September)

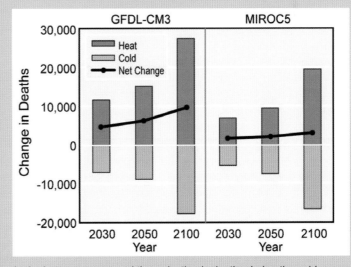

increase. Overall, the additional deaths from the warming in the hot season exceed the reduction in deaths during the cold season, resulting in a net increase in deaths attributable to temperature over time as a result of climate change. The baseline and future reporting years are based on 30-year periods where possible, with the exception of 2100: 1990 (1976–2005), 2030 (2016–2045), 2050 (2036–2065), and 2100 (2086–2100). (Figure source: adapted from Schwartz et al. 2015)[27]

be more common because of the additional demand on the electricity grid due to high air-conditioning usage.[98] Another potential effect is that if current trends of population growth and migration into large urban areas continue, there may be an increasing urban heat island effect which will magnify the rate of warming locally, possibly leading to more heat-related deaths and fewer cold-related deaths.

Projected changes in future health outcomes associated with extreme temperatures can be difficult to quantify. Projections can depend on 1) the characterization of population sensitivity to temperature event characteristics such as magnitude, duration, and humidity; 2) differences in population sensitivity depending on the timing and location of an extreme event; 3) future changes in baseline rates of death and illness as well as human tolerance and adaptive capacity; 4) the changing proportions of vulnerable populations, including the elderly, in the future; and 5) uncertainty in climate projections.

2.8 Populations of Concern for Death and Illness from Extreme Temperatures

Impacts of temperature extremes are geographically varied and disproportionally affect certain populations of concern (see also Ch. 9: Populations of Concern).[41] Certain populations are more at risk for experiencing detrimental consequences of exposure to extreme temperatures due to their sensitivity to hot and cold temperatures and limitations to their capacity for adapting to new climate conditions.

Older adults are a rapidly growing population in the United States, and heat impacts are projected to occur in places where older adults are heavily concentrated and therefore most exposed.[99] Older adults are at higher risk for temperature-related mortality and morbidity, particularly those who have preexisting diseases, those who take certain medications that affect thermoregulation or block nerve impulses (for example, beta-blockers, major tranquilizers, and diuretics), those who are living alone, or those with limited mobility (see also Ch. 9: Populations of Concern).[17, 24, 42, 45, 100] The relationship between increased temperatures and death in older adults is well-understood with strong evidence of heat-related vulnerability for adults over 65 and 75 years old.[101] An increased risk for respiratory and cardiovascular death is observed in older adults during temperature extremes due to reduced thermoregulation.[17, 42, 45, 65] Morbidity studies have also identified links between increased temperatures and respiratory and cardiovascular hospitalizations in older adults.[65]

Children are particularly vulnerable because they must rely on others to help keep them safe. This is especially true in environments that may lack air conditioning, including homes, schools, or cars (see also Ch. 9: Populations of Concern).[102] The primary health complications observed in children exposed to extreme heat include dehydration, electrolyte imbalance, fever, renal disease, heat stress, and hyperthermia.[64] Infec-

Physiological factors and participation in vigorous outdoor activities make children particularly vulnerable to extreme heat.

tious and respiratory diseases in children are affected by both hot and cold temperatures.[64] Inefficient thermoregulation, reduced cardiovascular output, and heightened metabolic rate are physiological factors driving vulnerability in children to extreme heat. Children also spend a considerable amount of time outdoors and participating in vigorous physical activities.[17, 42, 64, 103] High-school football players are especially vulnerable to heat illness (see also Ch. 9: Populations of Concern).[104] A limited number of studies show evidence of cold-related mortality in children. However, no study has examined the relationship between cold temperature and cause-specific mortality.[64] Pregnant women are also vulnerable to temperature extremes as preterm birth has been associated with extreme heat.[42, 105, 106] Elevated heat exposure can increase dehydration, leading to the release of labor-inducing hormones.[107] Extreme heat events are also associated with adverse birth outcomes, such as low birth weight and infant mortality (see Ch. 9: Populations of Concern).

Where a person lives, works, or goes to school can also make them more vulnerable to health impacts from extreme temperatures. Of particular concern for densely populated cities is the urban heat island effect, where manmade surfaces absorb sunlight during the day and then radiate the stored energy at night as heat. This process will exacerbate any warming from climate change and limit the potential relief of cooler nighttime temperatures in urban areas.[81] In addition to the urban heat island effect, land cover characteristics and poor air quality combine to increase the impacts of high ambient temperatures for city dwellers and further increase the burden on populations of concern within the urban area.[12, 17, 45, 108] The homeless are often more exposed to heat and cold extremes, while also sharing many risk factors with other populations of concern such as social isolation, psychiatric illness, and other health issues.[109]

Race, ethnicity, and socioeconomic status can affect vulnerability to temperature extremes. Non-Hispanic Black persons have been identified as being more vulnerable than other racial and ethnic groups to detrimental consequences of exposure to temperature extremes.[17, 42, 45, 103, 110, 111] One study found that non-Hispanic Blacks were 2.5 times more likely to experience heat-related mortality compared to non-Hispanic Whites, and non-Hispanic Blacks had a two-fold risk of dying from a heat-related event compared to Hispanics.[17] Evidence of racial differences in heat tolerance due to genetic differences is inconclusive.[110] However, other factors may contribute to increased vulnerability of Black populations, including comorbidities (co-existing chronic conditions) that increase susceptibility to higher temperatures, disparities in the availability and use of air conditioning and in heat risk-related land cover characteristics (for example, living in urban areas prone to heat-island effects), and environmental justice issues.[17, 42, 108, 110, 112] Overall, the link between temperature extremes, race, ethnicity, and socioeconomic status is multidimensional and dependent on the outcome being studied. Education level, income, safe housing, occupational risks, access to health care, and baseline health and nutrition status can further distort the association between temperature extremes, race, and ethnicity.[45, 110]

Outdoor workers spend a great deal of time exposed to temperature extremes, often while performing vigorous activities. Certain occupational groups such as agricultural workers, construction workers, and electricity and pipeline utility workers are at increased risk for heat- and cold-related illness, especially where jobs involve heavy exertion.[100, 113, 114] One study found failure of employers to provide for acclimatization to be the factor most clearly associated with heat-related death in workers.[113]

Mental, behavioral, and cognitive disorders can be triggered or exacerbated by heat waves. Specific illnesses impacted by heat include dementia, mood disorders, neurosis and stress, and substance abuse.[100, 115, 116, 117] Some medications interfere with thermoregulation, thereby increasing vulnerability to heat.[116] One study in Australia found that hospital admissions for mental and behavioral disorders increased by 7.3% during heat waves above 80°F.[115] Studies have also linked extreme heat and increased aggressive behavior. (See also Ch. 8: Mental Health).

2.9 Emerging and Cross-Cutting Issues

Emerging and cross-cutting issues include 1) disparate ways that extreme temperature and health are related, 2) urban and rural differences, 3) interactions between impacts and future changes in adaptation, and 4) projections of extreme temperature events.

The health effects addressed in this chapter are not the only ways in which heat and health are related. For example, research indicates that hotter temperatures may lead to an increase in violent crime[118] and could negatively affect the labor

force, especially occupational health for outdoor sectors.[119, 120] Extreme temperatures also interact with air quality, which can complicate estimating how extreme temperature events impact human health in the absence of air quality changes (see "Confounding Factors and Effect Modifiers" on page 49). In addition, increased heat may also increase vulnerability to poor air quality and allergens, leading to potential non-linear health outcome responses. Extreme temperature events, as well as other impacts from climate change, can also be associated with changes in electricity supply and distribution that can have important implications for the availability of heating and air conditioning, which are key adaptive measures.

Though the estimates of the health impact from extreme heat discussed in the "Research Highlight" were produced only for urban areas (which provided a large sample size for statistical validity), there is also emerging evidence regarding high rates of heat-related illness in rural areas.[6, 62] Occupational exposure and a lack of access to air conditioning are some of the factors that may make rural populations particularly susceptible to extreme heat. There are quantitative challenges to using statistical methods to estimate mortality impacts of temperatures in rural areas due to lower population density and more dispersed weather stations, but rural residents have also demonstrated vulnerability to heat events.[121]

Other changes in human behavior will also have implications for the linkage between climate and heat-related illness. Changes in building infrastructure as a response to changes in temperature can have impacts on indoor air quality. Similarly, changes in behavior as a result of temperature changes, for example, seeking access to air conditioning, can change exposure to indoor and outdoor pollution and vectorborne diseases (see Ch. 3: Air Quality Impacts; Ch. 5: Vectorborne Diseases).

Finally, projecting climate variability and the most extreme temperature events can be more challenging than projecting average warming. Extreme temperatures may rise faster than average temperatures,[122] with the coldest days warming faster than average for much of the twentieth century, and the warmest days warming faster than average temperatures in the past 30 years.[123] Extremely high temperatures in the future may also reach levels outside of past experience, in which case statistically based relationships may no longer hold for those events. There have been suggestive links between rapid recent Arctic sea ice loss[124] and an increased frequency of cold[125] and warm extremes,[126] but this is an active area of research with conflicting results.[127, 128] In regions where temperature variability increases, mortality will be expected to increase; mortality is expected to decrease in regions where variability decreases.[129]

2.10   Research Needs

In addition to the emerging issues identified above, the authors highlight the following potential areas for additional scientific and research activity on temperature-related illness and death based on their review of the literature. Improved modeling and more robust projections of climate variability and extreme temperatures will enhance the modeling of health impacts associated with extremes of heat and cold. While the surveillance for temperature-related deaths is relatively robust, understanding the impacts of future changes in heat waves and extreme temperatures can be improved with better surveillance and documentation of non-fatal illnesses, including hospitalizations and emergency room visits, for temperature-associated reasons. With growing implementation of heat early warning systems around the country, there is also a need for the development of evaluation methods and associated collection of data to be able to assess effectiveness of such systems and other means of health adaptation.

Future assessments can benefit from research activities that:

• further explore the associations between exposure to a range of high and low temperatures and exacerbation of illnesses across locations and healthcare settings;

• improve understanding of how genetic factors and social determinants contribute to vulnerability to illness and death from extreme temperature exposures;

• analyze the combined health effects of temperature and other discrete climate-sensitive stressors, such as changing air quality, smoke from wildfires, or impacts of extreme weather events;

• attribute changes in observed mortality to a changing climate;

• develop effective adaptive responses to reduce the potential adverse health outcomes attributable to changing temperatures; and

• explore how future adaptive measures and behaviors can be included in quantitative models of health impacts associated with extreme temperatures

Supporting Evidence

The chapter was developed through technical discussions of relevant evidence and expert deliberation by the report authors at several workshops, teleconferences, and email exchanges. The authors considered inputs and comments submitted by the public, the National Academies of Sciences, and Federal agencies. For additional information on the overall report process, see Appendices 2 and 3.

The content of this chapter was determined after reviewing the collected literature. The authors determined that there was substantial literature available to characterize both observed and projected mortality from elevated temperatures, with sufficient literature available to also characterize mortality from cold as well as cold-related hospitalizations and illness. Populations of concern were also considered to be a high priority for this chapter. As discussed in the chapter, there were limitations in terms of the state of the literature on understanding how future adaptation will influence climate-related changes in temperature-related mortality, addressing the impact of temperature on rural populations, and examining health-related endpoints other than mortality and morbidity.

KEY FINDING TRACEABLE ACCOUNTS

Future Increases in Temperature-Related Deaths

Key Finding 1: Based on present-day sensitivity to heat, an increase of thousands to tens of thousands of premature heat-related deaths in the summer *[Very Likely, High Confidence]* and a decrease of premature cold-related deaths in the winter *[Very Likely, Medium Confidence]* are projected each year as a result of climate change by the end of the century. Future adaptation will very likely reduce these impacts (see Changing Tolerance to Extreme Heat Finding). The reduction in cold-related deaths is projected to be smaller than the increase in heat-related deaths in most regions *[Likely, Medium Confidence]*.

Description of evidence base

An extensive literature examines projections of mortality due to increasing temperatures. In particular, nine studies were identified that provide heat mortality projections in the United States for at least 10% of the U.S. population.[22, 27, 38, 54, 75, 76, 77, 78, 79] Each of these studies projected an increase in heat-related mortality due to projections of future warming, though several noted the potential modification effect of adaptation (discussed in Key Finding #3). In general, the magnitude of projected increases in annual premature deaths in these studies was in the hundreds to thousands by mid-century, and thousands to tens of thousands by the end of the century, when scaled to the total U.S. population. These conclusions are further supported by studies at the city, county, and state level.[18, 26, 39, 80, 81, 82, 83, 84, 85]

The Third National Climate Assessment (2014 NCA) found that "While deaths and injuries related to cold events are projected to decline due to climate change, these reductions are not expected to compensate for the increase in heat-related deaths,"[41] and studies published since that time have further supported this finding. Of those studies that examine both heat and cold at the national scale, only Barreca found that the reductions in cold deaths would more than compensate for the increase in heat deaths.[22, 27, 38, 75, 77] Barreca's study was novel in terms of its treatment of humidity, finding that weather that was both cold and dry, or both hot and humid, was associated with higher mortality. However, this treatment of humidity was not the cause of the difference with other studies, as leaving out humidity actually showed a greater benefit from future climate change. Instead, the author stated that the reduction in net deaths was a result of relying on counties with over 100,000 inhabitants, and that using a state-level model covering all U.S. deaths would lead to a prediction of an increase of 1.7% in mortality rates rather than a decrease of 0.1%. The finding by the majority of studies at a national scale that heat deaths will increase more than cold deaths will decrease is consistent with studies at smaller spatial scales.[26] Moreover, several studies provide rationales for why heat mortality is expected to outpace cold mortality,[12, 22, 27] and some studies suggest that cold mortality may not be responsive to warming.[49, 93, 94] Barnett et al. (2012) showed that cold waves were not generally associated with an increase in deaths beyond the mortality already associated with cold weather, in contrast to heat waves.[2]

Major uncertainties

The largest remaining uncertainties concern questions of future adaptation, which are discussed in Key Finding #3. A related uncertainty involves the link between the temperatures measured at weather stations and the temperatures experienced by individuals. As long as the relationship between the weather station and the microclimate or indoor/outdoor difference remains constant, this should not impair projections. However, as microclimates, building construction, or behavior change, the relationship between recorded weather station temperature and actual temperature exposure will change. This is related to, but broader than, the question of adaptation. Additionally, there are uncertainties regarding the non-linearities of heat response with increasing temperatures.

Assessment of confidence and likelihood based on evidence

There is **high confidence** that heat deaths will **very likely** increase in the future compared to a future without climate change, based on high agreement and a large number of studies as well as consistency across scenarios and regions. Because there are fewer studies on winter mortality, and because studies exist that suggest that winter mortality is not strongly linked to temperatures, there is **medium confidence** that deaths due to extreme cold will **very likely** decrease. The majority of the studies that examine both heat and cold deaths find that the increase in heat deaths due to climate change will **likely** be larger than the decrease in cold deaths in most regions, but there are a limited number of such studies, leading to an assessment of **medium confidence**.

Even Small Differences from Seasonal Average Temperatures Result in Illness and Death

Key Finding 2: Days that are hotter than usual in the summer or colder than usual in the winter are both associated with increased illness and death *[Very High Confidence]*. Mortality effects are observed even for small differences from seasonal average temperatures *[High Confidence]*. Because small temperature differences occur much more frequently than large temperature differences, not accounting for the effect of these small differences would lead to underestimating the future impact of climate change *[Likely, High Confidence]*.

Description of evidence base

Two well-recognized conclusions from the literature are that extreme temperatures lead to illness and premature death and that these extreme temperatures are best described in relation to local average seasonal temperatures rather than absolute temperature values. Epidemiological studies find an increase in mortality at temperatures that are high related to the local average.[9, 10, 12, 15, 17, 38] Based on absolute temperatures, Anderson and Bell 2011 found that cities in the South and Southeast were the least sensitive to heat, demonstrating acclimatization.[9]

Illness has been linked with hot daily average temperature[4, 6, 51, 69, 71] and apparent temperature, among other metrics.[3, 66, 68, 87] Across studies, adverse health episodes were most strongly associated with exposures to high temperatures that occurred on the same day or the previous day.[3, 51] However, a cumulative effect of heat was also observed at periods of up to one week after exposure, tapering off beyond seven days.[69, 105] Cardiovascular and respiratory illness has been most commonly examined in relation to extreme heat, but the association is more varied for illness than for mortality due to effects across age groups[69, 70] and differences in morbidity risk associated with emergency room records versus hospital admissions.[4, 6, 24, 51, 66, 67, 68, 69, 70]

The evidence for mortality is clearest for extreme temperatures, as addressed in threshold-based studies,[12] but studies that account for smaller changes in temperature found mortality changes even for small deviations of temperature.[11, 27] This is consistent with studies showing a U-shaped relationship of temperature and mortality—while there may be some plateau near the optimal temperature, the plateau is often small, and not always coincident with the seasonal average temperature.[11, 40] However, some of the individuals who die in response to elevated temperatures were already near death, and so the temperature event is sometimes considered to have "displaced" the death by a matter of days rather than created an additional death. Studies have found that this effect is generally below 50% of the total deaths, and is much smaller than that (10% or less) for the most extreme events, such as the 2003 European heat wave.[12, 48, 49, 50] In contrast, one recent study found that in seven U.S. cities mortality displacement was greater than 80% for small temperature deviations and around 50% even for the 3% of warmest events in the study sample.[7]

Major uncertainties

This finding reflects consideration of a number of recent studies[17, 54] not referenced in the recent 2014 NCA.[41] There is a consensus of studies linking extreme temperatures and mortality, and a growing body of literature demonstrating that smaller differences in temperature are also linked with mortality. However, the mortality displacement effect, and the fact that deaths that do not occur during an identified heat wave are less likely to be directly attributed to extreme heat, contribute to continuing uncertainty about the magnitude of the effect of temperature on mortality.

Assessment of confidence and likelihood based on evidence

There is **very high confidence** in the relationship between extreme temperatures and premature deaths due to the consistency and strength of the literature, particularly given the different study designs that produce this result. There is **high confidence** that small temperature deviations from normal temperatures contribute to premature mortality due to high agreement among those studies that have examined the issue. Though some studies indicate that for these small temperature differences, mortality displacement may play a larger role than for more extreme temperatures. Fewer studies have examined the role of these smaller temperature differences in projections, but the directionality of the effect is clear, so the determination of the authors was that not including this effect would likely lead to an underestimate of future mortality, with **high confidence**.

Changing Tolerance to Extreme Heat

Key Finding 3: An increase in population tolerance to extreme heat has been observed over time *[Very High Confidence]*. Changes in this tolerance have been associated with increased use of air conditioning, improved social responses, and/or physiological acclimatization, among other factors *[Medium Confidence]*. Expected future increases in this tolerance will reduce the projected increase in deaths from heat *[Very Likely, Very High Confidence]*.

Description of evidence base

The increasing tolerance of the U.S. population to extreme heat has been shown by a number of studies.[9, 10, 54] However, there is less confidence in attributing this increase in tolerance: increased prevalence and use of air conditioning, physiological adaptation, available green space, and improved social responses have all been proposed as explanatory factors. There have been some indications (Sheridan et al. 2009)[97] that tolerance improvements in the United States might have plateaued, but Bobb et al. 2014 found continuing improvements through 2005.[54]

Several approaches to including adaptation have been used in temperature mortality projection studies. For example, two studies used an "analog city" approach, where the response of the population to future temperatures in a given city is assumed to be equal to that of a city with a hotter present-day climate.[22, 81] Another approach is to assume that critical temperature thresholds change by some quantity over time.[18, 91] A third approach is to calculate sensitivity to air conditioning prevalence in the present, and make assumptions about air conditioning in the future.[85] In general, inclusion of adaptation limits the projected increase in deaths, sometimes modestly, other times dramatically. However, approaches used to account for adaptation may be optimistic. Historically, adaptive measures have occurred as a response to extreme events, and therefore could be expected to lag warming.[39, 96] While the increase in mortality projected in these studies is reduced, the studies generally found that mortality increases compared to present day even under optimistic adaptation assumptions.[18, 22, 81, 85] A limit to adaptation may be seen in that even in cities with nearly 100% air conditioning penetration, heat deaths are observed today.

Major uncertainties
While studies have been published in recent years that include adaptation in sensitivity analyses,[22] this remains a challenging area of research. Difficulties in attributing observed increases in tolerance make it challenging to project future changes in tolerance, whether due to autonomous adaptation by individuals or planned adjustments by governments. Extrapolation of acclimatization is limited as there must be an increase in temperature beyond which acclimatization will not be possible.

Assessment of confidence and likelihood based on evidence
There is **very high confidence** that a decrease in sensitivity to heat events has occurred based on high agreement between studies, but only **medium confidence** that this decrease is due to some specific combination of air conditioning prevalence, physiological adaptation, presence of green space, and improved social responses because of the challenges involved in attribution. There is **very high confidence** that mortality due to heat will **very likely** be reduced compared to a no-adaptation scenario when adaptation is included, because all studies examined were in agreement with this conclusion, though the magnitude of this reduction is poorly constrained.

Some Populations at Greater Risk

Key Finding 4: Older adults and children have a higher risk of dying or becoming ill due to extreme heat *[Very High Confidence]*. People working outdoors, the socially isolated and economically disadvantaged, those with chronic illnesses, as well as some communities of color, are also especially vulnerable to death or illness *[Very High Confidence]*.

Description of evidence base
The relationship between increased temperatures and deaths in elderly populations is well-understood. An increased risk of respiratory and cardiovascular death is observed in elderly populations during temperature extremes due to reduced thermoregulation.[17, 42, 45, 65]

Studies cite dehydration, electrolyte imbalance, fever, heat stress, hyperthermia, and renal disease as the primary health conditions in children exposed to heat waves. Causes of heat-related illness in children include inefficient thermoregulation, reduced cardiovascular output, and heightened metabolic

rate. Children also spend a considerable amount of time outdoors and participating in vigorous activities.[17, 42, 64, 103] A limited number of studies found evidence of cold-related mortality in children; however, no study has examined the relationship between cold temperature and cause-specific mortality.[64]

Certain occupational groups that spend a great deal of time exposed to extreme temperatures, such as agricultural workers, construction workers, and electricity and pipeline utility workers, are at increased risk for heat- and cold-related illness, especially where jobs involve heavy exertion.[100, 113, 114] Lack of heat-illness-prevention programs in the workplace that include provisions for acclimatization was found to be a factor strongly associated with extreme temperature-related death.[113]

Race, ethnicity, and socioeconomic status have been shown to impact vulnerability to temperature extremes. Several studies have identified non-Hispanic Black populations to be more vulnerable than other racial and ethnic groups for experiencing detrimental consequences of exposure to temperature extremes.[17, 42, 45, 103, 110] Studies suggest comorbidities that enhance susceptibility to higher temperatures, availability and use of air conditioning, disparities in heat risk-related land cover characteristics, and other environmental justice issues contribute to increased vulnerability of non-Hispanic Blacks.[17, 42, 108, 110, 112]

Dementia, mood disorders, neurosis and stress-related illnesses, and substance abuse are shown to be impacted by extreme heat.[100, 115, 116, 117] Some medications interfere with thermoregulation, increasing vulnerability to heat.[116]

Major uncertainties
The literature available at the time of the development of the 2014 NCA had identified a number of vulnerable populations that were disproportionately at risk during heat waves, and literature since that time has only strengthened the understanding of the elevated risks for these populations. There continues to be a need for better understanding of the relative importance of genetics and environmental justice issues with regards to the observed higher risk for non-Hispanic Blacks, more work on understanding the risks to pregnant women from extreme temperature events, and a better understanding of the relationship between extreme cold vulnerabilities in populations of concern.

Assessment of confidence and likelihood based on evidence
Although some details regarding causation and identifying the most vulnerable subpopulations still require research, there is a large body of literature that demonstrates the increased vulnerability to extreme heat of a number of groups, and therefore there is **very high confidence** that the listed populations of concern are at greater risk of temperature-related death and illness.

DOCUMENTING UNCERTAINTY

This assessment relies on two metrics to communicate the degree of certainty in Key Findings. See Appendix 4: Documenting Uncertainty for more on assessments of likelihood and confidence.

Confidence Level
Very High
Strong evidence (established theory, multiple sources, consistent results, well documented and accepted methods, etc.), high consensus
High
Moderate evidence (several sources, some consistency, methods vary and/or documentation limited, etc.), medium consensus
Medium
Suggestive evidence (a few sources, limited consistency, models incomplete, methods emerging, etc.), competing schools of thought
Low
Inconclusive evidence (limited sources, extrapolations, inconsistent findings, poor documentation and/or methods not tested, etc.), disagreement or lack of opinions among experts

Likelihood
Very Likely
≥ 9 in 10
Likely
≥ 2 in 3
As Likely As Not
≈ 1 in 2
Unlikely
≤ 1 in 3
Very Unlikely
≤ 1 in 10

PHOTO CREDITS

Pg. 43–Construction worker: © Fotosearch

Pg. 44–Large Crowd: © iStockImages.com/Ints Vikmanis

Pg. 49–Snowstorm: © iStockImages.com/Dreef

Pg. 50–Construction worker: © Fotosearch

Pg. 54–Young baseball catcher: © iStockImages.com/jpbcpa

References

1. Walsh, J., D. Wuebbles, K. Hayhoe, J. Kossin, K. Kunkel, G. Stephens, P. Thorne, R. Vose, M. Wehner, J. Willis, D. Anderson, S. Doney, R. Feely, P. Hennon, V. Kharin, T. Knutson, F. Landerer, T. Lenton, J. Kennedy, and R. Somerville, 2014: Ch. 2: Our changing climate. *Climate Change Impacts in the United States: The Third National Climate Assessment.* Melillo, J.M., T.C. Richmond, and G.W. Yohe, Eds. U.S. Global Change Research Program, Washington, D.C., 19-67. http://dx.doi.org/10.7930/J0KW5CXT

2. Barnett, A.G., S. Hajat, A. Gasparrini, and J. Rocklöv, 2012: Cold and heat waves in the United States. *Environmental Research,* **112,** 218-224. http://dx.doi.org/10.1016/j.envres.2011.12.010

3. Gronlund, C.J., A. Zanobetti, J.D. Schwartz, G.A. Wellenius, and M.S. O'Neill, 2014: Heat, heat waves, and hospital admissions among the elderly in the United States, 1992–2006. *Environmental Health Perspectives,* **122,** 1187-1192. http://dx.doi.org/10.1289/ehp.1206132

4. Lavigne, E., A. Gasparrini, X. Wang, H. Chen, A. Yagouti, M.D. Fleury, and S. Cakmak, 2014: Extreme ambient temperatures and cardiorespiratory emergency room visits: Assessing risk by comorbid health conditions in a time series study. *Environmental Health,* **13,** 5. http://dx.doi.org/10.1186/1476-069x-13-5

5. Lin, S., M. Luo, R.J. Walker, X. Liu, S.-A. Hwang, and R. Chinery, 2009: Extreme high temperatures and hospital admissions for respiratory and cardiovascular diseases. *Epidemiology,* **20,** 738-746. http://dx.doi.org/10.1097/EDE.0b013e3181ad5522

6. Lippmann, S.J., C.M. Fuhrmann, A.E. Waller, and D.B. Richardson, 2013: Ambient temperature and emergency department visits for heat-related illness in North Carolina, 2007-2008. *Environmental Research,* **124,** 35-42. http://dx.doi.org/10.1016/j.envres.2013.03.009

7. Saha, M.V., R.E. Davis, and D.M. Hondula, 2014: Mortality displacement as a function of heat event strength in 7 US cities. *American Journal of Epidemiology,* **179,** 467-474. http://dx.doi.org/10.1093/aje/kwt264

8. Smith, T.T., B.F. Zaitchik, and J.M. Gohlke, 2013: Heat waves in the United States: Definitions, patterns and trends. *Climatic Change,* **118,** 811-825. http://dx.doi.org/10.1007/s10584-012-0659-2

9. Anderson, G.B. and M.L. Bell, 2011: Heat waves in the United States: Mortality risk during heat waves and effect modification by heat wave characteristics in 43 U.S. communities. *Environmental Health Perspectives,* **119,** 210-218. http://dx.doi.org/10.1289/ehp.1002313

10. Guo, Y., A.G. Barnett, and S. Tong, 2012: High temperatures-related elderly mortality varied greatly from year to year: Important information for heat-warning systems. *Scientific Reports,* **2.** http://dx.doi.org/10.1038/srep00830

11. Lee, M., F. Nordio, A. Zanobetti, P. Kinney, R. Vautard, and J. Schwartz, 2014: Acclimatization across space and time in the effects of temperature on mortality: A time-series analysis. *Environmental Health,* **13,** 89. http://dx.doi.org/10.1186/1476-069X-13-89

12. Medina-Ramón, M. and J. Schwartz, 2007: Temperature, temperature extremes, and mortality: A study of acclimatisation and effect modification in 50 US cities. *Occupational and Environmental Medicine,* **64,** 827-833. http://dx.doi.org/10.1136/oem.2007.033175

13. Sherwood, S.C. and M. Huber, 2010: An adaptability limit to climate change due to heat stress. *Proceedings of the National Academy of Sciences,* **107,** 9552-9555. http://dx.doi.org/10.1073/pnas.0913352107

14. Kent, S.T., L.A. McClure, B.F. Zaitchik, T.T. Smith, and J.M. Gohlke, 2014: Heat waves and health outcomes in Alabama (USA): The importance of heat wave definition. *Environmental Health Perspectives,* **122,** 151–158. http://dx.doi.org/10.1289/ehp.1307262

15. Kalkstein, L.S., S. Greene, D.M. Mills, and J. Samenow, 2011: An evaluation of the progress in reducing heat-related human mortality in major US cities. *Natural Hazards,* **56,** 113-129. http://dx.doi.org/10.1007/s11069-010-9552-3

16. Zhang, K., R.B. Rood, G. Michailidis, E.M. Oswald, J.D. Schwartz, A. Zanobetti, K.L. Ebi, and M.S. O'Neill, 2012: Comparing exposure metrics for classifying 'dangerous heat' in heat wave and health warning systems. *Environment International,* **46,** 23-29. http://dx.doi.org/10.1016/j.envint.2012.05.001

17. Berko, J., D.D. Ingram, S. Saha, and J.D. Parker, 2014: Deaths Attributed to Heat, Cold, and Other Weather Events in the United States, 2006–2010. National Health Statistics Reports No. 76, July 30, 2014, 15 pp. National Center for Health Statistics, Hyattsville, MD. http://www.cdc.gov/nchs/data/nhsr/nhsr076.pdf

18. Gosling, S.N., J.A. Lowe, G.R. McGregor, M. Pelling, and B.D. Malamud, 2009: Associations between elevated atmospheric temperature and human mortality: A critical review of the literature. *Climatic Change,* **92**, 299-341. http://dx.doi.org/10.1007/s10584-008-9441-x

19. WHO, 2004: *International Statistical Classification of Diseases and Related Health Problems, 10th Revision (ICD–10),* 2nd ed. World Health Organization, Geneva, Switzerland. http://www.who.int/classifications/icd/ICD-10_2nd_ed_volume2.pdf

20. CDC, 2005: Extreme Cold: A Prevention Guide to Promote Your Personal Health and Safety. 13 pp. U.S. Department of Health and Human Services, Centers for Disease Control and Prevention, Atlanta, GA. http://www.bt.cdc.gov/disasters/winter/pdf/extreme-cold-guide.pdf

21. CDC, 2006: Hypothermia-related deaths – United States, 1999-2002 and 2005. *MMWR. Morbidity and Mortality Weekly Report,* **55**, 282-284. http://www.cdc.gov/mmwr/preview/mmwrhtml/mm5510a5.htm

22. Mills, D., J. Schwartz, M. Lee, M. Sarofim, R. Jones, M. Lawson, M. Duckworth, and L. Deck, 2015: Climate change impacts on extreme temperature mortality in select metropolitan areas in the United States. *Climatic Change,* **131**, 83-95. http://dx.doi.org/10.1007/s10584-014-1154-8

23. Bustinza, R., G. Lebel, P. Gosselin, D. Bélanger, and F. Chebana, 2013: Health impacts of the July 2010 heat wave in Quebec, Canada. *BMC Public Health,* **13**, 56. http://dx.doi.org/10.1186/1471-2458-13-56

24. Knowlton, K., M. Rotkin-Ellman, G. King, H.G. Margolis, D. Smith, G. Solomon, R. Trent, and P. English, 2009: The 2006 California heat wave: Impacts on hospitalizations and emergency department visits. *Environmental Health Perspectives,* **117**, 61-67. http://dx.doi.org/10.1289/ehp.11594

25. Ye, X., R. Wolff, W. Yu, P. Vaneckova, X. Pan, and S. Tong, 2012: Ambient temperature and morbidity: A review of epidemiological evidence. *Environmental Health Perspectives,* **120**, 19-28. http://dx.doi.org/10.1289/ehp.1003198

26. Li, T., R.M. Horton, and P.L. Kinney, 2013: Projections of seasonal patterns in temperature-related deaths for Manhattan, New York. *Nature Climate Change,* **3**, 717-721. http://dx.doi.org/10.1038/nclimate1902

27. Schwartz, J.D., M. Lee, P.L. Kinney, S. Yang, D. Mills, M. Sarofim, R. Jones, R. Streeter, A. St. Juliana, J. Peers, and R.M. Horton, 2015: Projections of temperature-attributable premature deaths in 209 U.S. cities using a cluster-based Poisson approach. *Environmental Health,* **14**. http://dx.doi.org/10.1186/s12940-015-0071-2

28. CDC, 1994: Heat-related deaths--Philadelphia and United States, 1993-1994. *MMWR. Morbidity and Mortality Weekly Report,* **43**, 453-455. http://www.cdc.gov/mmwr/preview/mmwrhtml/00031773.htm

29. CDC, 1995: Heat-related mortality – Chicago, July 1995. *MMWR. Morbidity and Mortality Weekly Report,* **44**, 577-579. http://www.cdc.gov/mmwr/preview/mmwrhtml/00038443.htm

30. Hoshiko, S., P. English, D. Smith, and R. Trent, 2010: A simple method for estimating excess mortality due to heat waves, as applied to the 2006 California heat wave. *International Journal of Public Health,* **55**, 133-137. http://dx.doi.org/10.1007/s00038-009-0060-8

31. Jones, T.S., A.P. Liang, E.M. Kilbourne, M.R. Griffin, P.A. Patriarca, S.G. Fite Wassilak, R.J. Mullan, R.F. Herrick, D. Donnell, Jr., K. Choi, and S.B. Thacker, 1982: Morbidity and mortality associated with the July 1980 heat wave in St Louis and Kansas City, Mo. *JAMA: The Journal of the American Medical Association,* **247**, 3327–3331. http://dx.doi.org/10.1001/jama.1982.03320490025030

32. Jones, S., M. Griffin, A. Liang, and P. Patriarca, 1980: The Kansas City Heat Wave, July 1980: Effects of Health, Preliminary Report. Centers for Disease Control, Atlanta, GA.

33. Karl, T.R. and R.W. Knight, 1997: The 1995 Chicago heat wave: How likely is a recurrence? *Bulletin of the American Meteorological Society,* **78**, 1107-1119. http://dx.doi.org/10.1175/1520-0477(1997)078%3C1107:tchwhl%3E2.0.co;2

34. EPA, 2014: Climate Change Indicators in the United States, 2014. 3rd edition. EPA 430-R-14-04, 107 pp. U.S. Environmental Protection Agency, Washington, D.C. http://www.epa.gov/climatechange/pdfs/climateindicators-full-2014.pdf

35. Kaiser, R., A. Le Tertre, J. Schwartz, C.A. Gotway, W.R. Daley, and C.H. Rubin, 2007: The effect of the 1995 heat wave in Chicago on all-cause and cause-specific mortality. *American Journal of Public Health,* **97**, S158-S162. http://dx.doi.org/10.2105/ajph.2006.100081

36. Harlan, S.L., J.H. Declet-Barreto, W.L. Stefanov, and D.B. Petitti, 2013: Neighborhood effects on heat deaths: Social and environmental predictors of vulnerability in Maricopa County, Arizona. *Environmental Health Perspectives,* **121,** 197-204. http://dx.doi.org/10.1289/ehp.1104625

37. Madrigano, J., M.A. Mittleman, A. Baccarelli, R. Goldberg, S. Melly, S. von Klot, and J. Schwartz, 2013: Temperature, myocardial infarction, and mortality: Effect modification by individual- and area-level characteristics. *Epidemiology,* **24,** 439-446. http://dx.doi.org/10.1097/EDE.0b013e3182878397

38. Deschênes, O. and M. Greenstone, 2011: Climate change, mortality, and adaptation: Evidence from annual fluctuations in weather in the US. *American Economic Journal: Applied Economics,* **3,** 152-185. http://dx.doi.org/10.1257/app.3.4.152

39. Hayhoe, K., S. Sheridan, L. Kalkstein, and S. Greene, 2010: Climate change, heat waves, and mortality projections for Chicago. *Journal of Great Lakes Research,* **36,** 65-73. http://dx.doi.org/10.1016/j.jglr.2009.12.009

40. Gasparrini, A., Y. Guo, M. Hashizume, E. Lavigne, A. Zanobetti, J. Schwartz, A. Tobias, S. Tong, J. Rocklöv, B. Forsberg, M. Leone, M. De Sario, M.L. Bell, Y.-L.L. Guo, C.-f. Wu, H. Kan, S.-M. Yi, M. de Sousa Zanotti Stagliorio Coelho, P.H.N. Saldiva, Y. Honda, H. Kim, and B. Armstrong, 2015: Mortality risk attributable to high and low ambient temperature: A multicountry observational study. *The Lancet,* **386,** 369-375. http://dx.doi.org/10.1016/S0140-6736(14)62114-0

41. Luber, G., K. Knowlton, J. Balbus, H. Frumkin, M. Hayden, J. Hess, M. McGeehin, N. Sheats, L. Backer, C.B. Beard, K.L. Ebi, E. Maibach, R.S. Ostfeld, C. Wiedinmyer, E. Zielinski-Gutiérrez, and L. Ziska, 2014: Ch. 9: Human health. *Climate Change Impacts in the United States: The Third National Climate Assessment.* Melillo, J.M., T.C. Richmond, and G.W. Yohe, Eds. U.S. Global Change Research Program, Washington, D.C., 220-256. http://dx.doi.org/10.7930/J0PN93H5

42. Basu, R. and J.M. Samet, 2002: Relation between elevated ambient temperature and mortality: A review of the epidemiologic evidence. *Epidemiologic Reviews,* **24,** 190-202. http://dx.doi.org/10.1093/epirev/mxf007

43. Vanos, J.K., L.S. Kalkstein, and T.J. Sanford, 2015: Detecting synoptic warming trends across the US midwest and implications to human health and heat-related mortality. *International Journal of Climatology,* **35,** 85-96. http://dx.doi.org/10.1002/joc.3964

44. Dixon, P.G., D.M. Brommer, B.C. Hedquist, A.J. Kalkstein, G.B. Goodrich, J.C. Walter, C.C. Dickerson, S.J. Penny, and R.S. Cerveny, 2005: Heat mortality versus cold mortality: A study of conflicting databases in the United States. *Bulletin of the American Meteorological Society,* **86,** 937-943. http://dx.doi.org/10.1175/bams-86-7-937

45. Anderson, B.G. and M.L. Bell, 2009: Weather-related mortality: How heat, cold, and heat waves affect mortality in the United States. *Epidemiology,* **20,** 205-213. http://dx.doi.org/10.1097/EDE.0b013e318190ee08

46. Analitis, A., P. Michelozzi, D. D'Ippoliti, F. de'Donato, B. Menne, F. Matthies, R.W. Atkinson, C. Iñiguez, X. Basagaña, A. Schneider, A. Lefranc, A. Paldy, L. Bisanti, and K. Katsouyanni, 2014: Effects of heat waves on mortality: Effect modification and confounding by air pollutants. *Epidemiology,* **25,** 15-22. http://dx.doi.org/10.1097/EDE.0b013e31828ac01b

47. Madrigano, J., D. Jack, G.B. Anderson, M.L. Bell, and P.L. Kinney, 2015: Temperature, ozone, and mortality in urban and non-urban counties in the Northeastern United States. *Environmental Health,* **14,** 3. http://dx.doi.org/10.1186/1476-069X-14-3

48. Kalkstein, L.S., 1998: Climate and human mortality: Relationships and mitigating measures. *Advances in Bioclimatology,* **5,** 161-177. http://dx.doi.org/10.1007/978-3-642-80419-9_7

49. Kinney, P.L., M. Pascal, R. Vautard, and K. Laaidi, 2012: Winter mortality in a changing climate: Will it go down? *Bulletin Epidemiologique Hebdomadaire,* **12-13,** 5-7.

50. Le Tertre, A., A. Lefranc, D. Eilstein, C. Declercq, S. Medina, M. Blanchard, B. Chardon, P. Fabre, L. Filleul, J.-F. Jusot, L. Pascal, H. Prouvost, S. Cassadou, and M. Ledrans, 2006: Impact of the 2003 heatwave on all-cause mortality in 9 French cities. *Epidemiology,* **17,** 75-79. http://dx.doi.org/10.1097/01.ede.0000187650.36636.1f

51. Anderson, G.B., F. Dominici, Y. Wang, M.C. McCormack, M.L. Bell, and R.D. Peng, 2013: Heat-related emergency hospitalizations for respiratory diseases in the Medicare population. *American Journal of Respiratory and Critical Care Medicine,* **187,** 1098-103. http://dx.doi.org/10.1164/rccm.201211-1969OC

52. Zanobetti, A., M.S. O'Neill, C.J. Gronlund, and J.D. Schwartz, 2012: Summer temperature variability and long-term survival among elderly people with chronic disease. *Proceedings of the National Academy of Sciences,* **109,** 6608-6613. http://dx.doi.org/10.1073/pnas.1113070109

53. Hanna, E.G. and P.W. Tait, 2015: Limitations to thermo-regulation and acclimatization challenge human adaptation to global warming. *International Journal of Environmental Research and Public Health,* **12,** 8034-8074. http://dx.doi.org/10.3390/ijerph120708034

54. Bobb, J.F., R.D. Peng, M.L. Bell, and F. Dominici, 2014: Heat-related mortality and adaptation to heat in the United States. *Environmental Health Perspectives,* **122,** 811-816. http://dx.doi.org/10.1289/ehp.1307392

55. Petkova, E.P., A. Gasparrini, and P.L. Kinney, 2014: Heat and mortality in New York City since the beginning of the 20th century. *Epidemiology,* **25,** 554-560. http://dx.doi.org/10.1097/ede.0000000000000123

56. Hondula, D.M., R.C. Balling, Jr., J.K. Vanos, and M. Georgescu, 2015: Rising temperatures, human health, and the role of adaptation. *Current Climate Change Reports,* **1,** 144-154. http://dx.doi.org/10.1007/s40641-015-0016-4

57. Stone, B.J., J. Vargo, P. Liu, D. Habeeb, A. DeLucia, M. Trail, Y. Hu, and A. Russell, 2014: Avoided heat-related mortality through climate adaptation strategies in three US cities. *PLoS ONE,* **9,** e100852. http://dx.doi.org/10.1371/journal.pone.0100852

58. CDC, 2015: Emergency Preparedness and Response: Extreme Heat. Centers for Disease Control and Prevention, Atlanta, GA. http://www.bt.cdc.gov/disasters/extremeheat/

59. EPA, 2015: Natural Disasters: Extreme Heat. U.S. Environmental Protection Agency, Washington, D.C. http://epa.gov/naturaldisasters/extremeheat.html

60. Fischer, E.M. and R. Knutti, 2015: Anthropogenic contribution to global occurrence of heavy-precipitation and high-temperature extremes. *Nature Climate Change,* **5,** 560-564. http://dx.doi.org/10.1038/nclimate2617

61. Davis, R.E., P.C. Knappenberger, W.M. Novicoff, and P.J. Michaels, 2003: Decadal changes in summer mortality in US cities. *International Journal of Biometeorology,* **47,** 166-175. http://dx.doi.org/10.1007/s00484-003-0160-8

62. Hess, J.J., S. Saha, and G. Luber, 2014: Summertime acute heat illness in U.S. emergency departments from 2006 through 2010: Analysis of a nationally representative sample. *Environmental Health Perspectives,* **122,** 1209-1215. http://dx.doi.org/10.1289/ehp.1306796

63. Noe, R.S., J.O. Jin, and A.F. Wolkin, 2012: Exposure to natural cold and heat: Hypothermia and hyperthermia medicare claims, United States, 2004–2005. *American Journal of Public Health,* **102,** e11-e18. http://dx.doi.org/10.2105/ajph.2011.300557

64. Xu, Z., R.A. Etzel, H. Su, C. Huang, Y. Guo, and S. Tong, 2012: Impact of ambient temperature on children's health: A systematic review. *Environmental Research,* **117,** 120-131. http://dx.doi.org/10.1016/j.envres.2012.07.002

65. Åström, D.O., F. Bertil, and R. Joacim, 2011: Heat wave impact on morbidity and mortality in the elderly population: A review of recent studies. *Maturitas,* **69,** 99-105. http://dx.doi.org/10.1016/j.maturitas.2011.03.008

66. Basu, R., D. Pearson, B. Malig, R. Broadwin, and R. Green, 2012: The effect of high ambient temperature on emergency room visits. *Epidemiology,* **23,** 813-820. http://dx.doi.org/10.1097/EDE.0b013e31826b7f97

67. Green, R.S., R. Basu, B. Malig, R. Broadwin, J.J. Kim, and B. Ostro, 2010: The effect of temperature on hospital admissions in nine California counties. *International Journal of Public Health,* **55,** 113-121. http://dx.doi.org/10.1007/s00038-009-0076-0

68. Ostro, B., S. Rauch, R. Green, B. Malig, and R. Basu, 2010: The effects of temperature and use of air conditioning on hospitalizations. *American Journal of Epidemiology,* **172,** 1053-1061. http://dx.doi.org/10.1093/aje/kwq231

69. Schwartz, J., J.M. Samet, and J.A. Patz, 2004: Hospital admissions for heart disease: The effects of temperature and humidity. *Epidemiology,* **15,** 755-761. http://dx.doi.org/10.1097/01.ede.0000134875.15919.0f

70. Turner, L.R., A.G. Barnett, D. Connell, and S. Tong, 2012: Ambient temperature and cardiorespiratory morbidity: A systematic review and meta-analysis. *Epidemiology,* **23,** 594-606. http://dx.doi.org/10.1097/EDE.0b013e3182572795

71. Fletcher, B.A., S. Lin, E.F. Fitzgerald, and S.A. Hwang, 2012: Association of summer temperatures with hospital admissions for renal diseases in New York State: A case-cross-over study. *American Journal of Epidemiology,* **175,** 907-916. http://dx.doi.org/10.1093/aje/kwr417

72. Li, B., S. Sain, L.O. Mearns, H.A. Anderson, S. Kovats, K.L. Ebi, M.Y.V. Bekkedal, M.S. Kanarek, and J.A. Patz, 2012: The impact of extreme heat on morbidity in Milwaukee, Wisconsin. *Climatic Change,* **110,** 959-976. http://dx.doi.org/10.1007/s10584-011-0120-y

73. Alessandrini, E., S. Zauli Sajani, F. Scotto, R. Miglio, S. Marchesi, and P. Lauriola, 2011: Emergency ambulance dispatches and apparent temperature: A time series analysis in Emilia–Romagna, Italy. *Environmental Research,* **111,** 1192-1200. http://dx.doi.org/10.1016/j.envres.2011.07.005

74. Huang, C., A.G. Barnett, X. Wang, P. Vaneckova, G. Fitz-Gerald, and S. Tong, 2011: Projecting future heat-related mortality under climate change scenarios: A systematic review. *Environmental Health Perspectives,* **119,** 1681-1690. http://dx.doi.org/10.1289/Ehp.1103456

75. Barreca, A.I., 2012: Climate change, humidity, and mortality in the United States. *Journal of Environmental Economics and Management,* **63,** 19-34. http://dx.doi.org/10.1016/j.jeem.2011.07.004

76. Greene, S., L.S. Kalkstein, D.M. Mills, and J. Samenow, 2011: An examination of climate change on extreme heat events and climate–mortality relationships in large U.S. cities. *Weather, Climate, and Society,* **3,** 281-292. http://dx.doi.org/10.1175/WCAS-D-11-00055.1

77. Honda, Y., M. Kondo, G. McGregor, H. Kim, Y.-L. Guo, Y. Hijioka, M. Yoshikawa, K. Oka, S. Takano, S. Hales, and R.S. Kovats, 2014: Heat-related mortality risk model for climate change impact projection. *Environmental Health and Preventive Medicine,* **19,** 56-63. http://dx.doi.org/10.1007/s12199-013-0354-6

78. Voorhees, A.S., N. Fann, C. Fulcher, P. Dolwick, B. Hubbell, B. Bierwagen, and P. Morefield, 2011: Climate change-related temperature impacts on warm season heat mortality: A proof-of-concept methodology using BenMAP. *Environmental Science & Technology,* **45,** 1450-1457. http://dx.doi.org/10.1021/es102820y

79. Wu, J., Y. Zhou, Y. Gao, J.S. Fu, B.A. Johnson, C. Huang, Y.-M. Kim, and Y. Liu, 2014: Estimation and uncertainty analysis of impacts of future heat waves on mortality in the eastern United States. *Environmental Health Perspectives,* **122,** 10-16. http://dx.doi.org/10.1289/ehp.1306670

80. Peng, R.D., J.F. Bobb, C. Tebaldi, L. McDaniel, M.L. Bell, and F. Dominici, 2011: Toward a quantitative estimate of future heat wave mortality under global climate change. *Environmental Health Perspectives,* **119,** 701-706. http://dx.doi.org/10.1289/ehp.1002430

81. Knowlton, K., B. Lynn, R.A. Goldberg, C. Rosenzweig, C. Hogrefe, J.K. Rosenthal, and P.L. Kinney, 2007: Projecting heat-related mortality impacts under a changing climate in the New York City region. *American Journal of Public Health,* **97,** 2028-2034. http://dx.doi.org/10.2105/Ajph.2006.102947

82. Petkova, E.P., R.M. Horton, D.A. Bader, and P.L. Kinney, 2013: Projected heat-related mortality in the U.S. urban northeast. *International Journal of Environmental Research and Public Health,* **10,** 6734-6747. http://dx.doi.org/10.3390/ijerph10126734

83. Isaksen, T.B., M. Yost, E. Hom, and R. Fenske, 2014: Projected health impacts of heat events in Washington State associated with climate change. *Reviews on Environmental Health,* **29,** 119-123. http://dx.doi.org/10.1515/reveh-2014-0029

84. Jackson, J.E., M.G. Yost, C. Karr, C. Fitzpatrick, B.K. Lamb, S.H. Chung, J. Chen, J. Avise, R.A. Rosenblatt, and R.A. Fenske, 2010: Public health impacts of climate change in Washington State: Projected mortality risks due to heat events and air pollution. *Climatic Change,* **102,** 159-186. http://dx.doi.org/10.1007/s10584-010-9852-3

85. Ostro, B., S. Rauch, and S. Green, 2011: Quantifying the health impacts of future changes in temperature in California. *Environmental Research,* **111,** 1258-1264. http://dx.doi.org/10.1016/j.envres.2011.08.013

86. Petkova, E.P., D.A. Bader, G.B. Anderson, R.M. Horton, K. Knowlton, and P.L. Kinney, 2014: Heat-related mortality in a warming climate: Projections for 12 U.S. cities. *International Journal of Environmental Research and Public Health,* **11,** 11371-11383. http://dx.doi.org/10.3390/ijerph111111371

87. Lin, S., W.-H. Hsu, A.R. Van Zutphen, S. Saha, G. Luber, and S.-A. Hwang, 2012: Excessive heat and respiratory hospitalizations in New York State: Estimating current and future public health burden related to climate change. *Environmental Health Perspectives,* **120,** 1571-1577. http://dx.doi.org/10.1289/ehp.1104728

88. Brikowski, T.H., Y. Lotan, and M.S. Pearle, 2008: Climate-related increase in the prevalence of urolithiasis in the United States. *Proceedings of the National Academy of Sciences*, **105**, 9841-9846. http://dx.doi.org/10.1073/pnas.0709652105

89. Fakheri, R.J. and D.S. Goldfarb, 2011: Ambient temperature as a contributor to kidney stone formation: Implications of global warming. *Kidney International*, **79**, 1178-1185. http://dx.doi.org/10.1038/ki.2011.76

90. Tasian, G.E., J.E. Pulido, A. Gasparrini, C.S. Saigal, B.P. Horton, J.R. Landis, R. Madison, and R. Keren, 2014: Daily mean temperature and clinical kidney stone presentation in five U.S. metropolitan areas: A time-series analysis. *Environmental Health Perspectives*, **122**, 1081-1087. http://dx.doi.org/10.1289/ehp.1307703

91. Watkiss, P. and A. Hunt, 2012: Projection of economic impacts of climate change in sectors of Europe based on bottom up analysis: Human health. *Climatic Change*, **112**, 101-126. http://dx.doi.org/10.1007/s10584-011-0342-z

92. Mercer, J.B., 2003: Cold—an underrated risk factor for health. *Environmental Research*, **92**, 8-13. http://dx.doi.org/10.1016/s0013-9351(02)00009-9

93. Ebi, K.L. and D. Mills, 2013: Winter mortality in a warming climate: A reassessment. *Wiley Interdisciplinary Reviews: Climate Change*, **4**, 203-212. http://dx.doi.org/10.1002/wcc.211

94. Kinney, P.L., J. Schwartz, M. Pascal, E. Petkova, A. Le Tertre, S. Medina, and R. Vautard, 2015: Winter season mortality: Will climate warming bring benefits? *Environmental Research Letters*, **10**, 064016. http://dx.doi.org/10.1088/1748-9326/10/6/064016

95. Davis, R.E., P.C. Knappenberger, W.M. Novicoff, and P.J. Michaels, 2002: Decadal changes in heat-related human mortality in the eastern United States. *Climate Research*, **22**, 175-184. http://dx.doi.org/10.3354/cr022175

96. Ebi, K.L., T.J. Teisberg, L.S. Kalkstein, L. Robinson, and R.F. Weiher, 2004: Heat watch/warning systems save lives: Estimated costs and benefits for Philadelphia 1995–98. *Bulletin of the American Meteorological Society*, **85**, 1067-1073. http://dx.doi.org/10.1175/bams-85-8-1067

97. Sheridan, S.C., A.J. Kalkstein, and L.S. Kalkstein, 2009: Trends in heat-related mortality in the United States, 1975–2004. *Natural Hazards*, **50**, 145-160. http://dx.doi.org/10.1007/s11069-008-9327-2

98. Kovats, R.S. and S. Hajat, 2008: Heat stress and public health: A critical review. *Annual Review of Public Health*, **29**, 41-55. http://dx.doi.org/10.1146/annurev.publhealth.29.020907.090843

99. Gamble, J.L., B.J. Hurley, P.A. Schultz, W.S. Jaglom, N. Krishnan, and M. Harris, 2013: Climate change and older Americans: State of the science. *Environmental Health Perspectives*, **121**, 15-22. http://dx.doi.org/10.1289/ehp.1205223

100. Balbus, J.M. and C. Malina, 2009: Identifying vulnerable subpopulations for climate change health effects in the United States. *Journal of Occupational and Environmental Medicine*, **51**, 33-37. http://dx.doi.org/10.1097/JOM.0b013e318193e12e

101. Benmarhnia, T., S. Deguen, J.S. Kaufman, and A. Smargiassi, 2015: Review article: Vulnerability to heat-related mortality: A systematic review, meta-analysis, and meta-regression analysis. *Epidemiology*, **26**, 781-793. http://dx.doi.org/10.1097/EDE.0000000000000375

102. CDC, 2011: Extreme Heat and Your Health: Heat and Infants and Children. Centers for Disease Control and Prevention, Atlanta, GA. http://www.cdc.gov/extremeheat/children.html

103. Wasilevich, E.A., F. Rabito, J. Lefante, and E. Johnson, 2012: Short-term outdoor temperature change and emergency department visits for asthma among children: A case-crossover study. *American Journal of Epidemiology*, **176**, S123-S130. http://dx.doi.org/10.1093/aje/kws326

104. Kerr, Z.Y., D.J. Casa, S.W. Marshall, and R.D. Comstock, 2013: Epidemiology of exertional heat illness among U.S. high school athletes. *American Journal of Preventive Medicine*, **44**, 8-14. http://dx.doi.org/10.1016/j.amepre.2012.09.058

105. Basu, R., B. Malig, and B. Ostro, 2010: High ambient temperature and the risk of preterm delivery. *American Journal of Epidemiology*, **172**, 1108-1117. http://dx.doi.org/10.1093/aje/kwq170

106. Carolan-Olah, M. and D. Frankowska, 2014: High environmental temperature and preterm birth: A review of the evidence. *Midwifery*, **30**, 50-59. http://dx.doi.org/10.1016/j.midw.2013.01.011

107. Beltran, A.J., J. Wu, and O. Laurent, 2014: Associations of meteorology with adverse pregnancy outcomes: A systematic review of preeclampsia, preterm birth and birth weight. *International Journal of Environmental Research and Public Health*, **11**, 91-172. http://dx.doi.org/10.3390/ijerph110100091

108. Uejio, C.K., O.V. Wilhelmi, J.S. Golden, D.M. Mills, S.P. Gulino, and J.P. Samenow, 2011: Intra-urban societal vulnerability to extreme heat: The role of heat exposure and the built environment, socioeconomics, and neighborhood stability. *Health & Place,* **17,** 498-507. http://dx.doi.org/10.1016/j.healthplace.2010.12.005

109. Ramin, B. and T. Svoboda, 2009: Health of the homeless and climate change. *Journal of Urban Health,* **86,** 654-664. http://dx.doi.org/10.1007/s11524-009-9354-7

110. Gronlund, C.J., 2014: Racial and socioeconomic disparities in heat-related health effects and their mechanisms: A review. *Current Epidemiology Reports,* **1,** 165-173. http://dx.doi.org/10.1007/s40471-014-0014-4

111. Hansen, A., L. Bi, A. Saniotis, and M. Nitschke, 2013: Vulnerability to extreme heat and climate change: Is ethnicity a factor? *Global Health Action,* **6**. http://dx.doi.org/10.3402/gha.v6i0.21364

112. O'Neill, M.S., A. Zanobetti, and J. Schwartz, 2005: Disparities by race in heat-related mortality in four US cities: The role of air conditioning prevalence. *Journal of Urban Health,* **82,** 191-197. http://dx.doi.org/10.1093/jurban/jti043

113. Arbury, S., B. Jacklitsch, O. Farquah, M. Hodgson, G. Lamson, H. Martin, and A. Profitt, 2014: Heat illness and death among workers – United States, 2012-2013. *MMWR. Morbidity and Mortality Weekly Report,* **63,** 661-665. http://www.cdc.gov/mmwr/preview/mmwrhtml/mm6331a1.htm

114. Lundgren, K., K. Kuklane, C. Gao, and I. Holmer, 2013: Effects of heat stress on working populations when facing climate change. *Industrial Health,* **51,** 3-15. http://dx.doi.org/10.2486/indhealth.2012-0089 <Go to ISI>://WOS:000314383700002

115. Hansen, A., P. Bi, M. Nitschke, P. Ryan, D. Pisaniello, and G. Tucker, 2008: The effect of heat waves on mental health in a temperate Australian city. *Environmental Health Perspectives,* **116,** 1369-1375. http://dx.doi.org/10.1289/ehp.11339

116. Martin-Latry, K., M.P. Goumy, P. Latry, C. Gabinski, B. Bégaud, I. Faure, and H. Verdoux, 2007: Psychotropic drugs use and risk of heat-related hospitalisation. *European Psychiatry,* **22,** 335-338. http://dx.doi.org/10.1016/j.eurpsy.2007.03.007

117. Page, L.A., S. Hajat, R.S. Kovats, and L.M. Howard, 2012: Temperature-related deaths in people with psychosis, dementia and substance misuse. *The British Journal of Psychiatry,* **200,** 485-490. http://dx.doi.org/10.1192/bjp.bp.111.100404

118. Ranson, M., 2014: Crime, weather, and climate change. *Journal of Environmental Economics and Management,* **67,** 274-302. http://dx.doi.org/10.1016/j.jeem.2013.11.008

119. Dunne, J.P., R.J. Stouffer, and J.G. John, 2013: Reductions in labour capacity from heat stress under climate warming. *Nature Climate Change,* **3,** 563-566. http://dx.doi.org/10.1038/nclimate1827

120. Graff Zivin, J. and M. Neidell, 2014: Temperature and the allocation of time: Implications for climate change. *Journal of Labor Economics,* **32,** 1-26. http://dx.doi.org/10.1086/671766

121. Sheridan, S.C. and T.J. Dolney, 2003: Heat, mortality, and level of urbanization: Measuring vulnerability across Ohio, USA. *Climate Research,* **24,** 255-265. http://dx.doi.org/10.3354/cr024255

122. Orlowsky, B. and S.I. Seneviratne, 2012: Global changes in extreme events: Regional and seasonal dimension. *Climatic Change,* **10,** 669-696. http://dx.doi.org/10.1007/s10584-011-0122-9

123. Robeson, S.M., C.J. Willmott, and P.D. Jones, 2014: Trends in hemispheric warm and cold anomalies. *Geophysical Research Letters,* **41,** 9065-9071. http://dx.doi.org/10.1002/2014gl062323

124. Liu, J., M. Song, R.M. Horton, and Y. Hu, 2013: Reducing spread in climate model projections of a September ice-free Arctic. *Proceedings of the National Academy of Sciences,* **110,** 12571-12576. http://dx.doi.org/10.1073/pnas.1219716110

125. Liu, J., J.A. Curry, H. Wang, M. Song, and R.M. Horton, 2012: Impact of declining Arctic sea ice on winter snowfall. *Proceedings of the National Academy of Sciences,* **109,** 4074-4079. http://dx.doi.org/10.1073/pnas.1114910109

126. Francis, J.A. and S.J. Vavrus, 2012: Evidence linking Arctic amplification to extreme weather in mid-latitudes. *Geophysical Research Letters,* **39,** L06801. http://dx.doi.org/10.1029/2012GL051000

127. Barnes, E.A., 2013: Revisiting the evidence linking Arctic amplification to extreme weather in midlatitudes. *Geophysical Research Letters,* **40,** 4734-4739. http://dx.doi.org/10.1002/grl.50880

128. Wallace, J.M., I.M. Held, D.W.J. Thompson, K.E. Trenberth, and J.E. Walsh, 2014: Global warming and winter weather. *Science,* **343,** 729-730. http://dx.doi.org/10.1126/science.343.6172.729

129. Gosling, S.N., G.R. McGregor, and J.A. Lowe, 2009: Climate change and heat-related mortality in six cities Part 2: Climate model evaluation and projected impacts from changes in the mean and variability of temperature with climate change. *International Journal of Biometeorology,* **53,** 31-51. http://dx.doi.org/10.1007/s00484-008-0189-9

3 AIR QUALITY IMPACTS

Lead Author
Neal Fann
U.S. Environmental Protection Agency

Contributing Authors
Terry Brennan
Camroden Associates, Inc.

Patrick Dolwick
U.S. Environmental Protection Agency

Janet L. Gamble
U.S. Environmental Protection Agency

Vito Ilacqua
U.S. Environmental Protection Agency

Laura Kolb
U.S. Environmental Protection Agency

Christopher G. Nolte
U.S. Environmental Protection Agency

Tanya L. Spero
U.S. Environmental Protection Agency

Lewis Ziska
U.S. Department of Agriculture

Acknowledgements: Susan Anenberg, U.S. Chemical Safety Board; **Amanda Curry Brown,** U.S. Environmental Protection Agency; **William Fisk,** Lawrence Berkeley National Laboratory; **Patrick Kinney,** Columbia University; **Daniel Malashock,*** U.S. Department of Health and Human Services, Public Health Service; **David Mudarri,** CADMUS; **Sharon Phillips,** U.S. Environmental Protection Agency; **Marcus C. Sarofim,*** U.S. Environmental Protection Agency;

Recommended Citation: Fann, N., T. Brennan, P. Dolwick, J.L. Gamble, V. Ilacqua, L. Kolb, C.G. Nolte, T.L. Spero, and L. Ziska, 2016: Ch. 3: Air Quality Impacts. *The Impacts of Climate Change on Human Health in the United States: A Scientific Assessment.* U.S. Global Change Research Program, Washington, DC, 69–98. http://dx.doi.org/10.10.7930/J0GQ6VP6

*Chapter Coordinators

3 AIR QUALITY IMPACTS

Key Findings

Exacerbated Ozone Health Impacts
Key Finding 1: Climate change will make it harder for any given regulatory approach to reduce ground-level ozone pollution in the future as meteorological conditions become increasingly conducive to forming ozone over most of the United States *[Likely, High Confidence]*. Unless offset by additional emissions reductions of ozone precursors, these climate-driven increases in ozone will cause premature deaths, hospital visits, lost school days, and acute respiratory symptoms *[Likely, High Confidence]*.

Increased Health Impacts from Wildfires
Key Finding 2: Wildfires emit fine particles and ozone precursors that in turn increase the risk of premature death and adverse chronic and acute cardiovascular and respiratory health outcomes *[Likely, High Confidence]*. Climate change is projected to increase the number and severity of naturally occurring wildfires in parts of the United States, increasing emissions of particulate matter and ozone precursors and resulting in additional adverse health outcomes *[Likely, High Confidence]*.

Worsened Allergy and Asthma Conditions
Key Finding 3: Changes in climate, specifically rising temperatures, altered precipitation patterns, and increasing concentrations of atmospheric carbon dioxide, are expected to contribute to increases in the levels of some airborne allergens and associated increases in asthma episodes and other allergic illnesses *[High Confidence]*.

3.1 Introduction

Changes in the climate affect the air we breathe, both indoors and outdoors. Taken together, changes in the climate affect air quality through three pathways—via outdoor air pollution, aeroallergens, and indoor air pollution. The changing climate has modified weather patterns, which in turn have influenced the levels and location of outdoor air pollutants such as ground-level ozone (O_3) and fine particulate matter.[1, 2, 3, 4] Increasing carbon dioxide (CO_2) levels also promote the growth of plants that release airborne allergens (aeroallergens). Finally, these changes to outdoor air quality and aeroallergens also affect indoor air quality as both pollutants and aeroallergens infiltrate homes, schools, and other buildings.

Climate change influences outdoor air pollutant concentrations in many ways (Figure 1). The climate influences temperatures, cloudiness, humidity, the frequency and intensity of precipitation, and wind patterns,[5] each of which can influence air quality. At the same time, climate-driven changes in meteorology can also lead to changes in naturally occurring emissions that influence air quality (for example, wildfires, wind-blown dust, and emissions from vegetation). Over longer time scales, human responses to climate change may also affect the amount of energy that humans use, as well as how land is used and where people live. These changes would in turn modify emissions (depending on the fuel source) and thus further influence air quality.[6, 7] Some air pollutants such as ozone, sulfates, and black carbon also cause changes in

Climate Change and Health—Outdoor Air Quality

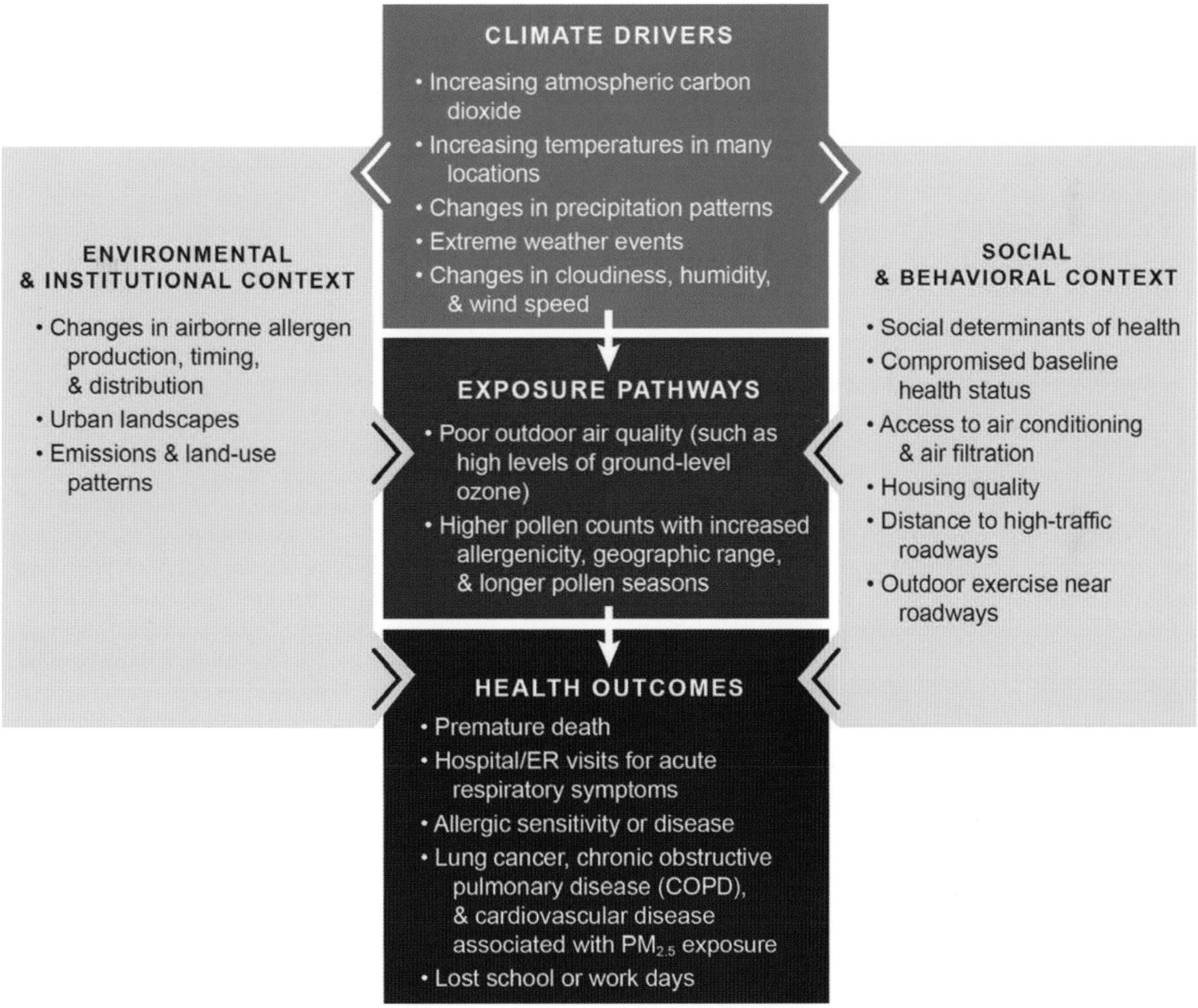

Figure 1: This conceptual diagram for an outdoor air quality example illustrates the key pathways by which humans are exposed to health threats from climate drivers, and potential resulting health outcomes (center boxes). These exposure pathways exist within the context of other factors that positively or negatively influence health outcomes (gray side boxes). Key factors that influence vulnerability for individuals are shown in the right box, and include social determinants of health and behavioral choices. Key factors that influence vulnerability at larger scales, such as natural and built environments, governance and management, and institutions, are shown in the left box. All of these influencing factors can affect an individual's or a community's vulnerability through changes in exposure, sensitivity, and adaptive capacity and may also be affected by climate change. See Chapter 1: Introduction for more information.

climate.[8] However, this chapter does not consider the climate effects of air pollutants, remaining focused on the health effects resulting from climate-related changes in air pollution exposure.

Poor air quality, whether outdoors or indoors, can negatively affect the human respiratory and cardiovascular systems. Outdoor ground-level ozone and particle pollution can have a range of adverse effects on human health. Current levels of ground-level ozone have been estimated to be responsible for tens of thousands of hospital and emergency room visits, millions of cases of acute respiratory symptoms and school absences, and thousands of premature deaths each year in the United States.[9, 10] Fine particle pollution has also been linked to even greater health consequences through harmful cardiovascular and respiratory effects.[11]

> *Human-caused climate change has the potential to increase ozone levels, may have already increased ozone pollution in some regions of the United States, and has the potential to affect future concentrations of ozone and fine particles.*

A changing climate can also influence the level of aeroallergens such as pollen, which in turn adversely affect human health. Rising levels of CO_2 and resulting climate changes alter the production, allergenicity (a measure of how much particular allergens, such as ragweed, affect people), distribution, and seasonal timing of aeroallergens. These changes increase the severity and prevalence of allergic diseases in humans. Higher pollen concentrations and longer pollen seasons can increase allergic sensitization and asthma episodes and thereby limit productivity at work and school.

Finally, climate change may alter the indoor concentrations of pollutants generated outdoors (such as ground-level ozone), particulate matter, and aeroallergens (such as pollen). Changes in the climate may also increase pollutants generated indoors,

Higher pollen concentrations and longer pollen seasons can increase allergic sensitization and asthma episodes.

such as mold and volatile organic compounds. Most of the air people breathe over their lifetimes will be indoors, since people spend the vast majority of their time in indoor environments. Thus, alterations in indoor air pollutant concentrations from climate change have important health implications.

3.2 Climate Impacts on Outdoor Air Pollutants and Health

Changes in the climate affect air pollution levels.[8, 12, 13, 14, 15, 16, 17, 18, 19, 20, 21, 22] Human-caused climate change has the potential to increase ozone levels,[1, 4] may have already increased ozone pollution in some regions of the United States,[3] and has the potential to affect future concentrations of ozone and fine particles (particulate matter smaller than 2.5 microns in diameter, referred to as $PM_{2.5}$).[2, 7] Climate change and air quality are both affected by, and influence, several factors; these include the levels and types of pollutants emitted, how land is used, the chemistry governing how these pollutants form in the atmosphere, and weather conditions.

Ground-Level Ozone

Ozone levels and subsequent ozone-related health impacts depend on 1) the amount of pollutants emitted that form ozone, and 2) the meteorological conditions that help determine the amount of ozone produced from those emissions. Both of these factors are expected to change in the future. The emissions of pollutants from anthropogenic (of human origin) sources that form ozone (that is, ozone "precursors") are expected to decrease over the next few decades in the United States.[23] However, irrespective of these changes in emissions, climate change will result in meteorological conditions more favorable to forming ozone. Consequently, attaining national air quality standards for ground-level ozone will also be more difficult, as climate changes offset some of the improvements that would otherwise be expected from emissions reductions. This effect is referred to as the "climate penalty."[7, 24]

Meteorological conditions influencing ozone levels include air temperatures, humidity, cloud cover, precipitation, wind trajectories, and the amount of vertical mixing in the atmosphere.[1, 2, 25, 26] Higher temperatures can increase the chemical rates at which ozone is formed and increase ozone precursor emissions from anthropogenic sources and biogenic (vegetative) sources. Lower relative humidity reduces cloud cover and rainfall, promoting the formation of ozone and extending ozone lifetime in the atmosphere. A changing climate will also modify wind patterns across the United States, which will influence local ozone levels. Over much of the country, the worst ozone episodes tend to occur when the local air mass does not change over a period of several days, allowing ozone and ozone precursor emissions

What is Ozone?

Ozone (O_3) is a compound that occurs naturally in Earth's atmosphere but is also formed by human activities. In the stratosphere (10–50 kilometers above the Earth's surface), O_3 prevents harmful solar ultraviolet radiation from reaching the Earth's surface. Near the surface, however, O_3 irritates the respiratory system. Ground-level O_3, a key component of smog, is formed by chemical interactions between sunlight and pollutants including nitrogen oxides (NO_x) and volatile organic compounds (VOCs). The emissions leading to O_3 formation can result from both human sources (for example, motor vehicles and electric power generation) and natural sources (for example, vegetation and wildfires). Occasionally, O_3 that is created naturally in the stratosphere can be mixed downward and contribute to O_3 levels near the surface. Once formed, O_3 can be transported by the wind before eventually being removed from the atmosphere via chemical reactions or by depositing on the surface.

At any given location, O_3 levels are influenced by complex interactions between emissions and meteorological conditions. Generally, higher temperatures, sunnier skies, and lighter winds lead to higher O_3 concentrations by increasing the rate of chemical reactions and by decreasing the extent to which pollutants are mixed with "clean" (less polluted) background air.

For a given level of emissions of O_3 precursors, climate change is generally expected to increase O_3 pollution in the future throughout much of the United States, in part due to higher temperatures and more frequent stagnant air conditions.[7] Unless offset by additional emissions reductions of ozone precursors, these climate-driven increases in O_3 will cause premature deaths, hospital visits, lost school days, and acute respiratory symptoms.[14]

to accumulate over time.[27, 28] Climate change is already increasing the frequency of these types of stagnation events over parts of the United States,[3] and further increases are projected.[29] Ozone concentrations near the ground are strongly influenced by upward and downward movement of air ("vertical mixing"). For example, high concentrations of ozone near the ground often occur in urban areas when there is downward movement of air associated with high pressure ("subsidence"), reducing the extent to which locally emitted pollutants are diluted in the atmosphere.[30] In addition, high concentrations of ozone can occur in some rural areas resulting from downward transport of ozone from the stratosphere or upper troposphere to the ground.[31]

Aside from the direct meteorological influences, there are also indirect impacts on U.S. ozone levels from other climate-influenced factors. For instance, higher water vapor concentrations due to increased temperatures will increase the natural rate of ozone depletion, particularly in remote areas,[32] thus decreasing the baseline level of ozone. Additionally, potential climate-driven increases in nitrogen oxides (NO_x) created by lightning or increased exchange of naturally produced ozone in the stratosphere to the troposphere could also affect ozone in those areas of the country most influenced by background ozone concentrations.[33] Increased occurrences of wildfires due to climate change can also lead to increased ozone concentrations near the ground.[34]

There is natural year-to-year variability in temperature and other meteorological factors that influence ozone levels.[7] While global average temperature over 30-year climatic timescales is expected to increase, natural interannual variability will continue to play a significant role in year-to-year changes in temperature.[35] Over the next several decades, the influence

of climate change on meteorological parameters affecting average levels of ozone is expected to be smaller than the natural interannual variability.[36]

To address these issues, most assessments of climate impacts on meteorology and associated ozone formation concurrently simulate global and regional chemical transport over multiple years using "coupled" models. This approach can isolate the influence of meteorology in forming ozone from the effect of changes in emissions. The consensus of these model-based assessments is that accelerated rates of photochemical reaction, increased occurrence of stagnation events, and other direct meteorological influences are likely to lead to higher levels of ozone over large portions of the United States.[8, 14, 16, 17] At the same time, ozone levels in certain regions are projected to decrease as a result of climate change, likely due to localized increases in cloud cover, precipitation, and/or increased dilution resulting from deeper mixed layers. These climate-driven changes in projected ozone vary by season and location, with climate and air quality models showing the most consistency in ozone increases due to climate change in the northeastern United States.[8, 37]

Generally, ozone levels will likely increase across the United States if ozone precursors are unchanged (see "Research Highlight: Ozone-Related Health Effects" on page 74).[4, 7, 8] This climate penalty for ozone will offset some of the expected health benefits that would otherwise result from the expected ongoing reductions of ozone precursor emissions, and could prompt the need for adaptive measures (for example, additional ozone precursor emissions reductions) to meet national air quality goals.

Research Highlight: Ozone-Related Health Effects

Los Angeles, California, May 22, 2012. Unless offset by additional emissions reductions of ozone precursors, climate-driven increases in ozone will cause premature deaths, hospital visits, lost school days, and acute respiratory symptoms.

Importance: Ozone is formed in the atmosphere by photochemical reactions of volatile organic compounds (VOCs) and nitrogen oxides (NO_x) in the presence of sunlight. Although U.S. air quality policies are projected to reduce VOC and NO_x emissions,[56] climate change will increase the frequency of regional weather patterns conducive to increasing ground-level ozone, partially offsetting the expected improvements in air quality.

Objective: Project the number and geographic distribution of additional ozone-related illnesses and premature deaths in the contiguous United States due to climate change between 2000 and 2030 under projected U.S. air quality policies.

Method: Climate scenarios from two global climate models (GCMs) using two different emissions pathways (RCP8.5 and RCP6.0) were dynamically downscaled following Otte et al. (2012)[57] and used with emissions projections for 2030 and a regional chemical transport model to simulate air quality in the contiguous United States. The resulting changes in ozone in each scenario were then used to compute regional ozone-related health effects attributable to climate change. Ozone-related health impacts were estimated using the environmental Benefits Mapping and Analysis Program–Community Edition (BenMAP–CE). Population exposure was estimated using projected population data from the Integrated Climate and Land Use Scenarios (ICLUS). Further details can be found in Fann et al. (2015).[14]

Results: The two downscaled GCM projections result in 1°C to 4°C (1.8°F to 7.2°F) increases in average daily maximum temperatures and 1 to 5 parts per billion increases in daily 8-hour maximum ozone in 2030 throughout the contiguous United States. As seen in previous modeling analyses of climate impacts on ozone, the air quality response to climate change can vary substantially by region and across scenarios.[22, 58] Unless reductions in ozone precursor emissions offset the influence of climate change, this climate penalty of increased ozone concentrations due to climate change would result in tens to thousands of additional ozone-related illnesses and premature deaths per year.

Research Highlight: Ozone-Related Health Effects, continued

Conclusions: Future climate change will result in higher ozone levels in polluted regions of the contiguous United States. This study isolates the effect of climate change on ozone by using the same emissions of ozone precursors for both 2000-era and 2030-era climate. In addition, this study uses the latest generation of GCM scenarios and represents the most comprehensive analysis of climate-related, ozone-attributable health effects in 2030, and includes not only deaths but also emergency department admissions for asthma, hospital visits for respiratory causes, acute respiratory symptoms, and missed days of school. These results are subject to important uncertainties and limitations. The ozone-climate modeling reflects two scenarios (based on two separate GCMs) considered. Several emissions categories that are important in the formation of ozone and that could be affected by climate, such as motor vehicles, electrical generating units, and wildfires, were left unchanged between the current and future periods. The analysis applied concentration–response relationships from epidemiology studies of historical air pollution episodes; this both implies that the relationship between air pollution and risk will remain constant into the future and that populations will not attempt to reduce their exposure to ozone.

Projected Change in Temperature, Ozone, and Ozone-Related Premature Deaths in 2030

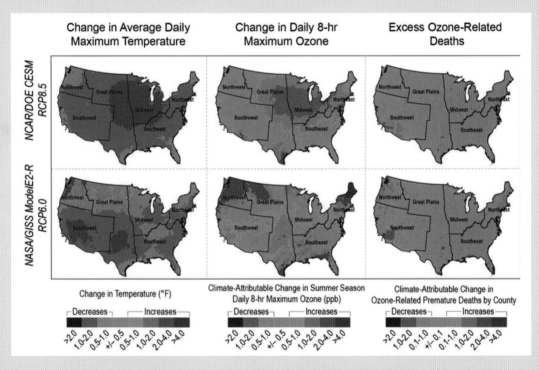

Figure 2. Projected changes in average daily maximum temperature (degrees Fahrenheit), summer average maximum daily 8-hour ozone (parts per billion), and excess ozone-related deaths (incidences per year by county) in the year 2030 relative to the year 2000, following two global climate models and two greenhouse gas concentration pathways, known as Representative Concentration Pathways, or RCPs (see van Vuuren et al. 2011[49]). Each year (2000 and 2030) is represented by 11 years of modeled data for May through September, the traditional ozone season in the United States.

The top panels are based on the National Center for Atmospheric Research/Department of Energy (NCAR/DOE) Community Earth System Model (CESM) following RCP8.5 (a higher greenhouse gas concentration pathway), and the bottom panels are based on the National Aeronautics and Space Administration (NASA) Goddard Institute for Space Studies (GISS) ModelE2-R following RCP6.0 (a moderate greenhouse gas concentration pathway).

The leftmost panels are based on dynamically downscaled regional climate using the NCAR Weather Research and Forecasting (WRF) model, the center panels are based on air quality simulations from the U.S. Environmental Protection Agency (EPA) Community Multiscale Air Quality (CMAQ) model, and the rightmost panels are based on the U.S. EPA Environmental Benefits and Mapping Program (BenMAP).

Fann et al. 2015 reports a range of mortality outcomes based on different methods of computing the mortality effects of ozone changes—the changes in the number of deaths shown in the rightmost panels were computed using the method described in Bell et al. 2004.[14, 38] (Figure source: adapted from Fann et al. 2015)[14]

Research Highlight: Ozone-Related Health Effects, continued

Projected Change in Ozone-Related Premature Deaths

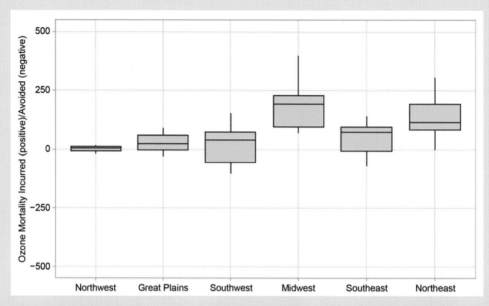

Figure 3. Projected change in ozone-related premature deaths from 2000 to 2030 by U.S. region and based on CESM/RCP8.5. Each year (2000 and 2030) is represented by 11 years of modeled data. Ozone-related premature deaths were calculated using the risk coefficient from Bell et al. (2004).[38] Boxes indicate 25th, 50th, and 75th percentile change over 11-year sample periods, and vertical lines extend to 1.5 times the interquartile range. U.S. regions follow geopolitical boundaries shown in Figure 2. (Figure source: Fann et al. 2015)[14]

Air pollution epidemiology studies describe the relationship between a population's historical exposure to air pollutants and the risk of adverse health outcomes. Populations exposed to ozone air pollution are at greater risk of dying prematurely, being admitted to the hospital for respiratory hospital admissions, being admitted to the emergency department, and suffering from aggravated asthma, among other impacts.[38, 39, 40]

Air pollution health impact assessments combine risk estimates from these epidemiology studies with modeled changes in future or historical air quality changes to estimate the number of air-pollution-related premature deaths and illness.[41] Future ozone-related human health impacts attributable to climate change are projected to lead to hundreds to thousands of premature deaths, hospital admissions, and cases of acute respiratory illnesses per year in the United States in 2030.[14, 42, 43, 44, 45, 46]

Health outcomes that can be attributed to climate change impacts on air pollution are sensitive to a number of factors noted above—including the climate models used to describe meteorological changes (including precipitation and cloud cover), the models simulating air quality levels (including wildfire incidence), the size and distribution of the population exposed, and the health status of that population (which influences their susceptibility to air pollution; see Ch. 1: Introduction).[42, 47, 48, 49] Moreover, there is emerging evidence that air pollution can interact with climate-related stressors such as

temperature to affect the human physiological response to air pollution.[39, 42, 50, 51, 52, 53, 54, 55] For example, the risk of dying from exposure to a given level of ozone may increase on warmer days.[51]

Particulate Matter

Particulate matter (PM) is a complex mixture of solid- or liquid-phase substances in the atmosphere that arise from both natural and human sources. Principal constituents of PM include sulfate, nitrate, ammonium, organic carbon, elemental carbon, sea salt, and dust. These particles (also known as aerosols) can either be directly emitted or can be formed in the atmosphere from gas-phase precursors. PM smaller than 2.5 microns in diameter ($PM_{2.5}$) is associated with serious chronic and acute health effects, including lung cancer, chronic obstructive pulmonary disease (COPD), cardiovascular disease, and asthma development and exacerbation.[11] The elderly are particularly sensitive to short-term particle exposure, with a higher risk of hospitalization and death.[59, 60]

As is the case for ozone, atmospheric $PM_{2.5}$ concentrations depend on emissions and on meteorology. Emissions of sulfur dioxide (SO_2), NO_x, and black carbon are projected to decline substantially in the United States over the next few decades due to regulatory controls,[56, 61, 62, 63] which will lead to reductions in sulfate and nitrate aerosols.

Climate change is expected to alter several meteorological factors that affect PM₂.₅, including precipitation patterns and humidity, although there is greater consensus regarding the effects of meteorological changes on ozone than on PM₂.₅.[2] Several factors, such as increased humidity, increased stag- nation events, and increased biogenic emissions are likely to increase PM₂.₅ levels, while increases in precipitation, en- hanced atmospheric mixing, and other factors could decrease PM₂.₅ levels.[2, 8, 37, 64] Because of the strong influence of changes in precipitation and atmospheric mixing on PM₂.₅ levels, and because there is more variability in projected changes to those variables, there is no consensus yet on whether meteorolog- ical changes will lead to a net increase or decrease in PM₂.₅ levels in the United States.[2, 8, 17, 21, 22, 64, 65]

As a result, while it is clear that PM₂.₅ accounts for most of the health burden of outdoor air pollution in the United States,[10] the health effects of climate-induced changes in PM₂.₅ are poorly quantified. Some studies have found that changes in PM₂.₅ will be the dominant driver of air quality-related health effects due to climate change,[44] while others have suggested a potentially more significant health burden from changes in ozone.[50]

PM resulting from natural sources (such as plants, wildfires, and dust) is sensitive to daily weather patterns, and those fluc- tuations can affect the intensity of extreme PM episodes (see also Ch. 4: Extreme Events, Section 4.6).[8] Wildfires are a major source of PM, especially in the western United States during summer.[66, 67, 68] Because winds carry PM₂.₅ and ozone precursor gases, air pollution from wildfires can affect people even far downwind from the fire location.[35, 69] PM₂.₅ from wildfires af- fects human health by increasing the risk of premature death and hospital and emergency department visits.[70, 71, 72]

Climate change has already led to an increased frequency of large wildfires, as well as longer durations of individual wildfires and longer wildfire seasons in the western United States.[73] Future climate change is projected to increase wild- fire risks[74, 75] and associated emissions, with harmful impacts on health.[76] The area burned by wildfires in North America is expected to increase dramatically over the 21st century due to climate change.[77, 78] By 2050, changes in wildfires in the west- ern United States are projected to result in 40% increases of organic carbon and 20% increases in elemental carbon aerosol concentrations.[79] Wildfires may dominate summertime PM₂.₅ concentrations, offsetting even large reductions in anthropo- genic PM₂.₅ emissions.[22]

Likewise, dust can be an important constituent of PM, espe- cially in the southwest United States. The severity and spatial extent of drought has been projected to increase as a result of climate change,[80] though the impact of increased aridity on airborne dust PM has not been quantified (see Ch. 4. Extreme Events).[2]

Nearly 6.8 million children in the United States are affected by asthma, making it a major chronic disease of childhood.

3.3 Climate Impacts on Aeroallergens and Respiratory Diseases

Climate change may alter the production, allergenicity, distri- bution, and timing of airborne allergens (aeroallergens). These changes contribute to the severity and prevalence of allergic disease in humans. The very young, those with compromised immune systems, and the medically uninsured bear the brunt of asthma and other allergic illnesses. While aeroallergen exposure is not the sole, or even necessarily the most signifi- cant factor associated with allergic illnesses, that relationship is part of a complex pathway that links aeroallergen expo- sure to the prevalence of allergic illnesses, including asthma episodes.[81, 82] On the other hand, climate change may reduce adverse allergic and asthmatic responses in some areas. For example, as some areas become drier, there is the potential for a shortening of the pollen season due to plant stress.

Aeroallergens and Rates of Allergic Diseases in the United States

Aeroallergens are substances present in the air that, once inhaled, stimulate an allergic response in sensitized individu- als. Aeroallergens include tree, grass, and weed pollen; indoor and outdoor molds; and other allergenic proteins associated with animal dander, dust mites, and cockroaches.[83] Ragweed is the aeroallergen that most commonly affects persons in the United States.[84]

Allergic diseases develop in response to complex and multi- ple interactions among both genetic and non-genetic factors, including a developing immune system, environmental expo- sures (such as ambient air pollution or weather conditions), and socioeconomic and demographic factors.[85, 86, 87] Aeroal- lergen exposure contributes to the occurrence of asthma episodes, allergic rhinitis or hay fever, sinusitis, conjunctivitis, urticaria (hives), atopic dermatitis or eczema, and anaphylaxis (a severe, whole-body allergic reaction that can be life-threat-

Ragweed Pollen Season Lengthens

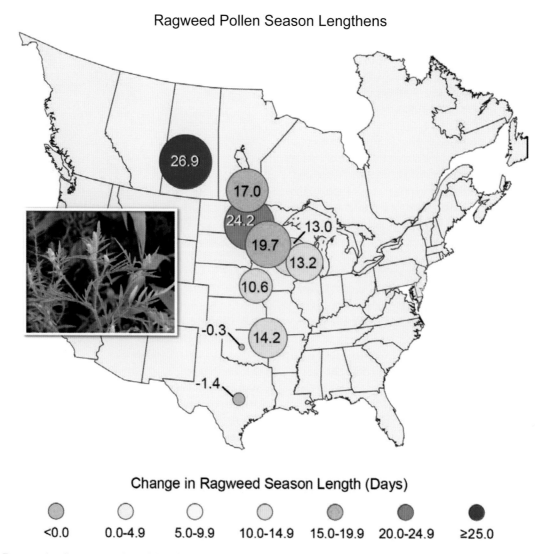

Change in Ragweed Season Length (Days)

| <0.0 | 0.0-4.9 | 5.0-9.9 | 10.0-14.9 | 15.0-19.9 | 20.0-24.9 | ≥25.0 |

Figure 4: Ragweed pollen season length has increased in central North America between 1995 and 2011 by as much as 11 to 27 days in parts of the United States and Canada, in response to rising temperatures. Increases in the length of this allergenic pollen season are correlated with increases in the number of days before the first frost. The largest increases have been observed in northern cities. (Figure source: Melillo et al. 2014. Photo credit: Lewis Ziska, USDA).[35]

ening).[84, 88] Allergic illnesses, including hay fever, affect about one-third of the U.S. population, and more than 34 million Americans have been diagnosed with asthma.[81] These diseases have increased in the United States over the past 30 years (see Ch. 1 Introduction). The prevalence of hay fever has increased from 10% of the population in 1970 to 30% in 2000.[84] Asthma rates have increased from approximately 8 to 55 cases per 1,000 persons to approximately 55 to 90 cases per 1,000 persons over that same time period;[89] however, there is variation in reports of active cases of asthma as a function of geography and demographics.[90]

Climate Impacts on Aeroallergen Characteristics

Climate change contributes to changes in allergic illnesses as greater concentrations of CO_2, together with higher temperatures and changes in precipitation, extend the start or duration of the growing season, increase the quantity and allergenicity of pollen, and expand the spatial distribution of pollens.[84, 91, 92, 93, 94]

Historical trends show that climate change has led to changes in the length of the growing season for certain allergenic pollens. For instance, the duration of pollen release for common ragweed (*Ambrosia artemisiifolia*) has been increasing as a function of latitude in recent decades in the midwestern region of North America (see Figure 4). Latitudinal effects on increasing season length were associated primarily with a delay in first frost during the fall season and lengthening of the frost-free period.[95] Studies in controlled indoor environments find that increases in temperature and CO_2 result in earlier flowering, greater floral numbers, greater pollen production, and increased allergenicity in common ragweed.[96, 97] In addition, studies using urban areas as proxies for both higher CO_2 and higher temperatures demonstrate earlier flowering of pollen species, which may lead to a longer total pollen season.[98, 99, 100]

For trees, earlier flowering associated with higher winter and spring temperatures has been observed over a 50-year period for oak.[101] Research on loblolly pine (*Pinus taeda*) also demonstrates that elevated CO_2 could induce earlier and greater seasonal pollen production.[102] Annual birch (*Betula*) pollen production and peak values from 2020 to 2100 are projected to be 1.3 to 2.3 times higher, relative to average values for 2000, with the start and peak dates of pollen release advancing by two to four weeks.[103]

Climate Variability and Effects on Allergic Diseases

Climate change related alterations in local weather patterns, including changes in minimum and maximum temperatures and rainfall, affect the burden of allergic diseases.[104, 105, 106] The role of weather on the initiation or exacerbation of allergic symptoms in sensitive persons is not well understood.[86, 107] So-called "thunderstorm asthma" results as allergenic particles are dispersed through osmotic rupture, a phenomenon where cell membranes burst. Pollen grains may, after contact with rain, release part of their cellular contents, including allergen-laced fine particles. Increases in the intensity and frequency of heavy rainfall and storminess over the coming decades is likely to be associated with spikes in aeroallergen concentrations and the potential for related increases in the number and severity of allergic illnesses.[108, 109]

Potential non-linear interactions between aeroallergens and ambient air pollutants (including ozone, nitrogen dioxide, sulfur dioxide, and fine particulate matter) may increase health risks for people who are simultaneously exposed.[87, 88, 106, 108, 110, 111, 112, 113, 114] In particular, pre-exposure to air pollution (especially ozone or fine particulate matter) may magnify the effects of aeroallergens, as prior damage to airways may increase the permeability of mucous membranes to the penetration of allergens, although existing evidence suggests greater sensitivity but not necessarily a direct link with ozone exposure.[115] A recent report noted remaining uncertainties across the epidemiologic, controlled human exposure, and toxicology studies on this emerging topic.[39]

3.4 Climate Impacts on Indoor Air Quality and Health: An Emerging Issue

Climate change may worsen existing indoor air problems and create new problems by altering outdoor conditions that affect indoor conditions and by creating more favorable conditions for the growth and spread of pests, infectious agents, and disease vectors that can migrate indoors.[116] Climate change can also lead to changes in the mixing of outdoor and indoor air. Reduced mixing of outdoor and indoor air limits penetration of outdoor pollutants into the indoors, but also leads to higher concentrations of pollutants generated indoors since their dilution by outdoor air is decreased.

Indoor air contains a complex mixture of chemical and biological pollutants or contaminants. Contaminants that can be found indoors include carbon monoxide (CO), fine particles (PM2.5), nitrogen dioxide, formaldehyde, radon, mold, and

pollen. Indoor air quality varies from building to building and over the course of a day in an individual building.

Public and environmental health professionals have known for decades that poor indoor air quality is associated with adverse respiratory and other health effects.[116, 117, 118, 119, 120, 121] Since most people spend about 90% of their time indoors,[122, 123, 124, 125, 126] much of their exposures to airborne pollutants (both those influenced by climate change and those driven by other factors) happen indoors.

Outdoor Air Changes Reflected in Indoor Air

Indoor air pollutants may come from indoor sources or may be transported into the building with outdoor air.[127, 128] Indoor pollutants of outdoor origin may include ozone, dust, pollen, and fine PM (PM2.5). Even if a building has an outdoor air intake, some air will enter the building through other openings, such as open windows or under doors, or through cracks in the buildings, bypassing any filters and bringing outdoor air pollutants inside.[129] If there are changes in airborne pollutants of outdoor origin, such as pollen and mold (see Section 3.3) and fine PM from wildfires (see "Particulate Matter" on page 76), there will be changes in indoor exposures to these contaminants. Although indoor fine PM levels from wildfires are typically lower than outdoors (about 50%), because people spend most of their time indoors, most of their exposure to and health effects from wildfire particles (about 80%) will come from particles inhaled

Dampness and mold in U.S. homes are linked to approximately 4.6 million cases of worsened asthma.

Research Highlight: Residential Infiltration and Indoor Air

Importance: Indoor and outdoor air are constantly mixing as air flows through small cracks and openings in buildings (infiltration) in addition to any open doors, windows, and vents. Infiltration or air exchange is driven by differences in barometric pressure, as a result of wind, and of the temperature difference between indoor and outdoor air. The greater this air exchange, the more similar the composition of indoor and outdoor air. Lower air exchange rates accentuate the impact of indoor sources while reducing that of some outdoor pollution. As climate change increases the average temperature of outdoor air, while indoor air continues to be maintained at the same comfortable temperatures, infiltration driven by temperature differences will change as well, modifying exposure to indoor and outdoor air pollution sources.

Objective: Project the relative change in infiltration and its effects on exposure to indoor and outdoor air pollution sources for different climates in the United States, between a late-20th century reference and the middle of the current century, in typical detached homes.

Method: The infiltration change projected for 2040–2070 compared to 1970–2000 was modeled for typical single-family residences in urban areas, using temperatures and wind speeds from eight global–regional model combinations for nine U.S. cities (Atlanta, Boston, Chicago, Houston, Los Angeles, Minneapolis, New York, Phoenix, and Seattle). This analysis compares a building to itself, removing the effects of individual building characteristics on infiltration. Indoor temperatures were assumed unchanged between these two periods. Further details can be found in Ilacqua et al. 2015.[131]

Results: Because current average yearly temperatures across the contiguous United States are generally below comfortable indoor temperatures, model results indicate that, under future warmer temperatures, infiltration is projected to decrease by about 5%, averaged across cities, seasons, and climate models. Exposure to some pollutants emitted indoors would correspondingly increase, while exposure to some outdoor air pollutants would decrease to some extent. Projections vary, however, among location, seasons, and climate models. In the warmer cities, infiltration during summer months would rise by up to 25% in some models, raising peak exposures to ozone and other related pollutants just when their concentrations are typically highest. Predictions of different models are less consistent for summer months, however, displaying more uncertainty (average modeling relative range of 14%) for summer than for the rest of the year, and in fact not all models predict summer infiltration increases. Modeling uncertainty for the rest of the year is lower than in the summer (relative range less than 6%).

Conclusions: This study shows the potential shifts in residential exposure to indoor and outdoor air pollution sources driven by a changing climate.[131] These conclusions can be applied to small buildings, including single-family homes, row houses, and small offices. Potential adaptations intended to promote energy efficiency by reducing the leakage area of buildings will enhance the effect of decreasing infiltration and increasing exposure to indoor sources. Because of its novelty and lack of additional evidence, the study results should be considered as suggestive of an emerging issue. If replicated by other studies, these findings would add to the evidence on the potential for climate change to alter indoor air quality and further emphasize the impact of indoor air sources on human health. The overall implications of these findings for exposure to ambient and indoor air pollution remain uncertain at present, as they need to be considered along with other determinants of air exchange, such as window-opening behavior, whose relationship with climate change remains poorly characterized.

indoors.[130] Climate-induced changes in indoor-outdoor temperature differences may somewhat reduce the overall intake of outdoor pollutants into buildings for certain regions and seasons (see "Research Highlight: Residential Infiltration and Indoor Air").[131]

Most exposures to high levels of ozone occur outdoors; however, indoor exposures, while lower, occur for much longer time periods. Indoors, ozone concentrations are usually about 10% to 50% of outdoor concentrations; however, since people spend most of their time indoors, most of their exposure to ozone is from indoor air.[130] Thus, about 45% to 75% of a person's overall exposure to ozone will occur indoors.[132] About half of the health effects resulting from any outdoor increases in ozone (see Section "Ground-Level Ozone" on page 72) will be due to indoor ozone exposures.[130] The elderly and children are particularly sensitive to short-term ozone exposure; however, they may spend even more time indoors than the general population and consequently their exposure to ozone is at lower levels for longer periods than the general public.[133, 134] In addition, ozone entering a building reacts with some organic compounds to produce secondary indoor air pollutants. These reactions lower indoor ozone concentrations but introduce new indoor air contaminants, including other respiratory irritants.[135]

Climate-related increases in droughts and dust storms may result in increases in indoor transmission of dust-borne pathogens, as the dust penetrates the indoor environment. Dust contains particles of biologic origin, including pollen and bacterial and fungal spores. Some of the particles are allergenic.[136] Pathogenic fungi and bacteria can be found in dust both indoors and outdoors.[137] For example, in the southwestern United States, spores from the fungi *Coccidioides*, which can cause valley fever, are found indoors.[138] The geographic range where *Coccidioides* is commonly found is increasing. Climate changes, including increases in droughts and temperatures, may be contributing to this spread and to a rise in valley fever (see Ch. 4: Extreme Events).

Legionnaires' disease is primarily contracted from aerosolized water contaminated with *Legionella* bacteria.[139] *Legionella* bacteria are naturally found outdoors in water and soil; they are also known to contaminate treated water systems in buildings,[140] as well as building cooling systems such as swamp coolers or cooling towers.[141] *Legionella* can also be found indoors inside plumbing fixtures such as showerheads, faucets, and humidifiers.[142, 143] *Legionella* can cause outbreaks of a pneumonia known as Legionnaire's disease, which is a potentially fatal infection.[144] Exposure can occur indoors when a spray or mist of contaminated water is inhaled, including mist or spray from showers and swamp coolers.[145] The spread of *Legionella* bacteria can be affected by regional environmental factors.[116] Legionnaires' disease is known to follow a seasonal pattern, with more cases in late summer and autumn, potentially due to warmer and damper conditions.[146, 147] Cases of Legionnaires' disease are rising in the United States, with an increase of 192% from 2000 to 2009.[148, 149] If climate change results in sustained higher temperatures and damper conditions in some areas, there could be increases in the spread and transmission of *Legionella*.

Contaminants Generated Indoors

Although research directly linking indoor dampness and climate change is not available, information on building science, climate change, and outdoor environmental factors that affect indoor air quality can be used to project how climate change may influence indoor environments.[130] Climate change could result in increased indoor dampness in at least two ways: 1) if there are more frequent heavy precipitation events and other severe weather events (including high winds, flooding, and winter storms) that result in damage to buildings, allowing water or moisture entry; and 2) if outdoor humidity rises with climate change, indoor humidity and the potential for condensation and dampness will likely rise. Outdoor humidity is usually the largest contributor to indoor dampness on a yearly basis.[127]

Increased indoor dampness and humidity will in turn increase indoor mold, dust mites, bacteria, and other bio-contamination indoors, as well as increase levels of volatile organic compounds (VOCs) and other chemicals resulting from the off-gassing of damp or wet building materials.[116, 119, 150] Dampness and mold in U.S. homes are linked to approximately 4.6 million cases of worsened asthma and between 8% and 20% of several common respiratory infections, such as acute bronchitis.[151, 152] If there are climate-induced rises in indoor dampness, there could be increases in adverse health effects related to dampness and mold, such as asthma exacerbation.

Additionally, power outages due to more frequent extreme weather events such as flooding could lead to a number of health effects (see Ch 4: Extreme Events). Heating, ventilation, and air conditioning (HVAC) systems will not function without power; therefore, many buildings could have difficulty maintaining indoor temperatures or humidity. Loss of ventilation, filtration, air circulation, and humidity control can lead to indoor mold growth and increased levels of indoor contaminants,[153] including VOCs such as formaldehyde.[119, 154, 155, 156] Power outages are also associated with increases in hospital visits from carbon monoxide (CO) poisoning, primarily due to the incorrect use of backup and portable generators that contaminate indoor air with carbon monoxide.[135] Following floods, CO poisoning is also associated with the improper indoor use of wood-burning appliances and other combustion appliances designed for use outdoors.[157] There were at least nine deaths from carbon monoxide poisoning related to power outages from 2000 to 2009.[158]

Climate factors can influence populations of rodents that produce allergens and can harbor pathogens such as hantaviruses, which can cause Hantavirus Pulmonary Syndrome. Hantaviruses can be spread to people by rodents that infest buildings,[159] and limiting indoor exposure is a key strategy to prevent the spread of hantavirus.[160] Climate change may increase rodent populations in some areas, including indoors, particularly when droughts are followed by periods of heavy rain (see Ch. 4: Extreme Events) and with increases in temperature and rainfall.[161] Also, extreme weather events such as heavy rains and flooding may drive some rodents to relocate indoors.[162] Increases in rodent populations may result in increased indoor exposures to rodent allergens and related health effects.[159, 163, 164] In addition, climate factors may also influence the prevalence of hantaviruses in rodents.[163, 164] This is a complex dynamic, because climate change may influence rodent populations, ranges, and infection rates.

3.5 Populations of Concern

Certain groups of people may be more susceptible to harm from air pollution due to factors including age, access to healthcare, baseline health status, or other characteristics.[60] In the contiguous United States, Blacks or African-Americans, women, and the elderly experience the greatest baseline risk from air pollution.[165] The young, older adults, asthmatics, and people whose immune systems are compromised are more vulnerable to indoor air pollutants than the general population.[166] Lower

socioeconomic status and housing disrepair have been associated with higher indoor allergen exposures, though higher-income populations may be more exposed to certain allergens such as dust mites.[167, 168]

Nearly 6.8 million children in the United States are affected by asthma, making it a major chronic disease of childhood.[169] It is also the main cause of school absenteeism and hospital admissions among children.[83] In 2008, 9.3% of American children age 2 to 17 years were reported to have asthma.[169] The onset of asthma in children has been linked to early allergen exposure and viral infections, which act in concert with genetic susceptibility.[170] Children can be particularly susceptible to allergens due to their immature respiratory and immune systems, as well as indoor or outdoor activities that contribute to aeroallergen exposure (see Table 1).[170, 171, 172, 173]

Minority adults and children also bear a disproportionate burden associated with asthma as measured by emergency department visits, lost work and school days, and overall poorer health status (see Table 1).[175, 176] Twice as many Black children had asthma-related emergency department visits and hospitalizations compared with White children. Fewer Black and Hispanic children reported using preventative medication like inhaled corticosteroids (ICS) as compared to White children. Black and Hispanic children also had more poorly controlled asthma symptoms, leading to increased emergency department visits and greater use of rescue medications rather than routine daily use of ICS, regardless of symptom control.[173, 177]

Children living in poverty were 1.75 times more likely to be hospitalized for asthma than their non-poor counterparts. When income is accounted for, no significant difference was observed in the rate of hospital admissions by race or ethnicity. This income effect may be related to access and use of health care and appropriate use of preventive medications such as ICS.[178]

Percentage of population with active asthma, by year and selected characteristics: United States, 2001 and 2010.		
Characteristic	**Year 2001 %**	**Year 2010 %**
Total	**7.3**	**8.4**
Gender		
Male	6.3	7.0
Female	8.3	9.8
Race		
White	7.2	7.8
Black	8.4	11.9
Other	7.2	8.1
Ethnicity		
Hispanic	5.8	7.2
Non-Hispanic	7.6	8.7
Age		
Children (0-17)	8.7	9.3
Adults (18 and older)	6.9	8.2
Age Group		
0-4 years	5.7	6.0
5-14 years	9.9	10.7
15-34 years	8.0	8.6
35-64 years	6.7	8.1
65 years and older	6.0	8.1
Region		
Northeast	8.3	8.8
Midwest	7.5	8.6
South	7.1	8.3
West	6.7	8.3
Federal Poverty Threshold		
Below 100%	9.9	11.2
100% to < 250%	7.7	8.7
250% to < 450%	6.8	8.2
450% or higher	6.6	7.1
Source: Moorman et al. 2012[174]		

Table 1: A recent study of children in California found that racial and ethnic minorities are more affected by asthma.[175] Among minority children, the prevalence of asthma varies with the highest rates among Blacks and American Indians/Alaska Natives (17%), followed by non-Hispanic or non-Latino Whites (10%), Hispanics (7%), and Asian Americans (7%).

People with preexisting medical conditions—including hypertension, diabetes, and chronic obstructive pulmonary disorder—are at greater risk for outdoor air pollution-related health effects than the general population.[179] Populations with irregular heartbeats (atrial fibrillation) who were exposed to air pollution and high temperatures experience increased risk.[165] People who live or work in buildings without air conditioning and other ventilation controls or in buildings that are unable to withstand extreme precipitation or flooding events are at greater risk of adverse health effects. Other health risks are related to exposures to poor indoor air quality from mold and other biological contaminants and chemical pollutants emitted from wet building materials. While the presence of air conditioning has been found to greatly reduce the risk of ozone-related deaths, communities with a higher percentage of unemployment and a greater population of Blacks are at greater risk.[59]

3.6 Research Needs

In addition to the emerging issues identified above, the authors highlight the following potential areas for additional scientific and research activity on air quality. Understanding of future air quality and the ability to model future health impacts associated with air quality changes—particularly $PM_{2.5}$ impacts—will be enhanced by improved modeling and projections of climate-dependent variables like wildfires and land-use patterns, as well as improved modeling of ecosystem responses to climate change. Improved collection of data on aeroallergen concentrations in association with other ecosystem variables will facilitate research and modeling of related health impacts.

Future assessments can benefit from research activities that:

• enhance understanding of how interactions among climate-related factors, such as temperature or relative humidity, aeroallergens, and air pollution, affect human health, and how to attribute health impacts to changes in these different risk factors;

• improve the ability to model and project climate change impacts on the formation and fate of air contaminants and quantify the compounded uncertainty in the projections; and

• identify the impacts of changes in indoor dampness, such as mold, other biological contaminants, volatile organic compounds, and indoor air chemistry on indoor air pollutants and health.

Supporting Evidence

PROCESS FOR DEVELOPING CHAPTER

The chapter was developed through technical discussions of relevant evidence and expert deliberation by the report authors at several workshops, teleconferences, and email exchanges. The authors considered inputs and comments submitted by the public, the National Academies of Sciences, and Federal agencies. For additional information on the overall report process, see Appendices 2 and 3.

In addition, the author team held an all-day meeting at the U.S. Environmental Protection Agency National Center for Environmental Assessment in Crystal City, Virginia, on October 15, 2014, to discuss the chapter and develop initial drafts of the Key Findings. A quorum of the authors participated and represented each of the three sections of the chapter—outdoor air quality, aeroallergens, and indoor air quality. These discussions were informed by the results of the literature review as well as the research highlights focused on outdoor air quality and indoor air quality. The team developed Key Finding 2 in response to comments from the National Research Council review panel and the general public.

The Key Findings for outdoor ozone, wildfires, and aeroallergen impacts reflect strong empirical evidence linking changes in climate to these outcomes. When characterizing the human health impacts from outdoor ozone, the team considered the strength of the toxicological, clinical, and epidemiological evidence evaluated in the Ozone Integrated Science Assessment.[39] Because there is increasing evidence that climate change will increase the frequency and intensity of wildfire events, this outcome was included as a key finding, despite the inability to quantify this impact with the available tools and data. Because altered patterns of precipitation and increasing levels of CO_2 are anticipated to promote the level of aeroallergens, this outcome is also included as a Key Finding. Finally, because the empirical evidence linking climate change to indoor air quality was more equivocal, we identified this topic as an emerging issue.

KEY FINDING TRACEABLE ACCOUNTS

Exacerbated Ozone Health Impacts

Key Finding 1: Climate change will make it harder for any given regulatory approach to reduce ground-level ozone pollution in the future as meteorological conditions become increasingly conducive to forming ozone over most of the United States *[Likely, High Confidence]*. Unless offset by additional emissions reductions of ozone precursors, these climate-driven increases in ozone will cause premature deaths, hospital visits, lost school days, and acute respiratory symptoms *[Likely, High Confidence]*.

Description of evidence base

The Intergovernmental Panel on Climate Change (IPCC) has concluded that warming of the global climate system is unequivocal and that continued increases in greenhouse gas emissions will cause further temperature increases.[5, 35] At the same time, there is a well-established relationship between measured temperature and monitored peak ozone levels in the United States.[1, 25] Numerous climate and air quality modeling studies have also confirmed that increasing temperatures, along with other changes in meteorological variables, are likely to lead to higher peak ozone levels in the future over the United States,[7, 37] if ozone precursor emissions remain unchanged.

Risk assessments using concentration–response relationships from the epidemiological literature and modeled air quality data have projected substantial health impacts associated with climate-induced changes in air quality.[14, 42, 43, 44, 46, 50] This literature reports a range of potential changes in ozone-related, non-accidental mortality due to modeled climate change between the present and 2030 or 2050, depending upon the scenario modeled, the climate and air quality models used, and assumptions about the concentration–response function and future populations. Many of the studies suggest that tens to thousands of premature deaths could occur in the future due to climate change impacts on air quality.[14, 42] At the same time, hundreds of thousands of days of missed school and hundreds of thousands to millions of cases of acute respiratory symptoms also result from the climate-driven ozone increases in the United States.[14]

Major uncertainties

Climate projections are driven by greenhouse gas emission scenarios, which vary substantially depending on assumptions for economic growth and climate change mitigation policies. There is significant internal variability in the climate system, which leads to additional uncertainties in climate projections, particularly on a regional basis. Ozone concentrations also depend on emissions that are influenced indirectly by climate change (for example, incidence of wildfires, changes in energy use, energy technology choices), which further compounds the uncertainty. Studies projecting human health impacts apply concentration–response relationships from existing epidemiological studies characterizing historical air quality changes; it is unclear how future changes in the relationship between air quality, population exposure, and baseline health may affect the concentration–response relationship. Finally, these studies do not account for the possibility of a physiological interaction between air pollutants and temperature, which could lead to increases or decreases in air pollution-related deaths and illnesses.

Assessment of confidence and likelihood based on evidence

Given the known relationship between temperature and ozone, as well as the numerous air quality modeling studies that suggest climate-driven meteorological changes will yield conditions more favorable for ozone formation in the future, there is **high confidence** that ozone levels will **likely** increase due to climate change, unless offset by reductions in ozone precursor emissions. Based on observed relationships between ozone concentrations and human health responses, there is **high confidence** that any climate-driven increases in ozone will **likely** cause additional cases of premature mortality, as well as increasingly frequent cases of hospital visits and lost school days due to respiratory impacts.

Increased Health Impacts from Wildfires

Key Finding 2: Wildfires emit fine particles and ozone precursors that in turn increase the risk of premature death and adverse chronic and acute cardiovascular and respiratory health outcomes [*Likely, High Confidence*]. Climate change is projected to increase the number and severity of naturally occurring wildfires in parts of the United States, increasing emissions of particulate matter and ozone precursors and resulting in additional adverse health outcomes [*Likely, High Confidence*].

Description of evidence base

The harmful effects of PM concentrations on human health have been well-documented, and there is equally strong evidence linking wildfires to higher PM concentrations regionally. Recent studies have established linkages between wildfire incidence and adverse health outcomes in the nearby population.[70, 71] Though projections of climate change impacts on precipitation patterns are less certain than those on temperature, there is greater agreement across models that precipitation will decrease in the western United States.[74] Rising temperatures, decreasing precipitation, and earlier springtime onset of snowmelt are projected to lead to increased frequency and severity of wildfires.[22, 75, 77]

Major uncertainties

Future climate projections, especially projections of precipitation, are subject to considerable uncertainty. Land management practices, including possible adaptive measures taken to mitigate risk, could alter the frequency and severity of wildfires, the emissions from wildfires, and the associated human exposure to smoke.

Assessment of confidence and likelihood based on evidence

Given the known association between PM and health outcomes and between wildfires and PM concentrations, there is **high confidence** that an increase in wildfire frequency and severity will **likely** lead to an increase in adverse respiratory and cardiac health outcomes. Based on the robustness of the projection by global climate models that precipitation amounts will decrease in parts of the United States, and that summer temperatures will increase, there is

high confidence that the frequency and severity of wildfire occurrence will **likely** increase, particularly in the western United States.

Worsened Allergy and Asthma Conditions

Key Finding 3: Changes in climate, specifically rising temperatures, altered precipitation patterns, and increasing concentrations of atmospheric carbon dioxide, are expected to contribute to increases in the levels of some airborne allergens and associated increases in asthma episodes and other allergic illnesses [*High Confidence*].

Description of evidence base

There is a large body of evidence supporting the observation that climate change will alter the production, allergenicity, distribution, and timing of aeroallergens. Historical trends show that climate change has led to changes in the length of the growing season for certain allergenic pollens. Climate change also contributes to changes in allergic illnesses as greater concentrations of CO_2, together with higher temperatures and changes in precipitation, extend the start or duration of the growing season, increase the quantity and allergenicity of pollen, and expand the spatial distribution of pollens.[84, 91, 92, 93, 94] While the role of weather on the initiation or exacerbation of allergic symptoms in sensitive persons is not entirely understood,[86, 107] increases in intensity and frequency of rainfall and storminess over the coming decades is expected to be associated with spikes in aeroallergen concentrations and the potential for related increases in the number and severity of allergic illnesses.[108, 109]

These changes in exposure to aeroallergens contribute to the severity and prevalence of allergic disease in humans. Given that aeroallergen exposure is not the sole, or even necessarily the most significant, factor associated with allergic illnesses, that relationship is part of a complex pathway that links exposure to aeroallergens to the prevalence of allergic illnesses.[81] There is consistent and robust evidence that aeroallergen exposure contributes significantly to the occurrence of asthma episodes, hay fever, sinusitis, conjunctivitis, hives, and anaphylaxis.[84, 88] There is also compelling evidence that allergic diseases develop in response to complex and multiple interactions among both genetic and non-genetic factors, including a developing immune system, environmental exposures (such as ambient air pollution or weather conditions), and socioeconomic and demographic factors.[85, 86, 87] Finally, there is evidence that potential non-linear interactions between aeroallergens and ambient air pollutants is likely to increase health risks for people who are simultaneously exposed.[87, 88, 106, 108, 110, 111, 112, 113, 114]

Major uncertainties

The interrelationships between climate variability and change and exposure to aeroallergens are complex. Where they exist, differences in findings from across the relevant scientific literature may be due to study designs, references to certain species of pollen, geographic characteristics, climate variables, and degree of allergy sensitization.[104] There are also uncertainties with respect to the role of climate change and the extent and nature of its effects as they contribute to aeroallergen-related diseases, especially asthma.[91] Existing uncertainties can be addressed through the development of standardized approaches for measuring exposures and tracking outcomes across a range of allergic illnesses, vulnerable populations, and geographic proximity to exposures.[82]

Assessment of confidence and likelihood based on evidence
The scientific literature suggests that there is **high confidence** that changes in climate, including rising temperatures and altered precipitation patterns, will affect the concentration, allergenicity, season length, and spatial distribution of a number of aeroallergens, and these changes are expected to impact the prevalence of some allergic diseases, including asthma attacks.

DOCUMENTING UNCERTAINTY

This assessment relies on two metrics to communicate the degree of certainty in Key Findings. See Appendix 4: Documenting Uncertainty for more on assessments of likelihood and confidence.

Confidence Level	Likelihood	
Very High	**Very Likely**	
Strong evidence (established theory, multiple sources, consistent results, well documented and accepted methods, etc.), high consensus	≥ 9 in 10	
High	**Likely**	
Moderate evidence (several sources, some consistency, methods vary and/or documentation limited, etc.), medium consensus	≥ 2 in 3	
Medium	**As Likely As Not**	
Suggestive evidence (a few sources, limited consistency, models incomplete, methods emerging, etc.), competing schools of thought	≈ 1 in 2	
	Unlikely	
	≤ 1 in 3	
Low	**Very Unlikely**	
Inconclusive evidence (limited sources, extrapolations, inconsistent findings, poor documentation and/or methods not tested, etc.), disagreement or lack of opinions among experts	≤ 1 in 10	

References

1. Bloomer, B.J., J.W. Stehr, C.A. Piety, R.J. Salawitch, and R.R. Dickerson, 2009: Observed relationships of ozone air pollution with temperature and emissions. *Geophysical Research Letters,* **36,** L09803. http://dx.doi.org/10.1029/2009gl037308

2. Dawson, J.P., B.J. Bloomer, D.A. Winner, and C.P. Weaver, 2014: Understanding the meteorological drivers of U.S. particulate matter concentrations in a changing climate. *Bulletin of the American Meteorological Society,* **95,** 521-532. http://dx.doi.org/10.1175/BAMS-D-12-00181.1

3. Leibensperger, E.M., L.J. Mickley, and D.J. Jacob, 2008: Sensitivity of US air quality to mid-latitude cyclone frequency and implications of 1980–2006 climate change. *Atmospheric Chemistry and Physics,* **8,** 7075-7086. http://dx.doi.org/10.5194/acp-8-7075-2008

4. Weaver, C.P., E. Cooter, R. Gilliam, A. Gilliland, A. Grambsch, D. Grano, B. Hemming, S.W. Hunt, C. Nolte, D.A. Winner, X.-Z. Liang, J. Zhu, M. Caughey, K. Kunkel, J.-T. Lin, Z. Tao, A. Williams, D.J. Wuebbles, P.J. Adams, J.P. Dawson, P. Amar, S. He, J. Avise, J. Chen, R.C. Cohen, A.H. Goldstein, R.A. Harley, A.L. Steiner, S. Tonse, A. Guenther, J.-F. Lamarque, C. Wiedinmyer, W.I. Gustafson, L.R. Leung, C. Hogrefe, H.-C. Huang, D.J. Jacob, L.J. Mickley, S. Wu, P.L. Kinney, B. Lamb, N.K. Larkin, D. McKenzie, K.-J. Liao, K. Manomaiphiboon, A.G. Russell, E. Tagaris, B.H. Lynn, C. Mass, E. Salathé, S.M. O'Neill, S.N. Pandis, P.N. Racherla, C. Rosenzweig, and J.-H. Woo, 2009: A preliminary synthesis of modeled climate change impacts on U.S. regional ozone concentrations. *Bulletin of the American Meteorological Society,* **90,** 1843-1863. http://dx.doi.org/10.1175/2009BAMS2568.1

5. IPCC, 2013: Climate Change 2013: The Physical Science Basis. Contribution of Working Group I to the Fifth Assessment Report of the Intergovernmental Panel on Climate Change. Stocker, T.F., D. Qin, G.-K. Plattner, M. Tignor, S.K. Allen, J. Boschung, A. Nauels, Y. Xia, V. Bex, and P.M. Midgley (Eds.), 1535 pp. Cambridge University Press, Cambridge, UK and New York, NY. http://www.climatechange2013.org

6. Bernard, S.M., J.M. Samet, A. Grambsch, K.L. Ebi, and I. Romieu, 2001: The potential impacts of climate variability and change on air pollution-related health effects in the United States. *Environmental Health Perspectives,* **109,** 199-209. PMC1240667

7. Jacob, D.J. and D.A. Winner, 2009: Effect of climate change on air quality. *Atmospheric Environment,* **43,** 51-63. http://dx.doi.org/10.1016/j.atmosenv.2008.09.051

8. Fiore, A.M., V. Naik, D.V. Spracklen, A. Steiner, N. Unger, M. Prather, D. Bergmann, P.J. Cameron-Smith, I. Cionni, W.J. Collins, S. Dalsøren, V. Eyring, G.A. Folberth, P. Ginoux, L.W. Horowitz, B. Josse, J.-F. Lamarque, I.A. MacKenzie, T. Nagashima, F.M. O'Connor, M. Righi, S.T. Rumbold, D.T. Shindell, R.B. Skeie, K. Sudo, S. Szopa, T. Takemura, and G. Zeng, 2012: Global air quality and climate. *Chemical Society Reviews,* **41,** 6663-6683. http://dx.doi.org/10.1039/C2CS35095E

9. Anenberg, S.C., I.J. West, A.M. Fiore, D.A. Jaffe, M.J. Prather, D. Bergmann, K. Cuvelier, F.J. Dentener, B.N. Duncan, M. Gauss, P. Hess, J.E. Jonson, A. Lupu, I.A. Mackenzie, E. Marmer, R.J. Park, M.G. Sanderson, M. Schultz, D.T. Shindell, S. Szopa, M.G. Vivanco, O. Wild, and G. Zeng, 2009: Intercontinental impacts of ozone pollution on human mortality. *Environmental Science & Technology,* **43,** 6482-6287. http://dx.doi.org/10.1021/es900518z

10. Fann, N., A.D. Lamson, S.C. Anenberg, K. Wesson, D. Risley, and B.J. Hubbell, 2012: Estimating the national public health burden associated with exposure to ambient PM2.5 and ozone. *Risk Analysis,* **32,** 81-95. http://dx.doi.org/10.1111/j.1539-6924.2011.01630.x

11. EPA, 2009: Integrated Science Assessment for Particulate Matter. EPA/600/R-08/139F. National Center for Environmental Assessment, Office of Research and Development, U.S. Environmental Protection Agency, Research Triangle Park, NC. http://cfpub.epa.gov/ncea/cfm/recordisplay.cfm?deid=216546

12. Chang, H.H., J. Zhou, and M. Fuentes, 2010: Impact of climate change on ambient ozone level and mortality in southeastern United States. *International Journal of Environmental Research and Public Health,* **7,** 2866-2880. http://dx.doi.org/10.3390/ijerph7072866

13. Chang, H.H., H. Hao, and S.E. Sarnat, 2014: A statistical modeling framework for projecting future ambient ozone and its health impact due to climate change. *Atmospheric Environment,* **89,** 290-297. http://dx.doi.org/10.1016/j.atmosenv.2014.02.037

14. Fann, N., C.G. Nolte, P. Dolwick, T.L. Spero, A. Curry Brown, S. Phillips, and S. Anenberg, 2015: The geographic distribution and economic value of climate change-related ozone health impacts in the United States in 2030. *Journal of the Air & Waste Management Association,* **65,** 570-580. http://dx.doi.org/10.1080/10962247.2014.996270

15. Jackson, J.E., M.G. Yost, C. Karr, C. Fitzpatrick, B.K. Lamb, S.H. Chung, J. Chen, J. Avise, R.A. Rosenblatt, and R.A. Fenske, 2010: Public health impacts of climate change in Washington State: Projected mortality risks due to heat events and air pollution. *Climatic Change,* **102,** 159-186. http://dx.doi.org/10.1007/s10584-010-9852-3

16. Kelly, J., P.A. Makar, and D.A. Plummer, 2012: Projections of mid-century summer air-quality for North America: Effects of changes in climate and precursor emissions. *Atmospheric Chemistry and Physics,* **12,** 5367-5390. http://dx.doi.org/10.5194/acp-12-5367-2012

17. Penrod, A., Y. Zhang, K. Wang, S.-Y. Wu, and L.R. Leung, 2014: Impacts of future climate and emission changes on U.S. air quality. *Atmospheric Environment,* **89,** 533-547. http://dx.doi.org/10.1016/j.atmosenv.2014.01.001

18. Pfister, G.G., S. Walters, J.F. Lamarque, J. Fast, M.C. Barth, J. Wong, J. Done, G. Holland, and C.L. Bruyère, 2014: Projections of future summertime ozone over the U.S. *Journal of Geophysical Research: Atmospheres,* **119,** 5559-5582. http://dx.doi.org/10.1002/2013JD020932

19. Sheffield, P.E., J.L. Carr, P.L. Kinney, and K. Knowlton, 2011: Modeling of regional climate change effects on ground-level ozone and childhood asthma. *American Journal of Preventive Medicine,* **41,** 251-257. http://dx.doi.org/10.1016/j.amepre.2011.04.017

20. Sheffield, P.E., K.R. Weinberger, and P.L. Kinney, 2011: Climate change, aeroallergens, and pediatric allergic disease. *Mount Sinai Journal of Medicine,* **78,** 78-84. http://dx.doi.org/10.1002/msj.20232

21. Tai, A.P.K., L.J. Mickley, and D.J. Jacob, 2012: Impact of 2000–2050 climate change on fine particulate matter (PM2.5) air quality inferred from a multi-model analysis of meteorological modes. *Atmospheric Chemistry and Physics,* **12,** 11329-11337. http://dx.doi.org/10.5194/acp-12-11329-2012

22. Val Martin, M., C.L. Heald, J.F. Lamarque, S. Tilmes, L.K. Emmons, and B.A. Schichtel, 2015: How emissions, climate, and land use change will impact mid-century air quality over the United States: A focus on effects at National Parks. *Atmospheric Chemistry and Physics,* **15,** 2805-2823. http://dx.doi.org/10.5194/acp-15-2805-2015

23. EPA, 2014: Regulatory Impact Analysis of the Proposed Revisions to the National Ambient Air Quality Standards for Ground-Level Ozone. EPA-452/P-14-006. U.S. Environmental Protection Agency, Office of Air and Radiation, Office of Air Quality Planning and Standards, Research Triangle Park, NC. http://epa.gov/ttn/ecas/regdata/RIAs/20141125ria.pdf

24. Wu, S., L.J. Mickley, E.M. Leibensperger, D.J. Jacob, D. Rind, and D.G. Streets, 2008: Effects of 2000–2050 global change on ozone air quality in the United States. *Journal of Geophysical Research: Atmospheres,* **113,** D06302. http://dx.doi.org/10.1029/2007JD008917

25. Camalier, L., W. Cox, and P. Dolwick, 2007: The effects of meteorology on ozone in urban areas and their use in assessing ozone trends. *Atmospheric Environment,* **41,** 7127-7137. http://dx.doi.org/10.1016/j.atmosenv.2007.04.061

26. Davis, J., W. Cox, A. Reff, and P. Dolwick, 2011: A comparison of CMAQ-based and observation-based statistical models relating ozone to meteorological parameters. *Atmospheric Environment,* **45,** 3481-3487. http://dx.doi.org/10.1016/j.atmosenv.2010.12.060

27. Leung, L.R. and W.I. Gustafson, Jr., 2005: Potential regional climate change and implications to U.S. air quality. *Geophysical Research Letters,* **32,** L16711. http://dx.doi.org/10.1029/2005GL022911

28. Zhu, J. and X.-Z. Liang, 2013: Impacts of the Bermuda high on regional climate and ozone over the United states. *Journal of Climate,* **26,** 1018-1032. http://dx.doi.org/10.1175/JCLI-D-12-00168.1

29. Horton, D.E., Harshvardhan, and N.S. Diffenbaugh, 2012: Response of air stagnation frequency to anthropogenically enhanced radiative forcing. *Environmental Research Letters,* **7,** 044034. http://dx.doi.org/10.1088/1748-9326/7/4/044034

30. Haman, C.L., E. Couzo, J.H. Flynn, W. Vizuete, B. Heffron, and B.L. Lefer, 2014: Relationship between boundary layer heights and growth rates with ground-level ozone in Houston, Texas. *Journal of Geophysical Research: Atmospheres,* **119,** 6230-6245. http://dx.doi.org/10.1002/2013JD020473

31. Zhang, L., D.J. Jacob, X. Yue, N.V. Downey, D.A. Wood, and D. Blewitt, 2014: Sources contributing to background surface ozone in the US Intermountain West. *Atmospheric Chemistry and Physics,* **14,** 5295-5309. http://dx.doi.org/10.5194/acp-14-5295-2014

32. Murazaki, K. and P. Hess, 2006: How does climate change contribute to surface ozone change over the United States? *Journal of Geophysical Research - Atmospheres,* **111,** D05301. http://dx.doi.org/10.1029/2005JD005873

33. Kirtman, B., S.B. Power, J.A. Adedoyin, G.J. Boer, R. Bojariu, I. Camilloni, F.J. Doblas-Reyes, A.M. Fiore, M. Kimoto, G.A. Meehl, M. Prather, A. Sarr, C. Schär, R. Sutton, G.J. van Oldenborgh, G. Vecchi, and H.J. Wang, 2013: Near-term climate change: Projections and predictability. *Climate Change 2013: The Physical Science Basis. Contribution of Working Group I to the Fifth Assessment Report of the Intergovernmental Panel on Climate Change.* Stocker, T.F., D. Qin, G.-K. Plattner, M. Tignor, S.K. Allen, J. Boschung, A. Nauels, Y. Xia, V. Bex, and P.M. Midgley, Eds. Cambridge University Press, Cambridge, UK and New York, NY, USA, 953–1028. http://dx.doi.org/10.1017/CBO9781107415324.023

34. Stavros, E.N., D. McKenzie, and N. Larkin, 2014: The climate-wildfire-air quality system: Interactions and feedbacks across spatial and temporal scales. *Wiley Interdisciplinary Reviews: Climate Change,* **5,** 719-733. http://dx.doi.org/10.1002/wcc.303

35. Melillo, J.M., T.C. Richmond, and G.W. Yohe, Eds., 2014: *Climate Change Impacts in the United States: The Third National Climate Assessment.* U.S. Global Change Research Program, Washington, D.C., 842 pp. http://dx.doi.org/10.7930/J0Z31WJ2

36. Rieder, H.E., A.M. Fiore, L.W. Horowitz, and V. Naik, 2015: Projecting policy-relevant metrics for high summertime ozone pollution events over the eastern United States due to climate and emission changes during the 21st century. *Journal of Geophysical Research: Atmospheres,* **120,** 784-800. http://dx.doi.org/10.1002/2014JD022303

37. Fiore, A.M., V. Naik, and E.M. Leibensperger, 2015: Air quality and climate connections. *Journal of the Air & Waste Management Association,* **65,** 645-686. http://dx.doi.org/10.1080/10962247.2015.1040526

38. Bell, M.L., A. McDermott, S.L. Zeger, J.M. Samet, and F. Dominici, 2004: Ozone and short-term mortality in 95 US urban communities, 1987-2000. *JAMA: The Journal of the American Medical Association,* **292,** 2372-2378. http://dx.doi.org/10.1001/jama.292.19.2372

39. EPA, 2013: Integrated Science Assessment for Ozone and Related Photochemical Oxidants. EPA 600/R-10/076F, 1251 pp. U.S. Environmental Protection Agency, National Center for Environmental Assessment, Office of Research and Development, Research Triangle Park, NC. http://cfpub.epa.gov/ncea/isa/recordisplay.cfm?deid=247492

40. Jerrett, M., R.T. Burnett, C.A. Pope, K. Ito, G. Thurston, D. Krewski, Y. Shi, E. Calle, and M. Thun, 2009: Long-term ozone exposure and mortality. *The New England Journal of Medicine,* **360,** 1085-1095. http://dx.doi.org/10.1056/NEJMoa0803894

41. Hubbell, B., N. Fann, and J. Levy, 2009: Methodological considerations in developing local-scale health impact assessments: Balancing national, regional, and local data. *Air Quality, Atmosphere & Health,* **2,** 99-110. http://dx.doi.org/10.1007/s11869-009-0037-z

42. Post, E.S., A. Grambsch, C. Weaver, P. Morefield, J. Huang, L.-Y. Leung, C.G. Nolte, P. Adams, X.-Z. Liang, J.-H. Zhu, and H. Mahone, 2012: Variation in estimated ozone-related health impacts of climate change due to modeling choices and assumptions. *Environmental Health Perspectives,* **120,** 1559-1564. http://dx.doi.org/10.1289/ehp.1104271

43. Selin, N.E., S. Wu, K.M. Nam, J.M. Reilly, S. Paltsev, R.G. Prinn, and M.D. Webster, 2009: Global health and economic impacts of future ozone pollution. *Environmental Research Letters,* **4,** 044014. http://dx.doi.org/10.1088/1748-9326/4/4/044014

44. Tagaris, E., K.-J. Liao, A.J. DeLucia, L. Deck, P. Amar, and A.G. Russell, 2009: Potential impact of climate change on air pollution-related human health effects. *Environmental Science & Technology,* **43,** 4979-4988. http://dx.doi.org/10.1021/es803650w

45. Garcia-Menendez, F., R.K. Saari, E. Monier, and N.E. Selin, 2015: U.S. air quality and health benefits from avoided climate change under greenhouse gas mitigation. *Environmental Science & Technology,* **49,** 7580-7588. http://dx.doi.org/10.1021/acs.est.5b01324

46. Bell, M.L., R. Goldberg, C. Hogrefe, P.L. Kinney, K. Knowlton, B. Lynn, J. Rosenthal, C. Rosenzweig, and J.A. Patz, 2007: Climate change, ambient ozone, and health in 50 US cities. *Climatic Change,* **82,** 61-76. http://dx.doi.org/10.1007/s10584-006-9166-7

47. Riahi, K., S. Rao, V. Krey, C. Cho, V. Chirkov, G. Fischer, G. Kindermann, N. Nakicenovic, and P. Rafaj, 2011: RCP 8.5—A scenario of comparatively high greenhouse gas emissions. *Climatic Change,* **109,** 33-57. http://dx.doi.org/10.1007/s10584-011-0149-y

48. Taylor, K.E., R.J. Stouffer, and G.A. Meehl, 2012: An overview of CMIP5 and the experiment design. *Bulletin of the American Meteorological Society,* **93,** 485-498. http://dx.doi.org/10.1175/BAMS-D-11-00094.1

49. van Vuuren, D.P., J. Edmonds, M. Kainuma, K. Riahi, A. Thomson, K. Hibbard, G.C. Hurtt, T. Kram, V. Krey, J.-F. Lamarque, T. Masui, M. Meinshausen, N. Nakicenovic, S.J. Smith, and S.K. Rose, 2011: The representative concentration pathways: An overview. *Climatic Change,* **109,** 5-31. http://dx.doi.org/10.1007/s10584-011-0148-z

50. Jacobson, M.Z., 2008: On the causal link between carbon dioxide and air pollution mortality. *Geophysical Research Letters,* **35,** L03809. http://dx.doi.org/10.1029/2007GL031101

51. Jhun, I., N. Fann, A. Zanobetti, and B. Hubbell, 2014: Effect modification of ozone-related mortality risks by temperature in 97 US cities. *Environment International,* **73,** 128-134. http://dx.doi.org/10.1016/j.envint.2014.07.009

52. Ren, C., G.M. Williams, K. Mengersen, L. Morawska, and S. Tong, 2008: Does temperature modify short-term effects of ozone on total mortality in 60 large eastern US communities? An assessment using the NMMAPS data. *Environment International,* **34,** 451-458. http://dx.doi.org/10.1016/j.envint.2007.10.001

53. Ren, C., G.M. Williams, L. Morawska, K. Mengersen, and S. Tong, 2008: Ozone modifies associations between temperature and cardiovascular mortality: Analysis of the NMMAPS data. *Occupational and Environmental Medicine,* **65,** 255-260. http://dx.doi.org/10.1136/oem.2007.033878

54. Ren, C., G.M. Williams, K. Mengersen, L. Morawska, and S. Tong, 2009: Temperature enhanced effects of ozone on cardiovascular mortality in 95 large US communities, 1987-2000: Assessment using the NMMAPS data. *Archives of Environmental & Occupational Health,* **64,** 177-184. http://dx.doi.org/10.1080/19338240903240749

55. Ren, C., S. Melly, and J. Schwartz, 2010: Modifiers of short-term effects of ozone on mortality in eastern Massachusetts-A casecrossover analysis at individual level. *Environmental Health,* **9,** Article 3. http://dx.doi.org/10.1186/1476-069X-9-3

56. EPA, 2015: Technical Support Document (TSD): Preparation of Emissions Inventories for the Version 6.2, 2011 Emissions Modeling Platform. U.S. Environmental Protection Agency, Office of Air and Radiation. http://www3.epa.gov/ttn/chief/emch/2011v6/2011v6_2_2017_2025_EmisMod_TSD_aug2015.pdf

57. Otte, T.L., C.G. Nolte, M.J. Otte, and J.H. Bowden, 2012: Does nudging squelch the extremes in regional climate modeling? *Journal of Climate,* **25,** 7046-7066. http://dx.doi.org/10.1175/JCLI-D-12-00048.1

58. Gao, Y., J.S. Fu, J.B. Drake, J.-F. Lamarque, and Y. Liu, 2013: The impact of emission and climate change on ozone in the United States under representative concentration pathways (RCPs). *Atmospheric Chemistry and Physics,* **13,** 9607-9621. http://dx.doi.org/10.5194/acp-13-9607-2013

59. Bell, M.L. and F. Dominici, 2008: Effect modification by community characteristics on the short-term effects of ozone exposure and mortality in 98 US communities. *American Journal of Epidemiology,* **167,** 986-97. http://dx.doi.org/10.1093/aje/kwm396

60. Sacks, J.D., L.W. Stanek, T.J. Luben, D.O. Johns, B.J. Buckley, J.S. Brown, and M. Ross, 2011: Particulate matter–induced health effects: Who is susceptible? *Environmental Health Perspectives,* **119,** 446-454. http://dx.doi.org/10.1289/ehp.1002255

61. EPA, 1999: Regulatory Impact Analysis - Control of Air Pollution from New Motor Vehicles: Tier 2 Motor Vehicle Emissions Standards and Gasoline Sulfur Control Requirements. EPA420-R-99-023, 522 pp. U.S. Environmental Protection Agency, Office of Transportation and Air Quality, Washington, D.C. http://www.epa.gov/tier2/documents/r99023.pdf

62. EPA, 2015: Cross-State Air Pollution Rule (CSAPR). U.S. Environmental Protection Agency, Washington, D.C. http://www.epa.gov/airtransport/CSAPR/

63. EPA, 2015: Tier 2 Vehicle and Gasoline Sulfur Program. U.S. Environmental Protection Agency, Office of Transportation and Air Quality, Washington, D.C. http://www.epa.gov/tier2/

64. Dawson, J.P., P.N. Racherla, B.H. Lynn, P.J. Adams, and S.N. Pandis, 2009: Impacts of climate change on regional and urban air quality in the eastern United States: Role of meteorology. *Journal of Geophysical Research: Atmospheres,* **114.** http://dx.doi.org/10.1029/2008JD009849

65. Trail, M., A.P. Tsimpidi, P. Liu, K. Tsigaridis, J. Rudokas, P. Miller, A. Nenes, Y. Hu, and A.G. Russell, 2014: Sensitivity of air quality to potential future climate change and emissions in the United States and major cities. *Atmospheric Environment,* **94,** 552-563. http://dx.doi.org/10.1016/j.atmosenv.2014.05.079

66. Delfino, R.J., S. Brummel, J. Wu, H. Stern, B. Ostro, M. Lipsett, A. Winer, D.H. Street, L. Zhang, T. Tjoa, and D.L. Gillen, 2009: The relationship of respiratory and cardiovascular hospital admissions to the southern California wildfires of 2003. *Occupational and Environmental Medicine,* **66,** 189-197. http://dx.doi.org/10.1136/oem.2008.041376

67. Künzli, N., E. Avol, J. Wu, W.J. Gauderman, E. Rappaport, J. Millstein, J. Bennion, R. McConnell, F.D. Gilliland, K. Berhane, F. Lurmann, A. Winer, and J.M. Peters, 2006: Health effects of the 2003 southern California wildfires on children. *American Journal of Respiratory and Critical Care Medicine,* **174,** 1221-1228. http://dx.doi.org/10.1164/rccm.200604-519OC

68. Park, R.J., 2003: Sources of carbonaceous aerosols over the United States and implications for natural visibility. *Journal of Geophysical Research,* **108,** D12, 4355. http://dx.doi.org/10.1029/2002JD003190

69. Sapkota, A., J.M. Symons, J. Kleissl, L. Wang, M.B. Parlange, J. Ondov, P.N. Breysse, G.B. Diette, P.A. Eggleston, and T.J. Buckley, 2005: Impact of the 2002 Canadian forest fires on particulate matter air quality in Baltimore City. *Environmental Science & Technology,* **39,** 24-32. http://dx.doi.org/10.1021/es035311z

70. Henderson, S.B., M. Brauer, Y.C. Macnab, and S.M. Kennedy, 2011: Three measures of forest fire smoke exposure and their associations with respiratory and cardiovascular health outcomes in a population-based cohort. *Environmental Health Perspectives,* **119,** 1266-1271. http://dx.doi.org/10.1289/ehp.1002288

71. Rappold, A.G., W.E. Cascio, V.J. Kilaru, S.L. Stone, L.M. Neas, R.B. Devlin, and D. Diaz-Sanchez, 2012: Cardio-respiratory outcomes associated with exposure to wildfire smoke are modified by measures of community health. *Environmental Health,* **11,** Article 71. http://dx.doi.org/10.1186/1476-069X-11-71

72. Liu, J.C., G. Pereira, S.A. Uhl, M.A. Bravo, and M.L. Bell, 2015: A systematic review of the physical health impacts from non-occupational exposure to wildfire smoke. *Environmental Research,* **136,** 120-132. http://dx.doi.org/10.1016/j.envres.2014.10.015

73. Westerling, A.L., H.G. Hidalgo, D.R. Cayan, and T.W. Swetnam, 2006: Warming and earlier spring increase western U.S. forest wildfire activity. *Science,* **313,** 940-943. http://dx.doi.org/10.1126/science.1128834

74. Walsh, J., D. Wuebbles, K. Hayhoe, J. Kossin, K. Kunkel, G. Stephens, P. Thorne, R. Vose, M. Wehner, J. Willis, D. Anderson, S. Doney, R. Feely, P. Hennon, V. Kharin, T. Knutson, F. Landerer, T. Lenton, J. Kennedy, and R. Somerville, 2014: Ch. 2: Our changing climate. *Climate Change Impacts in the United States: The Third National Climate Assessment.* Melillo, J.M., T.C. Richmond, and G.W. Yohe, Eds. U.S. Global Change Research Program, Washington, D.C., 19-67. http://dx.doi.org/10.7930/J0KW5CXT

75. Garfin, G., G. Franco, H. Blanco, A. Comrie, P. Gonzalez, T. Piechota, R. Smyth, and R. Waskom, 2014: Ch. 20: Southwest. *Climate Change Impacts in the United States: The Third National Climate Assessment.* Melillo, J.M., T.C. Richmond, and G.W. Yohe, Eds. U.S. Global Change Research Program, Washington, D.C., 462-486. http://dx.doi.org/10.7930/J08G8HMN

76. Luber, G., K. Knowlton, J. Balbus, H. Frumkin, M. Hayden, J. Hess, M. McGeehin, N. Sheats, L. Backer, C.B. Beard, K.L. Ebi, E. Maibach, R.S. Ostfeld, C. Wiedinmyer, E. Zielinski-Gutiérrez, and L. Ziska, 2014: Ch. 9: Human health. *Climate Change Impacts in the United States: The Third National Climate Assessment.* Melillo, J.M., T.C. Richmond, and G.W. Yohe, Eds. U.S. Global Change Research Program, Washington, D.C., 220-256. http://dx.doi.org/10.7930/J0PN93H5

77. Westerling, A.L., M.G. Turner, E.A.H. Smithwick, W.H. Romme, and M.G. Ryan, 2011: Continued warming could transform Greater Yellowstone fire regimes by mid-21st century. *Proceedings of the National Academy of Sciences,* **108,** 13165-13170. http://dx.doi.org/10.1073/pnas.1110199108

78. Keywood, M., M. Kanakidou, A. Stohl, F. Dentener, G. Grassi, C.P. Meyer, K. Torseth, D. Edwards, A.M. Thompson, U. Lohmann, and J. Burrows, 2013: Fire in the air: Biomass burning impacts in a changing climate. *Critical Reviews in Environmental Science and Technology,* **43,** 40-83. http://dx.doi.org/10.1080/10643389.2011.604248

79. Spracklen, D.V., L.J. Mickley, J.A. Logan, R.C. Hudman, R. Yevich, M.D. Flannigan, and A.L. Westerling, 2009: Impacts of climate change from 2000 to 2050 on wildfire activity and carbonaceous aerosol concentrations in the western United States. *Journal of Geophysical Research: Atmospheres,* **114,** D20301. http://dx.doi.org/10.1029/2008JD010966

80. Seager, R., M. Ting, I. Held, Y. Kushnir, J. Lu, G. Vecchi, H.-P. Huang, N. Harnik, A. Leetmaa, N.-C. Lau, C. Li, J. Velez, and N. Naik, 2007: Model projections of an imminent transition to a more arid climate in southwestern North America. *Science,* **316,** 1181-1184. http://dx.doi.org/10.1126/science.1139601

81. Reid, C.E. and J.L. Gamble, 2009: Aeroallergens, allergic disease, and climate change: Impacts and adaptation. *EcoHealth,* **6,** 458-470. http://dx.doi.org/10.1007/s10393-009-0261-x

82. Selgrade, M.K., R.F. Lemanske, Jr., M.I. Gilmour, L.M. Neas, M.D.W. Ward, P.K. Henneberger, D.N. Weissman, J.A. Hoppin, R.R. Dietert, P.D. Sly, A.M. Geller, P.L. Enright, G.S. Backus, P.A. Bromberg, D.R. Germolec, and K.B. Yeatts, 2006: Induction of asthma and the environment: What we know and need to know. *Environmental Health Perspectives,* **114,** 615-619. http://dx.doi.org/10.1289/ehp.8376

83. Kinney, P.L., 2008: Climate change, air quality, and human health. *American Journal of Preventive Medicine,* **35,** 459-467. http://dx.doi.org/10.1016/j.amepre.2008.08.025

84. Bielory, L., K. Lyons, and R. Goldberg, 2012: Climate change and allergic disease. *Current Allergy and Asthma Reports,* **12,** 485-494. http://dx.doi.org/10.1007/s11882-012-0314-z

85. AAAAI, Undated: Allergy Statistics. American Academy of Allergy, Asthma and Immunology. http://www.aaaai.org/about-the-aaaai/newsroom/allergy-statistics.aspx

86. Breton, M.-C., M. Garneau, I. Fortier, F. Guay, and J. Louis, 2006: Relationship between climate, pollen concentrations of *Ambrosia* and medical consultations for allergic rhinitis in Montreal, 1994–2002. *Science of The Total Environment,* **370,** 39-50. http://dx.doi.org/10.1016/j.scitotenv.2006.05.022

87. Bartra, J., J. Mullol, A. del Cuvillo, I. Davila, M. Ferrer, I. Jauregui, J. Montoro, J. Sastre, and A. Valero, 2007: Air pollution and allergens. *Journal of Investigational Allergology and Clinical Immunology,* **17 Suppl 2,** 3-8. http://www.jiaci.org/issues/vol17s2/vol17s2-2.htm

88. Blando, J., L. Bielory, V. Nguyen, R. Diaz, and H.A. Jeng, 2012: Anthropogenic climate change and allergic diseases. *Atmosphere,* **3,** 200-212. http://dx.doi.org/10.3390/atmos3010200

89. EPA, 2008: Review of the Impacts of Climate Variability and Change on Aeroallergens and Their Associated Effects. EPA/600/R-06/164F, 125 pp. U.S. Environmental Protection Agency, Washington, D.C. http://ofmpub.epa.gov/eims/eimscomm.getfile?p_download_id=490474

90. Lawson, J.A. and A. Senthilselvan, 2005: Asthma epidemiology: Has the crisis passed? *Current Opinion in Pulmonary Medicine,* **11,** 79-84.

91. Beggs, P.J., 2004: Impacts of climate change on aeroallergens: Past and future. *Clinical & Experimental Allergy,* **34,** 1507-1513. http://dx.doi.org/10.1111/j.1365-2222.2004.02061.x

92. Beggs, P.J. and H.J. Bambrick, 2005: Is the global rise of asthma an early impact of anthropogenic climate change? *Environmental Health Perspectives,* **113,** 915-919. http://dx.doi.org/10.1289/ehp.7724

93. D'Amato, G., C.E. Baena-Cagnani, L. Cecchi, I. Annesi-Maesano, C. Nunes, I. Ansotegui, M. D'Amato, G. Liccardi, M. Sofia, and W.G. Canonica, 2013: Climate change, air pollution and extreme events leading to increasing prevalence of allergic respiratory diseases. *Multidisciplinary Respiratory Medicine,* **8,** 1-9. http://dx.doi.org/10.1186/2049-6958-8-12

94. Albertine, J.M., W.J. Manning, M. DaCosta, K.A. Stinson, M.L. Muilenberg, and C.A. Rogers, 2014: Projected carbon dioxide to increase grass pollen and allergen exposure despite higher ozone levels. *PLoS ONE,* **9,** e111712. http://dx.doi.org/10.1371/journal.pone.0111712

95. Ziska, L., K. Knowlton, C. Rogers, D. Dalan, N. Tierney, M.A. Elder, W. Filley, J. Shropshire, L.B. Ford, C. Hedberg, P. Fleetwood, K.T. Hovanky, T. Kavanaugh, G. Fulford, R.F. Vrtis, J.A. Patz, J. Portnoy, F. Coates, L. Bielory, and D. Frenz, 2011: Recent warming by latitude associated with increased length of ragweed pollen season in central North America. *Proceedings of the National Academy of Sciences,* **108,** 4248-4251. http://dx.doi.org/10.1073/pnas.1014107108

96. Rogers, C.A., P.M. Wayne, E.A. Macklin, M.L. Muilenberg, C.J. Wagner, P.R. Epstein, and F.A. Bazzaz, 2006: Interaction of the onset of spring and elevated atmospheric CO_2 on ragweed (*Ambrosia artemisiifolia* L.) pollen production. *Environmental Health Perspectives,* **114,** 865-869. http://dx.doi.org/10.1289/ehp.8549

97. Singer, B.D., L.H. Ziska, D.A. Frenz, D.E. Gebhard, and J.G. Straka, 2005: Increasing Amb a 1 content in common ragweed (*Ambrosia artemisiifolia*) pollen as a function of rising atmospheric CO_2 concentration. *Functional Plant Biology,* **32,** 667-670. http://dx.doi.org/10.1071/fp05039

98. Neil, K. and J. Wu, 2006: Effects of urbanization on plant flowering phenology: A review. *Urban Ecosystems,* **9,** 243-257. http://dx.doi.org/10.1007/s11252-006-9354-2

99. George, K., L.H. Ziska, J.A. Bunce, and B. Quebedeaux, 2007: Elevated atmospheric CO_2 concentration and temperature across an urban–rural transect. *Atmospheric Environment,* **41,** 7654-7665. http://dx.doi.org/10.1016/j.atmosenv.2007.08.018

100. Roetzer, T., M. Wittenzeller, H. Haeckel, and J. Nekovar, 2000: Phenology in central Europe: Differences and trends of spring phenophases in urban and rural areas. *International Journal of Biometeorology,* **44,** 60-66. http://dx.doi.org/10.1007/s004840000062

101. Garcia-Mozo, H., C. Galán, V. Jato, J. Belmonte, C.D. de la Guardia, D. Fernández, M. Gutiérrez, M.J. Aira, J.M. Roure, L. Ruiz, M.M. Trigo, and E. Domínguez-Vilches, 2006: *Quercus* pollen season dynamics in the Iberian peninsula: Response to meteorological parameters and possible consequences of climate change. *Annals of Agricultural and Environmental Medicine,* **13,** 209-224. http://www.uco.es/aerobiologia/publicaciones/modelling/climate_change/Quercus_AAEM_def.pdf

102. LaDeau, S.L. and J.S. Clark, 2006: Elevated CO_2 and tree fecundity: The role of tree size, interannual variability, and population heterogeneity. *Global Change Biology,* **12,** 822-833. http://dx.doi.org/10.1111/j.1365-2486.2006.01137.x

103. Zhang, R., T. Duhl, M.T. Salam, J.M. House, R.C. Flagan, E.L. Avol, F.D. Gilliland, A. Guenther, S.H. Chung, B.K. Lamb, and T.M. VanReken, 2013: Development of a regional-scale pollen emission and transport modeling framework for investigating the impact of climate change on allergic airway disease. *Biogeosciences,* **10,** 3977-4023. http://dx.doi.org/10.5194/bgd-10-3977-2013

104. D'Amato, G. and L. Cecchi, 2008: Effects of climate change on environmental factors in respiratory allergic diseases. *Clinical & Experimental Allergy,* **38,** 1264-1274. http://dx.doi.org/10.1111/j.1365-2222.2008.03033.x

105. D'Amato, G., M. Rottem, R. Dahl, M.S. Blaiss, E. Ridolo, L. Cecchi, N. Rosario, C. Motala, I. Ansotegui, and I. Annesi-Maesano, 2011: Climate change, migration, and allergic respiratory diseases: An update for the allergist. *World Allergy Organization Journal,* **4,** 121-125. http://dx.doi.org/10.1097/WOX.0b013e3182260a57

106. Shea, K.M., R.T. Truckner, R.W. Weber, and D.B. Peden, 2008: Climate change and allergic disease. *Journal of Allergy and Clinical Immunology,* **122,** 443-453. http://dx.doi.org/10.1016/j.jaci.2008.06.032

107. D'Amato, G., L. Cecchi, M. D'Amato, and G. Liccardi, 2010: Urban air pollution and climate change as environmental risk factors of respiratory allergy: An update. *Journal of Investigational Allergology and Clinical Immunology,* **20,** 95-102. http://www.jiaci.org/issues/vol20issue2/1.pdf

108. Cecchi, L., G. D'Amato, J.G. Ayres, C. Galan, F. Forastiere, B. Forsberg, J. Gerritsen, C. Nunes, H. Behrendt, C. Akdis, R. Dahl, and I. Annesi-Maesano, 2010: Projections of the effects of climate change on allergic asthma: The contribution of aerobiology. *Allergy,* **65,** 1073-1081. http://dx.doi.org/10.1111/j.1398-9995.2010.02423.x

109. D'Amato, G., G. Liccardi, and G. Frenguelli, 2007: Thunderstorm-asthma and pollen allergy. *Allergy,* **62,** 11-16. http://dx.doi.org/10.1111/j.1398-9995.2006.01271.x

110. D'Amato, G., G. Liccardi, M. D'Amato, and M. Cazzola, 2001: The role of outdoor air pollution and climatic changes on the rising trends in respiratory allergy. *Respiratory Medicine,* **95,** 606-611. http://dx.doi.org/10.1053/rmed.2001.1112

111. D'Amato, G., 2002: Environmental urban factors (air pollution and allergens) and the rising trends in allergic respiratory diseases. *Allergy,* **57,** 30-33. http://dx.doi.org/10.1034/j.1398-9995.57.s72.5.x

112. D'Amato, G., G. Liccardi, M. D'Amato, and S. Holgate, 2005: Environmental risk factors and allergic bronchial asthma. *Clinical & Experimental Allergy,* **35,** 1113-1124. http://dx.doi.org/10.1111/j.1365-2222.2005.02328.x

113. Atkinson, R.W. and D.P. Strachan, 2004: Role of outdoor aeroallergens in asthma exacerbations: Epidemiological evidence. *Thorax,* **59,** 277-278. http://dx.doi.org/10.1136/thx.2003.019133

114. Zhong, W., L. Levin, T. Reponen, G.K. Hershey, A. Adhikari, R. Shukla, and G. LeMasters, 2006: Analysis of short-term influences of ambient aeroallergens on pediatric asthma hospital visits. *Science of The Total Environment,* **370,** 330-336. http://dx.doi.org/10.1016/j.scitotenv.2006.06.019

115. Cakmak, S., R.E. Dales, and F. Coates, 2012: Does air pollution increase the effect of aeroallergens on hospitalization for asthma? *Journal of Allergy and Clinical Immunology,* **129,** 228-231. http://dx.doi.org/10.1016/j.jaci.2011.09.025

116. IOM, 2011: *Climate Change, the Indoor Environment, and Health.* Institute of Medicine. The National Academies Press, Washington, D.C., 286 pp. http://dx.doi.org/10.17226/13115

117. IOM, 1993: *Indoor Allergens: Assessing and Controlling Adverse Health Effects.* Institute of Medicine. The National Academies Press, Washington, D.C., 350 pp. http://dx.doi.org/10.17226/2056

118. IOM, 2000: *Clearing the Air: Asthma and Indoor Air Exposures*. Institute of Medicine. The National Academies Press, Washington, D.C., 456 pp. http://www.nap.edu/catalog/9610/clearing-the-air-asthma-and-indoor-air-exposures

119. IOM, 2004: *Damp Indoor Spaces and Health*. Institute of Medicine. The National Academies Press, Washington, D.C., 370 pp. http://dx.doi.org/10.17226/11011

120. WHO, 2009: *WHO Handbook on Indoor Radon: A Public Health Perspective*. Zeeb, H. and F. Shannoun, Eds. World Health Organization, Geneva, Switzerland, 108 pp. http://whqlibdoc.who.int/publications/2009/9789241547673_eng.pdf

121. WHO, 2010: WHO Guidelines for Indoor Air Quality: Selected Pollutants. 484 pp. World Health Organization, Geneva. http://www.euro.who.int/__data/assets/pdf_file/0009/128169/e94535.pdf

122. Ott, W.R., 1989: Human activity patterns: A review of the literature for estimating time spent indoors, outdoors, and in transit. *Proceedings of the Research Planning Conference on Human Activity Patterns*. EPA National Exposure Research Laboratory, Las Vegas, NV, 3-1 to 3-38.

123. Klepeis, N.E., W.C. Nelson, W.R. Ott, J.P. Robinson, A.M. Tsang, P. Switzer, J.V. Behar, S.C. Hern, and W.H. Engelmann, 2001: The National Human Activity Pattern Survey (NHAPS): A resource for assessing exposure to environmental pollutants. *Journal of Exposure Analysis and Environmental Epidemiology*, **11**, 231-252. http://dx.doi.org/10.1038/sj.jea.7500165

124. Riley, W.J., T.E. McKone, A.C. Lai, and W.W. Nazaroff, 2002: Indoor particulate matter of outdoor origin: Importance of size-dependent removal mechanisms. *Environmental Science & Technology*, **36**, 200-207. http://dx.doi.org/10.1021/es010723y

125. Bennett, D.H. and P. Koutrakis, 2006: Determining the infiltration of outdoor particles in the indoor environment using a dynamic model. *Journal of Aerosol Science*, **37**, 766-785. http://dx.doi.org/10.1016/j.jaerosci.2005.05.020

126. Wallace, L., 1996: Indoor particles: A review. *Journal of the Air & Waste Management Association*, **46**, 98-126. http://dx.doi.org/10.1080/10473289.1996.10467451

127. Brennan, T., J.B. Cummings, and J. Lstiburek, 2002: Unplanned airflows & moisture problems. *ASHRAE journal*, **44**, 44-49. https://64.94.228.53/File Library/docLib/Public/2002102583440_266.pdf

128. Stephens, B. and J.A. Siegel, 2012: Penetration of ambient submicron particles into single-family residences and associations with building characteristics. *Indoor Air*, **22**, 501-513. http://dx.doi.org/10.1111/j.1600-0668.2012.00779.x

129. Persily, A., A. Musser, and S.J. Emmerich, 2010: Modeled infiltration rate distributions for U.S. housing. *Indoor Air*, **20**, 473-485. http://dx.doi.org/10.1111/j.1600-0668.2010.00669.x

130. Fisk, W.J., 2015: Review of some effects of climate change on indoor environmental quality and health and associated no-regrets mitigation measures. *Building and Environment*, **86**, 70-80. http://dx.doi.org/10.1016/j.buildenv.2014.12.024

131. Ilacqua, V., J. Dawson, M. Breen, S. Singer, and A. Berg, 2015: Effects of climate change on residential infiltration and air pollution exposure. *Journal of Exposure Science and Environmental Epidemiology*, Published online 27 May 2015. http://dx.doi.org/10.1038/jes.2015.38

132. Weschler, C.J., 2006: Ozone's impact on public health: Contributions from indoor exposures to ozone and products of ozone-initiated chemistry. *Environmental Health Perspectives*, **114**, 1489-1496. PMC1626413

133. Bell, M.L., A. Zanobetti, and F. Dominici, 2014: Who is more affected by ozone pollution? A systematic review and meta-analysis. *American Journal of Epidemiology*, **180**, 15-28. http://dx.doi.org/10.1093/aje/kwu115

134. Lin, S., X. Liu, L.H. Le, and S.-A. Hwang, 2008: Chronic exposure to ambient ozone and asthma hospital admissions among children. *Environmental Health Perspectives*, **116**, 1725-1730. http://dx.doi.org/10.1289/ehp.11184

135. Nazaroff, W.W., 2013: Exploring the consequences of climate change for indoor air quality. *Environmental Research Letters*, **8**. http://dx.doi.org/10.1088/1748-9326/8/1/015022

136. Bowers, R.M., N. Clements, J.B. Emerson, C. Wiedinmyer, M.P. Hannigan, and N. Fierer, 2013: Seasonal variability in bacterial and fungal diversity of the near-surface atmosphere. *Environmental Science & Technology*, **47**, 12097-12106. http://dx.doi.org/10.1021/es402970s

137. Hardin, B.D., B.J. Kelman, and A. Saxon, 2003: Adverse human health effects associated with molds in the indoor environment. *Journal of Occupational and Environmental Medicine*, **45**, 470-478. http://dx.doi.org/10.1097/00043764-200305000-00006

138. Walzer, P.D., 2013: The ecology of pneumocystis: Perspectives, personal recollections, and future research opportunities. *The Journal of Eukaryotic Microbiology*, **60**, 634-645. http://dx.doi.org/10.1111/jeu.12072

139. Parr, A., E.A. Whitney, and R.L. Berkelman, 2014: Legionellosis on the rise: A review of guidelines for prevention in the United States. *Journal of Public Health Management and Practice*, **21**, E17-E26. http://dx.doi.org/10.1097/phh.0000000000000123

140. Decker, B.K. and T.N. Palmore, 2014: Hospital water and opportunities for infection prevention. *Curr Infect Dis Rep*, **16**, 432. http://dx.doi.org/10.1007/s11908-014-0432-y

141. Cunha, B.A., A. Burillo, and E. Bouza, 2015: Legionnaires' disease. *Lancet*, Published online 28 July 2015. http://dx.doi.org/10.1016/S0140-6736(15)60078-2

142. Hines, S.A., D.J. Chappie, R.A. Lordo, B.D. Miller, R.J. Janke, H.A. Lindquist, K.R. Fox, H.S. Ernst, and S.C. Taft, 2014: Assessment of relative potential for *Legionella* species or surrogates inhalation exposure from common water uses. *Water Research*, **56**, 203-213. http://dx.doi.org/10.1016/j.watres.2014.02.013

143. Whiley, H., S. Giglio, and R. Bentham, 2015: Opportunistic pathogens Mycobacterium Avium Complex (MAC) and *Legionella* spp. colonise model shower *Pathogens*, **4**, 590-598. http://dx.doi.org/10.3390/pathogens4030590

144. Phin, N., F. Parry-Ford, T. Harrison, H.R. Stagg, N. Zhang, K. Kumar, O. Lortholary, A. Zumla, and I. Abubakar, 2014: Epidemiology and clinical management of Legionnaires' disease. *The Lancet Infectious Diseases*, **14**, 1011-1021. http://dx.doi.org/10.1016/s1473-3099(14)70713-3

145. Falkinham, J.O., III,, E.D. Hilborn, M.J. Arduino, A. Pruden, and M.A. Edwards, 2015: Epidemiology and ecology of opportunistic premise plumbing pathogens: *Legionella pneumophila, Mycobacterium avium,* and *Pseudomonas aeruginosa*. *Environmental Health Perspectives*, **123**, 749-758. http://dx.doi.org/10.1289/ehp.1408692

146. Halsby, K.D., C.A. Joseph, J.V. Lee, and P. Wilkinson, 2014: The relationship between meteorological variables and sporadic cases of Legionnaires' disease in residents of England and Wales. *Epidemiology and Infection*, **142**, 2352-2359. http://dx.doi.org/10.1017/S0950268813003294

147. Cunha, B.A., J. Connolly, and E. Abruzzo, 2015: Increase in pre-seasonal community-acquired Legionnaire's disease due to increased precipitation. *Clinical Microbiology and Infection*, **21**, e45-e46. http://dx.doi.org/10.1016/j.cmi.2015.02.015

148. Hicks, L.A., L.E. Garrison, G.E. Nelson, and L.M. Hampton, 2011: Legionellosis---United States, 2000-2009. *MMWR. Morbidity and Mortality Weekly Report*, **60**, 1083-1086. http://www.cdc.gov/mmwr/preview/mmwrhtml/mm6032a3.htm

149. Farnham, A., L. Alleyne, D. Cimini, and S. Balter, 2014: Legionnaires' disease incidence and risk factors, New York, New York, USA, 2001-2011. *Emerging Infectious Diseases*, **20**, 1795-1802. http://dx.doi.org/10.3201/eid2011.131872

150. Johanning, E., P. Auger, P.R. Morey, C.S. Yang, and E. Olmsted, 2014: Review of health hazards and prevention measures for response and recovery workers and volunteers after natural disasters, flooding, and water damage: Mold and dampness. *Environmental Health and Preventive Medicine*, **19**, 93-99. http://dx.doi.org/10.1007/s12199-013-0368-0

151. Mudarri, D. and W.J. Fisk, 2007: Public health and economic impact of dampness and mold. *Indoor Air*, **17**, 226-235. http://dx.doi.org/10.1111/j.1600-0668.2007.00474.x

152. Fisk, W.J., E.A. Eliseeva, and M.J. Mendell, 2010: Association of residential dampness and mold with respiratory tract infections and bronchitis: A meta-analysis. *Environmental Health*, **9**, Article 72. http://dx.doi.org/10.1186/1476-069x-9-72

153. Seltenrich, N., 2012: Healthier tribal housing: Combining the best of old and new. *Environmental Health Perspectives*, **120**, A460-A469. http://dx.doi.org/10.1289/ehp.120-a460

154. Parthasarathy, S., R.L. Maddalena, M.L. Russell, and M.G. Apte, 2011: Effect of temperature and humidity on formaldehyde emissions in temporary housing units. *Journal of the Air & Waste Management Association*, **61**, 689-695. http://dx.doi.org/10.3155/1047-3289.61.6.689

155. Norbäck, D., G. Wieslander, K. Nordström, and R. Wålinder, 2000: Asthma symptoms in relation to measured building dampness in upper concrete floor construction, and 2-ethyl-1-hexanol in indoor air. *International Journal of Tuberculosis and Lung Disease*, **4**, 1016-1025. http://www.nchh.org/portals/0/contents/article0877.pdf

156. Markowicz, P. and L. Larsson, 2015: Influence of relative humidity on VOC concentrations in indoor air. *Environmental Science and Pollution Research*, **22**, 5772-5779. http://dx.doi.org/10.1007/s11356-014-3678-x

157. Waite, T., V. Murray, and D. Baker, 2014: Carbon monoxide poisoning and flooding: Changes in risk before, during and after flooding require appropriate public health interventions. *Plos Currents Disasters,* July 3, Edition 1. http://dx.doi.org/10.1371/currents.dis.2b2eb9e-15f9b982784938803584487f1

158. Bronstein, A., J.H. Clower, S. Iqbal, F.Y. Yip, C.A. Martin, A. Chang, A.F. Wolkin, and J. Bell, 2011: Carbon monoxide exposures---United States, 2000--2009. *MMWR. Morbidity and Mortality Weekly Report,* **60,** 1014-1017. http://www.cdc.gov/mmwr/preview/mmwrhtml/mm6030a2.htm

159. Watson, D.C., M. Sargianou, A. Papa, P. Chra, I. Starakis, and G. Panos, 2014: Epidemiology of Hantavirus infections in humans: A comprehensive, global overview. *Critical Reviews in Microbiology,* **40,** 261-272. http://dx.doi.org/10.3109/1040841x.2013.783555

160. Jonsson, C.B., L.T.M. Figueiredo, and O. Vapalahti, 2010: A global perspective on hantavirus ecology, epidemiology, and disease. *Clinical Microbiology Reviews,* **23,** 412-441. http://dx.doi.org/10.1128/cmr.00062-09

161. Dearing, M.D. and L. Dizney, 2010: Ecology of hantavirus in a changing world. *Annals of the New York Academy of Sciences,* **1195,** 99-112. http://dx.doi.org/10.1111/j.1749-6632.2010.05452.x

162. Bezirtzoglou, C., K. Dekas, and E. Charvalos, 2011: Climate changes, environment and infection: Facts, scenarios and growing awareness from the public health community within Europe. *Anaerobe,* **17,** 337-340. http://dx.doi.org/10.1016/j.anaerobe.2011.05.016

163. Klein, S.L. and C.H. Calisher, 2007: Emergence and persistence of hantaviruses. *Wildlife and Emerging Zoonotic Diseases: The Biology, Circumstances and Consequences of Cross-Species Transmission.* Childs, J.E., J.S. Mackenzie, and J.A. Richt, Eds. Springer-Verlag, Berlin, 217-252. http://dx.doi.org/10.1007/978-3-540-70962-6_10

164. Reusken, C. and P. Heyman, 2013: Factors driving hantavirus emergence in Europe. *Current Opinion in Virology,* **3,** 92-99. http://dx.doi.org/10.1016/j.coviro.2013.01.002

165. Medina-Ramón, M. and J. Schwartz, 2007: Temperature, temperature extremes, and mortality: A study of acclimatisation and effect modification in 50 US cities. *Occupational and Environmental Medicine,* **64,** 827-833. http://dx.doi.org/10.1136/oem.2007.033175

166. Akinbami, L.J., J.E. Moorman, C. Bailey, H.S. Zahran, M. King, C.A. Johnson, and X. Liu, 2012: Trends in Asthma Prevalence, Health Care Use, and Mortality in the United States, 2001–2010. NCHS Data Brief No. 94, May 2012, 8 pp. National Center for Health Statistics, Hyattsville, MD. http://www.cdc.gov/nchs/data/databriefs/db94.pdf

167. Adamkiewicz, G., A.R. Zota, P. Fabian, T. Chahine, R. Julien, J.D. Spengler, and J.I. Levy, 2011: Moving environmental justice indoors: Understanding structural influences on residential exposure patterns in low-income communities. *American Journal of Public Health,* **101,** S238-S245. http://dx.doi.org/10.2105/AJPH.2011.300119

168. Kitch, B.T., G. Chew, H.A. Burge, M.L. Muilenberg, S.T. Weiss, T.A. Platts-Mills, G. O'Connor, and D.R. Gold, 2000: Socioeconomic predictors of high allergen levels in homes in the greater Boston area. *Environmental Health Perspectives,* **108,** 301-307. PMC1638021

169. Bloom, B., L.I. Jones, and G. Freeman, 2013: Summary Health Statistics for U.S. Children: National Health Interview Survey, 2012. Vital and Health Statistics 10(258), 73 pp. National Center for Health Statistics, Hyattsville, MD. http://www.cdc.gov/nchs/data/series/sr_10/sr10_258.pdf

170. Gelfand, E.W., 2009: Pediatric asthma: A different disease. *Proceedings of the American Thoracic Society,* **6,** 278-282. http://dx.doi.org/10.1513/pats.200808-090RM

171. DellaValle, C.T., E.W. Triche, B.P. Leaderer, and M.L. Bell, 2012: Effects of ambient pollen concentrations on frequency and severity of asthma symptoms among asthmatic children. *Epidemiology,* **23,** 55-63. http://dx.doi.org/10.1097/EDE.0b013e31823b66b8

172. Ebi, K.L. and J.A. Paulson, 2007: Climate change and children. *Pediatric Clinics of North America,* **54,** 213-226. http://dx.doi.org/10.1016/j.pcl.2007.01.004

173. Schmier, J.K. and K.L. Ebi, 2009: The impact of climate change and aeroallergens on children's health. *Allergy and Asthma Proceedings,* **30,** 229-237. http://dx.doi.org/10.2500/aap.2009.30.3229

174. Moorman, J.E., L.J. Akinbami, C.M. Bailey, H.S. Zahran, M.E. King, C.A. Johnson, and X. Liu, 2012: National Surveillance of Asthma: United States, 2001-2010. Vital and Health Statistics 3(35). National Center for Health Statistics. http://www.cdc.gov/nchs/data/series/sr_03/sr03_035.pdf

175. Meng, Y.Y., S.H. Babey, T.A. Hastert, and E.R. Brown, 2007: California's racial and ethnic minorities more adversely affected by asthma. *Policy Brief UCLA Center for Health Policy Research* 1-7. https://escholarship.org/uc/item/4k45v3xt

176. Brim, S.N., R.A. Rudd, R.H. Funk, and D.B. Callahan, 2008: Asthma prevalence among US children in under-represented minority populations: American Indian/Alaska Native, Chinese, Filipino, and Asian Indian. *Pediatrics,* **122,** e217-e222. http://dx.doi.org/10.1542/peds.2007-3825

177. Crocker, D., C. Brown, R. Moolenaar, J. Moorman, C. Bailey, D. Mannino, and F. Holguin, 2009: Racial and ethnic disparities in asthma medication usage and health-care utilization: Data from the National Asthma Survey. *Chest,* **136,** 1063-1071. http://dx.doi.org/10.1378/chest.09-0013

178. Miller, J.E., 2000: The effects of race/ethnicity and income on early childhood asthma prevalence and health care use. *American Journal of Public Health,* **90,** 428-430. Pmc1446167

179. Peel, J.L., K.B. Metzger, M. Klein, W.D. Flanders, J.A. Mulholland, and P.E. Tolbert, 2007: Ambient air pollution and cardiovascular emergency department visits in potentially sensitive groups. *American Journal of Epidemiology,* **165,** 625-633. http://dx.doi.org/10.1093/aje/kwk051

4 IMPACTS OF EXTREME EVENTS ON HUMAN HEALTH

Lead Authors

Jesse E. Bell
Cooperative Institute for Climate and Satellites–North Carolina

Stephanie C. Herring
National Oceanic and Atmospheric Administration

Lesley Jantarasami*
U.S. Environmental Protection Agency

Contributing Authors

Carl Adrianopoli
U.S. Department of Health and Human Services

Kaitlin Benedict
Centers for Disease Control and Prevention

Kathryn Conlon
Centers for Disease Control and Prevention

Vanessa Escobar
National Aeronautics and Space Administration

Jeremy Hess
University of Washington

Jeffrey Luvall
National Aeronautics and Space Administration

Carlos Perez Garcia-Pando
Columbia University

Dale Quattrochi
National Aeronautics and Space Administration

Jennifer Runkle*
Cooperative Institute for Climate and Satellites–North Carolina

Carl J. Schreck, III
Cooperative Institute for Climate and Satellites–North Carolina

Acknowledgements: Mark Keim, formerly of the Centers for Disease Control and Prevention; **Andrea Maguire***, U.S. Environmental Protection Agency

Recommended Citation: Bell, J.E., S.C. Herring, L. Jantarasami, C. Adrianopoli, K. Benedict, K. Conlon, V. Escobar, J. Hess, J. Luvall, C.P. Garcia-Pando, D. Quattrochi, J. Runkle, and C.J. Schreck, III, 2016: Ch. 4: Impacts of Extreme Events on Human Health. *The Impacts of Climate Change on Human Health in the United States: A Scientific Assessment.* U.S. Global Change Research Program, Washington, DC, 99–128. http://dx.doi.org/10.7930/J0BZ63ZV

On the web: health2016.globalchange.gov   *Chapter Coordinators

4 IMPACTS OF EXTREME EVENTS ON HUMAN HEALTH

Key Findings

Increased Exposure to Extreme Events

Key Finding 1: Health impacts associated with climate-related changes in exposure to extreme events include death, injury, or illness; exacerbation of underlying medical conditions; and adverse effects on mental health *[High Confidence]*. Climate change will increase exposure risk in some regions of the United States due to projected increases in the frequency and/or intensity of drought, wildfires, and flooding related to extreme precipitation and hurricanes *[Medium Confidence]*.

Disruption of Essential Infrastructure

Key Finding 2: Many types of extreme events related to climate change cause disruption of infrastructure, including power, water, transportation, and communication systems, that are essential to maintaining access to health care and emergency response services and safeguarding human health *[High Confidence]*.

Vulnerability to Coastal Flooding

Key Finding 3: Coastal populations with greater vulnerability to health impacts from coastal flooding include persons with disabilities or other access and functional needs, certain populations of color, older adults, pregnant women and children, low-income populations, and some occupational groups *[High Confidence]*. Climate change will increase exposure risk to coastal flooding due to increases in extreme precipitation and in hurricane intensity and rainfall rates, as well as sea level rise and the resulting increases in storm surge *[High Confidence]*.

4.1 Introduction

Some regions of the United States have already experienced costly impacts—in terms of both lives lost and economic damages—from observed changes in the frequency, intensity, or duration of certain extreme events (Figure 1). Climate change projections show that there will be continuing increases in the occurrence and severity of some extreme events by the end of the century, while for other extremes the links to climate change are more uncertain (Table 1). (See also Ch. 1: Introduction)

Four categories of extreme events with important health impacts in the United States are addressed in this chapter: 1) flooding related to extreme precipitation, hurricanes, and coastal storms, 2) droughts, 3) wildfires, and 4) winter storms and severe thunderstorms. The health impacts of extreme heat and extreme cold are discussed in Chapter 2: Temperature-Related Death and Illness. For each event type, the chapter integrates discussion of populations of concern that have greater vulnerability to adverse health outcomes. The air quality impacts of wildfires are discussed below and also in Chapter 3: Air Quality Impacts. Although mental health effects are noted briefly here and in later sections of this chapter, in-depth discussion of the impacts of extreme events on mental health is presented in Chapter 8: Mental Health.

While it is intuitive that extremes can have health impacts such as death or injury during an event (for example, drowning during floods), health impacts can also occur before or after an extreme event as individuals may be involved in activities that put their health at risk, such as disaster preparation and post-event cleanup.[1] Health risks may also arise long after the event, or in places outside the area where the event took place, as a result of damage to property, destruction of assets, loss of infrastructure and public services, social and economic impacts, environmental degradation, and other factors. Extreme events also pose unique health risks if multiple events occur simultaneously or in succession in a given location, but these issues of cumulative or compounding impacts are still emerging in the literature (see Front Matter and Ch. 1: Introduction).

Dynamic interactions between extreme events, their physical impacts, and population vulnerability and response can make it difficult to quantitatively measure all the health impacts that may be associated with an extreme event type, particularly those that are distributed over longer periods of time (See "Emerging Issues," Section 4.8). These complexities make it difficult to integrate human health outcomes into climate impact models, and thus projections of future health burdens due to extreme events under climate change are not available

Estimated Deaths and Billion Dollar Losses from Extreme Events in the United States 2004–2013

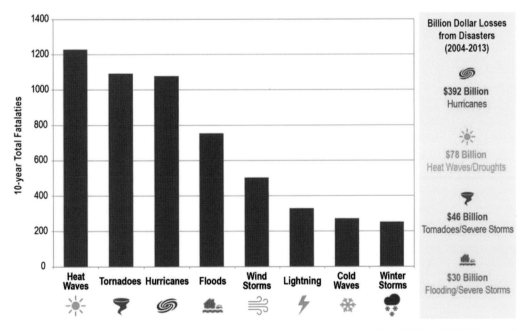

Figure 1: This figure provides 10-year estimates of fatalities related to extreme events from 2004 to 2013,[204] as well as estimated economic damages from 58 weather and climate disaster events with losses exceeding $1 billion (see Smith and Katz 2013 to understand how total losses were calculated).[205] These statistics are indicative of the human and economic costs of extreme weather events over this time period. Climate change will alter the frequency, intensity, and geographic distribution of some of these extremes,[2] which has consequences for exposure to health risks from extreme events. Trends and future projections for some extremes, including tornadoes, lightning, and wind storms are still uncertain.

Table 1: Health Impacts of Extreme Events

Event Type	Example Health Risks and Impacts (not a comprehensive list)	Observed and Projected Impacts of Climate Change on Extreme Events from 2014 NCA[2]
Flooding Related to Extreme Precipitation, Hurricanes, Coastal Storms	• Traumatic injury and death (drowning) • Mental health impacts • Preterm birth and low birth weight • Infrastructure disruptions and post-event disease spread • Carbon monoxide poisoning related to power outages	Heavy downpours are increasing nationally, especially over the last three to five decades, with the largest increases in the Midwest and Northeast. Increases in the frequency and intensity of extreme precipitation events are projected for all U.S. regions. *[High Confidence]*. The intensity, frequency, and duration of North Atlantic hurricanes, as well as the frequency of the strongest hurricanes, have all increased since the 1980s *[High Confidence]*. Hurricane intensity and rainfall are projected to increase as the climate continues to warm *[Medium Confidence]*. Increasing severity and frequency of flooding have been observed throughout much of the Mississippi and Missouri River Basins. Increased flood frequency and severity are projected in the Northeast and Midwest regions *[Low Confidence]*. In the Western United States, increasing snowmelt and rain-on-snow events (increased runoff when rain falls onto existing snowpack) will increase flooding in some mountain watersheds *[Medium Confidence]*. In the next several decades, storm surges and high tides could combine with sea level rise and land subsidence to further increase coastal flooding in many regions. The U.S. East and Gulf Coasts, Hawaii, and the U.S.-affiliated Pacific Islands are particularly at risk.
Droughts	• Reduced water quality and quantity • Respiratory impacts related to reduced air quality • Mental health impacts	Over the last several decades, drought patterns and trends have been changing, but patterns vary regionally across the United States. Droughts in the Southwest are projected to become more intense *[High Confidence]*.
Wildfires	• Smoke inhalation • Burns and other traumatic injury • Asthma exacerbations • Mental health impacts	Increased warming, drought, and insect outbreaks, all caused by or linked to climate change, have increased wildfires and impacts to people and ecosystems in the Southwest *[High Confidence]*. Rising temperatures and hotter, drier summers are projected to increase the frequency and intensity of large wildfires, particularly in the western United States and Alaska.
Winter Storms & Severe Thunderstorms	• Traumatic injury and death • Carbon monoxide poisoning related to power outages • Hypothermia and frostbite • Mental health impacts	Winter storms have increased in frequency and intensity since the 1950s, and their tracks have shifted northward *[Medium Confidence]*. Future trends in severe storms, including the intensity and frequency of tornadoes, hail, and damaging thunderstorm winds, are uncertain and are being studied intensively *[Low Confidence]*.

Climate Change and Health—Flooding

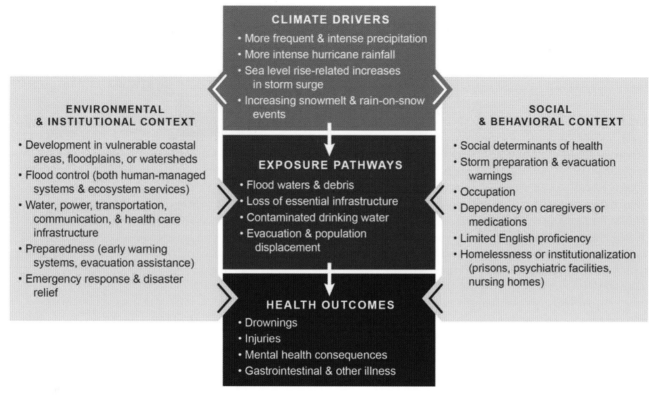

Figure 2: This conceptual diagram for a flooding event illustrates the key pathways by which humans are exposed to health threats from climate drivers, and potential resulting health outcomes (center boxes). These exposure pathways exist within the context of other factors that positively or negatively influence health outcomes (gray side boxes). Key factors that influence health outcomes and vulnerability for individuals are shown in the right box, and include social determinants of health and behavioral choices. Key factors that influence health outcomes and vulnerability at larger community or societal scales, such as natural and built environments, governance and management, and institutions, are shown in the left box. All of these influencing factors may also be affected by climate change. See Chapter 1: Introduction for more information.

in the literature. Instead, this chapter focuses on explaining the physical processes and pathways that scientists know contribute to human exposure and identifying overarching conclusions regarding the risk of adverse health impacts as a result of changing extreme weather and climate.

4.2 Complex Factors Determine Health Impacts

The severity and extent of health effects associated with extreme events depend on the physical impacts of the extreme events themselves as well as the unique human, societal, and environmental circumstances at the time and place where events occur. This complex set of factors can moderate or exacerbate health outcomes and vulnerability in the affected people and communities (Figure 2). Vulnerability is the tendency or predisposition to be adversely affected by climate-related health effects. It encompasses three elements—*exposure*, *sensitivity*, and *adaptive capacity*—that also interact with and are influenced by the social determinants of health (See Ch. 1: Introduction and Ch. 9: Populations of Concern for additional discussion and definitions of these terms.)

Exposure is contact between a person and one or more biological, psychosocial, chemical, or physical stressors, including stressors affected by climate change. Contact may occur in a single instance or repeatedly over time, and may occur in one location or over a wider geographic area. Demographic shifts and population migration may change exposure to public health impacts. For example, since 1970, coastal population growth (39%) has substantially increased compared to population growth for the United States as a whole (about 13%).[3] In the future, this coastal migration in conjunction with rising sea levels has the potential to result in increased vulnerability to storm surge events for a greater proportion of the U.S. population concentrated in these coastal areas. Choices by individuals and governments can reduce or increase some exposure risk to extreme events.[4] As shown in Figure 2, such choices can include whether to build or allow development in floodplains and coastal areas subject to extreme high tides and sea level rise. Individuals' responses to evacuation orders and other emergency warnings also affect their exposure to health threats. Factors such as income have been linked to how people perceive the risks to which they are exposed and choose to respond, as well as their ability to evacuate or relocate to a less risk-prone location.[5] The condition of the built environment also affects exposure to extreme events, and those living in low-quality, poorly maintained, or high-density housing may have greater risks of health impacts.[6]

Sensitivity is the degree to which people or communities are affected, either adversely or beneficially, by climate variability and change. It is determined, at least in part, by biologically based traits such as age. For example, older adults (generally defined as age 65 and older) are physiologically more sensitive to health impacts from extreme events because of normal aging processes; they are generally more frail, more likely to have chronic medical conditions that make them more dependent on medications, and require more assistance in activities of daily living.[7, 8] In addition, social determinants of health affect disparities in the prevalence of medical conditions that contribute to biological sensitivity.[9, 10] Health disparities are more prevalent in low-income populations, as well as in some communities of color, and are frequently exacerbated during extreme events.[11] For example, Black or African American populations have higher rates of chronic conditions such as asthma, decreased lung function, and cardiovascular issues, all of which are known to increase sensitivity to health effects of smoke from wildfires (Ch. 3: Air Quality Impacts).[12]

Adaptive capacity is the ability of communities, institutions, or people to adjust to potential hazards, to take advantage of opportunities, or to respond to consequences. Having strong adaptive capacity contributes to resilience—the ability to prepare and plan for, absorb, recover from, and more successfully adapt to adverse events.[13] In the context of extreme events, people with low adaptive capacity have difficulty responding, evacuating, or relocating when necessary, and recovering from event-related health impacts.

For individuals, health outcomes are strongly influenced by the social determinants of health that affect a person's adaptive capacity. Poverty is a key risk factor, and the poor are disproportionately affected by extreme events.[4, 9, 14] Low-income individuals may have fewer financial resources and social capital (such as human networks and relationships) to help them prepare for, respond to, and recover from an extreme event.[15, 16] In many urban, low-income neighborhoods, adaptive capacity is reduced where physical and social constructs, such as community infrastructure, neighborhood cohesion, and social patterns, promote social isolation.[17, 18, 19] Those with higher income possess a much higher level of resilience and availability of resources to increase their adaptive capacity.[20, 21] Other attributes of individuals that contribute to lower adaptive capacity include their age (very young or very old) and associated dependency on caregivers,

Family affected by Hurricane Sandy prepares to take shelter in Morristown, New Jersey, October 31, 2012.

disabilities such as mobility or cognitive impairments, having specific access and functional needs, medical or chemical dependence, limited English proficiency, social or cultural isolation, homelessness, and institutionalization (prisons, psychiatric facilities, nursing homes).[1, 8, 22]

At a larger community or societal level, adaptive capacity is heavily influenced by governance, management, and institutions.[23] Governments and non-governmental organizations provide essential extreme-event preparedness, coordination, emergency response, and recovery functions that increase adaptive capacity at the local, state, tribal, and federal levels—for example, in providing early warning systems where possible, evacuation assistance, and disaster relief.[13, 24] Risk sharing, management, and recovery schemes such as insurance can also play a significant role in building resilience in the context of extreme events and climate change.[25, 26] For instance, lack of health insurance has been associated with greater risk of hospital admission after exposure to certain weather events.[27] Public health actions or interventions that maintain or strengthen the adaptive capacity of communities, institutions, or people could help mediate certain health impacts due to extreme events.[28] On the other hand, climate change—particularly its effect on extreme events—has the potential to create unanticipated public health stressors that could overwhelm the U.S. public health system's adaptive capacity and could require new approaches.[28]

4.3 Disruption of Essential Infrastructure

When essential infrastructure and related services are disrupted during and after an extreme event, a population's exposure to health hazards can increase, and losses related to the event can reduce adaptive capacity.[4] Disruptions can include reduced functionality, such as poor road conditions that limit travel, or complete loss of infrastructure, such as roads and bridges being washed away. Serious health risks can arise from infrastructure and housing damage and disruption or loss of access to electricity, sanitation, safe food and water supplies, health care, com-

> *Having strong adaptive capacity contributes to resilience—the ability to prepare and plan for, absorb, recover from, and more successfully adapt to adverse events.*

munication, and transportation.[1, 29, 30, 31, 32] Identifying vulnerable infrastructure and investing in strategies to reduce vulnerability, including redundancy (having additional or alternate systems in place as backup) and ensuring a certain standard of condition and performance can reduce the likelihood of significant adverse impacts to infrastructure from extreme weather events.[33]

Health Risks Related to Infrastructure

Existing infrastructure is generally designed to perform at its engineered capacity assuming historical weather patterns, and these systems could be more vulnerable to failure in response to weather-related stressors under future climate scenarios.[4, 34, 35] Shifts in the frequency or intensity of extreme events outside their historical range pose infrastructure risks, which may be compounded by the fact that much of the existing critical infrastructure in the United States, like water and sewage systems, roads, bridges, and power plants, are aging and in need of repair or replacement.[4, 36] For example, the 2013 American Society of Civil Engineer's Report Card assigned a letter grade of D+ to the condition and performance of the Nation's infrastructure.[37]

In addition, recurrent weather-related stressors, such as "nuisance flooding" (frequent coastal flooding that is increasing in frequency due to sea level rise), contribute to overall deterioration of infrastructure like stormwater drainage systems and roads (see Ch. 6: Water-Related Illness).[38] These systems are important in the context of health because drainage helps to avoid sewage overflows and maintain water quality,[39] and roads are vital for evacuations and emergency response during and after extreme events.[40]

Energy infrastructure that relies on environmental inputs, such as water for cooling in power generation or for hydroelectric dams, is also vulnerable to changes in extreme events due to climate change.[34, 41] Power generation accounts for one of the largest withdrawals of freshwater in the United States.[42] Longer or more intense droughts that are projected for some regions of the United States (see Table 1) will contribute to reduced energy production in those regions, which may lead to supply interruptions of varying lengths and magnitudes and adverse impacts to other infrastructure that depends on energy supply.[34]

Power Outages

Electricity is fundamental to much modern infrastructure, and power outages are commonly associated with the types of extreme events highlighted in this chapter.[43] During power outages, observed health impacts include increased deaths from accidental and natural causes,[44] increased cases of foodborne diarrheal illness from consuming food spoiled by lack of refrigeration (see Ch. 7: Food Safety),[1] and increased rates of hospitalization.[45] In addition, extreme-event-related power outages are associated with increased injuries and deaths

from carbon monoxide poisoning after floods, hurricanes, severe winter storms, and ice storms.[1, 31, 46, 47, 48, 49] This is due to increased use of gasoline-powered generators, charcoal grills, and kerosene and propane heaters or stoves inside the home or other areas without proper ventilation (see also Ch. 3: Air Quality Impacts). Populations considered especially vulnerable to the health impacts of power outages include older adults, young children, those reliant on electrically powered medical equipment like ventilators and oxygen, those with preexisting health conditions, and those with disabilities (see Ch. 9: Populations of Concern).[1, 43, 44] In rural communities, power and communications can take longer to restore after damage from an extreme event.[50]

Transportation, Communication, and Access

Damage to transportation infrastructure or difficult road conditions may delay first responders, potentially delaying treatment of acute injuries and requiring more serious intervention or hospitalization.[40] Extreme events can disrupt access to health care services via damage to or loss of transportation infrastructure, evacuation, and population displacement.[32] For chronically ill people, treatment interruptions and lack of access to medication can exacerbate health conditions both during and after the extreme event.[1, 51] Surveys of patients after Hurricane Katrina showed that those with cancer, hypertension, kidney disease requiring dialysis, cardiovascular disease, and respiratory illnesses were particularly affected.[51, 52, 53] Evacuations also pose health risks to older adults—especially those who are frail, medically incapacitated, or residing in nursing or assisted living facilities—and may be complicated by the need for concurrent transfer of medical records, medications, and medical equipment.[1, 54] Some individuals with disabilities may also be disproportionally affected during evacuations if they are unable to access evacuation routes, have difficulty in understanding or receiving warnings of impending danger, or have limited ability to communicate their needs.[55]

Power lines damaged by Hurricane Isaac's wind and surge in Plaquemines Parish, Louisiana, September 3, 2012.

In addition, persons with limited English proficiency are less likely to understand or have timely access to emergency information, which may lead to delayed evacuation.[56, 57] Health risks increase if evacuation is delayed until after a storm hits; loss of power and damage to communications and transportation infrastructure can hinder health system operations.[1]

Water Infrastructure

Extreme precipitation events and storms can overwhelm or damage stormwater and wastewater treatment infrastructure, increasing the risk of exposure to contaminated water (see Ch. 6: Water-Related Illness). Risk of post-flood gastrointestinal illness outbreaks are considered to be low in the United States, but risk increases for displaced populations—especially young children and infants with immature immune systems—where shelter conditions are crowded or have poor sanitation.[1, 29] There is potential for post-flood mold and fungi growth inside houses to worsen allergic and asthmatic symptoms, but these types of health impacts have not been documented following floods or storms.[1, 29, 58, 59] In general, however, adverse health effects from dampness and mold in homes are well known and studied.[60, 61, 62]

Cascading Failures

Many infrastructure systems are reliant on one another, and disruption or failure of one system or at any place in the system can lead to the disruption of interconnected systems—a phenomenon referred to as a *cascading failure*. For example, electricity is essential to multiple systems, and a failure in the electrical grid can have cascading effects on water and sewage treatment, transportation, and health care systems.[36, 43] Extreme events can simultaneously strain single or multiple components of interconnected infrastructure and related facilities and equipment, which increases the risk of cascading infrastructure failure.[63, 64] This risk to interconnected systems has been particularly notable in the context of urban areas (especially cities for which the design or maintenance of critical infrastructure needs improvement) and industrial sites containing chemicals or hazardous materials that rely on specific equipment—such as holding tanks, pipelines, and electricity-dependent safety mechanisms like automatic shut-off valves—to prevent releases.[4, 65, 66] Dramatic infrastructure system failures are rare, but such cascading failures can lead to public health consequences when they do occur, including shifts in disease incidence.[67]

The 2003 blackout in the northeastern United States, caused indirectly by surging electrical demand during a heat wave, is an illustrative example of how climate change could introduce or exacerbate health threats from cascading infrastructure failures related to extreme weather. During this 31-hour event, lack of electricity compromised traffic control, health care and emergency services, wastewater treatment, solid waste collection, and a host of other critical infrastructure operations.[68, 69, 70, 71] New York City health officials responded to failure of

hospital emergency generators and interruptions in electrically powered medical equipment, contamination of recreational water and beaches with untreated sewage, pest control issues, and loss of refrigeration leading to potential impacts on food and vaccine spoilage.[72] Increased incidence of gastrointestinal illness from contaminated food or water, and a large increase in accidental and non-accidental deaths and hospitalizations in New York City were attributed to the blackout.[44, 45, 72] See Chapter 6: Water-Related Illness for other examples of health impacts when interconnected wastewater, stormwater, and drinking water infrastructure fails, such as during the 1993 Milwaukee *Cryptosporidium* outbreak.

Flood Terminology

Coastal floods – predominately caused by storm surges that are exacerbated by sea level rise. Coastal floods can destroy buildings and infrastructure, cause severe coastal erosion, and submerge large areas of the coast.

Flash and urban floods – occur in smaller inland natural or urban watersheds and are closely tied to heavy rainfall. Flash floods develop within minutes or hours after a rainfall event, and can result in severe damage and loss of life due to high water velocity, heavy debris load, and limited warning.

River floods – occur in large watersheds like the Mississippi and Missouri River Basins. River floods depend on many factors including precipitation, preexisting soil moisture conditions, river basin topography, and human factors like land-use change and flood control infrastructure (dams, levees).

Adapted from Georgakakos et al. (2014).[75]

4.4 Flooding Related to Extreme Precipitation, Hurricanes, and Coastal Storms

Floods are the primary health hazard associated with extreme precipitation events, hurricanes, and coastal storms. Risk of exposure to floods varies by region in the United States and by type of flooding that occurs in that location (see Table 1 and "Flood Terminology"). People in flood-prone regions are expected to be at greater risk of exposure to flood hazards due to climate change (Table 1),[9, 73, 74] which may result in various types of health impacts described below.

Most flood deaths in the United States are due to drowning associated with flash flooding.[1, 29, 58] The majority of these deaths are associated with becoming stranded or swept away when driving or walking near or through floodwaters.[58, 76, 77, 78] Flash floods in the United States occurred more frequently from 2006 to 2012 and were associated with more deaths and injuries in rural areas compared to urban areas.[78] Contributing factors include the following: 1) small, rural basins develop flash flood conditions much more quickly, providing less time to notify rural residents with emergency procedures like warnings, road closures, and evacuations; 2) more rural roads intersect low-water crossings without bridge infrastructure and rural areas have fewer alternative transportation options when roads are closed; and 3) rural areas have fewer emergency response units and slower response times.[78] Although flash floods are less frequent in urban areas, a single urban event is likely to result in more deaths and injuries than a rural event.[78]

Drowning in floodwaters was the leading cause of death (estimated 2,544 persons) among people directly exposed to hazards associated with hurricanes and coastal storms from 1963 to 2012.[79] Hurricanes are typically associated with coastal flooding, but they can also cause substantial inland flooding before, during, and after landfall, even when far from the storm's center (Figure 3).[80, 81, 82]

Hurricane-Induced Flood Effects in Eastern and Central United States

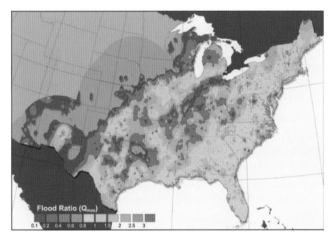

Figure 3: Composite map of floods associated with landfalling hurricanes over the past 31 years, based on stream gauge data. The Flood Ratio (Q) refers to maximum hurricane-related flood peaks compared to 10-year flood peaks (expected to occur, on average, once every 10 years and corresponds to the 90th percentile of the flood peak distribution) calculated for the same area. See Villarini et al. 2014 for a detailed description of how Q values are calculated.[80]

Q values between 0.6 and 1 (light blue and yellow) generally indicate minor to moderate flooding, while values above 1 (orange and red) generally indicate major flooding larger than the 10-year flood peak. The dark gray areas of the map represent the extent of the 500-km buffer around the center of circulation of the hurricanes included during the study period (the light gray areas of the map fall outside of the study area).

Figure 3 shows that hurricanes are important contributors to flooding in the eastern United States, as well as large areas of the central United States. Land use/land cover properties and soil moisture conditions are also important factors for flooding. (Figure source: adapted from Villarini et al. 2014)[80]

A truck gets stuck in the storm surge covering Highway 90 in Gulfport, Mississippi, during Hurricane Isaac, August 29, 2012.

The deadliest U.S. storms of this century to date were Hurricane Katrina and Superstorm Sandy. Katrina was a very large and powerful Category 3 storm that hit the Gulf Coast region in 2005. Hurricane Katrina was responsible for almost half of the hurricane-related deaths over the past 50 years,[79] with the majority of deaths directly related to the storm in Louisiana (an estimated 971 to 1,300 deaths) due to drowning or flood-related physical trauma due to the failure of the levees in New Orleans.[83, 84] Sandy was a historically rare storm that affected a large portion of the country in October 2012, with particularly significant human health and infrastructure impacts in New Jersey and the greater New York City area. Superstorm Sandy is estimated to have caused between 117 and 147 direct deaths across the Atlantic basin, also with drowning and flood-related physical trauma as the leading cause of death.[85, 86]

Both fatal and non-fatal flood-related injuries can occur in any phase of the event: before (preparation or evacuation), during, and after (cleanup and recovery). Common flood-related injuries include blunt trauma from falling debris or objects moving quickly in floodwater, electrocution, falls, and motor vehicle accidents from wet, damaged, or obstructed roads.[1, 29, 58] Other common, generally non-fatal injuries include cuts, puncture wounds, sprains/strains, burns, hypothermia, and animal bites.[1, 29, 58] Exposure to floodwaters or to contaminated drinking water can cause gastrointestinal illness; wound infections; skin irritations and infections; and eye, ear, nose, and throat infections.[1, 29] Many of these injuries have been observed in occupational settings [31] and in rural areas.[78]

In the United States, populations with greater vulnerability to flood-related injuries and illnesses include older adults, the immunocompromised and others with existing illness (especially if dependent on routine medical treatments or drug prescriptions), certain racial/ethnic groups (Black and Hispanic or Latino), people with limited English proficiency, and people with lower socioeconomic status (especially if uninsured, unemployed, or living in poor-quality housing).[1, 73] Differences in exposure, sensitivity, and adaptive capacity lead to a dispropor-tionate number of flood-related fatalities among older adults, males, and some low-income communities of color.[29] For exam-ple, almost half of deaths from Hurricane Katrina were people over age 75, while for Superstorm Sandy almost half were over age 65.[1, 29] The Black adult mortality rate from Hurricane Katrina was 1.7 to 4 times higher than that of whites.[29, 84] Floods and storms also create numerous occupational health risks, with most storm-related fatalities associated with cleanup activities (44%), construction (26%), public utilities restoration (8%), and security/policing (6%).[1] First responders and other emergency workers face greater health and safety risks when working in conditions with infrastructure disruptions, communication interruptions, and social unrest or violence following floods and storms.[73, 87, 88]

Pregnant women and newborns are uniquely vulnerable to flood health hazards. Flood exposure was associated with adverse birth outcomes (preterm birth, low birth weight) after Hurricane Katrina and the 1997 floods in North Dakota.[89, 90] Floods and storms can also create conditions in which chil-dren can become separated from their parents or caregivers, which—particularly for children with disabilities or special health care needs—increases their vulnerability to a range of health threats, including death, injury, disease, psychological trauma, and abuse.[91, 92, 93] Flood-related mental health impacts are associated with direct and longer-term losses, social im-pacts, stress, and economic hardship.[1, 29, 58] Women, children, older adults, low-income populations, and those in poor

Farmer in drought-stressed peanut field in Georgia. Health implications of drought include contamination and depletion of water sources.

health, with prior mental health issues, or with weak social networks may be especially vulnerable to the mental health impacts of floods (Ch. 8: Mental Health).

4.5 Droughts

Drought may be linked to a broad set of health hazards, including wildfires, dust storms, extreme heat events, flash flooding, degraded air and water quality, and reduced water quantity.[74] Exposure risk to potential drought health hazards is expected to vary widely across the nation, depending on several localized variables, such as characteristics of the built environment, loss of livelihoods, local demand for water, and changes in ecosystems.[94, 95] Researching the health effects of drought poses unique challenges given multiple definitions of the beginning and end of a drought, and because health effects tend to accumulate over time. In addition, health im-pacts do not occur in isolation. For example, droughts intensify heat waves by reducing evaporative cooling,[2] further compli-cating efforts to attribute specific health outcomes to specific drought conditions.

A primary health implication of drought arises from the contamination and depletion of water sources,[95] but there are few studies documenting specific health consequences in the United States.[96] Drought in coastal areas can increase saltwater intrusion (the movement of ocean water into fresh groundwater), reducing the supply and quality of potable wa-ter.[97, 98, 99] In addition to reducing water quantity, drought can decrease water quality when low flow or stagnant conditions increase concentrations of pollutants or contaminants (such as chemicals and heavy metals) and when higher temperatures encourage pathogen growth.[95, 96, 100, 101, 102, 103] Heavy rain follow-ing drought can flush accumulated pathogens or contaminants into water bodies.[104, 105] Reduced surface and groundwater quality can increase risk of water-related illness as well as foodborne illness if pathogens or contaminants enter the food chain (see Ch. 6: Water-Related Illness and Ch. 7: Food Safety).

In some regions of the United States, drought has been associ-ated with increased incidence of West Nile virus disease.[106, 107, 108, 109, 110] Human exposure risk to West Nile virus may increase during drought conditions due to a higher prevalence of the virus in mosquito and bird populations as a result of closer contact between birds (virus hosts) and mosquitoes (vectors) as they congregate around remaining water sources (see Ch. 5: Vector-Borne Diseases) .[111] Primarily in the Southwest, droughts followed by periods of heavy rainfall have been associated with an increase in rodent populations.[112, 113, 114] This could lead to increased exposures to rodent allergens and rodent-borne diseases, such as hantavirus.[115, 116, 117]

Wind Erosion and Dust Storms

Drought may increase the potential for wind erosion to cause soil dust to become airborne, and there is evidence from past trends showing regional increases in dust activity due to drought cycles, but there is large uncertainty about future projections of climate impacts on frequency or intensity of dust storms.[119, 128, 129] Wind erosion can also be exacerbated by human activities that disturb the soil, including growing crops, livestock grazing, recreation and suburbanization, and water diversion for irrigation.[119, 128, 130] Major dust activity in the United States is centered in the Southwest, where sources are mostly natural, and the Great Plains,

April 14, 2013. Dust storm on Interstate Highway 10 California USA.

extending from Montana to southern Texas, where sources are mainly from human activities associated with land use, such as agriculture.[131] These are also regions where climate change is expected to affect drought patterns.[2]

In the United States, dust exposure has been linked to increased incidence in respiratory disease, including asthma, acute bronchitis, and pneumonia.[27, 132, 133] However, the dust characteristics (such as composition and particle size), exposure levels, and biological mechanisms responsible for the observed health effects of dust are not completely understood. In part, this is because observations are generally unavailable in areas where dust exposure is greatest, including drylands and agricultural areas.[122] Apart from illness, intense dust storms are also associated with impaired visibility, which can cause road traffic accidents resulting in injury and death.[134, 135]

"In the United States, dust exposure has been linked to increased incidence in respiratory disease, including asthma, acute bronchitis, and pneumonia."

Fungal Diseases and Climate Change

Fungi growth and dispersal are sensitive to changes in temperature, moisture, and wind.[136] Illnesses or allergic reactions related to fungal toxins and superficial or invasive fungal infections can cause serious illness, permanent disability, or death. People generally become infected by breathing in fungal spores directly from the environment or having spores enter the skin at sites of injury. Coccidioidomycosis, also called "Valley Fever," is an infection caused by *Coccidioides*, a fungus found mainly in the southwestern United States. Reports of these infections are on the rise.[137] The fungus appears to grow best in soil after heavy rainfall and then becomes airborne most effectively during hot, dry conditions.[138] Several studies in Arizona and California, where most reported cases in the United States occur, suggest that climate likely plays a role in seasonal and yearly infection patterns.[139, 140] Recently, Coccidioides was found in soil in south-central Washington, far north of where it was previously known to exist.[141] Climate factors such as drought and increased temperature may be contributing to Coccidioides' expanded geographic range.[142] Thus, more prolonged or intense droughts resulting from climate change could lead to improved conditions for the spread of Coccidioides.[143] Understanding the impact of climate change on fungal infections (such as Coccidioidomycosis, Crypotcoccos gattii, and Mucormycosis) would require comprehensive epidemiologic surveillance, better methods to detect disease-causing fungi in the environment, and ongoing multidisciplinary collaboration.

Drought conditions also tend to reduce air quality and exacerbate respiratory illness by way of several mechanisms associated with soil drying, loss of vegetation, airborne particulate matter, and the creation of conditions conducive for dust storms and wildfires.[118, 119] In addition, air pollutants such as soluble trace gases and particles remain suspended in the air when there is a lack of precipitation (see Ch. 3: Air Quality Impacts).[120] Inhalation of particles can irritate bronchial passages and lungs, resulting in exacerbated chronic respiratory illnesses.[95] The size of particles is directly linked to their potential health effects. Exposure to fine particles is associated with cardiovascular illness (for example, heart attacks and strokes) and premature death, and is likely associated with adverse respiratory effects.[121] There is greater uncertainty regarding the health effects of inhaling coarse particles (often found in soil dust), but some evidence indicates an association with premature death and cardiovascular and respiratory effects.[121, 122, 123]

Mental health issues have also been observed during drought periods through research primarily conducted in Australia (see also Ch. 8: Mental Health).[94] Rural areas, in particular, can experience a rise in mental health issues related to economic insecurity from drought.[94, 124, 125, 126, 127]

Projected Increases in Very Large Fires

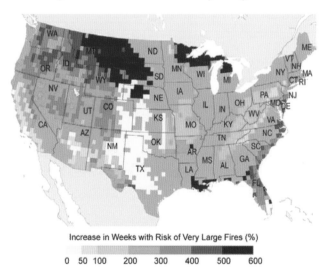

Increase in Weeks with Risk of Very Large Fires (%)

0 50 100 200 300 400 500 600

Figure 4: Based on 17 climate model simulations for the continental United States using a higher emissions pathway (RCP8.5), the map shows projected percentage increases in weeks with risk of very large fires by mid-century (2041–2070) compared to the recent past (1971–2000). The darkest shades of red indicated that up to a 6-fold increase in risk is projected for parts of the West. This area includes the Great Basin, Northern Rockies, and parts of Northern California. Gray represents areas within the continental United States where there is either no data or insufficient historical observations on very large fires to build robust models. The potential for very large fire events is also expected to increase along the southern coastline and in areas around the Great Lakes. (Figure source: adapted from Barbero et al. 2015 by NOAA)[206]

Exposure to smoke-related air pollutants from wildfires has been associated with a wide range of human health effects.

4.6 Wildfires

Climate change is projected to increase the frequency and intensity of large wildfires (Figure 4), with associated health risks projected to increase in many regions.[74, 144] Wildfire can have health impacts well beyond the perimeter of the fire. Populations near the fire or even thousands of miles downwind may be exposed to a complex smoke mixture containing various substances including carbon monoxide, ozone, toxic chemicals, and both fine and coarse particles,[145, 146] presenting a serious health risk for the exposed populations (see Ch. 3: Air Quality Impacts).[147, 148] For example, the 2002 forest fires in Quebec resulted in up to a 30-fold increase in airborne fine particulate concentrations in Baltimore, Maryland, a city nearly 1,000 miles downwind.[74] Exposure times can range from a few days to several weeks.[145, 149, 150]

Exposure to smoke-related air pollutants from wildfires has been associated with a wide range of human health effects, including early deaths and low infant birth weight, with the strongest evidence for acute respiratory illness.[145, 146, 151, 152, 153, 154, 155] Inhalation of smoke from wildfire has been linked to exacerbated respiratory problems, such as shortness of breath, asthma, and chronic obstructive pulmonary disease (COPD).[154, 156, 157, 158] While the association between smoke exposure and cardiovascular outcomes is uncertain,[154] exposure to fine particles contributes to risk of cardiovascular disease and premature death.[159, 160, 161, 162]

Wildfires can also affect indoor air quality for those living near affected areas by increasing particulate matter concentrations within homes, leading to many of the adverse health impacts already discussed.[149, 163] For example, during the 2007 San Diego wildfires, health monitoring showed excess emergency room visits for asthma, respiratory problems, chest pain, and COPD. During times of peak fire particulate matter concentrations, the odds of a person seeking emergency care increased by 50% when compared to non-fire conditions.[164] Smoke from wildfires can also impair driving visibility, increasing risks of motor vehicle deaths and injuries.[134, 165, 166, 167]

Freezing rain, snow, and ice have been linked to increased injuries associated with treacherous road conditions and impaired driving visibility.

Pregnant women, children, and the elderly are more sensitive to the harmful health effects of wildfire smoke exposure (see also Ch. 9: Populations of Concern).[12, 156, 168, 169] Firefighters are exposed to significantly higher levels and longer periods of exposure to combustion products from fires, leading to health risks that include decreased lung function, inflammation, and respiratory system problems, as well as injuries from burns and falling trees.[145, 168, 170, 171, 172, 173]

Wildfires can also create an increased burden on the health care system and public health infrastructure. For example, wildfires near populated areas often necessitate large evacuations, requiring extensive public health resources, including shelter, and treatment of individuals for injuries, smoke inhalation, and mental health impacts.[67, 166, 174, 175] Housing development in or near the wildland–urban interface has expanded over the last several decades and is expected to continue to expand.[176] These changing development patterns in combination with a changing climate are increasing the vulnerability of these areas to wildfires.[177, 178, 179]

Following wildfire, increased soil erosion rates and changes to runoff generation may contaminate water-supply reservoirs and disrupt downstream drinking water supplies.[180, 181] Post-wildfire erosion and runoff has been linked to increased flooding and debris flow hazards, depending on the severity of the fire, seasonal rainfall patterns, watershed characteristics, and the size of the burn area.[182, 183, 184, 185] Wildfires have a range of short- and long-term effects on watersheds that have the potential to change water quality, quantity, availability, and treatability downstream from the burned area.[186, 187, 188]

4.7 Winter Storms and Severe Thunderstorms

The primary health hazards of severe thunderstorms are from lightning and high winds, while the principal winter storm hazards include extreme cold temperatures (see Ch. 2: Temperature-related Deaths and Illness), frozen precipitation, and associated dangerous road and other conditions. Future health

impacts associated with these types of storms are uncertain and will depend on how climate change affects storm trends.

During the period 1956 to 2006, lightning caused an estimated 101.2 deaths per year,[189] while thunderstorm winds are estimated to have caused approximately 26 deaths per year from 1977 to 2007.[190] Thunderstorm precipitation and winds can damage structures, fell trees, and create hazardous road conditions and impair driving visibility, increasing risks of motor vehicle deaths and injuries.[134, 191, 192] Thunderstorm winds can cause blunt trauma or injuries, such as from being struck by falling trees and other flying debris,[46] and were responsible for an estimated 4,366 injuries during the period 1993 to 2003.[192]

Winter storms can be accompanied by freezing winds and frigid temperatures that can cause frostbite and hypothermia (see also Ch. 2: Temperature-Related Deaths and Illness).[193, 194] Individuals that lack proper clothing and shelter (for example, the homeless) are more at risk of injuries from direct exposure to weather conditions associated with winter storms and severe thunderstorms.[195] Low-income populations have increased exposure risk to severe winter weather conditions because they are more likely to live in low-quality, poorly insulated housing; be unable to afford sufficient domestic heating; or need to make tradeoffs between food and heating expenditures.[196, 197] Freezing rain, snow, and ice have been linked to increased injuries associated with falling[198] as well as motor vehicle deaths and injuries due to treacherous road conditions and impaired driving visibility.[134, 199]

After severe thunderstorms, individuals can suffer injuries during debris removal and cleanup activities[192, 200] as well as exposure to hazards if flooding occurs (see Section 4.4 of this chapter). Mental health issues and stress are also possible after storms (see Ch. 8: Mental Health). This is especially true of thunderstorms associated with tornadoes, as the aftermath of the storm can involve dealing with the loss of property, displacement, or loss of life.[201] After winter storms, snow removal can be strenuous work and can increase the likelihood of illness and death for individuals with preexisting cardiovascular or pulmonary conditions.[202]

4.8 Emerging Issues

Climate change and changing patterns of extreme weather have the potential to strain the capacity of public health systems. However, few comprehensive or systematic studies have examined the human health impacts of such health-system strain.[203] Particularly in the context of floods and hurricanes, the impacts on health systems from short- and long-term population displacement are not fully understood or well quantified.[67] In addition, the role of future population migration and demographic changes is just beginning to be elucidated in assessments of local adaptive capacity or resilience to the effects of future extreme events. Methodological challenges remain for accurately quantifying and attributing delayed mortality associated with, but not caused directly by, extreme event exposure—for example, elevated mortality associated with heart disease, cancer, diabetes, and infections and other complications from injuries in populations exposed to hurricanes.[30, 31]

4.9 Research Needs

In addition the emerging issues identified above, the authors highlight the following potential areas for additional scientific and research activity on extreme events based on their review of the literature. Current understanding is limited by a lack of systematic surveillance for the range of health impacts, both short and long term, associated with a wider range of extreme events, including prolonged events like droughts and other extremes that do not currently trigger post-event health surveillance.

Future assessments can benefit from multidisciplinary research activities that:

- better define the health implications associated with particular extreme events where longer-term impacts, as well as regional differences in health outcomes, are currently not well understood, such as droughts and floods;

- enhance understanding of how specific attributes that contribute to individual and community level vulnerability to health impacts after extreme events, including social and behavioral characteristics, interact and contribute to or mitigate risks of adverse health outcomes; and

- examine how health outcomes can be impacted by other cumulative, compounding, or secondary effects of extreme events, such as access to or disruption of healthcare services and damages to and cascading failures of infrastructure.

Supporting Evidence

PROCESS FOR DEVELOPING CHAPTER

The chapter was developed through technical discussions of relevant evidence and expert deliberation by the report authors at several workshops, teleconferences, and email exchanges. Authors considered inputs and comments submitted by the public, the National Academies of Sciences, and Federal agencies. For additional information on the overall report process, please see Appendices 2 and 3.

The health outcomes selected and prioritized for the chapter were based primarily on those that had substantial peer-reviewed literature to support statements. While many connections between changes in extreme events due to climate change and human health impacts appear intuitive, in some cases there may not be a robust body of peer-reviewed literature to support statements about direct effects. For example, while it is believed that droughts have the ability to impact water quality, which could in turn impact health, there are few studies documenting specific health consequences in the United States.[96]

In addition, due to space constraints, the authors did not intend to exhaustively identify all possible health impacts from every type of extreme event addressed in this chapter. Instead, the authors have provided an overview of possible impacts from different types of extreme events and provided a framework for understanding what additional factors (for example, population vulnerability, existing quality of infrastructure, etc.) can exacerbate or reduce adverse health outcomes.

Due to limited space and the uncertainty around future projections of tornadoes, we do not include detailed discussion of this topic in this chapter. We recognize that tornadoes can cause significant infrastructure damage and significant health impacts, and understanding how climate change will impact tornado intensity, frequency, and geographic distribution is an area of active scientific investigation.

KEY FINDING TRACEABLE ACCOUNTS

Increased Exposure to Extreme Events

Key Finding 1: Health impacts associated with climate-related changes in exposure to extreme events include death, injury, or illness; exacerbation of underlying medical conditions; and adverse effects on mental health *[High Confidence]*. Climate change will increase exposure risk in some regions of the United States due to projected increases in the frequency and/or intensity of drought, wildfires, and flooding related to extreme precipitation and hurricanes *[Medium Confidence]*.

Description of evidence base

The Third National Climate Assessment (2014 NCA) provides the most recent, peer-reviewed assessment conclusions for projected increases in the frequency and/or intensity of extreme precipitation, hurricanes, coastal inundation, drought, and wildfires in the United States.[2] To the extent that these extreme events are projected to increase in some regions of the United States, people are expected to be at greater risk of exposure to health hazards.

Flooding associated with extreme precipitation, hurricanes, and coastal storms is expected to increase in some regions of the United States due to climate change, thereby increasing exposure to a variety of health hazards.[9, 73, 74] The health impacts of floods and storms include death, injury, and illness; exacerbation of underlying medical conditions; and adverse effects on mental health.[1, 29, 31, 46, 51, 52, 53, 58]

Climate change is projected to lengthen or intensify droughts, especially in the Southwest,[2, 144] which may increase exposure to a broad set of health hazards.[9, 74] The potential health impacts of drought include: illness associated with reduced water quality and quantity[96, 100, 101, 102, 103] and reduced air quality,[95, 118, 119] associations with increased rates of some infectious diseases,[106, 107, 108, 109, 110] and adverse mental health impacts.[94, 124, 125, 126, 127]

Large, intense wildfires will occur more frequently in some regions of the United States, particularly in the western United States and Alaska,[2] and this is expected to increase exposure to wildfire-related health risks.[74, 144] The health impacts of wildfire include death, injury, and illness,[134, 145, 146, 151, 152, 153, 154, 155, 165, 166, 167, 168, 170, 172, 173] including exacerbation of underlying medical conditions.[154, 156, 157]

Major uncertainties

The role of climate change in observed shifts in and future projections of the frequency, intensity, geographic distribution, and duration of certain extreme events is an ongoing, active area of research. For example, although the 2014 NCA[2] concluded that extreme events will increase in some regions of the United States, uncertainties remain with respect to projections of climate impacts at smaller, more local scales and the timing of such impacts (see Table 1). Climate change related projections of winter storms and severe storms, including tornadoes, hail, and thunderstorms, are still uncertain.

The human health implications of the changes in extreme events have not received as much research attention to date, and there are currently no published, national-scale, quantitative projections of changes in exposure risks for the four categories of extreme events addressed in this chapter. Relevant health surveillance and epidemiological data for extreme events are limited by underreporting, underestimation, and lack of a common definition of what constitutes an adverse health impact from an extreme event.[30, 31] For drought in particular, there are few studies documenting specific health consequences in the United States.[96] Challenges to quantitatively estimating future human health risks for the four types of extreme events addressed in this chapter include limited data availability and lack of comprehensive modeling methods. For winter storms and severe storms especially, scientists need a better understanding of how climate change will affect future storm trends before they can make projections of future health impacts.

Assessment of confidence and likelihood based on evidence
There is **high confidence** that the types of health impacts associated with climate-related changes in extremes include death, injury, or illness; exacerbation of underlying medical conditions; and adverse effects on mental health (see Table 1). Based on the evidence presented in the peer-reviewed literature, there is **medium confidence** regarding increases in exposure to health hazards associated with projected increases in the frequency and/or intensity of extreme precipitation, hurricanes, coastal inundation, drought, and wildfires in some regions of the United States. Many qualitative studies have been published about the potential or expected health hazards from these events, but few draw strong or definitive conclusions that exposure to health hazards will increase due to climate change. Thus, the evidence is suggestive and supports a **medium confidence** level that, to the extent that these extreme events are projected to increase in some regions of the United States, people are expected to be at greater risk of exposure to health hazards. There is no quantitative information on which to base probability estimates of the likelihood of increasing exposure to health hazards associated with extreme precipitation, hurricanes, coastal inundation, drought, and wildfires.

Disruption of Essential Infrastructure

Key Finding 2: Many types of extreme events related to climate change cause disruption of infrastructure, including power, water, transportation, and communication systems, that are essential to maintaining access to health care and emergency response services and safeguarding human health *[High Confidence]*.

Description of evidence base

The frequency, intensity, and duration of extreme events determines their physical impacts and the extent to which essential infrastructure is disrupted. There is strong, consistent evidence from multiple studies that infrastructure can either exacerbate or moderate the physical impacts of extreme events, influencing the ultimate nature and severity of health impacts. Projections of increasing frequency and/or intensity of some extreme events suggest that they pose threats to essential infrastructure, such as water, transportation, and power systems.[4, 34, 36, 43] Disruption of essential infrastructure and services after extreme events can increase population exposure to health hazards and reduce their adaptive capacity.[4] There is substantial, high-quality literature supporting a finding that serious health risks can arise from utility outages; infrastructure and housing damage; and disruption or loss of access to sanitation, safe food and water supplies, health care, communication, and transportation.[1, 29, 30, 31, 40, 43, 44, 45, 46, 47, 48, 49, 51, 52, 53, 54, 58, 87] Infrastructure disruptions can have more or less impact on human health depending on the underlying vulnerability of the affected people and communities.[4] Urban populations face unique exposure risks due to their dependence on complex, often interdependent infrastructure systems that can be severely disrupted during extreme events.[2, 65] Rural communities also have vulnerabilities that are different from those faced by urban communities. For example, power and communications can take longer to restore after an outage.[50]

Existing infrastructure is generally designed to perform at its engineered capacity assuming historical weather patterns, and these systems could be more vulnerable to failure in response to weather-related stressors under future climate scenarios.[4, 34, 35] Shifts in the frequency or intensity of extreme events outside their historical range pose infrastructure risks that may be compounded by the fact that much of the existing critical infrastructure in the United States, including water and sewage systems, roads, bridges, and power plants, are aging and in need of repair or replacement.[4, 36]

Major uncertainties
Many of the uncertainties are similar to those of the previous key finding. There are few studies directly linking infrastructure impacts to health outcomes, and most are not longitudinal. Health impacts may occur after the event as a result of loss of infrastructure and public services. These impacts can be distributed over longer periods of time, making them harder to observe and quantify. Thus, the actual impact is likely underreported.

Uncertainties remain with respect to projecting how climate change will affect the severity of the physical impacts, including on infrastructure, of extreme events at smaller, more local scales and the timing of such impacts. Therefore, the subsequent impact on infrastructure also has a great deal of uncertainty. Thus, the key finding does not make any statements about future impacts. Instead the focus is on impacts that have occurred to date because there is supporting peer-reviewed literature. The extent to which infrastructure is exposed to extreme events, and the adaptive capacity of a community to repair infrastructure in a timely manner both influence the extent of the health outcomes. Thus, while the chapter makes general statements about trends in impacts due to extremes, there are uncertainties in the extent to which any specific location or infrastructure system could be impacted and the resulting health outcomes.

Assessment of confidence and likelihood based on evidence

There is **high confidence** that many types of extreme events can cause disruption of essential infrastructure (such as water, transportation, and power systems), and that such disruption can adversely affect human health. Many qualitative studies have been published about the effects of these factors on health impacts from an extreme event (noted above), and the evidence is of good quality and consistent.

Vulnerability to Coastal Flooding

Key Finding 3: Coastal populations with greater vulnerability to health impacts from coastal flooding include persons with disabilities or other access and functional needs, certain populations of color, older adults, pregnant women and children, low-income populations, and some occupational groups *[High Confidence].* Climate change will increase exposure risk to coastal flooding due to increases in extreme precipitation and in hurricane intensity and rainfall rates, as well as sea level rise and the resulting increases in storm surge *[High Confidence].*

Description of evidence base

The evidence in the peer-reviewed literature that climate change will increase coastal flooding in the future is very robust.[2, 4] Global sea level has risen by about 8 inches since reliable record keeping began in 1880 and it is projected to rise another 1 to 4 feet by 2100.[2] Rates of sea level rise are not uniform along U.S. coasts and can be exacerbated locally by land subsidence or reduced by uplift. In the next several decades, storm surges and high tides could combine with sea level rise and land subsidence to further increase coastal flooding in many regions. The U.S. East and Gulf coasts, Hawai'i, and the U.S.-affiliated Pacific Islands are particularly at risk.

In addition, recurrent weather-related stressors, such as "nuisance flooding" (frequent coastal flooding causing public inconveniences), contribute to overall deterioration of infrastructure like stormwater drainage systems and roads (see Ch. 6: Water-Related Illness).[38] These systems are important in the context of health because drainage helps to avoid sewage overflows and maintain water quality,[39] and roads are vital for evacuations and emergency response during and after extreme events.[40]

There is strong, consistent evidence in the literature that coastal flooding will increase exposure to a variety of health hazards—for example, direct physical impacts and impacts associated with disruption of essential infrastructure— which can result in death, injury, or illness; exacerbation of underlying medical conditions; and adverse effects on mental health.[1, 29, 31, 46, 51, 52, 53, 58] Multiple studies also consistently identify certain populations as especially vulnerable to the health impacts of coastal flooding. These populations include older adults (especially those who are frail, medically incapacitated, or residing in nursing or assisted living facilities), children, those reliant on electrically powered medical equipment like ventilators and oxygen supplies, those with preexisting health conditions, and people with disabilities.[1, 8, 22, 43, 44, 53, 54, 195, 196, 197] In addition, differences in exposure, sensitivity, and adaptive capacity lead to a disproportionate number of flood-related fatalities among older adults, males, and some low-income communities of color.[29] Floods and storms also create occupational health risks to first responders and other emergency workers and to people involved in cleanup activities, construction, public utilities restoration, and security/policing.[1, 73, 87, 88,]

Major uncertainties

It is nearly certain that coastal flooding will increase in the United States. There are varying estimates regarding the exact degree of flooding at any particular location along the coast. Modeling does provide estimated ranges with varying levels of confidence depending on the location. There is greater uncertainty about how coastal flooding will impact the health of specific populations. There are various ways in which these key risk factors interact with and contribute to the vulnerability (comprised of exposure, sensitivity, and adaptive capacity) of a population. Some uncertainties exist regarding the relative importance of each of these factors in determining a population's vulnerability to health impacts from extreme events. In addition, there is some uncertainty regarding how future demographic and population changes may affect the relative importance of each of these factors.

Assessment of confidence based on evidence

Based on the evidence presented in the peer-reviewed literature, there is **high confidence** that coastal flooding will increase in the United States, and that age, health status, socioeconomic status, race/ethnicity, and occupation are key risk factors that individually and collectively affect a population's vulnerability to health impacts from coastal flooding. Many qualitative studies have been published regarding how these key risk factors interact with and contribute to the exposure, sensitivity, and adaptive capacity of a population, and this evidence is of good quality and consistent.

DOCUMENTING UNCERTAINTY

This assessment relies on two metrics to communicate the degree of certainty in Key Findings. See Appendix 4: Documenting Uncertainty for more on assessments of likelihood and confidence.

Confidence Level
Very High
Strong evidence (established theory, multiple sources, consistent results, well documented and accepted methods, etc.), high consensus
High
Moderate evidence (several sources, some consistency, methods vary and/or documentation limited, etc.), medium consensus
Medium
Suggestive evidence (a few sources, limited consistency, models incomplete, methods emerging, etc.), competing schools of thought
Low
Inconclusive evidence (limited sources, extrapolations, inconsistent findings, poor documentation and/or methods not tested, etc.), disagreement or lack of opinions among experts

Likelihood
Very Likely
≥ 9 in 10
Likely
≥ 2 in 3
As Likely As Not
≈ 1 in 2
Unlikely
≤ 1 in 3
Very Unlikely
≤ 1 in 10

PHOTO CREDITS

Pg. 99–Firefighters battling fire: © Erich Schlegel/Corbis

Pg. 100–Family escaping flood waters: © Greg Vote/Corbis

Pg. 104–Young family in shelter: © © Robert Sciarrino/The Star-Ledger/Corbis

Pg. 105–Damaged power lines: © Julie Dermansky/Corbis

Pg. 107–Truck stuck in flood waters: © Mike Theiss/National Geographic Creative/Corbis

Pg. 108–Farmer in drought-stressed peanut field: © ERIK S. LESSER/epa/Corbis

Pg. 109–Dust storm: © Martyn Goddard/Corbis

Pg. 110–Firefighters battling fire: © Erich Schlegel/Corbis

Pg. 111–Gridlocked traffic: © Robin Nelson/ZUMA Press/Corbis

References

1. Lane, K., K. Charles-Guzman, K. Wheeler, Z. Abid, N. Graber, and T. Matte, 2013: Health effects of coastal storms and flooding in urban areas: A review and vulnerability assessment. *Journal of Environmental and Public Health,* **2013,** Article ID 913064. http://dx.doi.org/10.1155/2013/913064

2. Melillo, J.M., T.C. Richmond, and G.W. Yohe, Eds., 2014: *Climate Change Impacts in the United States: The Third National Climate Assessment.* U.S. Global Change Research Program, Washington, D.C., 842 pp. http://dx.doi.org/10.7930/J0Z31WJ2

3. Moser, S.C., M.A. Davidson, P. Kirshen, P. Mulvaney, J.F. Murley, J.E. Neumann, L. Petes, and D. Reed, 2014: Ch. 25: Coastal zone development and ecosystems. *Climate Change Impacts in the United States: The Third National Climate Assessment.* Melillo, J.M., T.C. Richmond, and G.W. Yohe, Eds. U.S. Global Change Research Program, Washington, D.C., 579-618. http://dx.doi.org/10.7930/J0MS3QNW

4. IPCC, 2012: Managing the Risks of Extreme Events and Disasters to Advance Climate Change Adaptation. A Special Report of Working Groups I and II of the Intergovernmental Panel on Climate Change. Field, C.B., V. Barros, T.F. Stocker, D. Qin, D.J. Dokken, K.L. Ebi, M.D. Mastrandrea, K.J. Mach, G.-K. Plattner, S.K. Allen, M. Tignor, and P.M. Midgley (Eds.), 582 pp. Cambridge University Press, Cambridge, UK and New York, NY. http://ipcc-wg2.gov/SREX/images/uploads/SREX-All_FINAL.pdf

5. Fothergill, A. and L.A. Peek, 2004: Poverty and disasters in the United States: A review of recent sociological findings. *Natural Hazards,* **32,** 89-110. http://dx.doi.org/10.1023/B:NHAZ.0000026792.76181.d9

6. Zoraster, R.M., 2010: Vulnerable populations: Hurricane Katrina as a case study. *Prehospital and Disaster Medicine,* **25,** 74-78. http://dx.doi.org/10.1017/s1049023x00007718

7. Aldrich, N. and W.F. Benson, 2008: Disaster preparedness and the chronic disease needs of vulnerable older adults. *Preventing Chronic Disease: Public Health Research, Practice, and Policy,* **5,** 1-7. http://www.cdc.gov/pcd//issues/2008/jan/07_0135.htm

8. Penner, S.J. and C. Wachsmuth, 2008: Disaster management and populations with special needs. *Disaster Management Handbook.* Pinkowski, J., Ed. CRC Press, Boca Raton, FL, 427-444.

9. Frumkin, H., J. Hess, G. Luber, J. Malilay, and M. McGeehin, 2008: Climate change: The public health response. *American Journal of Public Health,* **98,** 435-445. http://dx.doi.org/10.2105/AJPH.2007.119362

10. Keppel, K.G., 2007: Ten largest racial and ethnic health disparities in the United States based on Healthy People 2010 objectives. *American Journal of Epidemiology,* **166,** 97-103. http://dx.doi.org/10.1093/aje/kwm044

11. Collins, T.W., A.M. Jimenez, and S.E. Grineski, 2013: Hispanic health disparities after a flood disaster: Results of a population-based survey of individuals experiencing home site damage in El Paso (Texas, USA). *Journal of Immigrant and Minority Health,* **15,** 415-426. http://dx.doi.org/10.1007/s10903-012-9626-2

12. Weinhold, B., 2011: Fields and forests in flames: Vegetation smoke and human health. *Environmental Health Perspectives,* **119,** a386-a393. http://dx.doi.org/10.1289/ehp.119-a386

13. NRC, 2012: *Disaster Resilience: A National Imperative.* National Academies Press, Washington, D.C., 244 pp.

14. Brouwer, R., S. Akter, L. Brander, and E. Haque, 2007: Socioeconomic vulnerability and adaptation to environmental risk: A case study of climate change and flooding in Bangladesh. *Risk Analysis,* **27,** 313-326. http://dx.doi.org/10.1111/j.1539-6924.2007.00884.x

15. Thomalla, F., T. Downing, E. Spanger-Siegfried, G. Han, and J. Rockström, 2006: Reducing hazard vulnerability: Towards a common approach between disaster risk reduction and climate adaptation. *Disasters,* **30,** 39-48. http://dx.doi.org/10.1111/j.1467-9523.2006.00305.x

16. Tapsell, S.M., E.C. Penning-Rowsell, S.M. Tunstall, and T.L. Wilson, 2002: Vulnerability to flooding: Health and social dimensions. *Philosophical Transactions of the Royal Society A: Mathematical, Physical and Engineering Sciences,* **360,** 1511-1525. http://dx.doi.org/10.1098/rsta.2002.1013

17. Srinivasan, S., L.R. O'Fallon, and A. Dearry, 2003: Creating healthy communities, healthy homes, healthy people: Initiating a research agenda on the built environment and public health. *American Journal of Public Health,* **93,** 1446-1450. http://dx.doi.org/10.2105/ajph.93.9.1446

18. Bashir, S.A., 2002: Home is where the harm is: Inadequate housing as a public health crisis. *American Journal of Public Health,* **92,** 733-738. http://dx.doi.org/10.2105/ajph.92.5.733

19. Cattell, V., 2001: Poor people, poor places, and poor health: The mediating role of social networks and social capital. *Social Science & Medicine,* **52,** 1501-1516. http://dx.doi.org/10.1016/S0277-9536(00)00259-8

20. Masozera, M., M. Bailey, and C. Kerchner, 2007: Distribution of impacts of natural disasters across income groups: A case study of New Orleans. *Ecological Economics,* **63,** 299-306. http://dx.doi.org/10.1016/j.ecolecon.2006.06.013

21. Keim, M.E., 2008: Building human resilience: The role of public health preparedness and response as an adaptation to climate change. *American Journal of Preventive Medicine,* **35,** 508-516. http://dx.doi.org/10.1016/j.amepre.2008.08.022

22. Kovats, R.S. and S. Hajat, 2008: Heat stress and public health: A critical review. *Annual Review of Public Health,* **29,** 41-55. http://dx.doi.org/10.1146/annurev.publhealth.29.020907.090843

23. Engle, N.L., 2011: Adaptive capacity an d its assessment. *Global Environmental Change,* **21,** 647-656. http://dx.doi.org/10.1016/j.gloenvcha.2011.01.019

24. Cutter, S.L., L. Barnes, M. Berry, C. Burton, E. Evans, E. Tate, and J. Webb, 2008: A place-based model for understanding community resilience to natural disasters. *Global Environmental Change,* **18,** 598-606. http://dx.doi.org/10.1016/j.gloenvcha.2008.07.013

25. Linnenluecke, M.K., A. Griffiths, and M. Winn, 2012: Extreme weather events and the critical importance of anticipatory adaptation and organizational resilience in responding to impacts. *Business Strategy and the Environment,* **21,** 17-32. http://dx.doi.org/10.1002/bse.708

26. Warner, K., N. Ranger, S. Surminski, M. Arnold, J. Linnerooth-Bayer, E. Michel-Kerjan, P. Kovacs, and C. Herweijer, 2009: Adaptation to Climate Change: Linking Disaster Risk Reduction and Insurance. 30 pp. United Nations International Strategy for Disaster Reduction Secretariat, Geneva. http://www.unisdr.org/files/9654_linkingdrrinsurance.pdf

27. Grineski, S.E., J.G. Staniswalis, P. Bulathsinhala, Y. Peng, and T.E. Gill, 2011: Hospital admissions for asthma and acute bronchitis in El Paso, Texas: Do age, sex, and insurance status modify the effects of dust and low wind events? *Environmental Research,* **111,** 1148-1155. http://dx.doi.org/10.1016/j.envres.2011.06.007

28. Hess, J.J., J.Z. McDowell, and G. Luber, 2012: Integrating climate change adaptation into public health practice: Using adaptive management to increase adaptive capacity and build resilience. *Environmental Health Perspectives,* **120,** 171-179. http://dx.doi.org/10.1289/ehp.1103515

29. Alderman, K., L.R. Turner, and S. Tong, 2012: Floods and human health: A systematic review. *Environment International,* **47,** 37-47. http://dx.doi.org/10.1016/j.envint.2012.06.003

30. Senkbeil, J.C., D.M. Brommer, and I.J. Comstock, 2011: Tropical cyclone hazards in the USA. *Geography Compass,* **5,** 544-563. http://dx.doi.org/10.1111/j.1749-8198.2011.00439.x

31. McKinney, N., C. Houser, and K. Meyer-Arendt, 2011: Direct and indirect mortality in Florida during the 2004 hurricane season. *International Journal of Biometeorology,* **55,** 533-546. http://dx.doi.org/10.1007/s00484-010-0370-9

32. Guenther, R. and J. Balbus, 2014: Primary Protection: Enhancing Health Care Resilience for a Changing Climate. U.S. Department of Health and Human Services. https://toolkit.climate.gov/sites/default/files/SCRHCFI%20Best%20Practices%20Report%20final2%202014%20Web.pdf

33. Deshmukh, A., E. Ho Oh, and M. Hastak, 2011: Impact of flood damaged critical infrastructure on communities and industries. *Built Environment Project and Asset Management,* **1,** 156-175. http://dx.doi.org/10.1108/20441241111180415

34. Dell, J., S. Tierney, G. Franco, R.G. Newell, R. Richels, J. Weyant, and T.J. Wilbanks, 2014: Ch. 4: Energy supply and use. *Climate Change Impacts in the United States: The Third National Climate Assessment.* Melillo, J.M., T.C. Richmond, and G.W. Yohe, Eds. U.S. Global Change Research Program, Washington, D.C., 113-129. http://dx.doi.org/10.7930/J0BG2KWD

35. GAO, 2015: Army Corps of Engineers Efforts to Assess the Impact of Extreme Weather Events. GAO-15-660. United States Government Accountability Office. http://www.gao.gov/assets/680/671591.pdf

36. Cutter, S.L., W. Solecki, N. Bragado, J. Carmin, M. Fragkias, M. Ruth, and T. Wilbanks, 2014: Ch. 11: Urban systems, infrastructure, and vulnerability. *Climate Change Impacts in the United States: The Third National Climate Assessment.* Melillo, J.M., T.C. Richmond, and G.W. Yohe, Eds. U.S. Global Change Research Program, Washington, D.C., 282-296. http://dx.doi.org/10.7930/J0F769GR

37. ASCE, 2013: Report Card for America's Infrastructure. American Society of Civil Engineers, Reston, VA. http://www.infrastructurereportcard.org/

38. NOAA, 2014: Sea Level Rise and Nuisance Flood Frequency Changes around the United States. NOAA Technical Report NOS CO-OPS 073, 58 pp. U.S. Department of Commerce, National Oceanic and Atmospheric Administration, National Ocean Service, Silver Spring, MD. http://tidesandcurrents.noaa.gov/publications/NOAA_Technical_Report_NOS_COOPS_073.pdf

39. Cann, K.F., D.R. Thomas, R.L. Salmon, A.P. Wyn-Jones, and D. Kay, 2013: Extreme water-related weather events and waterborne disease. *Epidemiology and Infection*, **141**, 671-86. http://dx.doi.org/10.1017/s0950268812001653

40. Skinner, M.W., N.M. Yantzi, and M.W. Rosenberg, 2009: Neither rain nor hail nor sleet nor snow: Provider perspectives on the challenges of weather for home and community care. *Social Science & Medicine*, **68**, 682-688. http://dx.doi.org/10.1016/j.socscimed.2008.11.022

41. Mooney, H., A. Larigauderie, M. Cesario, T. Elmquist, O. Hoegh-Guldberg, S. Lavorel, G.M. Mace, M. Palmer, R. Scholes, and T. Yahara, 2009: Biodiversity, climate change, and ecosystem services. *Current Opinion in Environmental Sustainability*, **1**, 46-54. http://dx.doi.org/10.1016/j.cosust.2009.07.006

42. Kenny, J.F., N.L. Barber, S.S. Hutson, K.S. Linsey, J.K. Lovelace, and M.A. Maupin, 2009: Estimated Use of Water in the United States in 2005. U.S. Geological Survey Circular 1344, 52 pp. U.S. Geological Survey, Reston, VA. http://pubs.usgs.gov/circ/1344/

43. Klinger, C., O. Landeg, and V. Murray, 2014: Power outages, extreme events and health: A systematic review of the literature from 2011-2012. *PLoS Currents*, **6**. http://dx.doi.org/10.1371/currents.dis.04eb1dc5e73dd1377e-05a10e9edde673

44. Anderson, G.B. and M.L. Bell, 2012: Lights out: Impact of the August 2003 power outage on mortality in New York, NY. *Epidemiology*, **23**, 189-193. http://dx.doi.org/10.1097/EDE.0b013e318245c61c

45. Lin, S., B.A. Fletcher, M. Luo, R. Chinery, and S.-A. Hwang, 2011: Health impact in New York City during the Northeastern blackout of 2003. *Public Health Reports*, **126**, 384-93. PMC3072860

46. Goldman, A., B. Eggen, B. Golding, and V. Murray, 2014: The health impacts of windstorms: A systematic literature review. *Public Health*, **128**, 3-28. http://dx.doi.org/10.1016/j.puhe.2013.09.022

47. Daley, W.R., A. Smith, E. Paz-Argandona, J. Malilay, and M. McGeehin, 2000: An outbreak of carbon monoxide poisoning after a major ice storm in Maine. *The Journal of Emergency Medicine*, **18**, 87-93. http://dx.doi.org/10.1016/S0736-4679(99)00184-5

48. Iqbal, S., J.H. Clower, S.A. Hernandez, S.A. Damon, and F.Y. Yip, 2012: A review of disaster-related carbon monoxide poisoning: Surveillance, epidemiology, and opportunities for prevention. *American Journal of Public Health*, **102**, 1957-1963. http://dx.doi.org/10.2105/ajph.2012.300674

49. Lutterloh, E.C., S. Iqbal, J.H. Clower, H.A. Spillerr, M.A. Riggs, T.J. Sugg, K.E. Humbaugh, B.L. Cadwell, and D.A. Thoroughman, 2011: Carbon monoxide poisoning after an ice storm in Kentucky, 2009. *Public Health Reports*, **126 (Suppl 1)**, 108-115. PMC3072909

50. Hales, D., W. Hohenstein, M.D. Bidwell, C. Landry, D. McGranahan, J. Molnar, L.W. Morton, M. Vasquez, and J. Jadin, 2014: Ch. 14: Rural Communities. *Climate Change Impacts in the United States: The Third National Climate Assessment*. Melillo, J.M., T.C. Richmond, and G.W. Yohe, Eds. U.S. Global Change Research Program, Washington, DC, 333-349. http://dx.doi.org/10.7930/J01Z429C

51. Arrieta, M.I., R.D. Foreman, E.D. Crook, and M.L. Icenogle, 2009: Providing continuity of care for chronic diseases in the aftermath of Katrina: From field experience to policy recommendations. *Disaster Medicine and Public Health Preparedness*, **3**, 174-182. http://dx.doi.org/10.1097/DMP.0b013e3181b66ae4

52. Anderson, A.H., A.J. Cohen, N.G. Kutner, J.B. Kopp, P.L. Kimmel, and P. Muntner, 2009: Missed dialysis sessions and hospitalization in hemodialysis patients after Hurricane Katrina. *Kidney International*, **75**, 1202-1208. http://dx.doi.org/10.1038/ki.2009.5

53. Kleinpeter, M.A., 2011: Disaster preparedness for dialysis patients. *Clinical Journal of the American Society of Nephrology*, **6**, 2337-2339. http://dx.doi.org/10.2215/cjn.08690811

54. Laditka, S.B., J.N. Laditka, S. Xirasagar, C.B. Cornman, C.B. Davis, and J.V.E. Richter, 2008: Providing shelter to nursing home evacuees in disasters: Lessons from Hurricane Katrina. *American Journal of Public Health*, **98**, 1288-1293. http://dx.doi.org/10.2105/ajph.2006.107748

55. Battle, D.E., 2015: Persons with communication disabilities in natural disasters, war, and/or conflict. *Communication Disorders Quarterly,* **36,** 231-240. http://dx.doi.org/10.1177/1525740114545980

56. Andrulis, D.P., N.J. Siddiqui, and J.L. Gantner, 2007: Preparing racially and ethnically diverse communities for public health emergencies. *Health Affairs,* **26,** 1269-1279. http://dx.doi.org/10.1377/hlthaff.26.5.1269

57. Lippmann, A.L., 2011: Disaster preparedness in vulnerable communities. *International Law and Policy Review,* **1,** 69-96.

58. Du, W., G.J. FitzGerald, M. Clark, and X.-Y. Hou, 2010: Health impacts of floods. *Prehospital and Disaster Medicine,* **25,** 265-272. http://dx.doi.org/10.1017/S1049023X00008141

59. Solomon, G.M., M. Hjelmroos-Koski, M. Rotkin-Ellman, and S.K. Hammond, 2006: Airborne mold and endotoxin concentrations in New Orleans, Louisiana, after flooding, October through November 2005. *Environmental Health Perspectives,* **114,** 1381-1386. http://dx.doi.org/10.1289/ehp.9198

60. Fisk, W.J., E.A. Eliseeva, and M.J. Mendell, 2010: Association of residential dampness and mold with respiratory tract infections and bronchitis: A meta-analysis. *Environmental Health,* **9,** Article 72. http://dx.doi.org/10.1186/1476-069x-9-72

61. Fisk, W.J., Q. Lei-Gomez, and M.J. Mendell, 2007: Meta-analyses of the associations of respiratory health effects with dampness and mold in homes. *Indoor Air,* **17,** 284-296. http://dx.doi.org/10.1111/j.1600-0668.2007.00475.x

62. Mendell, M.J., A.G. Mirer, K. Cheung, and J. Douwes, 2011: Respiratory and allergic health effects of dampness, mold, and dampness-related agents: A review of the epidemiologic evidence. *Environmental Health Perspectives,* **119,** 748-756. http://dx.doi.org/10.1289/ehp.1002410

63. Leavitt, W.M. and J.J. Kiefer, 2006: Infrastructure interdependency and the creation of a normal disaster: The case of Hurricane Katrina and the City of New Orleans. *Public Works Management & Policy* **10,** 303-314. http://dx.doi.org/10.1177/1087724X06289055

64. Wilbanks, T.J. and S.J. Fernandez, Eds., 2014: *Climate Change and Infrastructure, Urban Systems, and Vulnerabilities. Technical Report for the U.S. Department of Energy in Support of the National Climate Assessment.* Island Press, Washington, D.C., 88 pp. http://www.cakex.org/sites/default/files/documents/Climate%20ChangeAndInfrastructureUrbanSystemsAndVulnerabilities.pdf

65. Kirshen, P., M. Ruth, and W. Anderson, 2008: Interdependencies of urban climate change impacts and adaptation strategies: A case study of Metropolitan Boston USA. *Climatic Change,* **86,** 105-122. http://dx.doi.org/10.1007/s10584-007-9252-5

66. Steinberg, L.J., H. Sengul, and A.M. Cruz, 2008: Natech risk and management: An assessment of the state of the art. *Natural Hazards,* **46,** 143-152. http://dx.doi.org/10.1007/s11069-007-9205-3

67. Knowlton, K., M. Rotkin-Ellman, L. Geballe, W. Max, and G.M. Solomon, 2011: Six climate change-related events in the United States accounted for about $14 billion in lost lives and health costs. *Health Affairs,* **30,** 2167-2176. http://dx.doi.org/10.1377/hlthaff.2011.0229

68. Kile, J.C., S. Skowronski, M.D. Miller, S.G. Reissman, V. Balaban, R.W. Klomp, D.B. Reissman, H.M. Mainzer, and A.L. Dannenberg, 2005: Impact of 2003 power outages on public health and emergency response. *Prehospital and Disaster Medicine,* **20,** 93-97. http://dx.doi.org/10.1017/s1049023x00002259

69. Prezant, D.J., J. Clair, S. Belyaev, D. Alleyne, G.I. Banauch, M. Davitt, K. Vandervoorts, K.J. Kelly, B. Currie, and G. Kalkut, 2005: Effects of the August 2003 blackout on the New York City healthcare delivery system: A lesson for disaster preparedness. *Critical Care Medicine,* **33,** S96-S101. http://dx.doi.org/10.1097/01.ccm.0000150956.90030.23

70. Freese, J., N.J. Richmond, R.A. Silverman, J. Braun, B.J. Kaufman, and J. Clair, 2006: Impact of citywide blackout on an urban emergency medical services system. *Prehospital and Disaster Medicine,* **21,** 372-378. http://dx.doi.org/10.1017/S1049023X00004064

71. Klein, K.R., P. Herzog, S. Smolinske, and S.R. White, 2007: Demand for poison control center services "surged" during the 2003 blackout. *Clinical Toxicology,* **45,** 248-254. http://dx.doi.org/10.1080/15563650601031676

72. Beatty, M.E., S. Phelps, C. Rohner, and I. Weisfuse, 2006: Blackout of 2003: Health effects and emergency responses. *Public Health Reports,* **121,** 36-44. PMC1497795

73. Lowe, D., K. Ebi, and B. Forsberg, 2013: Factors increasing vulnerability to health effects before, during and after floods. *International Journal of Environmental Research and Public Health,* **10,** 7015-7067. http://dx.doi.org/10.3390/ijerph10127015

74. Luber, G., K. Knowlton, J. Balbus, H. Frumkin, M. Hayden, J. Hess, M. McGeehin, N. Sheats, L. Backer, C.B. Beard, K.L. Ebi, E. Maibach, R.S. Ostfeld, C. Wiedinmyer, E. Zielinski-Gutiérrez, and L. Ziska, 2014: Ch. 9: Human health. *Climate Change Impacts in the United States: The Third National Climate Assessment.* Melillo, J.M., T.C. Richmond, and G.W. Yohe, Eds. U.S. Global Change Research Program, Washington, D.C., 220-256. http://dx.doi.org/10.7930/J0PN93H5

75. Georgakakos, A., P. Fleming, M. Dettinger, C. Peters-Lidard, T.C. Richmond, K. Reckhow, K. White, and D. Yates, 2014: Ch. 3: Water resources. *Climate Change Impacts in the United States: The Third National Climate Assessment.* Melillo, J.M., T.C. Richmond, and G.W. Yohe, Eds. U.S. Global Change Research Program, Washington, D.C., 69-112. http://dx.doi.org/10.7930/J0G44N6T

76. Kellar, D.M.M. and T.W. Schmidlin, 2012: Vehicle-related flood deaths in the United States, 1995-2005. *Journal of Flood Risk Management, 5,* 153-163. http://dx.doi.org/10.1111/j.1753-318X.2012.01136.x

77. Sharif, H.O., T.L. Jackson, M.M. Hossain, and D. Zane, 2014: Analysis of flood fatalities in Texas. *Natural Hazards Review, 16,* 04014016. http://dx.doi.org/10.1061/(asce)nh.1527-6996.0000145

78. Špitalar, M., J.J. Gourley, C. Lutoff, P.-E. Kirstetter, M. Brilly, and N. Carr, 2014: Analysis of flash flood parameters and human impacts in the US from 2006 to 2012. *Journal of Hydrology, 519,* 863-870. http://dx.doi.org/10.1016/j.jhydrol.2014.07.004

79. Rappaport, E.N., 2014: Fatalities in the United States from Atlantic tropical cyclones: New data and interpretation. *Bulletin of the American Meteorological Society, 95,* 341-346. http://dx.doi.org/10.1175/bams-d-12-00074.1

80. Villarini, G., R. Goska, J.A. Smith, and G.A. Vecchi, 2014: North Atlantic tropical cyclones and U.S. flooding. *Bulletin of the American Meteorological Society, 95,* 1381-1388. http://dx.doi.org/10.1175/bams-d-13-00060.1

81. Rowe, S.T. and G. Villarini, 2013: Flooding associated with predecessor rain events over the Midwest United States. *Environmental Research Letters, 8,* 024007. http://dx.doi.org/10.1088/1748-9326/8/2/024007

82. Schumacher, R.S. and T.J. Galarneau, 2012: Moisture transport into midlatitudes ahead of recurving tropical cyclones and its relevance in two predecessor rain events. *Monthly Weather Review, 140,* 1810-1827. http://dx.doi.org/10.1175/mwr-d-11-00307.1

83. Beven, J.L., II, L.A. Avila, E.S. Blake, D.P. Brown, J.L. Franklin, R.D. Knabb, R.J. Pasch, J.R. Rhome, and S.R. Stewart, 2008: Atlantic hurricane season of 2005. *Monthly Weather Review, 136,* 1109-1173. http://dx.doi.org/10.1175/2007MWR2074.1

84. Brunkard, J., G. Namulanda, and R. Ratard, 2008: Hurricane Katrina deaths, Louisiana, 2005. *Disaster Medicine and Public Health Preparedness, 2,* 215-223. http://dx.doi.org/10.1097/DMP.0b013e31818aaf55

85. Casey-Lockyer, M., R.J. Heick, C.E. Mertzlufft, E.E. Yard, A.F. Wolkin, R.S. Noe, and M. Murti, 2013: Deaths associated with Hurricane Sandy-October-November 2012. *MMWR. Morbidity and Mortality Weekly Report, 62,* 393-392. http://www.cdc.gov/mmwr/preview/mmwrhtml/mm6220a1.htm

86. NOAA, 2013: Service Assessment: Hurricane/Post-Tropical Cyclone Sandy, October 22–29, 2012. 66 pp. U.S. Department of Commerce, National Oceanic and Atmospheric Administration, National Weather Service, Silver Spring, MD. http://www.nws.noaa.gov/os/assessments/pdfs/Sandy13.pdf

87. Schulte, P.A. and H. Chun, 2009: Climate change and occupational safety and health: Establishing a preliminary framework. *Journal of Occupational and Environmental Hygiene, 6,* 542-554. http://dx.doi.org/10.1080/15459620903066008

88. Osofsky, H.J., J.D. Osofsky, J. Arey, M.E. Kronenberg, T. Hansel, and M. Many, 2011: Hurricane Katrina's first responders: The struggle to protect and serve in the aftermath of the disaster. *Disaster Medicine and Public Health Preparedness, 5,* S214-S219. http://dx.doi.org/10.1001/dmp.2011.53

89. Tong, V.T., M.E. Zotti, and J. Hsia, 2011: Impact of the Red River catastrophic flood on women giving birth in North Dakota, 1994–2000. *Maternal and Child Health Journal, 15,* 281-288. http://dx.doi.org/10.1007/s10995-010-0576-9

90. Xiong, X., E.W. Harville, D.R. Mattison, K. Elkind-Hirsch, G. Pridjian, and P. Buekens, 2008: Exposure to Hurricane Katrina, post-traumatic stress disorder and birth outcomes. *The American Journal of the Medical Sciences, 336,* 111-115. http://dx.doi.org/10.1097/MAJ.0b013e318180f21c

91. Peek, L. and L.M. Stough, 2010: Children with disabilities in the context of disaster: A social vulnerability perspective. *Child Development, 81,* 1260-1270. http://dx.doi.org/10.1111/j.1467-8624.2010.01466

92. Brandenburg, M.A., S.M. Watkins, K.L. Brandenburg, and C. Schieche, 2007: Operation Child-ID: Reunifying children with their legal guardians after Hurricane Katrina. *Disasters,* **31,** 277-287. http://dx.doi.org/10.1111/j.0361-3666.2007.01009.x

93. Thomas, D.S.K., B.D. Phillips, W.E. Lovekamp, and A. Fothergill, 2013: *Social Vulnerability to Disasters,* 2nd ed. CRC Press, Boca Raton, FL, 514 pp.

94. Stanke, C., M. Kerac, C. Prudhomme, J. Medlock, and V. Murray, 2013: Health effects of drought: A systematic review of the evidence. *Plos Currents Disasters.* http://dx.doi.org/10.1371/currents.dis.7a2cee9e980f91ad7697b570bc-c4b004

95. CDC, EPA, NOAA, and AWWA, 2010: When Every Drop Counts: Protecting Public Health During Drought Conditions—A Guide for Public Health Professionals. 56 pp. U.S. Department of Health and Human Services, Centers for Disease Control and Prevention, Atlanta, GA. http://www.cdc.gov/nceh/ehs/Docs/When_Every_Drop_Counts.pdf

96. Whitehead, P.G., R.L. Wilby, R.W. Battarbee, M. Kernan, and A.J. Wade, 2009: A review of the potential impacts of climate change on surface water quality. *Hydrological Sciences Journal,* **54,** 101-123. http://dx.doi.org/10.1623/hysj.54.1.101

97. Barlow, P.M., 2003: Ground Water in Freshwater-Saltwater Environments of the Atlantic Coast. U.S. Geological Survey Circular 1262, 113 pp., Reston, VA. http://pubs.usgs.gov/circ/2003/circ1262/pdf/circ1262.pdf

98. Renken, R.A., J. Dixon, J. Koehmstedt, S. Ishman, A.C. Lietz, R.L. Marella, P. Telis, J. Rogers, and S. Memberg, 2005: Impact of Anthropogenic Development on Coastal Ground-Water Hydrology in Southeastern Florida, 1900-2000. U.S. Geological Survey Circular 1275, 77 pp. U.S. Geological Survey, Reston, VA. http://sofia.usgs.gov/publications/circular/1275/cir1275_renken.pdf

99. Dausman, A. and C.D. Langevin, 2005: Movement of the Saltwater Interface in the Surficial Aquifer System in Response to Hydrologic Stresses and Water-Management Practices, Broward County, Florida. U.S. Geological Survey Scientific Investigations Report 2004-5256, 73 pp., Reston, VA. http://pubs.usgs.gov/sir/2004/5256/pdf/sir20045256.pdf

100. Olds, B.P., B.C. Peterson, K.D. Koupal, K.M. Farnsworth-Hoback, C.W. Schoenebeck, and W.W. Hoback, 2011: Water quality parameters of a Nebraska reservoir differ between drought and normal conditions. *Lake and Reservoir Management,* **27,** 229-234. http://dx.doi.org/10.1080/07438141.2011.601401

101. Zwolsman, J.J.G. and A.J. van Bokhoven, 2007: Impact of summer droughts on water quality of the Rhine River - a preview of climate change? *Water Science & Technology,* **56,** 45-55. http://dx.doi.org/10.2166/wst.2007.535

102. Delpla, I., A.-V. Jung, E. Baures, M. Clement, and O. Thomas, 2009: Impacts of climate change on surface water quality in relation to drinking water production. *Environment International,* **35,** 1225-1233. http://dx.doi.org/10.1016/j.envint.2009.07.001

103. van Vliet, M.T.H. and J.J.G. Zwolsman, 2008: Impact of summer droughts on the water quality of the Meuse river. *Journal of Hydrology,* **353,** 1-17. http://dx.doi.org/10.1016/j.jhydrol.2008.01.001

104. Charron, D.F., M.K. Thomas, D. Waltner-Toews, J.J. Aramini, T. Edge, R.A. Kent, A.R. Maarouf, and J. Wilson, 2004: Vulnerability of waterborne diseases to climate change in Canada: A review. *Journal of Toxicology and Environmental Health, Part A: Current Issues,* **67,** 1667-1677. http://dx.doi.org/10.1080/15287390490492313

105. Lal, A., M.G. Baker, S. Hales, and N.P. French, 2013: Potential effects of global environmental changes on cryptosporidiosis and giardiasis transmission. *Trends in Parasitology,* **29,** 83-90. http://dx.doi.org/10.1016/j.pt.2012.10.005

106. Epstein, P.R. and C. Defilippo, 2001: West Nile virus and drought. *Global Change and Human Health,* **2,** 105-107. http://dx.doi.org/10.1023/a:1015089901425

107. Johnson, B.J. and M.V.K. Sukhdeo, 2013: Drought-induced amplification of local and regional West Nile virus infection rates in New Jersey. *Journal of Medical Entomology,* **50,** 195-204. http://dx.doi.org/10.1603/me12035

108. Shaman, J., J.F. Day, and N. Komar, 2010: Hydrologic conditions describe West Nile virus risk in Colorado. *International Journal of Environmental Research and Public Health,* **7,** 494-508. http://dx.doi.org/10.3390/ijerph7020494

109. Landesman, W.J., B.F. Allan, R.B. Langerhans, T.M. Knight, and J.M. Chase, 2007: Inter-annual associations between precipitation and human incidence of West Nile Virus in the United States. *Vector-Borne and Zoonotic Diseases,* **7,** 337-343. http://dx.doi.org/10.1089/vbz.2006.0590

110. Wang, G., R.B. Minnis, J.L. Belant, and C.L. Wax, 2010: Dry weather induces outbreaks of human West Nile virus infections. *BMC Infectious Diseases,* **10,** 38. http://dx.doi.org/10.1186/1471-2334-10-38

111. Shaman, J., J.F. Day, and M. Stieglitz, 2005: Drought-induced amplification and epidemic transmission of West Nile virus in southern Florida. *Journal of Medical Entomology,* **42,** 134-141. http://dx.doi.org/10.1093/jmedent/42.2.134

112. Hjelle, B. and F. Torres-Pérez, 2010: Hantaviruses in the Americas and their role as emerging pathogens. *Viruses,* **2,** 2559-2586. http://dx.doi.org/10.3390/v2122559

113. Clement, J., J. Vercauteren, W.W. Verstraeten, G. Ducoffre, J.M. Barrios, A.-M. Vandamme, P. Maes, and M. Van Ranst, 2009: Relating increasing hantavirus incidences to the changing climate: The mast connection. *International Journal of Health Geographics,* **8,** 1. http://dx.doi.org/10.1186/1476-072x-8-1

114. Kuenzi, A.J., M.L. Morrison, N.K. Madhav, and J.N. Mills, 2007: Brush mouse (*Peromyscus boylii*) population dynamics and hantavirus infection during a warm, drought period in southern Arizona. *Journal of Wildlife Diseases,* **43,** 675-683. http://dx.doi.org/10.7589/0090-3558-43.4.675

115. Klein, S.L. and C.H. Calisher, 2007: Emergence and persistence of hantaviruses. *Wildlife and Emerging Zoonotic Diseases: The Biology, Circumstances and Consequences of Cross-Species Transmission.* Childs, J.E., J.S. Mackenzie, and J.A. Richt, Eds. Springer-Verlag, Berlin, 217-252. http://dx.doi.org/10.1007/978-3-540-70962-6_10

116. Reusken, C. and P. Heyman, 2013: Factors driving hantavirus emergence in Europe. *Current Opinion in Virology,* **3,** 92-99. http://dx.doi.org/10.1016/j.coviro.2013.01.002

117. Watson, D.C., M. Sargianou, A. Papa, P. Chra, I. Starakis, and G. Panos, 2014: Epidemiology of Hantavirus infections in humans: A comprehensive, global overview. *Critical Reviews in Microbiology,* **40,** 261-272. http://dx.doi.org/10.3109/1040841x.2013.783555

118. Peterson, T.C., T.R. Karl, J.P. Kossin, K.E. Kunkel, J.H. Lawrimore, J.R. McMahon, R.S. Vose, and X. Yin, 2014: Changes in weather and climate extremes: State of knowledge relevant to air and water quality in the United States. *Journal of the Air & Waste Management Association,* **64,** 184-197. http://dx.doi.org/10.1080/10962247.2013.851044

119. Munson, S.M., J. Belnap, and G.S. Okin, 2011: Responses of wind erosion to climate-induced vegetation changes on the Colorado Plateau. *Proceedings of the National Academy of Sciences,* **108,** 3854-3859. http://dx.doi.org/10.1073/pnas.1014947108

120. Neu, J.L. and M.J. Prather, 2012: Toward a more physical representation of precipitation scavenging in global chemistry models: Cloud overlap and ice physics and their impact on tropospheric ozone. *Atmospheric Chemistry and Physics,* **12,** 3289-3310. http://dx.doi.org/10.5194/acp-12-3289-2012

121. EPA, 2009: Integrated Science Assessment for Particulate Matter. EPA/600/R-08/139F. National Center for Environmental Assessment, Office of Research and Development, U.S. Environmental Protection Agency, Research Triangle Park, NC. http://cfpub.epa.gov/ncea/cfm/recordisplay.cfm?deid=216546

122. Morman, S.A. and G.S. Plumlee, 2013: The role of airborne mineral dusts in human disease. *Aeolian Research,* **9,** 203-212. http://dx.doi.org/10.1016/j.aeolia.2012.12.001

123. Puett, R.C., J. Schwartz, J.E. Hart, J.D. Yanosky, F.E. Speizer, H. Suh, C.J. Paciorek, L.M. Neas, and F. Laden, 2008: Chronic particulate exposure, mortality, and coronary heart disease in the Nurses' Health Study. *American Journal of Epidemiology,* **168,** 1161-1168. http://dx.doi.org/10.1093/aje/kwn232

124. Fritze, J.G., G.A. Blashki, S. Burke, and J. Wiseman, 2008: Hope, despair and transformation: Climate change and the promotion of mental health and wellbeing. *International Journal of Mental Health Systems,* **2,** 13. http://dx.doi.org/10.1186/1752-4458-2-13

125. Obrien, L.V., H.L. Berry, C. Coleman, and I.C. Hanigan, 2014: Drought as a mental health exposure. *Environmental Research,* **131,** 181-187. http://dx.doi.org/10.1016/j.envres.2014.03.014

126. Carnie, T.-L., H.L. Berry, S.A. Blinkhorn, and C.R. Hart, 2011: In their own words: Young people's mental health in drought-affected rural and remote NSW. *Australian Journal of Rural Health,* **19,** 244-248. http://dx.doi.org/10.1111/j.1440-1584.2011.01224.x

127. Hanigan, I.C., C.D. Butler, P.N. Kokic, and M.F. Hutchinson, 2012: Suicide and drought in New South Wales, Australia, 1970–2007. *Proceedings of the National Academy of Sciences,* **109,** 13950-13955. http://dx.doi.org/10.1073/pnas.1112965109

128. Brahney, J., A.P. Ballantyne, C. Sievers, and J.C. Neff, 2013: Increasing Ca2+ deposition in the western US: The role of mineral aerosols. *Aeolian Research,* **10,** 77-87. http://dx.doi.org/10.1016/j.aeolia.2013.04.003

129. Reheis, M.C. and F.E. Urban, 2011: Regional and climatic controls on seasonal dust deposition in the southwestern U.S. *Aeolian Research,* **3,** 3-21. http://dx.doi.org/10.1016/j.aeolia.2011.03.008

130. Belnap, J., B.J. Walker, S.M. Munson, and R.A. Gill, 2014: Controls on sediment production in two U.S. deserts. *Aeolian Research,* **14,** 15-24. http://dx.doi.org/10.1016/j.aeolia.2014.03.007

131. Ginoux, P., J.M. Prospero, T.E. Gill, N.C. Hsu, and M. Zhao, 2012: Global-scale attribution of anthropogenic and natural dust sources and their emission rates based on MODIS Deep Blue aerosol products. *Reviews of Geophysics,* **50,** RG3005. http://dx.doi.org/10.1029/2012rg000388

132. Hefflin, B.J., B. Jalaludin, E. McClure, N. Cobb, C.A. Johnson, L. Jecha, and R.A. Etzel, 1994: Surveillance for dust storms and respiratory diseases in Washington State, 1991. *Archives of Environmental Health: An International Journal,* **49,** 170-174. http://dx.doi.org/10.1080/00039896.1994.9940378

133. Panikkath, R., C.A. Jumper, and Z. Mulkey, 2013: Multilobar lung infiltrates after exposure to dust storm: The haboob lung syndrome. *The American Journal of Medicine,* **126,** e5-e7. http://dx.doi.org/10.1016/j.amjmed.2012.08.012

134. Ashley, W.S., S. Strader, D.C. Dziubla, and A. Haberlie, 2015: Driving blind: Weather-related vision hazards and fatal motor vehicle crashes. *Bulletin of the American Meteorological Society,* **96,** 755-778. http://dx.doi.org/10.1175/BAMS-D-14-00026.1

135. Ashley, W.S. and A.W. Black, 2008: Fatalities associated with nonconvective high-wind events in the United States. *Journal of Applied Meteorology and Climatology,* **47,** 717-725. http://dx.doi.org/10.1175/2007jamc1689.1

136. Beggs, P.J., 2004: Impacts of climate change on aeroallergens: Past and future. *Clinical & Experimental Allergy,* **34,** 1507-1513. http://dx.doi.org/10.1111/j.1365-2222.2004.02061.x

137. Pfaller, M.A., P.G. Pappas, and J.R. Wingard, 2006: Invasive fungal pathogens: Current epidemiological trends. *Clinical Infectious Diseases,* **43 Supp 1,** S3-S14. http://dx.doi.org/10.1086/504490

138. Smith, C.E., 1946: Effect of season and dust control on Coccidioidomycosis. *Journal of the American Medical Association,* **132,** 833-838. http://dx.doi.org/10.1001/jama.1946.02870490011003

139. Park, B.J., K. Sigel, V. Vaz, K. Komatsu, C. McRill, M. Phelan, T. Colman, A.C. Comrie, D.W. Warnock, J.N. Galgiani, and R.A. Hajjeh, 2005: An epidemic of Coccidioidomycosis in Arizona associated with climatic changes, 1998–2001. *Journal of Infectious Diseases,* **191,** 1981-1987. http://dx.doi.org/10.1086/430092

140. Zender, C.S. and J. Talamantes, 2006: Climate controls on valley fever incidence in Kern County, California. *International Journal of Biometeorology,* **50,** 174-182. http://dx.doi.org/10.1007/s00484-005-0007-6

141. Marsden-Haug, N., H. Hill, A.P. Litvintseva, D.M. Engelthaler, E.M. Driebe, C.C. Roe, C. Ralston, S. Hurst, M. Goldoft, L. Gade, R. Wohrle, G.R. Thompson, M.E. Brandt, and T. Chiller, 2014: Coccidioides immitis identified in soil outside of its known range - Washington, 2013. *MMWR. Morbidity and Mortality Weekly Report,* **63,** 450. PMID:24848217 http://www.cdc.gov/mmwr/preview/mmwrhtml/mm6320a3.htm

142. Litvintseva, A.P., N. Marsden-Haug, S. Hurst, H. Hill, L. Gade, E.M. Driebe, C. Ralston, C. Roe, B.M. Barker, M. Goldoft, P. Keim, R. Wohrle, G.R. Thompson, D.M. Engelthaler, M.E. Brandt, and T. Chiller, 2015: Valley fever: Finding new places for an old disease: *Coccidioides immitis* found in Washington State soil associated with recent human infection. *Clinical Infectious Diseases,* **60,** e1-e3. http://dx.doi.org/10.1093/cid/ciu681

143. Garcia-Solache, M.A. and A. Casadevall, 2010: Global warming will bring new fungal diseases for mammals. *mBio,* **1,** e00061-10. http://dx.doi.org/10.1128/mBio.00061-10

144. Garfin, G., G. Franco, H. Blanco, A. Comrie, P. Gonzalez, T. Piechota, R. Smyth, and R. Waskom, 2014: Ch. 20: Southwest. *Climate Change Impacts in the United States: The Third National Climate Assessment.* Melillo, J.M., T.C. Richmond, and G.W. Yohe, Eds. U.S. Global Change Research Program, Washington, D.C., 462-486. http://dx.doi.org/10.7930/J08G8HMN

145. Naeher, L.P., M. Brauer, M. Lipsett, J.T. Zelikoff, C.D. Simpson, J.Q. Koenig, and K.R. Smith, 2007: Woodsmoke health effects: A review. *Inhalation Toxicology,* **19,** 67-106. http://dx.doi.org/10.1080/08958370600985875

146. Stefanidou, M., S. Athanaselis, and C. Spiliopoulou, 2008: Health impacts of fire smoke inhalation. *Inhalation Toxicology*, **20**, 761-766. http://dx.doi.org/10.1080/08958370801975311

147. Wegesser, T.C., K.E. Pinkerton, and J.A. Last, 2009: California wildfires of 2008: Coarse and fine particulate matter toxicity. *Environmental Health Perspectives*, **117**, 893-897. http://dx.doi.org/10.1289/ehp.0800166

148. Viswanathan, S., L. Eria, N. Diunugala, J. Johnson, and C. McClean, 2006: An analysis of effects of San Diego wildfire on ambient air quality. *Journal of the Air & Waste Management Association*, **56**, 56-67. http://dx.doi.org/10.1080/10473289.2006.10464439

149. Sapkota, A., J.M. Symons, J. Kleissl, L. Wang, M.B. Parlange, J. Ondov, P.N. Breysse, G.B. Diette, P.A. Eggleston, and T.J. Buckley, 2005: Impact of the 2002 Canadian forest fires on particulate matter air quality in Baltimore City. *Environmental Science & Technology*, **39**, 24-32. http://dx.doi.org/10.1021/es035311z

150. Youssouf, H., C. Liousse, L. Roblou, E.M. Assamoi, R.O. Salonen, C. Maesano, S. Banerjee, and I. Annesi-Maesano, 2014: Quantifying wildfires exposure for investigating health-related effects. *Atmospheric Environment*, **97**, 239-251. http://dx.doi.org/10.1016/j.atmosenv.2014.07.041

151. Holstius, D.M., C.E. Reid, B.M. Jesdale, and R. Morello-Frosch, 2012: Birth weight following pregnancy during the 2003 southern California wildfires. *Environmental Health Perspectives*, **120**, 1340-1345. http://dx.doi.org/10.1289/ehp.1104515

152. Johnston, F., I. Hanigan, S. Henderson, G. Morgan, and D. Bowman, 2011: Extreme air pollution events from bushfires and dust storms and their association with mortality in Sydney, Australia 1994–2007. *Environmental Research*, **111**, 811-816. http://dx.doi.org/10.1016/j.envres.2011.05.007

153. Elliott, C., S. Henderson, and V. Wan, 2013: Time series analysis of fine particulate matter and asthma reliever dispensations in populations affected by forest fires. *Environmental Health*, **12**, 11. http://dx.doi.org/10.1186/1476-069X-12-11

154. Henderson, S.B., M. Brauer, Y.C. Macnab, and S.M. Kennedy, 2011: Three measures of forest fire smoke exposure and their associations with respiratory and cardiovascular health outcomes in a population-based cohort. *Environmental Health Perspectives*, **119**, 1266-1271. http://dx.doi.org/10.1289/ehp.1002288

155. Henderson, S.B. and F.H. Johnston, 2012: Measures of forest fire smoke exposure and their associations with respiratory health outcomes. *Current Opinion in Allergy and Clinical Immunology*, **12**, 221-227. http://dx.doi.org/10.1097/ACI.0b013e328353351f

156. Delfino, R.J., S. Brummel, J. Wu, H. Stern, B. Ostro, M. Lipsett, A. Winer, D.H. Street, L. Zhang, T. Tjoa, and D.L. Gillen, 2009: The relationship of respiratory and cardiovascular hospital admissions to the southern California wildfires of 2003. *Occupational and Environmental Medicine*, **66**, 189-197. http://dx.doi.org/10.1136/oem.2008.041376

157. Rappold, A.G., S.L. Stone, W.E. Cascio, L.M. Neas, V.J. Kilaru, M.S. Carraway, J.J. Szykman, A. Ising, W.E. Cleve, J.T. Meredith, H. Vaughan-Batten, L. Deyneka, and R.B. Devlin, 2011: Peat bog wildfire smoke exposure in rural North Carolina is associated with cardiopulmonary emergency department visits assessed through syndromic surveillance. *Environmental Health Perspectives*, **119**, 1415-1420. http://dx.doi.org/10.1289/ehp.1003206

158. Sutherland, E.R., B.J. Make, S. Vedal, L. Zhang, S.J. Dutton, J.R. Murphy, and P.E. Silkoff, 2005: Wildfire smoke and respiratory symptoms in patients with chronic obstructive pulmonary disease. *Journal of Allergy and Clinical Immunology*, **115**, 420-422. http://dx.doi.org/10.1016/j.jaci.2004.11.030

159. Garcia, C.A., P.-S. Yap, H.-Y. Park, and B.L. Weller, 2015: Association of long-term PM2.5 exposure with mortality using different air pollution exposure models: Impacts in rural and urban California. *International Journal of Environmental Health Research*, Published online 17 July 2015. http://dx.doi.org/10.1080/09603123.2015.1061113

160. Gold, D.R., A. Litonjua, J. Schwartz, E. Lovett, A. Larson, B. Nearing, G. Allen, M. Verrier, R. Cherry, and R. Verrier, 2000: Ambient pollution and heart rate variability. *Circulation*, **101**, 1267-1273. http://dx.doi.org/10.1161/01.CIR.101.11.1267

161. Pope, C.A., III, M.C. Turner, R.T. Burnett, M. Jerrett, S.M. Gapstur, W.R. Diver, D. Krewski, and R.D. Brook, 2015: Relationships between fine particulate air pollution, cardiometabolic disorders, and cardiovascular mortality. *Circulation Research*, **116**, 108-115. http://dx.doi.org/10.1161/CIRCRESAHA.116.305060

162. Pope, C.A., III, R.T. Burnett, G.D. Thurston, M.J. Thun, E.E. Calle, D. Krewski, and J.J. Godleski, 2004: Cardiovascular mortality and long-term exposure to particulate air pollution: Epidemiological evidence of general pathophysiological pathways of disease. *Circulation*, **109**, 71-77. http://dx.doi.org/10.1161/01.CIR.0000108927.80044.7F

163. Phuleria, H.C., P.M. Fine, Y. Zhu, and C. Sioutas, 2005: Air quality impacts of the October 2003 southern California wildfires. *Journal of Geophysical Research*, **110**, D07S20. http://dx.doi.org/10.1029/2004jd004626

164. Thelen, B., N.H.F. French, B.W. Koziol, M. Billmire, R.C. Owen, J. Johnson, M. Ginsberg, T. Loboda, and S. Wu, 2013: Modeling acute respiratory illness during the 2007 San Diego wildland fires using a coupled emissions-transport system and generalized additive modeling. *Environmental Health*, **12**, 94. http://dx.doi.org/10.1186/1476-069x-12-94

165. Olsen, C.S., D.K. Mazzotta, E. Toman, and A.P. Fischer, 2014: Communicating about smoke from wildland fire: Challenges and opportunities for managers. *Environmental Management*, **54**, 571-582. http://dx.doi.org/10.1007/s00267-014-0312-0

166. Richardson, L.A., P.A. Champ, and J.B. Loomis, 2012: The hidden cost of wildfires: Economic valuation of health effects of wildfire smoke exposure in Southern California. *Journal of Forest Economics*, **18**, 14-35. http://dx.doi.org/10.1016/j.jfe.2011.05.002

167. McMeeking, G.R., S.M. Kreidenweis, M. Lunden, J. Carrillo, C.M. Carrico, T. Lee, P. Herckes, G. Engling, D.E. Day, J. Hand, N. Brown, W.C. Malm, and J.L. Collett, 2006: Smoke-impacted regional haze in California during the summer of 2002. *Agricultural and Forest Meteorology*, **137**, 25-42. http://dx.doi.org/10.1016/j.agrformet.2006.01.011

168. Youssouf, H., C. Liousse, L. Roblou, E.-M. Assamoi, R.O. Salonen, C. Maesano, S. Banerjee, and I. Annesi-Maesano, 2014: Non-accidental health impacts of wildfire smoke. *International Journal of Environmental Research and Public Health*, **11**, 11772-11804. http://dx.doi.org/10.3390/ijerph111111772

169. Künzli, N., E. Avol, J. Wu, W.J. Gauderman, E. Rappaport, J. Millstein, J. Bennion, R. McConnell, F.D. Gilliland, K. Berhane, F. Lurmann, A. Winer, and J.M. Peters, 2006: Health effects of the 2003 southern California wildfires on children. *American Journal of Respiratory and Critical Care Medicine*, **174**, 1221-1228. http://dx.doi.org/10.1164/rccm.200604-519OC

170. Leonard, S.S., V. Castranova, B.T. Chen, D. Schwegler-Berry, M. Hoover, C. Piacitelli, and D.M. Gaughan, 2007: Particle size-dependent radical generation from wildland fire smoke. *Toxicology*, **236**, 103-113. http://dx.doi.org/10.1016/j.tox.2007.04.008

171. Ghilarducci, D.P. and R.S. Tjeerdema, 1995: Fate and effects of acrolein. *Reviews of Environmental Contamination and Toxicology*, **144**, 95-146. http://dx.doi.org/10.1007/978-1-4612-2550-8_2

172. Squire, B., C. Chidester, and S. Raby, 2011: Medical events during the 2009 Los Angeles County Station fire: Lessons for wildfire EMS planning. *Prehospital Emergency Care*, **15**, 464-472. http://dx.doi.org/10.3109/10903127.2011.598607

173. Booze, T.F., T.E. Reinhardt, S.J. Quiring, and R.D. Ottmar, 2004: A screening-level assessment of the health risks of chronic smoke exposure for wildland firefighters. *Journal of Occupational and Environmental Hygiene*, **1**, 296-305. http://dx.doi.org/10.1080/15459620490442500

174. Caamano-Isorna, F., A. Figueiras, I. Sastre, A. Montes-Martinez, M. Taracido, and M. Pineiro-Lamas, 2011: Respiratory and mental health effects of wildfires: An ecological study in Galician municipalities (north-west Spain). *Environmental Health*, **10**, Article 48. http://dx.doi.org/10.1186/1476-069X-10-48

175. Jones, R.T., D.P. Ribbe, P.B. Cunningham, J.D. Weddle, and A.K. Langley, 2002: Psychological impact of fire disaster on children and their parents. *Behavior Modification*, **26**, 163-186. http://dx.doi.org/10.1177/0145445502026002003

176. Mann, M.L., P. Berck, M.A. Moritz, E. Batllori, J.G. Baldwin, C.K. Gately, and D.R. Cameron, 2014: Modeling residential development in California from 2000 to 2050: Integrating wildfire risk, wildland and agricultural encroachment. *Land Use Policy*, **41**, 438-452. http://dx.doi.org/10.1016/j.landusepol.2014.06.020

177. Thomas, D.S. and D.T. Butry, 2014: Areas of the U.S. wildland–urban interface threatened by wildfire during the 2001–2010 decade. *Natural Hazards*, **71**, 1561-1585. http://dx.doi.org/10.1007/s11069-013-0965-7

178. Peters, M.P., L.R. Iverson, S.N. Matthews, and A.M. Prasad, 2013: Wildfire hazard mapping: Exploring site conditions in eastern US wildland–urban interfaces. *International Journal of Wildland Fire*, **22**, 567-578. http://dx.doi.org/10.1071/WF12177

179. Radeloff, V.C., R.B. Hammer, S.I. Stewart, J.S. Fried, S.S. Holcomb, and J.F. McKeefry, 2005: The wildland-urban interface in the United States. *Ecological Applications,* **15,** 799-805. http://dx.doi.org/10.1890/04-1413

180. Smith, H.G., G.J. Sheridan, P.N.J. Lane, P. Nyman, and S. Haydon, 2011: Wildfire effects on water quality in forest catchments: A review with implications for water supply. *Journal of Hydrology,* **396,** 170-192. http://dx.doi.org/10.1016/j.jhydrol.2010.10.043

181. Emelko, M.B., U. Silins, K.D. Bladon, and M. Stone, 2011: Implications of land disturbance on drinking water treatability in a changing climate: Demonstrating the need for "source water supply and protection" strategies. *Water Research,* **45,** 461-472. http://dx.doi.org/10.1016/j.watres.2010.08.051

182. Moody, J.A., R.A. Shakesby, P.R. Robichaud, S.H. Cannon, and D.A. Martin, 2013: Current research issues related to post-wildfire runoff and erosion processes. *Earth-Science Reviews,* **122,** 10-37. http://dx.doi.org/10.1016/j.earscirev.2013.03.004

183. Cannon, S.H., J.E. Gartner, R.C. Wilson, J.C. Bowers, and J.L. Laber, 2008: Storm rainfall conditions for floods and debris flows from recently burned areas in southwestern Colorado and southern California. *Geomorphology,* **96,** 250-269. http://dx.doi.org/10.1016/j.geomorph.2007.03.019

184. Cannon, S.H. and J. DeGraff, 2009: The increasing wildfire and post-fire debris-flow threat in western USA, and implications for consequences of climate change. *Landslides – Disaster Risk Reduction.* Sassa, K. and P. Canuti, Eds. Springer, Berlin, 177-190. http://dx.doi.org/10.1007/978-3-540-69970-5_9

185. Jordan, P., K. Turner, D. Nicol, and D. Boyer, 2006: Developing a risk analysis procedure for post-wildfire mass movement and flooding in British Columbia. *1st Specialty Conference on Disaster Medicine,* May 23-26, Calgary, Alberta, Canada. http://www.for.gov.bc.ca/hfd/pubs/rsi/fsp/Misc/Misc071.pdf

186. Sham, C.H., M.E. Tuccillo, and J. Rooke, 2013: Effects of Wildfire on Drinking Water Utilities and Best Practices for Wildfire Risk Reduction and Mitigation. Web Report #4482, 119 pp. Water Research Foundation, Denver, CO. http://www.waterrf.org/publicreportlibrary/4482.pdf

187. USGS, 2012: Wildfire Effects on Source-Water Quality: Lessons from Fourmile Canyon Fire, Colorado, and Implications for Drinking-Water Treatment. U.S. Geological Survey Fact Sheet 2012-3095, 4 pp. http://pubs.usgs.gov/fs/2012/3095/FS12-3095.pdf

188. Rhoades, C.C., D. Entwistle, and D. Butler, 2012: Water quality effects following a severe fire. *Fire Management Today,* **72,** (2):35-39. http://www.fs.fed.us/fire/fmt/fmt_pdfs/FMT72-2.pdf

189. Ashley, W.S. and C.W. Gilson, 2009: A reassessment of U.S. lightning mortality. *Bulletin of the American Meteorological Society,* **90,** 1501-1518. http://dx.doi.org/10.1175/2009bams2765.1

190. Black, A.W. and W.S. Ashley, 2010: Nontornadic convective wind fatalities in the United States. *Natural Hazards,* **54,** 355-366. http://dx.doi.org/10.1007/s11069-009-9472-2

191. Schmidlin, T.W., 2009: Human fatalities from wind-related tree failures in the United States, 1995–2007. *Natural Hazards,* **50,** 13-25. http://dx.doi.org/10.1007/s11069-008-9314-7

192. Ashley, W.S. and T.L. Mote, 2005: Derecho hazards in the United States. *Bulletin of the American Meteorological Society,* **86,** 1577-1592. http://dx.doi.org/10.1175/BAMS-86-11-1577

193. Jahromi, A.H., R. Wigle, and A.M. Youssef, 2011: Are we prepared yet for the extremes of weather changes? Emergence of several severe frostbite cases in Louisiana. *The American Surgeon,* **77,** 1712-1713.

194. Lim, C. and J. Duflou, 2008: Hypothermia fatalities in a temperate climate: Sydney, Australia. *Pathology,* **40,** 46-51. http://dx.doi.org/10.1080/00313020701716466

195. Ramin, B. and T. Svoboda, 2009: Health of the homeless and climate change. *Journal of Urban Health,* **86,** 654-664. http://dx.doi.org/10.1007/s11524-009-9354-7

196. Liddell, C. and C. Morris, 2010: Fuel poverty and human health: A review of recent evidence. *Energy Policy,* **38,** 2987-2997. http://dx.doi.org/10.1016/j.enpol.2010.01.037

197. Bhattacharya, J., T. DeLeire, S. Haider, and J. Currie, 2003: Heat or eat? Cold-weather shocks and nutrition in poor american families. *American Journal of Public Health,* **93,** 1149-1154. http://dx.doi.org/10.2105/AJPH.93.7.1149

198. Dey, A.N., P. Hicks, S. Benoit, and J.I. Tokars, 2010: Automated monitoring of clusters of falls associated with severe winter weather using the BioSense system. *Injury Prevention,* **16,** 403-407. http://dx.doi.org/10.1136/ip.2009.025841

199. Eisenberg, D. and K.E. Warner, 2005: Effects of snowfalls on motor vehicle collisions, injuries, and fatalities. *American Journal of Public Health,* **95,** 120-124. http://dx.doi.org/10.2105/AJPH.2004.048926

200. Fayard, G.M., 2009: Fatal work injuries involving natural disasters, 1992–2006. *Disaster Medicine and Public Health Preparedness,* **3,** 201-209. http://dx.doi.org/10.1097/DMP.0b013e3181b65895

201. Adams, Z.W., J.A. Sumner, C.K. Danielson, J.L. McCauley, H.S. Resnick, K. Grös, L.A. Paul, K.E. Welsh, and K.J. Ruggiero, 2014: Prevalence and predictors of PTSD and depression among adolescent victims of the Spring 2011 tornado outbreak. *Journal of Child Psychology and Psychiatry,* **55,** 1047-1055. http://dx.doi.org/10.1111/jcpp.12220

202. Conlon, K.C., N.B. Rajkovich, J.L. White-Newsome, L. Larsen, and M.S. O'Neill, 2011: Preventing cold-related morbidity and mortality in a changing climate. *Maturitas,* **69,** 197-202. http://dx.doi.org/10.1016/j.maturitas.2011.04.004

203. Bowles, D.C., C.D. Butler, and S. Friel, 2014: Climate change and health in Earth's future. *Earth's Future,* **2,** 60-67. http://dx.doi.org/10.1002/2013ef000177

204. NOAA, 2015: Natural Hazard Statistics: Weather Fatalities. National Oceanic and Atmospheric Administration, National Weather Service, Office of Climate, Water, and Weather Services. www.nws.noaa.gov/om/hazstats.shtml

205. Smith, A.B. and R.W. Katz, 2013: US billion-dollar weather and climate disasters: Data sources, trends, accuracy and biases. *Natural Hazards,* **67,** 387-410. http://dx.doi.org/10.1007/s11069-013-0566-5

206. Barbero, R., J.T. Abatzoglou, N.K. Larkin, C.A. Kolden, and B. Stocks, 2015: Climate change presents increased potential for very large fires in the contiguous United States. *International Journal of Wildland Fire.* http://dx.doi.org/10.1071/WF15083

THE IMPACTS OF CLIMATE CHANGE ON HUMAN HEALTH IN THE UNITED STATES
A Scientific Assessment

5 VECTOR-BORNE DISEASES

Lead Authors

Charles B. Beard
Centers for Disease Control and Prevention

Rebecca J. Eisen
Centers for Disease Control and Prevention

Contributing Authors

Christopher M. Barker
University of California, Davis

Jada F. Garofalo*
Centers for Disease Control and Prevention

Micah Hahn
Centers for Disease Control and Prevention

Mary Hayden
National Center for Atmospheric Research

Andrew J. Monaghan
National Center for Atmospheric Research

Nicholas H. Ogden
Public Health Agency of Canada

Paul J. Schramm
Centers for Disease Control and Prevention

Recommended Citation: Beard, C.B., R.J. Eisen, C.M. Barker, J.F. Garofalo, M. Hahn, M. Hayden, A.J. Monaghan, N.H. Ogden, and P.J. Schramm, 2016: Ch. 5: Vectorborne Diseases. *The Impacts of Climate Change on Human Health in the United States: A Scientific Assessment.* U.S. Global Change Research Program, Washington, DC, 129–156. http://dx.doi.org/10.7930/J0765C7V

*Chapter Coordinator

5 VECTOR-BORNE DISEASES

Key Findings

Changing Distributions of Vectors and Vector-Borne Diseases
Key Finding 1: Climate change is expected to alter the geographic and seasonal distributions of existing vectors and vector-borne diseases *[Likely, High Confidence]*.

Earlier Tick Activity and Northward Range Expansion
Key Finding 2: Ticks capable of carrying the bacteria that cause Lyme disease and other pathogens will show earlier seasonal activity and a generally northward expansion in response to increasing temperatures associated with climate change *[Likely, High Confidence]*. Longer seasonal activity and expanding geographic range of these ticks will increase the risk of human exposure to ticks *[Likely, Medium Confidence]*.

Changing Mosquito-Borne Disease Dynamics
Key Finding 3: Rising temperatures, changing precipitation patterns, and a higher frequency of some extreme weather events associated with climate change will influence the distribution, abundance, and prevalence of infection in the mosquitoes that transmit West Nile virus and other pathogens by altering habitat availability and mosquito and viral reproduction rates *[Very Likely, High Confidence]*. Alterations in the distribution, abundance, and infection rate of mosquitoes will influence human exposure to bites from infected mosquitoes, which is expected to alter risk for human disease *[Very Likely, Medium Confidence]*.

Emergence of New Vector-Borne Pathogens
Key Finding 4: Vector-borne pathogens are expected to emerge or reemerge due to the interactions of climate factors with many other drivers, such as changing land-use patterns *[Likely, High Confidence]*. The impacts to human disease, however, will be limited by the adaptive capacity of human populations, such as vector control practices or personal protective measures *[Likely, High Confidence]*.

5.1 Introduction

Vector-borne diseases are illnesses that are transmitted by vectors, which include mosquitoes, ticks, and fleas. These vectors can carry infective pathogens such as viruses, bacteria, and protozoa, which can be transferred from one host (carrier) to another. In the United States, there are currently 14 vector-borne diseases that are of national public health concern. These diseases account for a significant number of human illnesses and deaths each year and are required to be reported to the National Notifiable Diseases Surveillance System at the Centers for Disease Control and Prevention (CDC). In 2013, state and local health departments reported 51,258 vector-borne disease cases to the CDC (Table 1).

The seasonality, distribution, and prevalence of vector-borne diseases are influenced significantly by climate factors, primarily high and low temperature extremes and precipitation patterns.[11] Climate change can result in modified weather patterns and an increase in extreme events (see Ch. 1: Introduc-

tion) that can affect disease outbreaks by altering biological variables such as vector population size and density, vector survival rates, the relative abundance of disease-carrying animal (zoonotic) reservoir hosts, and pathogen reproduction rates. Collectively, these changes may contribute to an increase in the risk of the pathogen being carried to humans.

Climate change is likely to have both short- and long-term effects on vector-borne disease transmission and infection patterns, affecting both seasonal risk and broad geographic changes in disease occurrence over decades. However, models for predicting the effects of climate change on vector-borne diseases are subject to a high degree of uncertainty, largely due to two factors: 1) vector-borne diseases are maintained in nature in complex transmission cycles that involve vectors, other intermediate zoonotic hosts, and humans; and 2) there are a number of other significant social and environmental drivers of vector-borne disease transmission in addition to cli-

Summary of Reported Case Counts of Notifiable[a] Vector-Borne Diseases in the United States.

Diseases	2013 Reported Cases	Median (range) 2004–2013[b]
Tick-Borne		
Lyme disease	36,307	30,495 (19,804–38,468)
Spotted Fever Rickettsia	3,359	2,255 (1,713–4,470)
Anaplasmosis/Ehrlichiosis	4,551	2,187 (875–4,551)
Babesiosis[b]	1,792	1,128 (940–1,792)
Tularemia	203	136 (93–203)
Powassan	15	7 (1–16)
Mosquito-Borne		
West Nile virus	2,469	1,913 (712–5,673)
Malaria[c]	1,594	1,484 (1,255–1,773)
Dengue[b,c]	843	624 (254–843)
California serogroup viruses	112	78 (55–137)
Eastern equine encephalitis	8	7 (4–21)
St. Louis encephalitis	1	10 (1–13)
Flea-Borne		
Plague	4	4 (2–17)

[a] State Health Departments are required by law to report regular, frequent, and timely information about individual cases to the CDC in order to assist in the prevention and control of diseases. Case counts are summarized based on annual reports of nationally notifiable infectious diseases.[1, 2, 3, 4, 5, 6, 7, 8, 9, 10]

[b] Babesiosis and dengue were added to the list of nationally notifiable diseases in 2011 and 2009, respectively. Median and range values encompass cases reported from 2011 to 2013 for babesiosis and from 2010 to 2013 for dengue.

[c] Primarily acquired outside of the United States and based on travel-related exposures.

Table 1: Vectors and hosts involved in the transmission of these infective pathogens are sensitive to climate change and other environmental factors which, together, affect vector-borne diseases by influencing one or more of the following: vector and host survival, reproduction, development, activity, distribution, and abundance; pathogen development, replication, maintenance, and transmission; geographic range of pathogens, vectors, and hosts; human behavior; and disease outbreak frequency, onset, and distribution.[11]

mate change. For example, while climate variability and climate change both alter the transmission of vector-borne diseases, they will likely interact with many other factors, including how pathogens adapt and change, the availability of hosts, changing ecosystems and land use, demographics, human behavior, and adaptive capacity.[12, 13] These complex interactions make it difficult to predict the effects of climate change on vector-borne diseases.

The risk of introducing exotic pathogens and vectors not currently present in the United States, while likely to occur, is similarly difficult to project quantitatively.[14, 15, 16] In recent years, several important vector-borne pathogens have been introduced or reintroduced into the United States. These include West Nile virus, dengue virus, and chikungunya virus. In the case of the 2009 dengue outbreak in southern Florida, climate change was not responsible for the reintroduction of the virus in this area, which arrived via infected travelers from disease-endemic regions of the Caribbean.[17] In fact, vector populations capable of transmitting dengue have been present for many years throughout much of the southern United States, including Florida.[18] Climate change has the potential to increase human exposure risk or disease transmission following shifts in extended spring and summer seasons as dengue becomes more established in the United States. Climate change effects, however, are difficult to quantify due to the adaptive capacity of a population that may reduce exposure to vector-borne pathogens through such means as air conditioning, screens on windows, vector control and public health practices.

This chapter presents case studies of Lyme disease and West Nile virus infection in relation to weather and climate. Although ticks and mosquitoes transmit multiple infectious pathogens to humans in the United States, Lyme disease and West Nile virus infection are the most commonly reported tick-borne and mosquito-borne diseases in this country (Table 1). In addition, a substantial number of studies have been conducted to elucidate the role of climate in the transmission of these infectious pathogens. These broad findings, together with the areas of uncertainty from these case studies, are generalizable to other vector-borne diseases.[11]

5.2 Lyme Disease

State of the Science

Lyme disease is a tick-borne bacterial disease that is endemic (commonly found) in parts of North America, Europe, and Asia. In the United States, Lyme disease is caused by the bacterium *Borrelia burgdorferi sensu stricto* (*B. burgdorferi;* one of the spiral-shaped bacteria known as spirochetes) and is the most commonly reported vector-borne illness. It is primarily transmitted to humans in the eastern United States by the tick species *Ixodes scapularis* (formerly *I. dammini*), known as blacklegged ticks or deer ticks, and in the far western United States by *I. pacificus*, commonly known as western blacklegged ticks.[19] Ill-

ness in humans typically presents with fever, headache, fatigue, and a characteristic skin rash called erythema migrans. If left untreated, infection can spread to joints, the heart, and the nervous system.[20] Since 1991, when standardized surveillance and reporting of Lyme disease began in the United States, case counts have increased steadily.[21] Since 2007, more than 25,000 Lyme disease cases have been reported annually.[22] The geographic distribution of the disease is limited to specific regions in the United States (Figure 2), transmission occurs seasonally, and year-to-year variation in case counts and in seasonal onset is considerable.[20, 21, 23] Each of these observations suggest that geographic location and seasonal climate variability may play a significant role in determining when and where Lyme disease cases are most likely to occur.

Although the reported incidence of Lyme disease is greater in the eastern United States compared with the westernmost United States,[20, 21] in both geographical regions, nymphs (small immature ticks) are believed to be the life stage that is most significant in pathogen transmission from infected hosts (primarily rodents) to humans (Figure 2, Figure 3).[24, 25] Throughout the United States, the majority of human cases report onset of clinical signs of infection during the months of June, July, and August. The summer is a period of parallel increased activity for both blacklegged and western blacklegged ticks in the nymphal life stage (the more infectious stage) and for human recreational activity outdoors.[21, 25]

Infection rates in humans vary significantly from year to year. From 1992 to 2006, variation in case counts of Lyme disease was as high as 57% from one year to the next.[21] Likewise, the precise week of onset of Lyme disease cases across states in the eastern United States, where Lyme disease is endemic, differed by as much as 10 weeks from 1992 to 2007. Much of this variation in timing of disease onset can be explained by geographic region (cases occurred earlier in warmer states in the mid-Atlantic region compared with cooler states in the North); however, the annual variation of disease onset within regions was notable and linked to winter and spring climate variability (see "Annual and Seasonal Variation in Lyme Disease" on page 136).[23]

The geographic and seasonal distributions of Lyme disease case occurrence are driven, in part, by the life cycle of vector ticks (Figure 3). Humans are only exposed to Lyme disease spirochetes (*B. burgdorferi*) in locations where both the vector tick populations and the infection-causing spirochetes are present.[27] Within these locations, the potential for contracting Lyme disease depends on three key factors: 1) tick vector abundance (the density of host-seeking nymphs being particularly important), 2) prevalence of *B. burgdorferi* infection in ticks (the prevalence in nymphs being particularly important), and 3) contact frequency between infected ticks and humans.[28] To varying degrees, climate change can affect all three of these factors.

Climate Change and Health—Lyme Disease

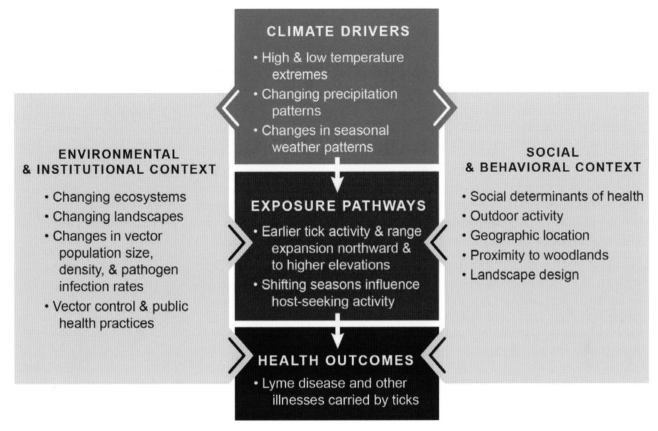

Figure 1: This conceptual diagram illustrates the key pathways by which climate change influences human exposure to Lyme disease and the potential resulting health outcomes (center boxes). These exposure pathways exist within the context of other factors that positively or negatively influence health outcomes (gray side boxes). Key factors that influence vulnerability for individuals are shown in the right box, and include social determinants of health and behavioral choices. Key factors that influence vulnerability at larger scales, such as natural and built environments, governance and management, and institutions, are shown in the left box. All of these influencing factors can affect an individual's or a community's vulnerability through changes in exposure, sensitivity, and adaptive capacity and may also be affected by climate change. See Ch. 1: Introduction for more information.

Changes in Lyme Disease Case Report Distribution

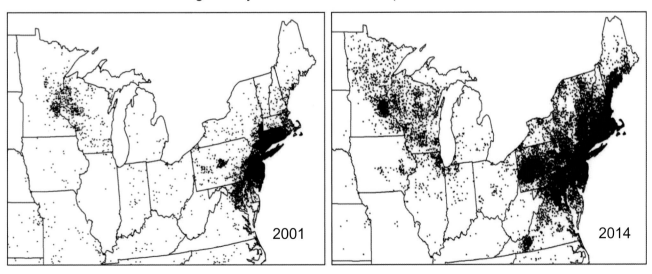

Figure 2: Maps show the reported cases of Lyme disease in 2001 and 2014 for the areas of the country where Lyme disease is most common (the Northeast and Upper Midwest). Both the distribution and the numbers of cases have increased. (Figure source: adapted from CDC 2015)[26]

Life Cycle of Blacklegged Ticks, *Ixodes scapularis*

The seasonal occurrence of Lyme disease cases is related, partially, to the timing of a blood meal (host-seeking activity) of ticks and the three-stage life cycle (larvae, nymph, and adult) of ticks.[48] Increasing temperatures and the accompanying changes in seasonal patterns are expected to result in earlier seasonal tick activity and an expansion in tick habitat range, increasing the risk of human exposure to ticks.

For blacklegged ticks and western blacklegged ticks, spirochete transmission from adult ticks to eggs is rare or does not occur.[49] Instead, immature ticks (larvae and nymphs) acquire infection-causing *B. burgdorferi* spirochetes by feeding on rodents, other small mammals, and birds during the spring and summer months. The spirochetes are maintained throughout the tick life cycle from larva to nymph and from nymph to adult. The spirochetes are primarily passed to humans from nymphs and less frequently by adults.

Prevalence of *B. burgdorferi* infection in nymphal ticks depends in part on the structure of the host community.[50, 51] Larval ticks are more likely to be infected in areas where they feed mostly on animals that can carry and transmit the disease-causing bacteria (such as white-footed mice), compared with areas where they feed mostly on hosts that cannot become infected and thus do not pass on the bacteria (such as certain lizards).

Natural variation in potential for rodents, birds, and reptiles to carry *B. burgdorferi* in the wild leads to large differences in infection rates in nymphal ticks, resulting in considerable geographic variation in the transmission cycles and in the opportunity for humans to contract Lyme disease.[52] Unlike nymphal or larval ticks, adult ticks feed mainly during the cooler months of the year, and primarily on deer, which are resistant to *B. burgdorferi* infection and thus play little role in increasing the abundance of infected ticks in the population. However, deer are important for tick reproduction and therefore influence the abundance of nymphs in subsequent generations.[19]

Life Cycle of Blacklegged Ticks, *Ixodes scapularis*

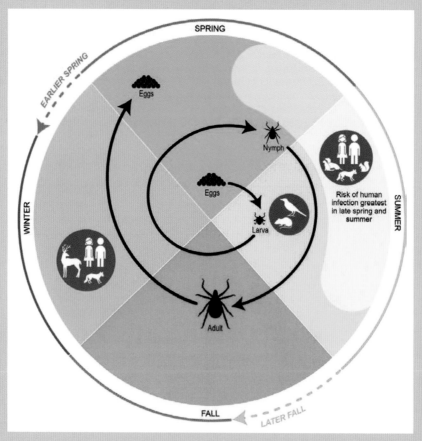

Figure 3: Figure depicts the life cycle of blacklegged ticks, including the phases in which humans can be exposed to Lyme disease, and some of the changes in seasonality expected with climate change. (Figure source: adapted from CDC 2015)[47]

Aside from short periods of time when they are feeding on hosts (less than three weeks of their two- to three-year life cycle), ticks spend most of their lives off of hosts in various natural landscapes (such as woodlands or grasslands) where weather factors including temperature, precipitation, and humidity affect their survival and host-seeking behavior. In general, both low and high temperatures increase tick mortality rates, although increasing humidity can increase their ability to tolerate higher temperatures.[29, 30, 31, 32, 33, 34, 35, 36, 37, 38] Within areas where tick vector populations are present, some studies have demonstrated an association among temperature, humidity, and tick abundance.[39, 40, 41] Factors that are less immediately dependent on climate (for example, landscape and the relative proportions within a community of zoonotic hosts that carry or do not carry Lyme disease-causing bacteria) may be more important in smaller geographic areas.[42, 43] Temperature and humidity also influence the timing of host-seeking activity,[32, 35, 36, 44] and can influence which seasons are of highest risk to the public.

In summary, weather-related variables can determine geographic distributions of ticks and seasonal activity patterns. However, the importance of these weather variables in Lyme disease transmission to humans compared with other important predictors is likely scale-dependent. In general, across the entire country, climate-related variables often play a significant role in determining the occurrence of tick vectors and Lyme disease incidence in the United States (for example, Lyme disease vectors are absent in the arid Intermountain West where climate conditions are not suitable for tick survival). However, within areas where conditions are suitable for tick survival, other variables (for example, landscape and the relative proportions within a community of zoonotic hosts that carry or do not carry Lyme disease-causing bacteria) are more important for determining tick abundance, infection rates in ticks, and ultimately human infection rates.[39, 45, 46]

Observed Trends and Measures of Human Risk

Geographic Distribution of Ticks

Because the presence of tick vectors is required for *B. burgdorferi* transmission to humans, information on where vector tick species live provides basic information on where Lyme disease risk occurs. Minimum temperature appears to be a key variable in defining the geographic distribution of blacklegged ticks.[39, 45, 53] Low minimum temperatures in winter may lead to environmental conditions that are unsuitable for tick population survival. The probability of a given geographic area being suitable for tick populations increases as minimum temperature rises.[45] In the case of the observed northward range expansion of blacklegged ticks into Canada, higher temperatures appear to be a key factor affecting where, and how fast, ticks are colonizing new localities.[54, 55, 56, 57, 58]

In the eastern United States, Lyme disease is transmitted to humans primarily by blacklegged (deer) ticks.

Maximum temperatures also significantly affect where blacklegged ticks live.[39, 45] Higher temperatures increase tick development and hatching rates, but reduce tick survival and egg-laying (reproduction) success.[30]

Declines in rainfall amount and humidity are also important in limiting the geographic distribution of blacklegged ticks. Ticks are more likely to reside in moister areas because increased humidity can increase tick survival.[35, 38, 39, 45, 53, 55]

Geographic Distribution of Infected Ticks

Climate variables have been shown to be strong predictors of geographic locations in which blacklegged ticks reside, but less important for determining how many nymphs live in a given area or what proportion of those ticks is infected.[39, 40] The presence of uninfected nymphs and infected nymphs can vary widely over small geographic areas experiencing similar temperature and humidity conditions, which supports the hypothesis that factors other than weather play a significant role in determining nymph survival and infection rates.[37, 39, 40, 41, 44] Additional studies that modeled nymphal density within small portions of the blacklegged tick range (north-central states and Hudson River Valley, NY), and modeling studies that include climate and other non-biological variables indicate only a weak relationship to nymphal density.[59, 60] Nonetheless, climate variables can be used to model nymphal density in some instances. For example, in a single county in northern coastal California with strong climate gradients, warmer areas with less variation between maximum and minimum monthly water vapor in the air were characteristic of areas with elevated concentrations of infected nymphs.[41] However, it is likely that differences in animal host community structure, which vary with climatic conditions (for example, relative abundances of hosts that carry or do not carry Lyme disease-causing bacteria), influenced the concentration of infected nymphs.[37, 61]

Geographic Distribution of Lyme Disease

Though there are links between climate and tick distribution, studies that look for links between weather and geographical differences in human infection rates do not show a clear or consistent link between temperature and Lyme disease incidence.[46, 62, 63]

Annual and Seasonal Variation in Lyme Disease

Temperature and precipitation both influence the host-seeking activity of ticks, which may result in year-to-year variation in the number of new Lyme disease cases and the timing of the season in which Lyme disease infections occur. However, identified associations between precipitation and Lyme disease incidence, or temperature and Lyme disease incidence, are limited or weak.[64, 65] Overall, the association between summer moisture and Lyme disease infection rates in humans remains inconsistent across studies.

The peak period when ticks are seeking hosts starts earlier in the warmer, more southern, states than in northern states.[44] Correspondingly, the onset of human Lyme disease cases occurs earlier as the growing degree days (a measurement of temperature thresholds that must be met for biological processes to occur) increases, yet, the timing of the end of the Lyme disease season does not appear to be determined by weather-related variables.[23] Rather, the number of potential carriers (for example, deer, birds, and humans) likely influences the timing of the end of the Lyme disease season.

The effects of temperature and humidity or precipitation on the seasonal activity patterns of nymphal western blacklegged ticks is more certain than the impacts of these factors on the timing of Lyme disease case occurrence.[36, 37] Peak nymphal activity is generally reached earlier in hotter and drier areas, but lasts for shorter durations. Host-seeking activity ceases earlier in the season in cooler and more humid conditions. The density of nymphal western blacklegged ticks in north-coastal California consistently begins to decline when average daily maximum temperatures are between 70°F (21°C) and 73.5°F (23°C), and when average maximum daily relative humidity decreases below 83%–85%.[36, 37]

Projected Impacts

Warmer winter and spring temperatures are projected to lead to earlier annual onset of Lyme disease cases in the eastern United States (see "Research Highlight" below) and in an earlier onset of nymphal host-seeking behavior.[66] Limited research shows that the geographic distribution of blacklegged ticks is expected to expand to higher latitudes and elevations in the future and retract in the southern United States.[67] Declines in subfreezing temperatures at higher latitudes may be responsible for improved survival of ticks. In many woodlands, ticks can find refuge from far-subzero winter air temperatures in the surface layers of the soil.[68, 69] However, a possibly important impact of climate change will be acceleration of the tick life cycles due to higher temperatures during the spring, summer, and autumn, which would increase the likelihood that ticks survive to reproduce.[58, 70] This prediction is consistent with recent observations of the spread of *I. scapularis* in Canada.[55, 71]

Research Highlight: Lyme Disease

Importance: Lyme disease occurrence is highly seasonal. The annual springtime onset of Lyme disease cases is regulated by climate variability in preceding months. Until now, the possible effects of climate change on the timing of Lyme disease infection in humans early and late in the 21st century have not been addressed for the United States, where Lyme disease is the most commonly reported vector-borne disease.

Objectives: Examine the potential impacts of 21st century climate change on the timing of the beginning of the annual Lyme disease season (annual onset week) in the eastern United States.

Methods: Downscaled future climate projections for four greenhouse gas (GHG) concentration trajectories from five atmosphere–ocean global climate models (AOGCMs) are input to the national-level empirical model of Moore et al. (2014)[23] to simulate the potential impact of 21st century climate change on the annual onset week of Lyme disease in the United States.[23] The four GHG trajectories in order of lowest to highest concentrations are RCP2.6, RCP4.5, RCP6.0, and RCP8.5 (see Appendix 1: Technical Support Document).

Results: Historical and future projections for the beginning of the Lyme disease season are shown in Figure 4. Historical results are for the period 1992–2007, where the national-average peak onset date for Lyme disease occurs on week 21.2 of the calendar year (mid-May). Future projections are for two time periods: 1) 2025–2040 and 2) 2065–2080. On average, the start of the Lyme disease season is projected to arrive a few days earlier for 2025–2040 (0.4–0.5 weeks), and approximately one to two weeks earlier for 2065–2080 (0.7–1.9 weeks) depending on the GHG trajectory. Winter and spring temperature increases are primarily responsible for the earlier peak onset of Lyme disease infections.

Research Highlight: Lyme Disease, continued

Conclusions: Results demonstrate that 21st century climate change will lead to environmental conditions suitable for earlier annual onset of Lyme disease cases in the United States, with possible implications for the timing of public health interventions. The end of the Lyme disease season is not strongly affected by climate variables; therefore, conclusions about the duration of the transmission season or changes in the annual number of new Lyme disease cases cannot be drawn from this study.

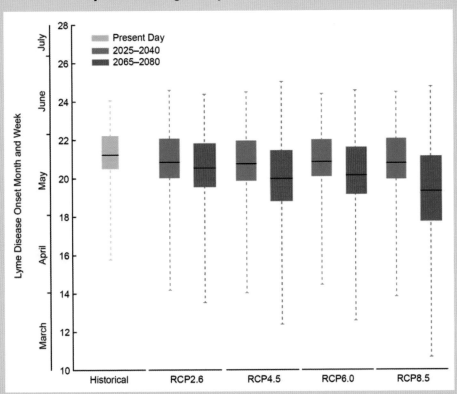

Projected Change in Lyme Disease Onset Week

Figure 4: Box plots comparing the distributions of the national-level historical observed data for annual Lyme disease onset week (1992–2007 in green) with the distributions of AOGCM multi-model mean projections of Lyme onset week for each of four Representative Concentration Pathways (RCP2.6, 4.5, 6.0, and 8.5) and two future time periods (2025–2040 in blue, 2065–2080 in red). Each box plot shows the values of Lyme disease onset week for the maximum (top of dashed line), 75th percentile (top of box), average (line through box), 25th percentile (bottom of box), and minimum (bottom of dashed line) of the distribution. All distributions are comprised of values for 12 eastern states and 16 years (N = 192). Additional details can be found in Monaghan et al. (2015). (Figure source: adapted from Monaghan et al. 2015).[72]

To project accurately the changes in Lyme disease risk in humans based on climate variability, long-term data collection on tick vector abundance and human infection case counts are needed to better understand the relationships between changing climate conditions, tick vector abundance, and Lyme disease case occurrence.

5.3 West Nile Virus

State of the Science

West Nile virus (WNV) is the leading cause of mosquito-borne disease in the United States. From 1999 to 2013, a total of 39,557 cases of WNV disease were reported in the United States.[73] Annual variation is substantial, both in terms of case counts and the geographic distribution of cases of human

infection (Figure 5).[73] Since the late summer of 1999, when an outbreak of WNV first occurred in New York City,[74] human WNV cases have occurred in the United States every year. After the introduction of the virus to the United States, WNV spread westward, and by 2004 WNV activity was reported throughout the contiguous United States.[75, 76] Annual human WNV incidence remained stable through 2007, decreased substantially through 2011, and increased again in 2012, raising questions about the factors driving year-to-year variation in disease transmission.[75] The locations of annual WNV outbreaks vary, but several states have reported consistently high rates of disease over the years, including Arizona, California, Colorado, Idaho, Illinois, Louisiana, New York, North Dakota, South Dakota, and Texas.[73, 75]

Incidence of West Nile Neuroinvasive Disease by County in the United States

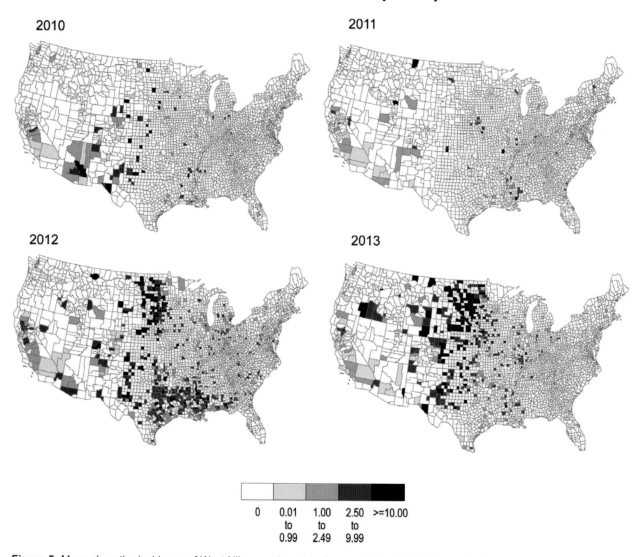

2010

2011

2012

2013

| 0 | 0.01 to 0.99 | 1.00 to 2.49 | 2.50 to 9.99 | >=10.00 |

Figure 5: Maps show the incidence of West Nile neuroinvasive disease in the United States for 2010 through 2013. Shown as cases per 100,000 people. (Data source: CDC 2014)[73]

The majority (70% to 80%) of people infected with WNV do not show symptoms of the disease. Of those infected, 20% to 30% develop acute systemic febrile illness, which may include headache, myalgias (muscle pains), rash, or gastrointestinal symptoms; fewer than 1% experience neuroinvasive disease, which may include meningitis (inflammation around the brain and spinal cord), encephalitis (inflammation of the brain), or myelitis (inflammation of the spinal cord) (see "5.4 Populations of Concern" on page 142).[77] Because most infected persons are asymptomatic (showing no symptoms), there is significant under-reporting of cases.[78, 79, 80] More than three million people were estimated to be infected with WNV in the United States from 1999 to 2010, resulting in about 780,000 illnesses.[77] However, only about 30,700 cases were reported during the same time span.[73]

West Nile virus is maintained in transmission cycles between birds (the natural hosts of the virus) and mosquitoes (Figure 6). The number of birds and mosquitoes infected with WNV increases as mosquitoes pass the virus from bird to bird starting in late winter or spring. Human infections can occur from a bite of a mosquito that has previously bitten an infected bird.[81] Humans do not pass on the virus to biting mosquitoes because they do not have sufficient concentrations of the virus in their bloodstreams.[82, 83] In rare instances, WNV can be transmitted through blood transfusions or organ transplants.[82, 84] Peak transmission of WNV to humans in the United States typically occurs between June and September, coinciding with the summer season when mosquitoes are most active and temperatures are highest.[85]

Climate Impacts on West Nile Virus Transmission

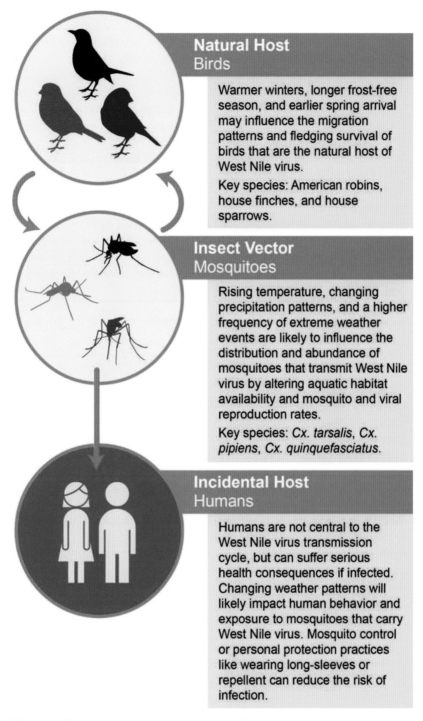

Natural Host
Birds

Warmer winters, longer frost-free season, and earlier spring arrival may influence the migration patterns and fledging survival of birds that are the natural host of West Nile virus.

Key species: American robins, house finches, and house sparrows.

Insect Vector
Mosquitoes

Rising temperature, changing precipitation patterns, and a higher frequency of extreme weather events are likely to influence the distribution and abundance of mosquitoes that transmit West Nile virus by altering aquatic habitat availability and mosquito and viral reproduction rates.

Key species: *Cx. tarsalis, Cx. pipiens, Cx. quinquefasciatus.*

Incidental Host
Humans

Humans are not central to the West Nile virus transmission cycle, but can suffer serious health consequences if infected. Changing weather patterns will likely impact human behavior and exposure to mosquitoes that carry West Nile virus. Mosquito control or personal protection practices like wearing long-sleeves or repellent can reduce the risk of infection.

Figure 6: Climate Impacts on West Nile Virus Transmission

Observed Impacts and Indicators

Mosquito vectors and bird hosts are required for WNV to persist, and the dynamics of both are strongly affected by climate in a number of ways. Geographical variation in average climate constrains the ranges of both vectors and hosts, while shorter-term climate variability affects many aspects of vector and host population dynamics. Unlike ticks, mosquitoes have short life cycles and respond more quickly to climate drivers over relatively short timescales of days to weeks. Impacts on bird abundance are often realized over longer timescales of months to years due to impacts on annual reproduction and migration cycles.

WNV has been detected in 65 mosquito species and more than 300 bird species in the United States,[85] although only a relatively small number of these species contribute substantively to human infections. Three *Culex* (*Cx.*) mosquito species are the primary vectors of the virus in different regions of the continental United States, and differences in their preferred breeding habitats mean that climate change will likely impact human WNV disease risk differently across these regions (Figure 5). Bird species that contribute to WNV transmission include those that develop sufficient viral concentrations in their blood to transmit the virus to feeding mosquitoes.[86, 87] As with mosquitoes, the bird species involved in the transmission cycle are likely to respond differently to climate change, increasing the complexity of projecting future WNV risk.

Impacts of Climate and Weather

Climate, or the long-term average weather, is important for defining WNV's transmission range limits because extreme conditions—too cold, hot, wet, or dry—can alter mosquito and bird habitat availability, increase mortality in mosquitoes or birds, and/or disrupt viral transmission. WNV is an invasive pathogen that was first detected in the United States just over 15 years ago, which is long enough to observe responses of WNV to key weather variables, but not long enough to observe responses to climate change trends.

Climate change may influence mosquito survival rates through changes in season length, although mosquitoes are also able to adapt to changing conditions. For example, mosquitoes that transmit WNV are limited to latitudes and altitudes where winters are short enough for them to survive.[88] However, newly emerged adult female mosquitoes have some ability to survive cold temperatures by entering a reproductive arrest called diapause as temperatures begin to cool and days grow shorter in late summer.[89, 90] These females will not seek a

Climate change has already begun to cause shifts in bird breeding and migration patterns, but it is unknown how these changes may affect West Nile virus transmission.

blood meal until temperatures begin to warm the following year. Even during diapause, very harsh winters may reduce mosquito populations, as temperatures near freezing have been shown to kill diapausing *Cx. tarsalis*.[91]

During the warmer parts of the year, *Culex* mosquitoes must have aquatic habitat available on a nearly continuous basis because their eggs hatch within a few days after they are laid and need moisture to remain viable. The breeding habitats of WNV vectors vary by species, ranging from fresh, sunlit water found in irrigated crops and wetlands preferred by *Cx. tarsalis* to stagnant, organically enriched water sources, such as urban storm drains, unmaintained swimming pools, or backyard containers, used by *Cx. pipiens* and *Cx. quinquefasciatus*.[92, 93, 94]

WNV has become endemic within a wide range of climates in the United States, but there is substantial geographic variation in the intensity of virus transmission. Part of this geographic variation can be attributed to the abundance and distributions of suitable bird hosts.[95] Important hosts, such as robins, migrate annually between summer breeding grounds and winter foraging areas.[86, 96] Migrating birds have shown potential as a vehicle for long-range virus movement.[97, 98] Although the timing of migration is driven by climate, the impact of climate change-driven migration changes on WNV transmission have not yet been documented by scientists. Climate change has already begun to cause shifts in bird breeding and migration patterns,[99] but it is unknown how these changes may affect WNV transmission.

Temperature is the most studied climate driver of the dynamics of WNV transmission. It is clear that warm temperatures accelerate virtually all of the biological processes that affect transmission: accelerating the mosquito life cycle,[100, 101, 102, 103, 104] increasing the mosquito biting rates that determine the frequency of contact between mosquitoes and hosts,[105, 106] and increasing viral replication rates inside the mosquito that decrease the time needed for a blood-fed mosquito to be able to pass on the virus.[107, 108, 109] These relationships between increasing temperatures and the biological processes that affect WNV transmission suggest a subsequent increase in risk of human disease.[110, 111, 112, 113] However, results from models have suggested that extreme high temperatures combined with decreased precipitation may decrease mosquito populations.[114]

Precipitation can create aquatic breeding sites for WNV vectors,[115, 116] and in some areas snowpack increases the amount of stored water available for urban or agricultural systems, which provide important habitat for WNV vectors,[117, 118] although effects depend on human water management

Birds such as the house finch are the natural host of West Nile virus.

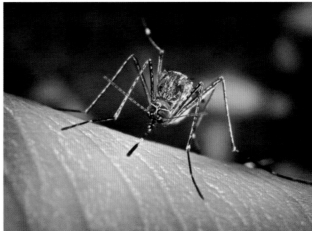

Humans can be infected from a bite of a mosquito that has previously bitten an infected bird.

decisions and vary spatially.[101] Droughts have been associated with increased WNV activity, but the association between decreased precipitation and WNV depends on location and the particular sequence of drought and wetting that precedes the WNV transmission season.[119, 120, 121, 122]

The impact of year-to-year changes in precipitation on mosquito populations varies among the regions of the United States and is affected by the typical climate of the area as well as other non-climate factors, such as land use or water infrastructure and management practices. In the northern Great Plains—a hotspot for WNV activity—increased precipitation has been shown to lead to higher *Cx. tarsalis* abundance a few weeks later.[116] In contrast, in the typically wet Pacific Northwest, weekly precipitation was found to be unrelated to subsequent mosquito abundance.[123] In urban areas, larvae (aquatic immature mosquitoes) may be washed out of their underground breeding habitats by heavy rainfall events, making drier conditions more favorable for WNV transmission.[110, 124, 125] In rural areas or drier regions, increased precipitation or agricultural irrigation may provide the moisture necessary for the development of breeding habitats.[121]

Impacts of Long-Term Climate Trends

The relatively short period of WNV's transmission in the United States prevents direct observation of the impacts of long-term climate trends on WNV incidence. However, despite the short history of WNV in the United States, there are some lessons to be learned from other mosquito-borne diseases with longer histories in the United States.

Western equine encephalomyelitis virus (WEEV) and St. Louis encephalitis virus (SLEV) were first identified in the 1930s and have been circulating in the United States since that time. Like WNV, both viruses are transmitted primarily by *Culex* mosquitoes and are climate-sensitive. WEEV outbreaks were associated with wet springs followed by warm summers.[118, 126] Outbreaks of SLEV were associated with hot, dry periods

when urban mosquito production increased due to stagnation of water in underground systems or when cycles of drought and wetting set up more complex transmission dynamics.[127, 128]

Despite climatic warming that would be expected to favor increased WEEV and SLEV transmission, both viruses have had sharply diminished incidence during the past 30 to 40 years.[129, 130] Although the exact reason for this decline is unknown, it is likely a result of non-climate factors, such as changes in human behavior or undetected aspects of viral evolution. Several other mosquito-borne pathogens, such as chikungunya and dengue, have grown in importance as global health threats during recent decades; however, a link to climate change induced disease expansion in the United States has not yet been confirmed. These examples demonstrate the variable impact that climate change can have on different mosquito-borne diseases and help to explain why the direction of future trends in risk for WNV remain unclear.

Projected Impacts

Given WNV's relatively short history in the United States, the described geographic variation in climate responses, and the complexity of transmission cycles, projecting the future distribution of WNV under climate change remains a challenge. Despite the growing body of work examining the connections between WNV and weather, climate-based seasonal forecasts of WNV outbreak risk are not yet available at a national scale. Forecasting the annual presence of WNV disease on the basis of climate and other ecological factors has been attempted for U.S. counties, with general agreement between modeled expectations and observed data, but more quantitative predictions of disease incidence or the risk for human exposure are needed.[131]

Longer-term projections of WNV under climate change scenarios are also rare. WNV is projected to increase in much of the northern and southeastern United States due to rising temperatures and declining precipitation, respectively, with the potential for decreased occurrence across the central United States.[132] Future projections show that the season when mosquitoes are most abundant will begin earlier and end later, possibly resulting in fewer mosquitoes in mid-summer in southern locations where extreme heat is predicted to coincide with decreased summer precipitation.[114]

5.4 Populations of Concern

Climate change will influence human vulnerability to vector-borne disease by influencing the seasonality and the location of exposures to pathogens and vectors. These impacts may influence future disease patterns; certain vector-borne diseases may emerge in areas where they had previously not been observed and other diseases may become less common in areas where they had previously been very common. As such, some segments of the U.S. population may be disproportionately affected by, or exposed to, vector-borne diseases in response to climate change (see also Ch. 9: Populations of Concern).

In addition to climate factors, multiple non-climate factors also influence human exposure to vector-borne pathogens.[17, 133, 134, 135, 136, 137] Some of these include factors from an environmental or institutional context (Figure 1), such as pathogen adaptation and change, changes in vector and host population and composition, changes in pathogen infection rates, and vector control or other public health practices (pesticide applications, integrated vector management, vaccines, and other disease interventions). Other non-climate factors that influence vulnerability to vector-borne disease include those from a social and behavioral context, such as outdoor activity, occupation, landscape design, proximity to vector habitat, and personal protective behaviors (applying repellents before spending time in tick habitat, performing tick checks, and bathing after being outdoors).[138] For Lyme disease, behavioral factors, especially the number of hours spent working or playing outdoors in tick habitat as well as proximity to dense shrubbery, can increase exposure to the ticks that transmit the bacteria that causes Lyme disease.[139] For example, outdoor workers in the northeastern United States are at higher risk for contact with blacklegged ticks and, therefore, are at a greater risk for contracting Lyme disease.[140, 141, 142] If outdoor workers are working in areas where there are infected mosquitoes, occupational exposures can also occur for WNV.[143]

Individual characteristics, such as age, gender, and immune function, may also affect vulnerability by influencing susceptibility to infection.[21, 80, 140, 143, 144, 145] Lyme disease is more frequently reported in children between 5 and 9 years of age and in adults between the ages of 55 and 59,[21] and advanced age and being male contribute to a higher risk for severe WNV infections.[79, 144, 145]

The impacts of climate change on human vulnerability to vector-borne disease may be minimized by individual- or community-level adaptive capacity, or the ability to reduce the potential exposures that may be caused by climate change. For example, socioeconomic status and domestic protective barriers, such as screens on windows and doors, can limit exposures to vector-borne pathogens.[17, 134, 135, 136, 137] From 1980 to 1999, the infected mosquito counts in Laredo, Texas, were significantly higher than in three adjoining Mexican states—yet, while there were only 64 cases of dengue fever reported in Texas, more than 62,000 dengue fever cases were reported in the Mexican states.[137] In Texas, socioeconomic factors and adaptive measures, including houses with air conditioning and intact screens, contributed to the significantly lower dengue incidence by reducing human–mosquito contact.[137] The adaptive capacity of a population may augment or limit the impacts of climate change to human vulnerability for vector-borne disease.[137]

> *Some segments of the U.S. population may be disproportionately affected by, or exposed to, vector-borne diseases in response to climate change.*

Climate factors are useful benchmarks to indicate seasonal risk and broad geographic changes in disease occurrence over decades. However, human vulnerability to vector-borne disease is more holistically evaluated by examining climate factors with non-climate factors (environmental or institutional context, social and behavioral context, and individual characteristics). Ultimately, a community's capacity to adapt to both the climate and non-climate factors will affect population vulnerability to vector-borne disease.

5.5 Emerging Issues

Some vector-borne diseases may be introduced or become re-established in the United States by a variety of mechanisms. In conjunction with trade and travel, climate change may contribute by creating habitats suitable for the establishment of disease-carrying vectors or for locally sustained transmission of vector-borne pathogens. Examples of emerging vector-borne diseases in the United States include the West Nile virus introduction described above, recent outbreaks of locally acquired dengue in Florida[17, 146] and southern Texas,[147] and chikungunya cases in the Caribbean and southern Florida,[148] all of which have raised public health concern about emergence and re-emergence of these mosquito-borne diseases in the United States. Collecting data on the spread of disease-causing insect vectors and the viruses that cause dengue and chikungunya is critical to understanding and predicting the threat of emergence or reemergence of these diseases. Understanding the role of climate change in disease emergence and reemergence would also require additional research.

5.6 Research Needs

In addition to the emerging issues identified above, and based on their review of the literature, the authors highlight the following areas for potential scientific research activity on vector-borne disease. Climate and non-climate factors interact to determine the burden of vector-borne diseases on humans, but the mechanisms of these processes are still poorly understood.[149] Evidence-based models that include vector–host interaction, host immunity, pathogen evolution, and land use, as well as socioeconomic drivers of transmission, human behavior, and adaptive capacity are needed to facilitate a better understanding of the mechanisms by which climate and non-climate factors drive vector-borne disease emergence. Socioeconomic and human behavioral factors, in particular, appear to limit vector-borne diseases, even in neighboring cities.[136, 137] This is a fertile area for future research, and one that is particularly relevant for increasing our adaptive capacity to address future vector-borne disease threats.

Numerous studies have identified associations between vector-borne diseases and weather or climate, but most have focused on risk mapping or estimating associations of broad aggregates of temperature and precipitation with disease-related outcomes. A move beyond correlative associations to a more mechanistic understanding of climate's impacts on the discrete events that give rise to transmission is needed. Models must also be accompanied by empirical research to inform their parameters. Climate effects are complex, and models frequently borrow information across vector species and pathogens or make simplifying assumptions that can lead to incorrect conclusions.[150]

The risk for vector-borne diseases is highly variable geographically and over time. Monitoring responses of pathogens to climate change at a continental scale requires coordinated, systematically collected long-term surveillance datasets to document changes in vector occurrence, abundance, and infection rates. Collecting these data will provide a clearer understanding of how external drivers work in conjunction with climate change to determine the risk for human exposure to vector-borne disease.

Future assessments can benefit from research activities that:

- evaluate how climatic variables, socioeconomic factors, and human behavior influence vector-borne disease occurrence and are expected to affect human adaptive capacity and the ability to respond to future disease threats;

- enhance long-term, systematic data collection on vector and pathogen distributions to detect changes over time. Such datasets must span a range of land-use types, including urban areas, and should be coupled with data on human disease;

- utilize mechanistic models that provide an evidence-based view of climate's impacts on vector-borne diseases by explicitly accounting for the series of discrete but intertwined events that give rise to transmission. Models should be supported and validated by data specific to the disease system and include a realistic assessment of parameter uncertainty and variability;

- study the natural maintenance cycles of vector-borne pathogen evolution, emergence, and transmission as well as how climatic variables influence these cycles.

Supporting Evidence

The chapter was developed through technical discussions of relevant evidence and expert deliberation by the report authors at several workshops, teleconferences, and email exchanges. The authors considered inputs and comments submitted by the public, the National Academies of Sciences, and Federal agencies. For additional information on the overall report process, see Appendices 2 and 3.

The approach and organization of this chapter was decided after conducting a comprehensive literature review. Two case studies, Lyme disease and West Nile virus, were chosen as representative examples of vector-borne diseases in the United States for this chapter because of their high incidence rates and the body of literature available on the association between climatic and meteorological variables and occurrence of these diseases.

Regarding human outcomes related to vector-borne diseases, there is a much greater volume of published literature available on meteorological and climatic influences on vectors. As a result, our certainty in how climate change is likely to influence the vectors far exceeds our certainty in how changing climatic conditions are likely to affect when, where, and how many cases of vector-borne diseases are likely to occur.

Although the topic of zoonotic diseases was included in the original prospectus, it was later removed due to space constraints. Additionally, since both West Nile virus infection and Lyme disease are zoonotic diseases, these case studies address concepts that are common to both vector-borne and zoonotic diseases.

KEY FINDING TRACEABLE ACCOUNTS

Changing Distributions of Vectors and Vector-Borne Diseases

Key Finding 1: Climate change is expected to alter the geographic and seasonal distributions of existing vectors and vector-borne diseases *[Likely, High Confidence].*

Description of evidence base
Vector-borne diseases result from complex interactions involving vectors, reservoirs, humans, and both climate and non-climate factors. Numerous studies explain how climate variables influence the relationships between vectors, animal reservoirs, humans, and other non-climate factors to ultimately influence the spatial and temporal distribution of vector-borne disease.[11, 39, 45, 53, 101, 104, 114, 116, 123, 135]

Major uncertainties
It is certain that climate change will alter the geographic and seasonal distribution of existing vectors, pathogens, and reservoirs; the influence of climate change on the timing, prevalence, and location of specific vector-borne disease outbreaks is likely to vary depending on the influence of other significant non-climate drivers of disease occurrence.

Assessment of confidence and likelihood based on evidence
Based on the evidence that climate change will influence the temporal and spatial distributions of vectors, pathogens, and animal reservoirs, there is **high confidence** that climate change is **likely** to alter the geographic and seasonal distributions of vectors and vector-borne diseases.

Earlier Tick Activity and Northward Range Expansion

Key Finding 2: Ticks capable of carrying the bacteria that cause Lyme disease and other pathogens will show earlier seasonal activity and a generally northward expansion in response to increasing temperatures associated with climate change *[Likely, High Confidence].* Longer seasonal activity and expanding geographic range of these ticks will increase the risk of human exposure to ticks *[Likely, Medium Confidence].*

Description of evidence base

There is strong evidence that temperature affects the geographical distribution of ticks,[39, 45, 53, 67] the timing of host-seeking activity of ticks,[36, 37, 44] and even the timing of Lyme disease case occurrence.[23] However, the abundance of ticks infected with Lyme disease spirochetes, which is considered a better predictor of human risk for Lyme disease compared with nymphal density alone, has rarely been found to be strongly associated with meteorological variables.[41] Studies aimed at identifying meteorological variables associated with the geographical distribution of human Lyme disease vary in their support for demonstrating positive associations between temperature and Lyme disease.[46, 62, 63]

Major uncertainties
While the effects of temperature, precipitation, and humidity on the spatial distribution of ticks and the timing of their host-seeking activity have been clearly established in both the eastern and western regions of the United States, where Lyme disease is common, the degree to which climate change will alter Lyme disease incidence remains uncertain. The observation that meteorological variables play a lesser role than other variables in predicting the density of nymphs infected with Lyme disease bacteria raises uncertainty in how climate change will affect the distribution and magnitude of Lyme disease incidence. This uncertainty is reflected in results from models aiming to associate meteorological variables with Lyme disease incidence that yielded inconsistent findings.[46, 62, 63]

Assessment of confidence and likelihood based on evidence

Based on the evidence, there is **high confidence** that climate change, especially temperature change, is **likely** to cause shifts in the geographical distribution of ticks capable of carrying *B. burgdorferi* to more northern latitudes, the timing of host-seeking activity of ticks, and the timing of Lyme disease case occurrence. While these changes are **likely** to influence human disease, due to the few sources with limited consistency, incomplete models with methods still emerging, and some competing schools of thought, there is **medium confidence** surrounding how, and how much, climate change will influence the risk of human exposure to ticks carrying *B. burgdorferi*.

Changing Mosquito-Borne Disease Dynamics

Key Finding 3: Rising temperatures, changing precipitation patterns, and a higher frequency of some extreme weather events associated with climate change will influence the distribution, abundance, and prevalence of infection in the mosquitoes that transmit West Nile virus and other pathogens by altering habitat availability and mosquito and viral reproduction rates [*Very Likely, High Confidence*]. Alterations in the distribution, abundance, and infection rate of mosquitoes will influence human exposure to bites from infected mosquitoes, which is expected to alter risk for human disease [*Very Likely, Medium Confidence*].

Description of evidence base

Higher temperatures affect the West Nile virus (WNV) system by accelerating mosquito development[102, 104] and virus reproduction rates,[101, 107, 108, 109] increasing egg-laying and biting frequency,[106] and affecting mosquito survival.[102, 126] Increased WNV activity has been associated with warm temperatures, mild winters, and drought.[101, 110, 116] Very few studies have used climate variables to predict the occurrence of human WNV cases in the United States in response to climate change (for example, Harrigan et al. 2014),[132] but available results suggest that areas of WNV transmission will expand in the northern latitudes and higher elevations driven by increasing temperature, while WNV transmission may decrease in the South if increasing temperatures reduce mosquito survival or limit availability of surface water, such as that provided by agricultural irrigation.

Major uncertainties

While the influence of temperature and precipitation on mosquito and WNV biology are fairly well-understood, these relationships vary across the United States depending on the local mosquito vector species, land use, and human activity.[112, 121] For mosquitoes in urban areas, droughts may lead to stagnation of water and increased mosquito populations that enhance WNV transmission,[110, 125] while in rural or agricultural areas, droughts may reduce mosquito populations by reducing available mosquito habitat for breeding,[101] except when irrigation compensates for drought conditions.[121] Long-term projections of human WNV risk

under climate change scenarios are still in the early stages of development and are impeded by the complexities of the disease transmission cycle. Evolution of the virus, improvements in mosquito control, and the potential for long-term changes in human behavior that may affect exposure to WNV are key sources of uncertainty. For this reason, short-term, seasonal forecasts of WNV may be more fruitful in the near term and may provide information for seasonal resource allocation and public health planning.

Assessment of confidence and likelihood based on evidence

Based on the evidence, there is **high confidence** that climate change is **very likely** to influence mosquito distribution, abundance, and infection prevalence by altering habitat availability and mosquito and viral reproduction rates. While this is **very likely** to influence human disease, due to the few sources with limited consistency, incomplete models with methods still emerging, and some competing schools of thought, there is **medium confidence** surrounding how, and how much, climate change will influence human incidence of disease.

Emergence of New Vector-Borne Pathogens

Key Finding 4: Vector-borne pathogens are expected to emerge or reemerge due to the interactions of climate factors with many other drivers, such as changing land-use patterns [*Likely, High Confidence*]. The impacts to human disease, however, will be limited by the adaptive capacity of human populations, such as vector control practices or personal protective measures [*Likely, High Confidence*].

Description of evidence base

The literature shows that climate change must be considered together with the many other non-climate factors of disease emergence[11, 12] and the availability of other mitigating factors, such as air conditioning, screens on windows, and vector control practices,[17, 134, 136, 137] in order to appropriately quantify the impact climate has on the risk of emerging or reemerging exotic pathogens and vectors.

Major uncertainties

It remains uncertain how climate interacts as a driver with travel-related exposures and evolutionary adaptation of invasive vectors and pathogens to affect human disease. Improved longitudinal datasets and empirical models that include vector–host interaction, host immunity, and pathogen evolution as well as socioeconomic drivers of transmission are needed to address these knowledge gaps in research on climate sensitive diseases.

Assessment of confidence and likelihood based on evidence

Based on the evidence, there is **high confidence** that a multitude of interacting factors, one of which being climate change, will **likely** influence the emergence or reemergence of vector-borne pathogens to the United States. Additionally, there is **high confidence** that the influence of climate change on human disease is **likely** to be limited by the adaptive capacity of a population.

DOCUMENTING UNCERTAINTY

See Appendix 4: Documenting Uncertainty for more information on assessments of confidence and likelihood used in this report.

Confidence Level	Likelihood
Very High	**Very Likely**
Strong evidence (established theory, multiple sources, consistent results, well documented and accepted methods, etc.), high consensus	≥ 9 in 10
High	**Likely**
Moderate evidence (several sources, some consistency, methods vary and/or documentation limited, etc.), medium consensus	≥ 2 in 3
Medium	**As Likely As Not**
Suggestive evidence (a few sources, limited consistency, models incomplete, methods emerging, etc.), competing schools of thought	≈ 1 in 2
Low	**Unlikely**
Inconclusive evidence (limited sources, extrapolations, inconsistent findings, poor documentation and/or methods not tested, etc.), disagreement or lack of opinions among experts	≤ 1 in 3
	Very Unlikely
	≤ 1 in 10

PHOTO CREDITS

Pg. 129–Mosquito: ©CDC/Science Faction/Corbis

Pg. 130–Woman applying repellent: © iStockPhoto.com/powerofforever

Pg. 135–Blacklegged tick: ©Science Stills/ARS/Visuals Unlimited/Corbis

Pg. 140–House finch: ©Pete Oxford/Minden Pictures/Corbis

Pg. 141–Mosquito warning sign: © iStockPhotos.com/leekris

Pg. 141–Mosquito: ©CDC/Science Faction/Corbis

References

1. CDC, 2005: Notice to readers: Final 2004 reports of Notifiable Diseases. *MMWR. Morbidity and Mortality Weekly Report,* **54,** 770-780. http://www.cdc.gov/mmwr/preview/mmwrhtml/mm5431a4.htm

2. CDC, 2006: Notice to readers: Final 2005 Reports of Notifiable Diseases. *MMWR. Morbidity and Mortality Weekly Report,* **55,** 880-881. http://www.cdc.gov/mmwr/preview/mmwrhtml/mm5532a4.htm

3. CDC, 2007: Notice to readers: Final 2006 Reports of Nationally Notifiable Infectious Diseases. *MMWR. Morbidity and Mortality Weekly Report,* **56,** 853-863. http://www.cdc.gov/mmwr/preview/mmwrhtml/mm5633a4.htm

4. CDC, 2008: Notice to readers: Final 2007 Reports of Nationally Notifiable Infectious Diseases. *MMWR. Morbidity and Mortality Weekly Report,* **57,** 901,903-913. http://www.cdc.gov/mmwr/preview/mmwrhtml/mm5733a6.htm

5. CDC, 2009: Notice to readers: Final 2008 Reports of Nationally Notifiable Infectious Diseases. *MMWR. Morbidity and Mortality Weekly Report,* **58,** 856-869. http://www.cdc.gov/mmwr/preview/mmwrhtml/mm5831a5.htm

6. CDC, 2010: Notice to readers: Final 2009 Reports of Nationally Notifiable Infectious Diseases. *MMWR. Morbidity and Mortality Weekly Report,* **59,** 1027-1039. http://www.cdc.gov/mmwr/preview/mmwrhtml/mm5932a5.htm?s_cid=mm5932a5_w

7. CDC, 2011: Notice to readers: Final 2010 Reports of Nationally Notifiable Infectious Diseases. *MMWR. Morbidity and Mortality Weekly Report,* **60,** 1088-1101. http://www.cdc.gov/mmwr/preview/mmwrhtml/mm6032a5.htm?s_cid=mm6032a5_w

8. CDC, 2012: Notice to readers: Final 2011 Reports of Nationally Notifiable Infectious Diseases. *MMWR. Morbidity and Mortality Weekly Report,* **61,** 624-637. http://www.cdc.gov/mmwr/preview/mmwrhtml/mm6132a8.htm?s_cid=mm6132a8_w

9. CDC, 2013: Notice to readers: Final 2012 Reports of Nationally Notifiable Infectious Diseases. *MMWR. Morbidity and Mortality Weekly Report,* **62,** 669-682. http://www.cdc.gov/mmwr/preview/mmwrhtml/mm6233a6.htm?s_cid=mm6233a6_w

10. CDC, 2014: Notice to readers: Final 2013 Reports of Nationally Notifiable Infectious Diseases. *MMWR. Morbidity and Mortality Weekly Report,* **63,** 702-715. http://www.cdc.gov/mmwr/preview/mmwrhtml/mm6332a6.htm?s_cid=mm6332a6_w

11. Gage, K.L., T.R. Burkot, R.J. Eisen, and E.B. Hayes, 2008: Climate and vector-borne diseases. *American Journal of Preventive Medicine,* **35,** 436-450. http://dx.doi.org/10.1016/j.amepre.2008.08.030

12. IOM, 2003: *Microbial Threats to Health: Emergence, Detection, and Response. Smolinski, M.S., M.A. Hamburg, and J. Lederberg, Eds. Institute of Medicine. The National Academies Press, Washington, D.C., 398 pp.* http://dx.doi.org/10.17226/10636

13. Allan, B.F., F. Keesing, and R.S. Ostfeld, 2003: Effect of forest fragmentation on Lyme disease risk. *Conservation Biology,* **17,** 267-272. http://dx.doi.org/10.1046/j.1523-1739.2003.01260.x

14. Jones, K.E., N.G. Patel, M.A. Levy, A. Storeygard, D. Balk, J.L. Gittleman, and P. Daszak, 2008: Global trends in emerging infectious diseases. *Nature,* **451,** 990-993. http://dx.doi.org/10.1038/nature06536

15. Rosenberg, R., M.A. Johansson, A.M. Powers, and B.R. Miller, 2013: Search strategy has influenced the discovery rate of human viruses. *PNAS,* **110,** 13961-13964. http://dx.doi.org/10.1073/pnas.1307243110

16. Kilpatrick, A.M. and S.E. Randolph, 2012: Drivers, dynamics, and control of emerging vector-borne zoonotic diseases. *Lancet,* **380,** 1946-1955. http://dx.doi.org/10.1016/s0140-6736(12)61151-9

17. Radke, E.G., C.J. Gregory, K.W. Kintziger, E.K. Sauer-Schatz, E. Hunsperger, G.R. Gallagher, J.M. Barber, B.J. Biggerstaff, D.R. Stanek, K.M. Tomashek, and C.G.M. Blackmore, 2012: Dengue outbreak in Key West, Florida, USA, 2009. *Emerging Infectious Diseases,* **18,** 135-137. http://dx.doi.org/10.3201/eid1801.110130

18. Dick, O.B., J.L. San Martín, R.H. Montoya, J. del Diego, B. Zambrano, and G.H. Dayan, 2012: The history of dengue outbreaks in the Americas. *American Journal of Tropical Medicine and Hygiene,* **87,** 584-593. http://dx.doi.org/10.4269/ajtmh.2012.11-0770

19. Lane, R.S., J. Piesman, and W. Burgdorfer, 1991: Lyme borreliosis: Relation of its causative agent to its vectors and hosts in North America and Europe. *Annual Review of Entomology*, **36**, 587-609. http://dx.doi.org/10.1146/annurev.en.36.010191.003103

20. Mead, P.S., 2015: Epidemiology of Lyme disease. *Infectious Disease Clinics of North America*, **29**, 187-210. http://dx.doi.org/10.1016/j.idc.2015.02.010

21. Bacon, R.M., K.J. Kugeler, and P.S. Mead, 2008: Surveillance for Lyme disease--United States, 1992-2006. *MMWR Surveillance Summaries*, **57(SS10)**, 1-9. http://www.cdc.gov/MMWR/PREVIEW/MMWRHTML/ss5710a1.htm

22. CDC, 2015: Reported Cases of Lyme Disease by Year, United States, 1995-2013. Centers for Disease Control and Prevention, Atlanta, GA. http://www.cdc.gov/lyme/stats/chartstables/casesbyyear.html

23. Moore, S.M., R.J. Eisen, A. Monaghan, and P. Mead, 2014: Meteorological influences on the seasonality of Lyme disease in the United States. *American Journal of Tropical Medicine and Hygiene*, **90**, 486-496. http://dx.doi.org/10.4269/ajtmh.13-0180

24. Clover, J.R. and R.S. Lane, 1995: Evidence implicating nymphal Ixodes pacificus (Acari, ixodidae) in the epidemiology of Lyme disease in California. *American Journal of Tropical Medicine and Hygiene*, **53**, 237-240.

25. Piesman, J., 1989: Transmission of Lyme disease spirochetes (*Borrelia-Burgdorferi*). *Experimental & Applied Acarology*, **7**, 71-80. http://dx.doi.org/10.1007/Bf01200454

26. CDC, 2015: Lyme Disease: Data and Statistics: Maps-Reported Cases of Lyme Disease – United States, 2001-2014. Centers for Disease Control and Prevention. http://www.cdc.gov/lyme/stats/

27. Dennis, D.T., T.S. Nekomoto, J.C. Victor, W.S. Paul, and J. Piesman, 1998: Forum: Reported distribution of *Ixodes scapularis* and in *Ixodes pacificus (Acari: Ixodidae)* in the United States. *Journal of Medical Entomology*, **35**, 629-638. http://dx.doi.org/10.1093/jmedent/35.5.629

28. Pepin, K.M., R.J. Eisen, P.S. Mead, J. Piesman, D. Fish, A.G. Hoen, A.G. Barbour, S. Hamer, and M.A. Diuk-Wasser, 2012: Geographic variation in the relationship between human Lyme disease incidence and density of infected host-seeking *Ixodes scapularis nymphs in the eastern United States. American Journal of Tropical Medicine and Hygiene*, **86**, 1062-1071. http://dx.doi.org/10.4269/ajtmh.2012.11-0630

29. Yuval, B. and A. Spielman, 1990: Duration and regulation of the developmental cycle of *Ixodes dammini (Acari: Ixodidae)*. *Journal of Medical Entomology*, **27**, 196-201. http://dx.doi.org/10.1093/jmedent/27.2.196

30. Needham, G.R. and P.D. Teel, 1991: Off-host physiological ecology of ixodid ticks. *Annual Review of Entomology*, **36**, 659-681. http://dx.doi.org/10.1146/annurev.en.36.010191.003303

31. Stafford III, K.C., 1994: Survival of immature *Ixodes scapularis (Acari: Ixodidae)* at different relative humidities. *Journal of Medical Entomology*, **31**, 310-314. http://dx.doi.org/10.1093/jmedent/31.2.310

32. Lane, R.S., J.E. Kleinjan, and G.B. Schoeler, 1995: Diel activity of nymphal *Dermacentor occidentalis and Ixodes pacificus (Acari: Ixodidae) in relation to meteorological factors and host activity periods. Journal of Medical Entomology*, **32**, 290-299. http://dx.doi.org/10.1093/jmedent/32.3.290

33. Bertrand, M.R. and M.L. Wilson, 1996: Microclimate-dependent survival of unfed adult *Ixodes scapularis (Acari: Ixodidae) in nature: Life cycle and study design implications. Journal of Medical Entomology*, **33**, 619-627. http://dx.doi.org/10.1093/jmedent/33.4.619

34. Mount, G.A., D.G. Haile, and E. Daniels, 1997: Simulation of management strategies for the blacklegged tick (Acari: Ixodidae) and the Lyme disease spirochete, *Borrelia burgdorferi. Journal of Medical Entomology*, **34**, 672-683. http://dx.doi.org/10.1093/jmedent/34.6.672

35. Vail, S.G. and G. Smith, 1998: Air temperature and relative humidity effects on behavioral activity of blacklegged tick (Acari: Ixodidae) nymphs in New Jersey. *Journal of Medical Entomology*, **35**, 1025-1028. http://dx.doi.org/10.1093/jmedent/35.6.1025

36. Eisen, L., R.J. Eisen, and R.S. Lane, 2002: Seasonal activity patterns of *Ixodes pacificus nymphs in relation to climatic conditions. Medical and Veterinary Entomology*, **16**, 235-244. http://dx.doi.org/10.1046/j.1365-2915.2002.00372.x

37. Eisen, R.J., L. Eisen, M.B. Castro, and R.S. Lane, 2003: Environmentally related variability in risk of exposure to Lyme disease spirochetes in northern California: Effect of climatic conditions and habitat type. *Environmental Entomology*, **32**, 1010-1018. http://dx.doi.org/10.1603/0046-225X-32.5.1010

38. Schulze, T.L. and R.A. Jordan, 2003: Meteorologically mediated diurnal questing of *Ixodes scapularis and Amblyomma americanum (Acari: Ixodidae) nymphs. Journal of Medical Entomology,* **40,** 395-402. http://dx.doi.org/10.1603/0022-2585-40.4.395

39. Diuk-Wasser, M.A., G. Vourc'h, P. Cislo, A.G. Hoen, F. Melton, S.A. Hamer, M. Rowland, R. Cortinas, G.J. Hickling, J.I. Tsao, A.G. Barbour, U. Kitron, J. Piesman, and D. Fish, 2010: Field and climate-based model for predicting the density of host-seeking nymphal *Ixodes scapularis, an important vector of tick-borne disease agents in the eastern United States. Global Ecology and Biogeography,* **19,** 504-514. http://dx.doi.org/10.1111/j.1466-8238.2010.00526.x

40. Diuk-Wasser, M.A., A.G. Hoen, P. Cislo, R. Brinkerhoff, S.A. Hamer, M. Rowland, R. Cortinas, G. Vourc'h, F. Melton, G.J. Hickling, J.I. Tsao, J. Bunikis, A.G. Barbour, U. Kitron, J. Piesman, and D. Fish, 2012: Human risk of infection with *Borrelia burgdorferi, the Lyme disease agent, in eastern United States. American Journal of Tropical Medicine and Hygiene,* **86,** 320-327. http://dx.doi.org/10.4269/ajtmh.2012.11-0395

41. Eisen, R.J., L. Eisen, Y.A. Girard, N. Fedorova, J. Mun, B. Slikas, S. Leonhard, U. Kitron, and R.S. Lane, 2010: A spatially-explicit model of acarological risk of exposure to *Borrelia burgdorferi-infected Ixodes pacificus nymphs in northwestern California based on woodland type, temperature, and water vapor. Ticks and Tick-Borne Diseases,* **1,** 35-43. http://dx.doi.org/10.1016/j.ttbdis.2009.12.002

42. Ostfeld, R.S., C.D. Canham, K. Oggenfuss, R.J. Winchcombe, and F. Keesing, 2006: Climate, deer, rodents, and acorns as determinants of variation in Lyme-disease risk. *Plos Biology,* **4,** 1058-1068. http://dx.doi.org/10.1371/Journal.Pbio.0040145

43. Schulze, T.L., R.A. Jordan, C.J. Schulze, and R.W. Hung, 2009: Precipitation and Temperature as Predictors of the Local Abundance of Ixodes scapularis (Acari: Ixodidae) Nymphs. *Journal of Medical Entomology,* **46,** 1025-1029. http://dx.doi.org/10.1603/033.046.0508

44. Diuk-Wasser, M.A., A.G. Gatewood, M.R. Cortinas, S. Yaremych-Hamer, J. Tsao, U. Kitron, G. Hickling, J.S. Brownstein, E. Walker, J. Piesman, and D. Fish, 2006: Spatiotemporal patterns of host-seeking *Ixodes scapularis nymphs (Acari: Iodidae) in the United States. Journal of Medical Entomology,* **43,** 166-176. http://dx.doi.org/10.1093/jmedent/43.2.166

45. Brownstein, J.S., T.R. Holford, and D. Fish, 2003: A climate-based model predicts the spatial distribution of the Lyme disease vector Ixodes scapularis in the United States. *Environmental Health Perspectives,* **111,** 1152-1157. http://dx.doi.org/10.1289/ehp.6052

46. Tran, P.M. and L. Waller, 2013: Effects of landscape fragmentation and climate on Lyme disease incidence in the northeastern United States. *Ecohealth,* **10,** 394-404. http://dx.doi.org/10.1007/s10393-013-0890-y

47. CDC, 2015: Ticks: Life Cycle of Hard Ticks that Spread Disease. Centers for Disease Control and Prevention, Atlanta, GA. http://www.cdc.gov/ticks/life_cycle_and_hosts.html

48. Falco, R.C., D.F. McKenna, T.J. Daniels, R.B. Nadelman, J. Nowakowski, D. Fish, and G.P. Wormser, 1999: Temporal relation between *Ixodes scapularis abundance and risk for Lyme disease associated with erythema migrans. American Journal of Epidemiology,* **149,** 771-776.

49. Rollend, L., D. Fish, and J.E. Childs, 2013: Transovarial transmission of *Borrelia spirochetes by Ixodes scapularis: A summary of the literature and recent observations. Ticks and Tick-Borne Diseases,* **4,** 46-51. http://dx.doi.org/10.1016/j.ttbdis.2012.06.008

50. LoGiudice, K., S.T. Duerr, M.J. Newhouse, K.A. Schmidt, M.E. Killilea, and R.S. Ostfeld, 2008: Impact of host community composition on Lyme disease risk. *Ecology,* **89,** 2841-2849. http://dx.doi.org/10.1890/07-1047.1

51. Mather, T.N., M.L. Wilson, S.I. Moore, J.M. Ribeiro, and A. Spielman, 1989: Comparing the relative potential of rodents as reservoirs of the Lyme disease spirochete (Borrelia burgdorferi). *American Journal of Epidemiology,* **130,** 143-50. http://www.ncbi.nlm.nih.gov/pubmed/2787105

52. Stromdahl, E.Y. and G.J. Hickling, 2012: Beyond Lyme: Aetiology of tick-borne human diseases with emphasis on the south-eastern United States. *Zoonoses and Public Health,* **59,** 48-64. http://dx.doi.org/10.1111/j.1863-2378.2012.01475.x

53. Estrada-Peña, A., 2002: Increasing habitat suitability in the United States for the tick that transmits Lyme disease: A remote sensing approach. *Environmental Health Perspectives,* **110,** 635-640. PMC1240908

54. Bouchard, C., G. Beauchamp, P.A. Leighton, R. Lindsay, D. Belanger, and N.H. Ogden, 2013: Does high biodiversity reduce the risk of Lyme disease invasion? *Parasites & Vectors,* **6,** 195. http://dx.doi.org/10.1186/1756-3305-6-195

55. Leighton, P.A., J.K. Koffi, Y. Pelcat, L.R. Lindsay, and N.H. Ogden, 2012: Predicting the speed of tick invasion: An empirical model of range expansion for the Lyme disease vector Ixodes scapularis in Canada. *Journal of Applied Ecology,* **49,** 457-464. http://dx.doi.org/10.1111/j.1365-2664.2012.02112.x

56. Ogden, N.H., L. St-Onge, I.K. Barker, S. Brazeau, M. Bigras-Poulin, D.F. Charron, C.M. Francis, A. Heagy, L.R. Lindsay, A. Maarouf, P. Michel, F. Milord, C.J. O'Callaghan, L. Trudel, and R.A. Thompson, 2008: Risk maps for range expansion of the Lyme disease vector, *Ixodes scapularis, in Canada now and with climate change. International Journal of Health Geographics,* **7, 24.** http://dx.doi.org/10.1186/1476-072X-7-24

57. Ogden, N.H., C. Bouchard, K. Kurtenbach, G. Margos, L.R. Lindsay, L. Trudel, S. Nguon, and F. Milord, 2010: Active and passive surveillance and phylogenetic analysis of *Borrelia burgdorferi elucidate the process of Lyme disease risk emergence in Canada. Environmental Health Perspectives,* **118,** 909-914. http://dx.doi.org/10.1289/ehp.0901766

58. Ogden, N.H., M. Radojević, X. Wu, V.R. Duvvuri, P.A. Leighton, and J. Wu, 2014: Estimated effects of projected climate change on the basic reproductive number of the Lyme disease vector *Ixodes scapularis. Environmental Health Perspectives,* **122,** 631-638. http://dx.doi.org/10.1289/ehp.1307799

59. Guerra, M., E. Walker, C. Jones, S. Paskewitz, M.R. Cortinas, A. Stancil, L. Beck, M. Bobo, and U. Kitron, 2002: Predicting the risk of Lyme disease: Habitat suitability for *Ixodes scapularis in the north central United States. Emerging Infectious Diseases,* **8,** 289-297. http://dx.doi.org/10.3201/eid0803.010166

60. Khatchikian, C.E., M. Prusinski, M. Stone, P.B. Backenson, I.N. Wang, M.Z. Levy, and D. Brisson, 2012: Geographical and environmental factors driving the increase in the Lyme disease vector *Ixodes scapularis. Ecosphere,* **3,** art85. http://dx.doi.org/10.1890/ES12-00134.1

61. Eisen, R.J., L. Eisen, and R.S. Lane, 2004: Habitat-related variation in infestation of lizards and rodents with *Ixodes ticks in dense woodlands in Mendocino County, California. Experimental and Applied Acarology,* **33,** 215-233. http://dx.doi.org/10.1023/B:Appa.0000032954.71165.9e

62. Ashley, S.T. and V. Meentemeyer, 2004: Climatic analysis of Lyme disease in the United States. *Climate Research,* **27,** 177-187. http://dx.doi.org/10.3354/cr027177

63. Tuite, A.R., A.L. Greer, and D.N. Fisman, 2013: Effect of latitude on the rate of change in incidence of Lyme disease in the United States. *CMAJ Open,* **1,** E43-E47. http://dx.doi.org/10.9778/cmajo.20120002

64. McCabe, G.J. and J.E. Bunnell, 2004: Precipitation and the occurrence of Lyme disease in the northeastern United States. *Vector-Borne and Zoonotic Diseases,* **4,** 143-148. http://dx.doi.org/10.1089/1530366041210765

65. Subak, S., 2003: Effects of climate on variability in Lyme disease incidence in the northeastern United States. *American Journal of Epidemiology,* **157,** 531-538. http://dx.doi.org/10.1093/aje/kwg014

66. Levi, T., F. Keesing, K. Oggenfuss, and R.S. Ostfeld, 2015: Accelerated phenology of blacklegged ticks under climate warming. *Philosophical Transactions of the Royal Society B: Biological Sciences,* **370.** http://dx.doi.org/10.1098/rstb.2013.0556

67. Brownstein, J.S., T.R. Holford, and D. Fish, 2005: Effect of climate change on Lyme disease risk in North America. *EcoHealth,* **2,** 38-46. http://dx.doi.org/10.1007/s10393-004-0139-x

68. Brunner, J.L., M. Killilea, and R.S. Ostfeld, 2012: Overwintering survival of nymphal *Ixodes scapularis (Acari: Ixodidae) under natural conditions. Journal of Medical Entomology,* **49,** 981-987. http://dx.doi.org/10.1603/me12060

69. Lindsay, L.R., I.K. Barker, G.A. Surgeoner, S.A. McEwen, T.J. Gillespie, and J.T. Robinson, 1995: Survival and development of Ixodes scapularis (Acari: Ixodidae) under various climatic conditions in Ontario, Canada. *Journal of Medical Entomology,* **32,** 143-152. http://dx.doi.org/10.1093/jmedent/32.2.143

70. Ogden, N.H., A. Maarouf, I.K. Barker, M. Bigras-Poulin, L.R. Lindsay, M.G. Morshed, J. O'Callaghan C, F. Ramay, D. Waltner-Toews, and D.F. Charron, 2006: Climate change and the potential for range expansion of the Lyme disease vector Ixodes scapularis in Canada. *International Journal for Parasitology,* **36,** 63-70. http://dx.doi.org/10.1016/j.ijpara.2005.08.016

71. Ogden, N.H., J.K. Koffi, Y. Pelcat, and L.R. Lindsay, 2014: Environmental risk from Lyme disease in central and eastern Canada: A summary of recent surveillance information. *Canada Communicable Disease Report,* **40,** 74-82. http://www.phac-aspc.gc.ca/publicat/ccdr-rmtc/14vol40/dr-rm40-05/dr-rm40-05-1-eng.php

72. Monaghan, A.J., S.M. Moore, K.M. Sampson, C.B. Beard, and R.J. Eisen, 2015: Climate change influences on the annual onset of Lyme disease in the United States. *Ticks and Tick-Borne Diseases,* **6,** 615-622. http://dx.doi.org/10.1016/j.ttbdis.2015.05.005

73. CDC, 2014: Surveillance Resources: ArboNET. Centers for Disease Control and Prevention, Arboviral Diseases Branch, Fort Collins, CO. http://www.cdc.gov/westnile/resourcepages/survResources.html

74. Lanciotti, R.S., J.T. Roehrig, V. Deubel, J. Smith, M. Parker, K. Steele, B. Crise, K.E. Volpe, M.B. Crabtree, J.H. Scherret, R.A. Hall, J.S. MacKenzie, C.B. Cropp, B. Panigrahy, E. Ostlund, B. Schmitt, M. Malkinson, C. Banet, J. Weissman, N. Komar, H.M. Savage, W. Stone, T. McNamara, and D.J. Gubler, 1999: Origin of the West Nile Virus responsible for an outbreak of encephalitis in the northeastern United States. *Science,* **286,** 2333-2337. http://dx.doi.org/10.1126/science.286.5448.2333

75. Beasley, D.W., A.D. Barrett, and R.B. Tesh, 2013: Resurgence of West Nile neurologic disease in the United States in 2012: What happened? What needs to be done? *Antiviral Research,* **99,** 1-5. http://dx.doi.org/10.1016/j.antiviral.2013.04.015

76. Petersen, L.R. and E.B. Hayes, 2004: Westward ho?—The spread of West Nile virus. *The New England Journal of Medicine,* **351,** 2257-2259. http://dx.doi.org/10.1056/NEJMp048261

77. Petersen, L.R., P.J. Carson, B.J. Biggerstaff, B. Custer, S.M. Borchardt, and M.P. Busch, 2013: Estimated cumulative incidence of West Nile virus infection in US adults, 1999-2010. *Epidemiology and Infection,* **141,** 591-595. http://dx.doi.org/10.1017/S0950268812001070

78. Busch, M.P., D.J. Wright, B. Custer, L.H. Tobler, S.L. Stramer, S.H. Kleinman, H.E. Prince, C. Bianco, G. Foster, L.R. Petersen, G. Nemo, and S.A. Glynn, 2006: West Nile virus infections projected from blood donor screening data, United States, 2003. *Emerging Infectious Diseases,* **12,** 395-402. http://dx.doi.org/10.3201/eid1203.051287

79. Carson, P.J., S.M. Borchardt, B. Custer, H.E. Prince, J. Dunn-Williams, V. Winkelman, L. Tobler, B.J. Biggerstaff, R. Lanciotti, L.R. Petersen, and M.P. Busch, 2012: Neuroinvasive disease and West Nile virus infection, North Dakota, USA, 1999–2008. *Emerging Infectious Diseases,* **18,** 684-686. http://dx.doi.org/10.3201/eid1804.111313

80. Mostashari, F., M.L. Bunning, P.T. Kitsutani, D.A. Singer, D. Nash, M.J. Cooper, N. Katz, K.A. Liljebjelke, B.J. Biggerstaff, A.D. Fine, M.C. Layton, S.M. Mullin, A.J. Johnson, D.A. Martin, E.B. Hayes, and G.L. Campbell, 2001: Epidemic West Nile encephalitis, New York, 1999: Results of a household-based seroepidemiological survey. *The Lancet,* **358,** 261-264. http://dx.doi.org/10.1016/S0140-6736(01)05480-0

81. Hayes, E.B., N. Komar, R.S. Nasci, S. Montgomery, D.R. O'Leary, and G.L. Campbell, 2005: Epidemiology and transmission dynamics of West Nile virus disease. *Emerging Infectious Diseases,* **11,** 1167-1173. http://dx.doi.org/10.3201/eid1108.050289a

82. Pealer, L.N., A.A. Marfin, L.R. Petersen, R.S. Lanciotti, P.L. Page, S.L. Stramer, M.G. Stobierski, K. Signs, B. Newman, H. Kapoor, J.L. Goodman, and M.E. Chamberland, 2003: Transmission of West Nile virus through blood transfusion in the United States in 2002. *The New England Journal of Medicine,* **349,** 1236-1245. http://dx.doi.org/10.1056/NEJMoa030969

83. Zou, S., G.A. Foster, R.Y. Dodd, L.R. Petersen, and S.L. Stramer, 2010: West Nile fever characteristics among viremic persons identified through blood donor screening. *The Journal of Infectious Diseases,* **202,** 1354-1361. http://dx.doi.org/10.1086/656602

84. Nett, R.J., M.J. Kuehnert, M.G. Ison, J.P. Orlowski, M. Fischer, and J.E. Staples, 2012: Current practices and evaluation of screening solid organ donors for West Nile virus. *Transplant Infectious Disease,* **14,** 268-277. http://dx.doi.org/10.1111/j.1399-3062.2012.00743.x

85. Petersen, L.R., A.C. Brault, and R.S. Nasci, 2013: West Nile virus: Review of the literature. *JAMA: The Journal of the American Medical Association,* **310,** 308-315. http://dx.doi.org/10.1001/jama.2013.8042

86. Komar, N., 2003: West Nile virus: Epidemiology and ecology in North America. *The Flaviviruses: Detection, Diagnosis and Vaccine Development.* Chambers, T. and T. Monath, Eds. Elsevier Academic Press, London, UK, 185-234. http://dx.doi.org/10.1016/s0065-3527(03)61005-5

87. Reisen, W.K., Y. Fang, and V.M. Martinez, 2005: Avian host and mosquito (Diptera: Culicidae) vector competence determine the efficiency of West Nile and St. Louis encephalitis virus transmission. *Journal of Medical Entomology,* **42,** 367-375. http://dx.doi.org/10.1603/0022-2585(2005)042%5B0367:ahamdc%5D2.0.co;2

88. Darsie, R.F. and R.A. Ward, 2005: *Identification and Geographical Distribution of the Mosquitos of North America, North of Mexico.* University Press of Florida, Gainesville, FL, 383 pp.

89. Eldridge, B.F., 1987: *Diapause and related phenomena in Culex mosquitoes: Their relation to arbovirus disease ecology.* Current Topics in Vector Research, **4,** 1-28. http://dx.doi.org/10.1007/978-1-4612-4712-8_1

90. Nelms, B.M., P.A. Macedo, L. Kothera, H.M. Savage, and W.K. Reisen, 2013: Overwintering biology of *Culex (Diptera: Culicidae) mosquitoes in the Sacramento Valley of California.* Journal of Medical Entomology, **50,** 773-790. http://dx.doi.org/10.1603/me12280

91. Mail, G. and R. McHugh, 1961: Relation of temperature and humidity to winter survival of Culex pipiens and Culex tarsalis. *Mosquito News,* **21,** 252-254.

92. DeGroote, J.P., R. Sugumaran, S.M. Brend, B.J. Tucker, and L.C. Bartholomay, 2008: Landscape, demographic, entomological, and climatic associations with human disease incidence of West Nile virus in the state of Iowa, USA. *International Journal of Health Geographics,* **7,** 19. http://dx.doi.org/10.1186/1476-072x-7-19

93. Eisen, L., C.M. Barker, C.G. Moore, W.J. Pape, A.M. Winters, and N. Cheronis, 2010: Irrigated agriculture is an important risk factor for West Nile virus disease in the hyperendemic Larimer-Boulder-Weld area of north central Colorado. *Journal of Medical Entomology,* **47,** 939-951. http://dx.doi.org/10.1093/jmedent/47.5.939

94. Gibney, K.B., J. Colborn, S. Baty, A.M. Bunko Patterson, T. Sylvester, G. Briggs, T. Stewart, C. Levy, K. Komatsu, K. MacMillan, M.J. Delorey, J.-P. Mutebi, M. Fischer, and J.E. Staples, 2012: Modifiable risk factors for West Nile virus infection during an outbreak—Arizona, 2010. *American Journal of Tropical Medicine and Hygiene,* **86,** 895-901. http://dx.doi.org/10.4269/ajtmh.2012.11-0502

95. Kilpatrick, A.M., P. Daszak, M.J. Jones, P.P. Marra, and L.D. Kramer, 2006: Host heterogeneity dominates West Nile virus transmission. *Proceedings of the Royal Society B: Biological Sciences,* **273,** 2327-2333. http://dx.doi.org/10.1098/rspb.2006.3575

96. Kilpatrick, A.M., 2011: Globalization, land use, and the invasion of West Nile virus. *Science,* **334,** 323-327. http://dx.doi.org/10.1126/science.1201010

97. Dusek, R.J., R.G. McLean, L.D. Kramer, S.R. Ubico, A.P. Dupuis, G.D. Ebel, and S.C. Guptill, 2009: Prevalence of West Nile virus in migratory birds during spring and fall migration. *American Journal of Tropical Medicine and Hygiene,* **81,** 1151-1158. http://dx.doi.org/10.4269/ajtmh.2009.09-0106

98. Owen, J., F. Moore, N. Panella, E. Edwards, R. Bru, M. Hughes, and N. Komar, 2006: Migrating birds as dispersal vehicles for West Nile virus. *EcoHealth,* **3,** 79-85. http://dx.doi.org/10.1007/s10393-006-0025-9

99. Parmesan, C., 2006: Ecological and evolutionary responses to recent climate change. *Annual Review of Ecology, Evolution, and Systematics,* **37,** 637-669. http://dx.doi.org/10.1146/annurev.ecolsys.37.091305.110100

100. Dodson, B.L., L.D. Kramer, and J.L. Rasgon, 2012: Effects of larval rearing temperature on immature development and West Nile virus vector competence of *Culex tarsalis. Parasites & Vectors,* **5,** 199. http://dx.doi.org/10.1186/1756-3305-5-199

101. Reisen, W.K., D. Cayan, M. Tyree, C.M. Barker, B. Eldridge, and M. Dettinger, 2008: Impact of climate variation on mosquito abundance in California. *Journal of Vector Ecology,* **33,** 89-98. http://dx.doi.org/10.3376/1081-1710(2008)33%5B89:iocvom%5D2.0.co;2

102. Rueda, L.M., K.J. Patel, R.C. Axtell, and R.E. Stinner, 1990: Temperature-dependent development and survival rates of *Culex quinquefasciatus and Aedes aegypti (Diptera: Culicidae). Journal of Medical Entomology,* **27,** 892-898. http://dx.doi.org/10.1093/jmedent/27.5.892

103. Walter, N.M. and C.S. Hacker, 1974: Variation in life table characteristics among three geographic strains of Culex pipiens quinquefasciatus. *Journal of Medical Entomology,* **11,** 541-550.

104. Reisen, W.K., 1995: Effect of temperature on *Culex tarsalis (Diptera: Culicidae) from the Coachella and San Joaquin Valleys of California. Journal of Medical Entomology,* **32,** 636-645. http://dx.doi.org/10.1093/jmedent/32.5.636

105. Garcia-Rejón, J.E., J.A. Farfan-Ale, A. Ulloa, L.F. Flores-Flores, E. Rosado-Paredes, C. Baak-Baak, M.A. Loroño-Pino, I. Fernández-Salas, and B.J. Beaty, 2008: Gonotrophic cycle estimate for *Culex quinquefasciatus in Mérida, Yucatán, México. Journal of the American Mosquito Control Association,* **24,** 344-348. http://dx.doi.org/10.2987/5667.1

106. Hartley, D.M., C.M. Barker, A. Le Menach, T. Niu, H.D. Gaff, and W.K. Reisen, 2012: Effects of temperature on emergence and seasonality of West Nile virus in California. *American Journal of Tropical Medicine and Hygiene,* **86,** 884-894. http://dx.doi.org/10.4269/ajtmh.2012.11-0342

107. Dohm, D.J., M.L. O'Guinn, and M.J. Turell, 2002: Effect of environmental temperature on the ability of *Culex pipiens (Diptera: Culicidae) to transmit West Nile virus. Journal of Medical Entomology,* **39,** 221-225. http://dx.doi.org/10.1603/0022-2585-39.1.221

108. Kilpatrick, A.M., M.A. Meola, R.M. Moudy, and L.D. Kramer, 2008: Temperature, viral genetics, and the transmission of West Nile virus by *Culex pipiens mosquitoes. PLoS Pathogens,* **4,** e1000092. http://dx.doi.org/10.1371/journal.ppat.1000092

109. Reisen, W.K., Y. Fang, and V.M. Martinez, 2006: Effects of temperature on the transmission of West Nile virus by *Culex tarsalis (Diptera: Culicidae). Journal of Medical Entomology,* **43,** 309-317. http://dx.doi.org/10.1093/jmedent/43.2.309

110. Ruiz, M.O., L.F. Chaves, G.L. Hamer, T. Sun, W.M. Brown, E.D. Walker, L. Haramis, T.L. Goldberg, and U.D. Kitron, 2010: Local impact of temperature and precipitation on West Nile virus infection in *Culex species mosquitoes in northeast Illinois, USA. Parasites & Vectors,* **3,** Article 19. http://dx.doi.org/10.1186/1756-3305-3-19

111. Soverow, J.E., G.A. Wellenius, D.N. Fisman, and M.A. Mittleman, 2009: Infectious disease in a warming world: How weather influenced West Nile virus in the United States (2001–2005). *Environmental Health Perspectives,* **117,** 1049-1052. http://dx.doi.org/10.1289/ehp.0800487

112. Wimberly, M.C., A. Lamsal, P. Giacomo, and T.-W. Chuang, 2014: Regional variation of climatic influences on West Nile virus outbreaks in the United States. *American Journal of Tropical Medicine and Hygiene,* **91,** 677-684. http://dx.doi.org/10.4269/ajtmh.14-0239

113. Winters, A.M., R.J. Eisen, S. Lozano-Fuentes, C.G. Moore, W.J. Pape, and L. Eisen, 2008: Predictive spatial models for risk of West Nile virus exposure in eastern and western Colorado. *American Journal of Tropical Medicine and Hygiene,* **79,** 581-590. PMC2581834

114. Morin, C.W. and A.C. Comrie, 2013: Regional and seasonal response of a West Nile virus vector to climate change. *Proceedings of the National Academy of Sciences,* **110,** 15620-15625. http://dx.doi.org/10.1073/pnas.1307135110

115. Calhoun, L.M., M. Avery, L. Jones, K. Gunarto, R. King, J. Roberts, and T.R. Burkot, 2007: Combined sewage overflows (CSO) are major urban breeding sites for *Culex quinquefasciatus in Atlanta, Georgia. American Journal of Tropical Medicine and Hygiene,* **77,** 478-484.

116. Chuang, T.W., M.B. Hildreth, D.L. Vanroekel, and M.C. Wimberly, 2011: Weather and land cover influences on mosquito populations in Sioux Falls, South Dakota. *Journal of Medical Entomology,* **48,** 669-79. http://dx.doi.org/10.1603/me10246

117. Barker, C.M., B.G. Bolling, W.C. Black, IV, C.G. Moore, and L. Eisen, 2009: Mosquitoes and West Nile virus along a river corridor from prairie to montane habitats in eastern Colorado. *Journal of Vector Ecology,* **34,** 276-293. http://dx.doi.org/10.1111/j.1948-7134.2009.00036.x

118. Wegbreit, J. and W.K. Reisen, 2000: Relationships among weather, mosquito abundance, and encephalitis virus activity in California: Kern County 1990-98. *Journal of the American Mosquito Control Association,* **16,** 22-27.

119. Landesman, W.J., B.F. Allan, R.B. Langerhans, T.M. Knight, and J.M. Chase, 2007: Inter-annual associations between precipitation and human incidence of West Nile Virus in the United States. *Vector-Borne and Zoonotic Diseases,* **7,** 337-343. http://dx.doi.org/10.1089/vbz.2006.0590

120. Shaman, J., J.F. Day, and M. Stieglitz, 2005: Drought-induced amplification and epidemic transmission of West Nile virus in southern Florida. *Journal of Medical Entomology,* **42,** 134-141. http://dx.doi.org/10.1093/jmedent/42.2.134

121. Shaman, J., J.F. Day, and N. Komar, 2010: Hydrologic conditions describe West Nile virus risk in Colorado. *International Journal of Environmental Research and Pulic Health,* **7,** 494-508. http://dx.doi.org/10.3390/ijerph7020494

122. Shaman, J., K. Harding, and S.R. Campbell, 2011: Meteorological and hydrological influences on the spatial and temporal prevalence of West Nile virus in *Culex mosquitos, Suffolk County, New York. Journal of Medical Entomology,* **48,** 867-875. http://dx.doi.org/10.1603/ME10269

123. Pecoraro, H.L., H.L. Day, R. Reineke, N. Stevens, J.C. Withey, J.M. Marzluff, and J.S. Meschke, 2007: Climatic and landscape correlates for potential West Nile virus mosquito vectors in the Seattle region. *Journal of Vector Ecology,* **32,** 22-28. http://dx.doi.org/10.3376/1081-1710(2007)32%5B22:-CALCFP%5D2.0.CO;2

124. Gardner, A.M., G.L. Hamer, A.M. Hines, C.M. Newman, E.D. Walker, and M.O. Ruiz, 2012: Weather variability affects abundance of larval *Culex (Diptera: Culicidae) in storm water catch basins in suburban Chicago. Journal of Medical Entomology,* **49,** 270-276. http://dx.doi.org/10.1603/ME11073

125. Johnson, B.J. and M.V.K. Sukhdeo, 2013: Drought-induced amplification of local and regional West Nile virus infection rates in New Jersey. *Journal of Medical Entomology,* **50,** 195-204. http://dx.doi.org/10.1603/me12035

126. Reeves, W.C., S.M. Asman, J.L. Hardy, M.M. Milby, and W.K. Reisen, 1990: *Epidemiology and Control of Mosquito-Borne Arboviruses in California, 1943-1987. California Mosquito and Vector Control Association, Sacramento, CA, 508 pp.*

127. *Reisen, W.K., R.P. Meyer, M.M. Milby, S.B. Presser, R.W. Emmons, J.L. Hardy, and W.C. Reeves, 1992: Ecological observations on the 1989 outbreak of St. Louis encephalitis virus in the southern San Joaquin Valley of California. Journal of Medical Entomology,* **29,** 472-482. http://dx.doi.org/10.1093/jmedent/29.3.472

128. Shaman, J., J.F. Day, and M. Stieglitz, 2004: The spatial-temporal distribution of drought, wetting, and human cases of St. Louis encephalitis in southcentral Florida. *American Journal of Tropical Medicine and Hygiene,* **71,** 251-261. http://www.ajtmh.org/content/71/3/251.long

129. Forrester, N.L., J.L. Kenney, E. Deardorff, E. Wang, and S.C. Weaver, 2008: Western equine encephalitis submergence: Lack of evidence for a decline in virus virulence. *Virology,* **380,** 170-172. http://dx.doi.org/10.1016/j.virol.2008.08.012

130. Reisen, W.K., Y. Fang, and A.C. Brault, 2008: Limited interdecadal variation in mosquito (Diptera: Culicidae) and avian host competence for Western equine encephalomyelitis virus (Togaviridae: Alphavirus). *American Journal of Tropical Medicine and Hygiene,* **78,** 681-686. http://www.ajtmh.org/content/78/4/681.full.pdf+html

131. Manore, C.A., J.K. Davis, R.C. Christofferson, D.M. Wesson, J.M. Hyman, and C.N. Mores, 2014: Towards an early warning system for forecasting human west nile virus incidence. *PLoS Currents,* **6.** http://dx.doi.org/10.1371/currents.outbreaks.f0b3978230599a56830ce30cb9ce0500

132. Harrigan, R.J., H.A. Thomassen, W. Buermann, and T.B. Smith, 2014: A continental risk assessment of West Nile virus under climate change. *Global Change Biology,* **20,** 2417-2425. http://dx.doi.org/10.1111/gcb.12534

133. Bennett, C.M. and A.J. McMichael, 2010: Non-heat related impacts of climate change on working populations. *Global Health Action,* **3,** 5640. http://dx.doi.org/10.3402/gha.v3i0.5640

134. Brunkard, J.M., E. Cifuentes, and S.J. Rothenberg, 2008: Assessing the roles of temperature, precipitation, and ENSO in dengue re-emergence on the Texas-Mexico border region. *Salud Publica Mex,* **50,** 227-234. http://dx.doi.org/10.1590/S0036-36342008000300006

135. Gubler, D.J., P. Reiter, K.L. Ebi, W. Yap, R. Nasci, and J.A. Patz, 2001: Climate variability and change in the United States: Potential impacts on vector- and rodent-borne diseases. *Environmental Health Perspectives,* **109,** 223-233. http://dx.doi.org/10.2307/3435012

136. Ramos, M.M., H. Mohammed, E. Zielinski-Gutierrez, M.H. Hayden, J.L.R. Lopez, M. Fournier, A.R. Trujillo, R. Burton, J.M. Brunkard, L. Anaya-Lopez, A.A. Banicki, P.K. Morales, B. Smith, J.L. Muñoz, and S.H. Waterman, 2008: Epidemic dengue and dengue hemorrhagic fever at the Texas–Mexico border: Results of a household-based seroepidemiologic survey, December 2005. *American Journal of Tropical Medicine and Hygiene,* **78,** 364-369. http://www.ajtmh.org/content/78/3/364.full.pdf+html

137. Reiter, P., S. Lathrop, M. Bunning, B. Biggerstaff, D. Singer, T. Tiwari, L. Baber, M. Amador, J. Thirion, J. Hayes, C. Seca, J. Mendez, B. Ramirez, J. Robinson, J. Rawlings, V. Vorndam, S. Waterman, D. Gubler, G. Clark, and E. Hayes, 2003: Texas lifestyle limits transmission of dengue virus. *Emerging Infectious Diseases,* **9,** 86-89. http://dx.doi.org/10.3201/eid0901.020220

138. CDC, 2015: Lyme Disease: Preventing Tick Bites on People. Centers for Disease Control and Prevention, Atlanta, GA. http://www.cdc.gov/lyme/prev/on_people.html

139. Finch, C., M.S. Al-Damluji, P.J. Krause, L. Niccolai, T. Steeves, C.F. O'Keefe, and M.A. Diuk-Wasser, 2014: Integrated assessment of behavioral and environmental risk factors for Lyme disease infection on Block Island, Rhode Island. *PLoS ONE,* **9,** e84758. http://dx.doi.org/10.1371/journal.pone.0084758

140. Bowen, G.S., T.L. Schulze, C. Hayne, and W.E. Parkin, 1984: A focus of Lyme disease in Monmouth County, New Jersey. *American Journal of Epidemiology,* **120,** 387-394. http://www.ncbi.nlm.nih.gov/pubmed/6475916

141. Schwartz, B.S. and M.D. Goldstein, 1990: Lyme disease in outdoor workers: Risk factors, preventive measures, and tick removal methods. *American Journal of Epidemiology,* **131,** 877-885.

142. Schwartz, B.S., M.D. Goldstein, and J.E. Childs, 1994: Longitudinal study of *Borrelia burgdorferi infection in New Jersey outdoor workers, 1988-1991. American Journal of Epidemiology,* **139,** 504-512.

143. NIOSH, 2005: Recommendations for Protecting Outdoor Workers from West Nile Virus Exposure. DHHS (NIOSH) Publication No. 2005–155, 16 pp. Department of Health and Human Services, Centers for Disease Control and Prevention, National Institute for Occupational Safety and Health. http://www.cdc.gov/niosh/docs/2005-155/

144. Brien, J.D., J.L. Uhrlaub, A. Hirsch, C.A. Wiley, and J. Nikolich-Žugich, 2009: Key role of T cell defects in age-related vulnerability to West Nile virus. *The Journal of Experimental Medicine,* **206,** 2735-2745. http://dx.doi.org/10.1084/jem.20090222

145. Weiss, D., D. Carr, J. Kellachan, C. Tan, M. Phillips, E. Bresnitz, M. Layton, and West Nile Virus Outbreak Response Working Group, 2001: Clinical findings of West Nile virus infection in hospitalized patients, New York and New Jersey, 2000. *Emerging Infectious Diseases,* **7,** 654-658. PMC2631758

146. WHO, 2015: Dengue and Severe Dengue. Fact Sheet No. 117. World Health Organization. http://www.who.int/mediacentre/factsheets/fs117/en/index.html

147. USGS, 2014: Dengue Fever (Locally Acquired) Human 2013. Cumulative data as of May 7, 2014. United States Geological Survey. http://diseasemaps.usgs.gov/2013/del_us_human.html

148. PAHO, 2014: Chikungunya. Pan American Health Organization, Washington, D.C. http://www.paho.org/hq/index.php?Itemid=40931

149. Parham, P.E., J. Waldock, G.K. Christophides, D. Hemming, F. Agusto, K.J. Evans, N. Fefferman, H. Gaff, A. Gumel, S. LaDeau, S. Lenhart, R.E. Mickens, E.N. Naumova, R.S. Ostfeld, P.D. Ready, M.B. Thomas, J. Velasco-Hernandez, and E. Michael, 2015: Climate, environmental and socio-economic change: Weighing up the balance in vector-borne disease transmission. *Philosophical Transactions of the Royal Society B: Biological Sciences,* **370.** http://dx.doi.org/10.1098/rstb.2013.0551

150. Reiner, R.C., T.A. Perkins, C.M. Barker, T. Niu, L.F. Chaves, A.M. Ellis, D.B. George, A. Le Menach, J.R.C. Pulliam, D. Bisanzio, C. Buckee, C. Chiyaka, D.A.T. Cummings, A.J. Garcia, M.L. Gatton, P.W. Gething, D.M. Hartley, G. Johnston, E.Y. Klein, E. Michael, S.W. Lindsay, A.L. Lloyd, D.M. Pigott, W.K. Keisen, N. Ruktanonchai, B.K. Singh, A.J. Tatem, U. Kitron, S.I. Hay, T.W. Scott, and D.L. Smith, 2013: A systematic review of mathematical models of mosquito-borne pathogen transmission: 1970-2010. *Journal of the Royal Society Interface,* **10.** http://dx.doi.org/10.1098/rsif.2012.0921

6 CLIMATE IMPACTS ON WATER-RELATED ILLNESS

Lead Authors

Juli Trtanj
National Oceanic and Atmospheric Administration

Lesley Jantarasami*
U.S. Environmental Protection Agency

Contributing Authors

Joan Brunkard
Centers for Disease Control and Prevention

Tracy Collier
National Oceanic and Atmospheric Administration and
University Corporation for Atmospheric Research

John Jacobs
National Oceanic and Atmospheric Administration

Erin Lipp
The University of Georgia

Sandra McLellan
University of Wisconsin-Milwaukee

Stephanie Moore
National Oceanic and Atmospheric Administration and
University Corporation for Atmospheric Research

Hans Paerl
The University of North Carolina at Chapel Hill

John Ravenscroft
U.S. Environmental Protection Agency

Mario Sengco
U.S. Environmental Protection Agency

Jeanette Thurston
U.S. Department of Agriculture

Acknowledgements: Sharon Nappier, U.S. Environmental Protection Agency

Recommended Citation: Trtanj, J., L. Jantarasami, J. Brunkard, T. Collier, J. Jacobs, E. Lipp, S. McLellan, S. Moore, H. Paerl, J. Ravenscroft, M. Sengco, and J. Thurston, 2016: Ch. 6: Climate Impacts on Water-Related Illness. *The Impacts of Climate Change on Human Health in the United States: A Scientific Assessment.* U.S. Global Change Research Program, Washington, DC, 157–188. http://dx.doi.org/10.7930/J03F4MH4

On the web: health2016.globalchange.gov   *Chapter Coordinator*

6 CLIMATE IMPACTS ON WATER-RELATED ILLNESS

Key Findings

Seasonal and Geographic Changes in Waterborne Illness Risk

Key Finding 1: Increases in water temperatures associated with climate change will alter the seasonal windows of growth and the geographic range of suitable habitat for freshwater toxin-producing harmful algae *[Very Likely, High Confidence]*, certain naturally occurring *Vibrio* bacteria *[Very Likely, Medium Confidence]*, and marine toxin-producing harmful algae *[Likely, Medium Confidence]*. These changes will increase the risk of exposure to waterborne pathogens and algal toxins that can cause a variety of illnesses *[Medium Confidence]*.

Runoff from Extreme Precipitation Increases Exposure Risk

Key Finding 2: Runoff from more frequent and intense extreme precipitation events will increasingly compromise recreational waters, shellfish harvesting waters, and sources of drinking water through increased introduction of pathogens and prevalence of toxic algal blooms *[High Confidence]*. As a result, the risk of human exposure to agents of water-related illness will increase *[Medium Confidence]*.

Water Infrastructure Failure

Key Finding 3: Increases in some extreme weather events and storm surges will increase the risk that infrastructure for drinking water, wastewater, and stormwater will fail due to either damage or exceedance of system capacity, especially in areas with aging infrastructure *[High Confidence]*. As a result, the risk of exposure to water-related pathogens, chemicals, and algal toxins will increase in recreational and shellfish harvesting waters and in drinking water where treatment barriers break down *[Medium Confidence]*.

6.1 Introduction

Across most of the United States, climate change is expected to affect fresh and marine water resources in ways that will increase people's exposure to water-related contaminants that cause illness. Water-related illnesses include waterborne diseases caused by pathogens, such as bacteria, viruses, and protozoa. Water-related illnesses are also caused by toxins produced by certain harmful algae and cyanobacteria (also known as blue-green algae) and by chemicals introduced into the environment by human activities. Exposure occurs through ingestion, inhalation, or direct contact with contaminated drinking or recreational water and through consumption of fish and shellfish.

Factors related to climate change—including temperature, precipitation and related runoff, hurricanes, and storm surge—affect the growth, survival, spread, and virulence or toxicity of agents (causes) of water-related illness. Heavy downpours are already on the rise and increases in the frequency and intensity of extreme precipitation events are projected for all U.S. regions.[1] Projections of temperature, precipitation, extreme events such as flooding and drought, and other climate factors vary by region of the United States, and thus the extent of climate health impacts will also vary by region.

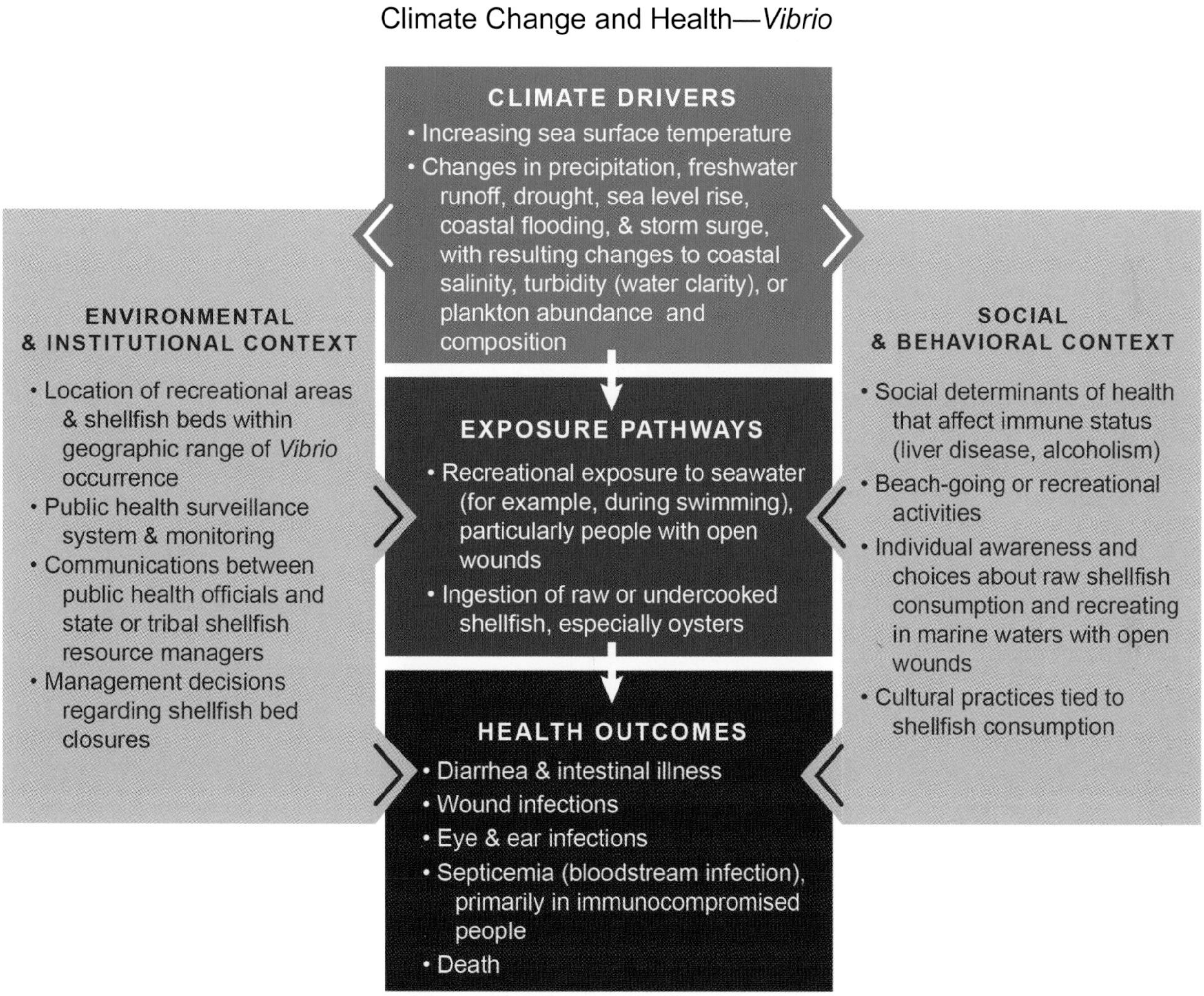

Climate Change and Health—*Vibrio*

Figure 1: This conceptual diagram for an example of infection by *Vibrio* species (*V. vulnificus*, *V. parahaemolyticus*, or *V. alginolyticus*) illustrates the key pathways by which humans are exposed to health threats from climate drivers. These climate drivers create more favorable growing conditions for these naturally occurring pathogens in coastal environments through their effects on coastal salinity, turbidity (water clarity), or plankton abundance and composition. Longer seasons for growth and expanding geographic range of occurrence increase the risk of exposure to *Vibrio*, which can result in various potential health outcomes (center boxes). These exposure pathways exist within the context of other factors that positively or negatively influence health outcomes (gray side boxes). Key factors that influence vulnerability for individuals are shown in the right box and include social determinants of health and behavioral choices. Key factors that influence vulnerability at larger scales, such as natural and built environments, governance and management, and institutions, are shown in the left box. All of these influencing factors can affect an individual's or a community's vulnerability through changes in exposure, sensitivity, and adaptive capacity and may also be affected by climate change. See Ch. 1: Introduction for more information.

Waterborne pathogens are estimated to cause 8.5% to 12% of acute gastrointestinal illness cases in the United States, affecting between 12 million and 19 million people annually.[2, 3, 4] Eight pathogens, which are all affected to some degree by climate, account for approximately 97% of all suspected waterborne illnesses in the United States: the enteric viruses norovirus, rotavirus, and adenovirus; the bacteria *Campylobacter jejuni*, *E. coli* O157:H7, and *Salmonella enterica*; and the protozoa *Cryptosporidium* and *Giardia*.[5]

Specific health outcomes are determined by different exposure pathways and multiple other social and behavioral factors, some of which are also affected by climate (Figure 1). Most research to date has focused on understanding how climate drivers affect physical and ecological processes that act as key exposure pathways for pathogens and toxins, as shown by the arrow moving from the top to the middle box in Figure 1. There is currently less information and fewer methods with which to measure actual human exposure and incidence of illness based on those physical and ecological metrics (arrow moving from middle to bottom box in Figure 1). Thus, it is often not possible to quantitatively project future health outcomes from water-related illnesses under climate change (bottom box in Figure 1).

This chapter covers health risks associated with changes in natural marine, coastal, and freshwater systems and water infrastructure for drinking water, wastewater, and stormwater (*Legionella* in aerosolized water is covered in Ch. 3: Air Quality Impacts). This chapter also includes fish and shellfish illnesses associated with the waters in which they grow and which are affected by the same climate factors that affect drinking and recreational waters (impacts related to handling and post-harvest processing of seafood are covered in Ch. 7: Food Safety). The framing of this chapter addresses sources of contaminations, exposure pathways, and health outcomes when available. Based on the available data and research, many of the examples are regionally focused and make evident that the impact of climate change on water-related illness is inherently regional. Table 1 lists various health outcomes that can result from exposure to agents of water-related illness as well as key climate-related changes affecting their occurrence.

Whether or not illness results from exposure to contaminated water, fish, or shellfish is dependent on a complex set of factors, including human behavior and social determinants of health that may affect a person's exposure, sensitivity, and adaptive capacity (Figure 1; see also Ch. 1: Introduction and

Table 1. Climate Sensitive Agents of Water Related Illness

Pathogen or Toxin Producer	Exposure Pathway	Selected Health Outcomes & Symptoms	Major Climate Correlation or Driver (strongest driver(s) listed first)
Algae: Toxigenic marine species of *Alexandrium, Pseudo-nitzschia, Dinophysis, Gambierdiscus; Karenia brevis*	**Shellfish Fish Recreational waters (aerosolized toxins)**	Gastrointestinal and neurologic illness caused by shellfish poisoning (paralytic, amnesic, diarrhetic, neurotoxic) or fish poisoning (ciguatera). Asthma exacerbations, eye irritations caused by contact with aerosolized toxins (*K. brevis*).	Temperature (increased water temperature), ocean surface currents, ocean acidification, hurricanes (*Gambierdiscus* spp. and *K. brevis*)
Cyanobacteria (multiple freshwater species producing toxins including microcystin)	**Drinking water Recreational waters**	Liver and kidney damage, gastroenteritis (diarrhea and vomiting), neurological disorders, and respiratory arrest.	Temperature, precipitation patterns
Enteric bacteria & protozoan parasites: *Salmonella enterica; Campylobacter* species; Toxigenic *Escherichia coli; Cryptosporidium; Giardia*	**Drinking water Recreational waters Shellfish**	Enteric pathogens generally cause gastroenteritis. Some cases may be severe and may be associated with long-term and recurring effects.	Temperature (air and water; both increase and decrease), heavy precipitation, and flooding
Enteric viruses: enteroviruses; rotaviruses; noroviruses; hepatitis A and E	**Drinking water Recreational waters Shellfish**	Most cases result in gastrointestinal illness. Severe outcomes may include paralysis and infection of the heart or other organs.	Heavy precipitation, flooding, and temperature (air and water; both increase and decrease)
Leptospira and *Leptonema* bacteria	**Recreational waters**	Mild to severe flu-like illness (with or without fever) to severe cases of meningitis, kidney, and liver failure.	Flooding, temperature (increased water temperature), heavy precipitation
Vibrio bacteria species	**Recreational waters Shellfish**	Varies by species but include gastroenteritis (*V. parahaemolyticus, V. cholerae*), septicemia (bloodstream infection) through ingestion or wounds (*V. vulnificus*), skin, eye, and ear infections (*V. alginolyticus*).	Temperature (increased water temperature), sea level rise, precipitation patterns (as it affects coastal salinity)

Ch. 9: Populations of Concern). Water resource, public health, and environmental agencies in the United States provide many public health safeguards to reduce risk of exposure and illness even if water becomes contaminated. These include water quality monitoring, drinking water treatment standards and practices, beach closures, and issuing advisories for boiling drinking water and harvesting shellfish.

Many water-related illnesses are either undiagnosed or unreported, and therefore the total incidence of waterborne disease is underestimated (see Ch. 1: Introduction for discussion of public health surveillance data limitations related to "reportable" and "nationally notifiable" diseases).[6, 7] On average, illnesses from pathogens associated with water are thought to be underestimated by as much as 43-fold, and may be underestimated by up to 143 times for certain *Vibrio* species.[7]

6.2 Sources of Water-Related Contaminants

The primary sources of water contamination are human and animal waste and agricultural activities, including the use of fertilizers. Runoff and flooding resulting from expected increases in extreme precipitation, hurricane rainfall, and storm surge (see Ch. 4: Extreme Events) may increase risks of contamination. Contamination occurs when agents of water-related illness and nutrients, such as nitrogen and phosphorus, are carried from

urban, residential, and agricultural areas into surface waters, groundwater, and coastal waters (Figure 2). The nutrient loading can promote growth of naturally occurring pathogens and algae. Human exposure occurs via contamination of drinking water sources (page 163), recreational waters (page 164), and fish and shellfish (page 165).

Water contamination by human waste is tied to failure of local urban or rural water infrastructure, including municipal wastewater, septic, and stormwater conveyance systems. Failure can occur either when rainfall and subsequent runoff overwhelm the capacity of these systems—causing, for example, sewer overflows, basement backups, or localized flooding—or when extreme events like flooding and storm surges damage water conveyance or treatment infrastructure and result in reduction or loss of performance and functionality. Many older cities in the Northeast and around the Great Lakes region of the United States have combined sewer systems (with stormwater and sewage sharing the same pipes), which are prone to discharging raw sewage directly into surface waters after moderate to heavy rainfall.[8] The amount of rain that causes combined sewer overflows is highly variable between cities because of differences in infrastructure capacity and design, and ranges from 5 mm (about 0.2 inches) to 2.5 cm (about 1 inch).[9, 10] Overall, combined sewer overflows are expected to increase,[11] but site-specific analysis is needed to predict the extent of these increases (see Case Study on page 164). Extreme precipitation events will exacerbate existing problems with inadequate, aging, or deteriorating wastewater infrastructure throughout the country.[12, 13] These problems include broken or leaking sewer pipes and failing septic systems that leach sewage into the ground. Runoff or contaminated groundwater discharge also carries pathogens and nutrients into surface water, including freshwater and marine coastal areas and beaches.[14, 15, 16, 17, 18, 19, 20, 21]

Water contamination from agricultural activities is related to the release of microbial pathogens or nutrients in livestock manure and inorganic fertilizers that can stimulate rapid and excessive growth or blooms of harmful algae. Agricultural land covers about 900 million acres across the United States,[22] comprising over 2 million farms, with livestock sectors concentrated in certain regions of the United States (Figure 3). Depending on the type and number of animals, a large livestock operation can produce between 2,800 and 1,600,000 tons of manure each year.[23, 24] With the projected increases in heavy precipitation for all U.S. regions,[1] agricultural sources of contamination can affect water quality across

Links between Climate Change, Water Quantity and Quality, and Human Exposure to Water-Related Illness.

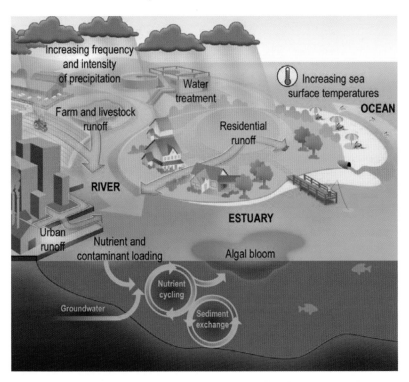

Figure 2: Precipitation and temperature changes affect fresh and marine water quantity and quality primarily through urban, rural, and agricultural runoff. This runoff in turn affects human exposure to water-related illnesses primarily through contamination of drinking water, recreational water, and fish and shellfish.

Locations of Livestock and Projections of Heavy Precipitation

Number of Broilers and Other Meat-Type Chickens per Square Mile, 2012

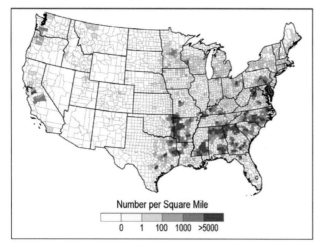

Number per Square Mile

0   1   100   1000   >5000

Number of Cattle and Calves per Square Mile, 2012

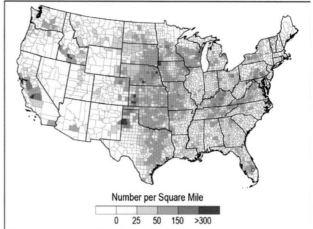

Number per Square Mile

0   25   50   150   >300

Number of Hogs and Pigs per Square Mile, 2012

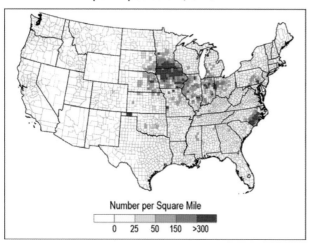

Number per Square Mile

0   25   50   150   >300

Projected Changes in Heavy Precipitation

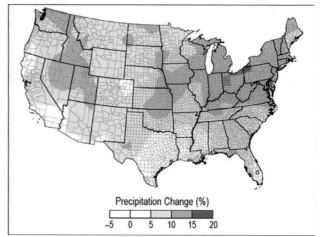

Precipitation Change (%)

−5   0   5   10   15   20

Figure 3: This figure compares the geographic distribution of chicken, cattle, and hog and pig densities to the projected change in annual maximum 5-day precipitation totals (2046–2065 compared to 1981–2000, multi-model average using RCP8.5) across the continental United States. Increasing frequency and intensity of precipitation and subsequent increases in runoff are key climate factors that increase the potential for pathogens associated with livestock waste to contaminate water bodies. (Figure sources: adapted from USDA 2014 and Sun et al. 2015).[26, 27]

the Nation. Runoff from lands where manure has been used as fertilizer or where flooding has caused wastewater lagoons to overflow can carry contamination agents directly from the land into water bodies.[23, 24, 25]

Management practices and technologies, such as better timing of manure application and improved animal feeds, help reduce or eliminate the risks of manure-borne contaminant transport to public water supplies and shellfish harvesting waters and reduce nutrients that stimulate harmful algal blooms.[23, 25, 28, 29] Drinking water treatment and monitoring practices also help to decrease or eliminate exposure to waterborne illness agents originating from agricultural environments.

Water contamination from wildlife (for example, rodents, birds, deer, and wild pigs) occurs via feces and urine of infected animals, which are reservoirs of enteric and other pathogens.[29, 30, 31] Warmer winters and earlier springs are expected to increase animal activity and alter the ecology and habitat of animals that may carry pathogens.[1] This may lengthen the exposure period for humans and expand the geographic ranges in which pathogens are transmitted.[1, 32]

6.3 Exposure Pathways and Health Risks

Humans are exposed to agents of water-related illness through several pathways, including drinking water (treated and untreated), recreational waters (freshwater, coastal, and marine), and fish and shellfish.

Drinking Water

Although the United States has one of the safest municipal drinking water supplies in the world, water-related outbreaks (more than one illness case linked to the same source) still occur.[33] Public drinking water systems provide treated water to approximately 90% of Americans at their places of residence, work, or schools.[34] However, about 15% of the population relies fully or in part on untreated private wells or other private sources for their drinking water.[35] These private sources are not regulated under the Safe Drinking Water Act.[36] The majority of drinking water outbreaks in the United States are associated with untreated or inadequately treated groundwater and distribution system deficiencies.[33, 37]

Pathogen and Algal Toxin Contamination

Between 1948 and 1994, 68% of waterborne disease outbreaks in the United States were preceded by extreme precipitation events,[38] and heavy rainfall and flooding continue to be cited as contributing factors in more recent outbreaks in multiple regions of the United States.[39] Extreme precipitation events have been statistically linked to increased levels of pathogens in treated drinking water supplies[40] and to an increased incidence of gastrointestinal illness in children.[21, 41] This established relationship suggests that extreme precipitation is a key climate factor for waterborne disease.[42, 43, 44, 45] The Milwaukee *Cryptosporidium* outbreak in 1993—the largest documented waterborne disease outbreak in U.S. history, causing an estimated 403,000 illnesses and more than 50 deaths[46]—was preceded by the heaviest rainfall event in 50 years in the adjacent watersheds.[10] Various treatment plant operational problems were also key contributing factors.[47] (See future projections in the Case Study on page 164). Observations in England and Wales also show waterborne disease outbreaks were preceded by weeks of low cumulative rainfall and then heavy precipitation events, suggesting that drought or periods of low rainfall may also be important climate-related factors.[48]

Small community or private groundwater wells or other drinking water systems where water is untreated or minimally treated are especially susceptible to contamination following extreme precipitation events.[49] For example, in May 2000, following heavy rains, livestock waste containing *E. coli* O157:H7 and *Campylobacter* was carried in runoff to a well that served as the primary drinking water source for the town of Walkerton, Ontario, Canada, resulting in 2,300 illnesses and 7 deaths.[43, 44, 50] High rainfall amounts were an important catalyst for the outbreak, although non-climate factors, such as well infrastructure, operational and maintenance problems, and lack of communication between public utilities staff and local health officials were also key factors.[44, 51]

Likewise, extreme precipitation events and subsequent increases in runoff are key climate factors that increase nutrient loading in drinking water sources, which in turn increases the likelihood of harmful cyanobacterial blooms that produce algal

Extreme precipitation events have been statistically linked to increased levels of pathogens in treated drinking water supplies.

toxins.[52] The U.S. Environmental Protection Agency has established health advisories for two algal toxins (microcystins and cylindrospermopsin) in drinking water.[53] Lakes and reservoirs that serve as sources of drinking water for between 30 million and 48 million Americans may be periodically contaminated by algal toxins.[54] Certain drinking water treatment processes can remove cyanobacterial toxins; however, efficacy of the treatment processes may vary from 60% to 99.9%. Ineffective treatment could compromise water quality and may lead to severe treatment disruption or treatment plant shutdown.[53, 54, 55, 56] Such an event occurred in Toledo, Ohio, in August 2014, when nearly 500,000 residents of the state's fourth-largest city lost access to their drinking water after tests revealed the presence of toxins from a cyanobacterial bloom in Lake Erie near the water plant's intake.[57]

Water Supply

Climate-related hydrologic changes such as those related to flooding, drought, runoff, snowpack and snowmelt, and saltwater intrusion (the movement of ocean water into fresh groundwater) have implications for freshwater management and supply (see also Ch. 4: Extreme Events).[58] Adequate freshwater supply is essential to many aspects of public health, including provision of drinking water and proper sanitation and personal hygiene. For example, following floods or storms, short-term loss of access to potable water has been linked to increased incidence of illnesses including gastroenteritis and respiratory tract and skin infections.[59] Changes in precipitation and runoff, combined with changes in consumption and withdrawal, have reduced surface and groundwater supplies in many areas, primarily in the western United States.[58] These trends are expected to continue under future climate change, increasing the likelihood of water shortages for many uses.[58]

Future climate-related water shortages may result in more municipalities and individuals relying on alternative sources for drinking water, including reclaimed water and roof-harvested rainwater.[60, 61, 62, 63] Water reclamation refers to the treatment of stormwater, industrial wastewater, and municipal wastewater for beneficial reuse.[64] States like California, Arizona, New Mexico, Texas, and Florida are already implementing wastewater reclamation and reuse practices as a means of conserving and adding to freshwater supplies.[65] However, no federal regulations or criteria for public health protection have been developed or proposed specifically for potable water reuse in the United States.[66] Increasing household rainwater collection has also been seen in some areas of the country (primarily Arizona, Colorado, and Texas), although in some cases, exposure to untreated rainwater has been found to pose health risks from bacterial or protozoan pathogens, such as *Salmonella enterica* and *Giardia lamblia*.[67, 68, 69]

Projected Changes

Runoff from more frequent and intense extreme precipitation events will contribute to contamination of drinking water sources with pathogens and algal toxins and place additional stresses on the capacity of drinking water treatment facilities and distribution systems.[10, 52, 59, 70, 71, 72, 73] Contamination of drinking water sources may be exacerbated or insufficiently addressed by treatment processes at the treatment plant or by breaches in the distribution system, such as during water main breaks or low-pressure events.[13] Untreated groundwater drawn from municipal and private wells is of particular concern.

Climate change is not expected to substantially increase the risk of contracting illness from drinking water for those people who are served by treated drinking water systems, if appropriate treatment and distribution is maintained. However, projections

of more frequent or severe extreme precipitation events, flooding, and storm surge suggest that drinking water infrastructure may be at greater risk of disruption or failure due to damage or exceedance of system capacity.[6, 58, 70, 74, 75] Aging drinking water infrastructure is one longstanding limitation in controlling waterborne disease, and may be especially susceptible to failure.[6, 13, 74] For example, there are more than 50,000 systems providing treated drinking water to communities in the United States, and most water distribution pipes in these systems are already failing or will reach their expected lifespan and require replacement within 30 years.[6] Breakdowns in drinking water treatment and distribution systems, compounded by aging infrastructure, could lead to more serious and frequent health consequences than those we experience now.

Recreational Waters

Humans are exposed to agents of water-related illness through recreation (such as swimming, fishing, and boating) in freshwater and marine or coastal waters. Exposure may occur directly (ingestion and contact with water) or incidentally (inhalation of aerosolized water droplets).

Pathogen and Algal Toxin Contamination

Enteric viruses, especially noroviruses, from human waste are a primary cause of gastrointestinal illness from exposure to contaminated recreational fresh and marine water (Table 1).[77] Although there are comparatively few reported illnesses and outbreaks of gastrointestinal illness from recreating in marine waters compared to freshwater, marine contamination still presents a significant health risk.[39, 78, 79, 80, 81] Illnesses from marine sources are less likely to be reported than those from freshwater beaches in part because the geographical residences of beachgoers are more widely distributed (for example, tourists may travel to marine beaches for vacation) and illnesses are less often attributed to marine exposure as a common source.[39, 77]

Key climate factors associated with risks of exposure to enteric pathogens in both freshwater and marine recreational waters include extreme precipitation events, flooding, and temperature. For example, *Salmonella* and *Campylobacter* concentrations in freshwater streams in the southeastern United States increase significantly in the summer months and following heavy rainfall.[82, 83, 84] In the Great Lakes—a freshwater system—changes in rainfall, higher lake temperatures, and low lake levels have been linked to increases in fecal bacteria levels.[10] The zoonotic bacteria *Leptospira* are introduced into water from the urine of animals,[85, 86] and increased illness rates in humans are linked to warm temperatures and flooding events.[87, 88, 89, 90, 91]

In marine waters, recreational exposure to naturally occurring bacterial pathogens (such as *Vibrio* species) may result in eye, ear, and wound infections, diarrheal illness, or death (Table 1).[92, 93, 94] Reported rates of illness for all *Vibrio* infections have tripled since 1996, with *V. alginolyticus* infections having increased by 40-fold.[92] *Vibrio* growth rates are highly responsive to rising sea

---

Case Study: Modeling Future Extreme Precipitation and Combined Sewer Overflows in Great Lakes Urban Coastal Areas

The Great Lakes contain 20% of the Earth's surface freshwater and provide drinking water to 40 million people. Milwaukee, WI, is typical of urban areas in the Great Lakes in that it has a combined sewer system that overflows during moderate or heavy rainfall. In 1994, unrelated to but shortly after the 1993 *Cryptosporidium* outbreak, the city completed a project to increase sewer capacity; reducing combined sewage overflows from 50 to 60 per year, to 2 to 3 per year.[10]

In order to assess how changing rainfall patterns might affect sewer capacity in the future, Milwaukee was one of the first cities to integrate regional climate projections into its detailed engineering models. Under a future climate scenario (for 2050) that had one of the largest projected increases in spring rain, a 37% increase in the number of combined sewage overflows in spring was projected, resulting in an overall 20% increase from the baseline in the volume of discharge each year.[76]

surface temperatures, particularly in coastal waters, which generally have high levels of the dissolved organic carbon required for *Vibrio* growth. The distribution of species changes with salinity patterns related to sea level rise and to changes in delivery of freshwater to coastal waters, which is affected by flooding and drought. For instance, *V. parahaeomolyticus* and *V. alginolyticus* favor higher salinities while *V. vulnificus* favors more moderate salinities.[95, 96, 97, 98, 99, 100]

Harmful algal blooms caused by cyanobacteria were responsible for nearly half of all reported outbreaks in untreated recreational freshwater in 2009 and 2010, resulting in approximately 61 illnesses (health effects included dermatologic, gastrointestinal, respiratory, and neurologic symptoms), primarily reported in children/young adults age 1–19.[101] Cyanobacterial blooms are strongly influenced by rising temperatures, altered precipitation patterns, and changes in freshwater discharge or flushing rates of water bodies (Table 1).[102, 103, 104, 105, 106, 107, 108] Higher temperatures (77°F and greater) favor surface-bloom-forming cyanobacteria over less harmful types of algae.[109] In marine water, the toxins associated with harmful "red tide" blooms of *Karenia brevis* can aerosolize in water droplets through wind and wave action and cause acute respiratory illness and eye irritation in recreational beachgoers.[110, 111] People with preexisting respiratory diseases, specifically asthma, are at increased risk of illness.[112, 113] Prevailing winds and storms are important climate factors influencing the accumulation of *K. brevis* cells in the water.[78, 114] For example, in 1996, Tropical Storm Josephine transported a Florida panhandle bloom as far west as Louisiana,[115] the first documented occurrence of *K. brevis* in that state.

Projected Changes

Overall, climate change will contribute to contamination of recreational waters and increased exposure to agents of water-related illness.[10, 82, 116, 117, 118, 119, 120] Increases in flooding, coastal inundation, and nuisance flooding (linked to sea level rise and storm surge from changing patterns of coastal storms and hurricanes) will negatively affect coastal infrastructure and increase chances for pathogen contamination, especially in populated areas (see also Ch. 4: Extreme Events).[70, 121] In areas

In areas where increasing temperatures lengthen recreational swimming seasons, exposure risks are expected to increase.

where increasing temperatures lengthen the seasons for recreational swimming and other water activities, exposure risks are expected to increase.[122, 123]

As average temperatures rise, the seasonal and geographic range of suitable habitat for cyanobacterial species is projected to expand.[124, 125, 126, 127, 128] For example, tropical and subtropical species like *Cylindrospermopsis raciborskii*, *Anabaena* spp., and *Aphanizomenon* spp. have already shown poleward expansion into mid-latitudes of Europe, North America, and South America.[107, 129, 130] Increasing variability in precipitation patterns and more frequent and intense extreme precipitation events (which will increase nutrient loading) will also affect cyanobacterial communities. If such events are followed by extended drought periods, the stagnant, low-flow conditions accompanying droughts will favor cyanobacterial dominance and bloom formation.[103, 131]

In recreational waters, projected increases in sea surface temperatures are expected to lengthen the seasonal window of growth and expand geographic range of *Vibrio* species,[96, 132] although the certainty of regional projections is affected by underlying model structure.[133] While the specific response of *Vibrio* and degree of growth may vary by species and locale, in general, longer seasons and expansion of *Vibrio* into areas where it had not previously been will increase the likelihood of exposure to *Vibrio* in recreational waters. Regional climate changes that affect coastal salinity (such as flooding, drought, and sea level rise) can also affect the population dynamics of these agents,[97, 99, 134] with implications for human exposure risk. Increases in hurricane intensity and rainfall are projected as the climate continues to warm (see Ch 4: Extreme Events). Such increases may redistribute toxic blooms of *K. brevis* ("red tide" blooms) into new geographic locations, which would change human exposure risk in newly affected areas.

Fish and Shellfish

Water-related contaminants as well as naturally occurring harmful bacteria and algae can be accumulated by fish or shellfish, providing a route of human exposure through consumption (see also Ch. 7: Food Safety).[135, 136, 137] Shellfish, including oysters, are often consumed raw or very lightly cooked, which increases the potential for ingestion of an infectious pathogen.[138]

Pathogens Associated with Fish and Shellfish

Enteric viruses (for example, noroviruses and hepatitis A virus) found in sewage are the primary causes of gastrointestinal illness due to shellfish consumption.[139, 140] Rainfall increases the load of contaminants associated with sewage delivered to shellfish harvesting waters and may also temporarily reduce salinity, which can increase persistence of many enteric bacteria and viruses.[141, 142, 143, 144] Many enteric viruses also exhibit seasonal patterns in infection rates and detection rates in the environment, which may be related to temperature.[145, 146, 147]

Among naturally occurring water-related pathogens, *Vibrio vulnificus* and *V. parahaemolyticus* are the species most often implicated in foodborne illness in the United States, accounting for more than 50% of reported shellfish-related illnesses annually.[140, 148, 149, 150, 151] Cases have increased significantly since 1996.[92, 148] Rising sea surface temperatures have contributed to an expanded geographic and seasonal range in outbreaks associated with shellfish.[96, 152, 153, 154, 155]

Precipitation is expected to be the primary climate driver affecting enteric pathogen loading to shellfish harvesting areas, although temperature also affects bioaccumulation rates of enteric viruses in shellfish. There are currently no national projections for the associated risk of illness from shellfish consumption. Many local and state agencies have developed plans for closing shellfish beds in the event of threshold-exceeding rain events that lead to loading of these contaminants and deterioration of water quality.[156]

Research Highlight: The Effect of Warming on Seasonal *Vibrio* Abundance and Distribution

Importance: *Vibrio* species are naturally occurring pathogens in coastal environments that cause illnesses ranging from gastroenteritis to septicemia (bloodstream infection) and death from both water contact and ingestion of raw or undercooked seafood, especially shellfish.[93] *Vibrio* are highly responsive to environmental conditions. For example, local nutrient availability can affect *Vibrio* abundance, though coastal waters generally have sufficient levels of the dissolved organic carbon required for *Vibrio* growth.[159]

Over longer timescales and larger geographic areas, key climate-related factors that increase *Vibrio* growth and abundance include rising sea surface temperatures and changes in precipitation, freshwater runoff, drought, sea level rise, coastal flooding, and storm surge, with resulting changes to coastal salinity patterns, turbidity (water clarity), and plankton abundance and composition (see Figure 1).[95, 96, 97, 98, 99, 100, 134, 160, 161, 162, 163]

Water temperature is a major contributor to *Vibrio* growth potential and, in turn, human exposure risk. The minimum water temperature threshold for the growth of most *Vibrio* species that cause illness in humans is 15°C (59°F), with growth rates increasing as temperature increases.[132, 152, 154, 157] Thus, it is projected that global ocean warming will increase risk of exposure by extending seasonal windows of growth and geographical range of occurrence.[132]

Projections of *Vibrio* Occurrence and Abundance in Chesapeake Bay

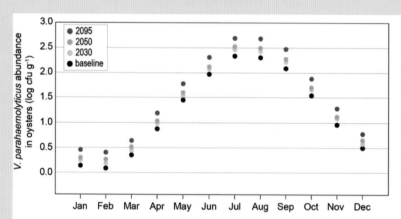

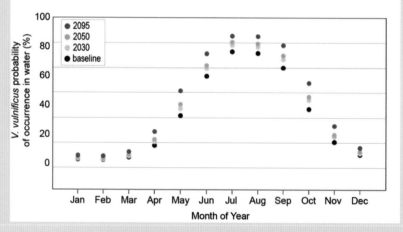

Figure 4: Seasonal and decadal projections of abundance *of V. parahaemolyticus* in oysters of Chesapeake Bay (top) and probability of occurrence of *V. vulnificus* in Chesapeake Bay surface waters (bottom). The circles show average values in the baseline period (1985–2000) and future years averaged by decadal period: 2030 (2025–2034), 2050 (2045–2054), and 2095 (2090–2099). (Figure source: adapted from Jacobs et al. 2015).[132]

Research Highlight: The Effect of Warming on Seasonal *Vibrio* Abundance and Distribution, continued

Objective: A quantitative projection of future shifts in *Vibrio* seasonal abundance and geographic range.

Method: Monthly average sea surface temperatures were projected for the 2030s, 2050s, and 2090s based on statistical downscaling of up to 21 global climate models for the Chesapeake Bay and Alaskan coastline. Previously published empirical models relating sea surface temperature and salinity to *Vibrio vulnificus* and *V. parahaemolyticus* were used to project probability of occurrence and abundance in Chesapeake Bay waters and oysters. Geographic information system (GIS) mapping of Alaskan coastal waters was used to project the distribution of monthly average water temperatures exceeding 15°C (59°F), considered to be the minimum temperature favorable for growth.[132]

Results and Conclusions: Modeling results find increases in abundance, geographical range, and seasonal extent of available habitat for *Vibrio*. In the Chesapeake Bay, the probability of occurrence of *V. vulnificus* is projected to increase by nearly 16% in the shoulder months of the growing season (May and September), with a similar increase in abundance of *V. parahaemolyticus* in oysters (Figure 4).

Analysis of temperature projections for Alaskan coastal waters based on an average of four climate models showed that habitat availability for *Vibrio* growth will increase to nearly 60% of the Alaskan shoreline in August by the 2090s (Figure 5).

Sources of uncertainty include different rates of warming associated with each model ensemble and other factors that affect growth and abundance, but all models used in this study project warming of coastal waters.

Changes in Suitable Coastal *Vibrio* Habitat in Alaska

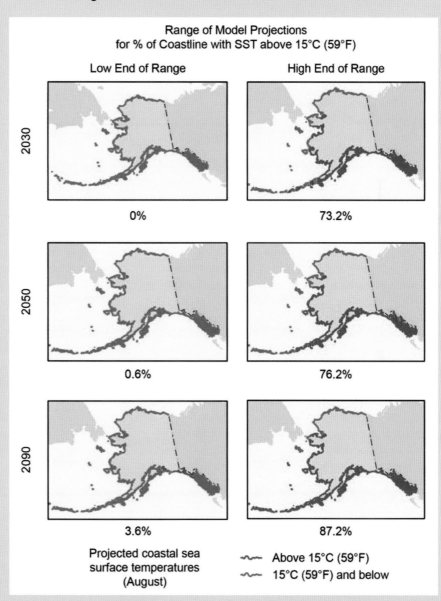

Figure 5: *Vibrio* growth increases in temperatures above 15°C (59°F). These maps show the low and high end of the ranges for projected area of Alaskan coastline with water temperature averages in August that are greater than this threshold. The projections were made for the following future time periods: 2030 (2026–2035), 2050 (2046–2055), and 2090 (2086–2095). On average, the models project that by 2090, nearly 60% of the Alaskan shoreline in August will become suitable *Vibrio* habitat. (Figure source: adapted from Jacobs et al. 2015)[132]

Increases in sea surface temperatures, changes in precipitation and freshwater delivery to coastal waters, and sea level rise will continue to affect *Vibrio* growth and are expected to increase human exposure risk.[96, 134, 152, 157] Regional models project increased abundance and extended seasonal windows of growth of *Vibrio* pathogens (see Research Highlight on page 166).[132] The magnitude of health impacts depends on the use of intervention strategies and on public and physician awareness.[158]

Harmful Algal Toxins

Harmful algal blooms (HABs) that contaminate seafood with toxins are becoming increasingly frequent and persistent in coastal marine waters, and some have expanded into new geographic locations.[164, 165, 166, 167, 168] Attribution of this trend has been complicated for some species, with evidence to suggest that human-induced changes (such as ballast water exchange, aquaculture, nutrient loading to coastal waters, and climate change) have contributed to this expansion.[167, 169]

Among HABs associated with seafood, ciguatera fish poisoning (CFP) is most strongly influenced by climate.[170, 171, 172] CFP is caused by toxins produced by the benthic algae *Gambierdiscus* (Table 1) and is the most frequently reported fish poisoning in humans.[173] There is a well-established link between warm sea surface temperatures and increased occurrences of CFP,[170, 171, 172] and in some cases, increases have also been linked to El Niño–Southern Oscillation events.[174] The frequency of tropical cyclones in the United States has also been associated with CFP, but with an 18-month lag period associated with the time required for a new *Gambierdiscus* habitat to develop.[170, 171]

Paralytic shellfish poisoning (PSP) is the most globally widespread shellfish poisoning associated with algal toxins,[175] and records of PSP toxins in shellfish tissues (an indicator of toxin-producing species of *Alexandrium*) provide the longest time series in the United States for evaluating climate impacts. Warm phases of the naturally occurring climate pattern known as the Pacific Decadal Oscillation co-occur with increased PSP toxins in Puget Sound shellfish on decadal timescales.[176] Further, it is very likely that the 20th century warming trend also contributed to the observed increase in shellfish toxicity since the 1950s.[177, 178] Warm spring temperatures also contributed to a bloom of *Alexandrium* in a coastal New York estuary in 2008.[179] Decadal patterns in PSP toxins in Gulf of Maine shellfish show no clear relationships with long-term trends in climate,[180, 181, 182] but ocean–climate interactions and changing oceanographic conditions are important factors for understanding *Alexandrium* bloom dynamics in this region.[183]

There is less agreement on the extent of climate impacts on other marine HAB-related diseases in the United States.

Increased abundances of *Pseudo-nitzschia* species, which can cause amnesic shellfish poisoning, have been attributed to nutrient enrichment in the Gulf of Mexico.[184] On the U.S. West Coast, increased abundances of at least some species of *Pseudo-nitzschia* occur during warm phases associated with El Niño events.[185] For *Dinophysis* species that can cause diarrhetic shellfish poisoning, data records are too short to evaluate potential relationships with climate in the United States,[164, 186] but studies in Sweden have found relationships with natural climate oscillations.[187]

The projected impacts of climate change on toxic marine harmful algae include geographic range changes in both warm- and cold-water species, changes in abundance and toxicity, and changes in the timing of the seasonal window of growth.[188, 189, 190, 191] These impacts will likely result from climate change related impacts on one or more of 1) water temperatures, 2) salinities, 3) enhanced surface stratification, 4) nutrient availability and supply to coastal waters (upwelling and freshwater runoff), and 5) altered winds and ocean currents.[188, 190, 191, 192, 193]

> *Climate change, especially continued warming, will dramatically increase the burden of some marine HAB-related diseases in some parts of the United States*

Limited understanding of the interactions among climate and non-climate stressors and, in some cases, limitations in the design of experiments for investigating decadal- or century-scale trends in phytoplankton communities, makes forecasting the direction and magnitude of change in toxic marine HABs challenging.[189, 191] Still, changes to the community composition of marine microalgae, including harmful species, will occur.[188, 194] Conditions for the growth of dinoflagellates—the algal group containing numerous toxic species—could potentially be increasingly favorable with climate change because these species possess certain physiological characteristics that allow them to take advantage of climatically-driven changes in the structure of the ocean (for example, stronger vertical stratification and reduced turbulence).[190, 193, 195, 196, 197]

Climate change, especially continued warming, will dramatically increase the burden of some marine HAB-related diseases in some parts of the United States, with strong implications for disease surveillance and public health preparedness. For example, the projected 4.5°F to 6.3°F increase in sea surface temperature in the Caribbean over the coming century is expected to increase the incidence of ciguatera fish poisoning by 200% to 400%.[171] In Puget Sound, warming is projected to increase the seasonal window of growth for *Alexandrium* by approximately 30 days by 2040, allowing blooms to begin earlier in the year and persist for longer.[177, 190, 198]

Research Highlight: Increased Risk of Ciguatera Fish Poisoning (CFP)

Importance: Ciguatera fish poisoning is caused by consumption of fish contaminated with toxins produced by dinoflagellates, such as those of the genus *Gambierdiscus*. There is a well-established link between warm sea surface temperatures and increased occurrence of CFP,[171] and thus concern that global ocean warming will affect the risk of illness.

Objective: A quantitative projection of future shifts in species of *Gambierdiscus*.

Method: Growth models developed for three Caribbean species of *Gambierdiscus* were run using 11 global climate model projections for specific buoy locations in the western Gulf of Mexico, Yucatan channel, and eastern Caribbean Sea through 2099. For further detail, see Kibler et al. 2015.[199]

Results and Conclusions: Modeling results suggest substantial changes in dominant species composition (Figure 6). Lower thermal tolerances of some species may result in geographic range shifts to more northern latitudes, particularly from the Yucatan and eastern Caribbean Sea. The projected shift in distribution is likely to mean that dominant CFP toxins enter the marine food web through different species, with increases of toxins in new areas where waters are warming and potential decreases in existing areas where waters are warming less rapidly.

Projected Changes in Caribbean *Gambierdiscus* Species

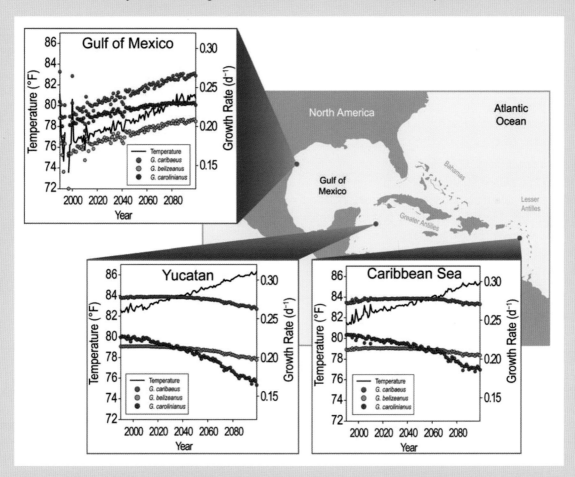

Figure 6: Water temperature data from 1990–2013 were collected or reconstructed for buoy sites in the western Gulf of Mexico, Yucatan channel, and eastern Caribbean Sea. These data were then used in calculations to project average annual water temperature and average growth rates for three Caribbean *Gambierdiscus* species (*G. caribaeus, G. belizeanus, G. carolinianus*) for the period 2014–2099. (Figure source: adapted from Kibler et al. 2015).[199]

Research Highlight: Expanded Seasonal Windows for Harmful Algal Blooms

Importance: When some harmful algae in the genus *Alexandrium* bloom, toxins that can accumulate in shellfish are produced. When these shellfish are consumed, gastrointestinal illness and neurological symptoms, known as paralytic shellfish poisoning (PSP), can occur. Death can result in extreme cases. Because growth of *Alexandrium* is regulated in part by water temperature, warm water conditions appropriate for bloom formation may expand seasonally, increasing the risk of illness.

Objective: A quantitative projection of future conditions appropriate for *Alexandrium* bloom formation in Puget Sound.

Method: Monthly average sea surface temperature was projected for Quartermaster Harbor, Puget Sound, for the 2030s, 2050s, and 2090s based on statistical downscaling of 21 global climate models. The projections were applied to previously published empirical models relating temperature and salinity to *Alexandrium* growth. For more detail, see Jacobs et al. 2015.[132]

Projections of Growth of *Alexandrium* in Puget Sound

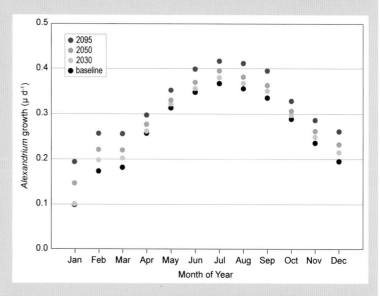

Figure 7: Seasonal and decadal projections of growth of *Alexandrium* in Puget Sound. The circles show average values in the baseline period (2006–2013) and future years averaged by decadal period: 2030 (2025–2035), 2050 (2045–2055), and 2095 (2090–2099). Growth rate values above 0.25µd⁻¹ constitute a bloom of *Alexandrium* (Figure source: adapted from Jacobs et al. 2015)[132]

Results and Conclusions: Modeling results indicate that *Alexandrium* blooms could develop up to two months earlier in the year and persist for up to two months longer by 2100 compared to the present day (Figure 7). All model projections indicate that the bloom season will expand by at least one month on either side of the present-day bloom season by 2100. Therefore, it is likely that the risk of *Alexandrium* blooms that can contaminate shellfish with potent toxins will increase. This may increase the risk of human exposure to the toxins, which can cause paralytic shellfish poisoning. Sources of uncertainty include different rates of warming associated with each model ensemble and other factors that affect growth and abundance, but all models used project warming of coastal waters.

6.4 Populations of Concern

Climate change impacts on the drinking water exposure pathway (see page 163) will act as an additional stressor on top of existing exposure disparities in the United States. Lack of consistent access to potable drinking water and inequities in exposure to contaminated water disproportionately affects the following populations: tribes and Alaska Natives, especially those in remote reservations or villages; residents of low-income rural subdivisions known as colonias along the U.S.–Mexico border; migrant farm workers; the homeless; and low-income communities not served by public water utilities—which can be urban, suburban, or rural, and some of which are predominately Hispanic or Latino and Black or African American communities in certain regions of the country.[200, 201, 202, 203, 204, 205, 206, 207, 208] In general, the heightened vulnerability of these populations primarily results from unequal access to adequate water and sewer

infrastructure, and various environmental, political, economic, and social factors jointly create these disparities.[201]

Children, older adults (primarily age 65 and older), pregnant women, and immunocompromised individuals have higher risk of gastrointestinal illness and severe health outcomes from contact with contaminated water.[4, 209, 210, 211, 212, 213] Pregnant women who develop severe gastrointestinal illness are at high risk for adverse pregnancy outcomes (pregnancy loss and preterm birth).[214] Because children swallow roughly twice as much water as adults while swimming, they have higher recreational exposure risk for both pathogens and freshwater HABs.[101, 120] Recent cryptosporidiosis and giardiasis cases have frequently been reported in children aged one to nine years, with onset of illness peaking during the summer months.[215] In addition, 40%

of swimming-related eye and ear infections from *Vibrio algino-lyticus* during the period 1997–2006 were reported in children (median age of 15).[93]

Traditional tribal consumption of fish and shellfish in the Pacific Northwest and Alaska can be on average 3 to 10 times higher than that of average U.S. consumers, or even up to 20 times higher.[216] Climate change will contribute to increased seafood contamination by toxins and potentially by chemical contaminants (see "6.5 Emerging Issues" below), with potential health risks and cultural implications for tribal communities. Those who continue to consume traditional diets may face increased health risks from contamination.[217] Alternatively, replacing these traditional nutrition sources may involve consuming less nutritious processed foods and the loss of cultural practices tied to fish and shellfish harvest.[218, 219]

6.5 Emerging Issues

A key emerging issue is the impact of climate on new and re-emerging pathogens. While cases of nearly-always-fatal primary amoebic meningoencephalitis due to the amoeba *Naegleria fowleri* and other related species remain relatively uncommon, a northward expansion of cases has been observed in the last five years.[220, 221] Evidence suggests that in addition to detection in source water (ground and surface waters), these amoebae may be harbored in biofilms associated with water distribution systems, where increased temperatures decrease efficacy of chlorine disinfection and support survival and potentially growth.[222, 223, 224]

Climate change may also alter the patterns or magnitude of chemical contamination of seafood, leading to altered effects on human health—most of which are chronic conditions. Rising temperatures and reduced ice cover are already linked to increasing burdens of mercury and organohalogens in arctic fish,[225] a sign of increasing contamination of the arctic food chain. Changes in hydrology resulting from climate change are expected to alter releases of chemical contaminants into the Nation's surface waters,[226] with as-yet-unknown effects on seafood contamination.

6.6 Research Needs

In addition to those identified in the emerging issues discussion above, the authors highlight the following potential areas for additional scientific and research activity on water-related illness, based on their review of the literature. Enhanced understanding of climate change impacts will be facilitated by improved public health surveillance for water-related infectious diseases and expanded monitoring and surveillance of surface and coastal water quality. In addition, improved understanding of how human behaviors affect the risk of waterborne diseases can facilitate the development of predictive models and effective adaptation measures. Predictive models can also help identify major areas of uncertainty and refine key research questions.

Water-related contamination of shellfish may reduce consumption and contribute to loss of tribal cultural practices tied to shellfish harvest.

Future assessments can benefit from research activities that

- assess the interactions among climate drivers, ecosystem changes, water quality and infectious pathogens, including *Vibrio spp., N. fowlerii,* chemical contaminants, and harmful algal blooms;

- increase understanding of how marine and terrestrial wildlife, including waterfowl, contribute to the distribution of pathogens and transmission of infectious disease and assess the role of climate;

- explore how ocean acidification affects toxin production and distribution of marine HABs and pathogens;

- analyze the hydrologic (discharge, flow-residence time, and mixing) thresholds for predicting HAB occurrences; and

- increase understanding of how the impacts of climate change on drinking water infrastructure, including the need for development of new and emerging technologies for provision of drinking water, affect the risks of waterborne diseases.

Supporting Evidence

PROCESS FOR DEVELOPING CHAPTER

The chapter was developed through technical discussions of relevant evidence and expert deliberation by the report authors at several workshops, teleconferences, and email exchanges. Authors considered inputs and comments submitted by the public, the National Academies of Sciences, and Federal agencies. For additional information on the overall report process, see Appendices 2 and 3.

Many water-related illnesses are of critical importance globally, such as cholera and hepatitis E virus, and they affect U.S. interests abroad, but the focus of this chapter is to address climate impacts on water-related illnesses of primary importance to human health within the United States. In addition, although climate change has the potential to impact national as well as global seafood supplies, this chapter does not cover these types of impacts because the peer-reviewed literature is not yet robust enough to make connections to human health outcomes in the United States. Even with those constraints, the impacts of climate on water-related illness are regionally or locally specific and may include increased risks as well as benefits. For example, the projected geographic range shifts of some *Gambieridiscus* species to more northern latitudes may mean that dominant ciguatera fish poisoning toxins enter the marine food web through different species, with increases of toxins in new areas where waters are warming and potential decreases in areas such as the Yucatan and eastern Caribbean Sea.[199]

KEY FINDING TRACEABLE ACCOUNTS

Seasonal and Geographic Changes in Waterborne Illness Risk

Key Finding 1: Increases in water temperatures associated with climate change will alter the seasonal windows of growth and the geographic range of suitable habitat for freshwater toxin-producing harmful algae [*Very Likely, High Confidence*], certain naturally occurring *Vibrio* bacteria [*Very Likely, Medium Confidence*], and marine toxin-producing harmful algae [*Likely, Medium Confidence*]. These changes will increase the risk of exposure to waterborne pathogens and algal toxins that can cause a variety of illnesses [*Medium Confidence*].

Description of evidence base

Vibrio, a genus of naturally occurring waterborne pathogens, thrives in water temperatures above a 15°C/59°F threshold.[132, 152, 154, 157] Rising sea surface temperatures have contributed to an expanded geographic and seasonal range in outbreaks of human illness associated with *Vibrio* in shellfish.[96, 152, 153, 154, 155] In recreational waters, projected increases in sea surface temperatures are expected to lengthen the seasonal window of growth and expand geographic range of *Vibrio*.[96, 132] Like

other heterotrophic bacteria, growth of *Vibrio* is ultimately limited by availability of carbon substrate, though the coastal areas where *Vibrio* exposure is most likely, either through recreation or consumption of shellfish, generally have sufficient dissolved organic carbon.[159] Reported rates of all *Vibrio* infections have tripled since 1996 in the United States, with *V. alginolyticus* infections having increased by 40-fold.[92] Increasing sea surface temperatures, changes in precipitation and freshwater delivery to coastal waters, and sea level rise will continue to affect *Vibrio* growth and are expected to increase human exposure.[96, 134, 152, 157]

Most harmful algae, including freshwater cyanobacteria that can contaminate drinking water and marine dinoflagellate species that can contaminate fish and shellfish with natural toxins, thrive during the warm summer season or when water temperatures are higher than usual. As the climate continues to warm, water temperatures will rise above thresholds that promote bloom development earlier in the spring and will persist longer into the fall and expand into higher latitudes. This will result in a longer seasonal window and expanded geographic range for human exposure into higher latitudes.[124, 125, 126, 127, 128, 188, 189, 190, 191, 192, 193] Climate change, especially continued warming, will increase the burden of some marine HAB-related diseases, particularly ciguatera fish poisoning, in some regions of the United States.

Major uncertainties

Uncertainty remains regarding the relative importance of additional factors that may also act on naturally occurring pathogens and harmful algae at local or regional levels to influence their growth, distribution, and toxicity. In many cases, it is uncertain how these multiple factors may interact with each other to influence the seasonal windows and geographic range for pathogens and harmful algae, especially in dynamic coastal marine environments. For example, changes in salinity, competition with other plankton, and presence of viruses or other organisms that consume plankton or bacteria can affect abundance.[162, 163] Changing distribution patterns for some marine species of harmful algae is not well understood and some regions may become too warm for certain species of harmful algae to grow, shifting (without changing in total size) or even shrinking their geographic range.

Additionally, there are limited studies on projections for changes in illness rates due to naturally occurring waterborne pathogens and harmful algae. Uncertainty remains regarding appropriate methods for projecting changes in illness rates, including how to integrate considerations of human behavior into modeling (current methods to assess exposure risk assume similar human behavior across time scales and geography). Methodological challenges are related to 1) underreporting and underdiagnosis of cases that affect the accuracy of baseline estimates of illness, 2) ability to project changes in strain virulence, 3) accounting for the effects of potential adaptation strategies/public health interventions (for example, public service announcements on how to avoid exposure), and 4) accounting for changes in public healthcare infrastructure and access that can reduce the risk of exposure or illness/death if exposed.

Assessment of confidence and likelihood based on evidence
Based on the evidence, there is **medium confidence** that, with changing climate, the annual seasonal and the geographic range for *Vibrio* and certain marine harmful algae will expand. The assessment of **medium confidence** is due to less certainty from modeling results regarding the magnitude of projected changes in abundance. The conclusions were deemed **very likely** to occur for *Vibrio* and **likely** for marine harmful algae based on good levels of agreement found in the published quantitative modeling projections for both *Vibrio* and marine harmful algae (*Alexandrium* and *Gambieridiscus*) cited above. This conclusion takes into consideration that for some marine algae (for example, *Gambieridiscus*), lower latitudes may become too warm and risk may decline in those areas as it increases at higher latitudes. For freshwater harmful algae, there is **high confidence** that annual season and geographic range will expand with changing climate, which will also prolong the time for exposure and the potential for public health impacts. Consistent and high-quality evidence from a limited number of laboratory studies, modeling efforts, field surveys, and comparisons of historic and contemporary conditions support this assessment. The conclusion was deemed **very likely** to occur for freshwater harmful algae with **high confidence** based on laboratory studies and field observations, as well as a greater fundamental understanding of inland hydrodynamics and bloom ecology as indicated in the literature cited in the chapter. There is **medium confidence** regarding increased risk to human health from a longer potential time for exposure to waterborne pathogens and algal toxins and potential exposure for a wider (or novel) population. This confidence level was chosen due to less certainty stemming from a relative lack of quantitative data and projections for future illness rates in the peer-reviewed literature.

Runoff from Extreme Precipitation Increases Exposure Risk

Key Finding 2: Runoff from more frequent and intense extreme precipitation events will increasingly compromise recreational waters, shellfish harvesting waters, and sources of drinking water through increased introduction of pathogens and prevalence of toxic algal blooms [*High Confidence*]. As a result, the risk of human exposure to agents of water-related illness will increase [*Medium Confidence*].

Description of evidence base
Extreme precipitation can mobilize pathogens, nutrients, and chemical contaminants from agricultural, wildlife, and urban sources. Waterborne illness and outbreaks from pathogens following heavy precipitation events have been well documented in multiple studies using both passive and active surveillance on a local and regional level.[38, 39, 40, 42, 43, 44, 45, 46, 47] Likewise, extreme precipitation events and subsequent increases in runoff are key climate factors that increase nutrient loading in freshwater and marine recreational waters, shellfish harvesting waters, and sources of drinking water, which in turn increases the likelihood of harmful cyanobacterial blooms that produce algal toxins.[56] The drinking water treatment process can remove cyanobacterial blooms; however, efficacy of the treatment processes may vary from 60% to 99.9%. Ineffective treatment could compromise water quality and may lead to severe treatment disruption or treatment plant shutdown.[53, 54, 55, 56] More frequent and intense extreme precipitation events are projected for many regions in the United States as climate changes. Consistent, high-quality evidence from multiple studies supports a finding that increased runoff and flooding events are expected to increase contamination of source waters (for drinking water supply) and surface waters used for recreation, which may increase people's exposure to pathogens and algal toxins that cause illness.[10, 52, 59, 70, 71, 72, 73, 76, 82, 116, 117, 118, 119, 120] Other factors may modify these risks, such as increased air or water temperatures, residence time in the environment, lower water levels, or dilution.

Major uncertainties
Changes in exposure and risk are attributable to many factors in addition to climate. While extreme precipitation and flooding events introduce contaminants and pathogens to water to varying degrees depending on the characteristics of each individual event, they may not always result in increases in exposure due to planning and adaptive actions. There are limited studies on actual projections for changes in illness rates due to increasing frequency or intensity of extreme precipitation events. Uncertainty remains regarding appropriate methods for projecting changes in illness rates, including how to integrate considerations of human behavior into modeling (current methods to assess exposure risk assume similar human behavior across time scales and geography). Methodological challenges are related to 1) baseline case reporting issues (underreporting and underdiagnosis), 2) accounting for the effects of potential adaptation strategies/public health interventions (for example, public service announcements about how to avoid exposure), and 3) accounting for changes in public healthcare infrastructure and access that can reduce the risk of exposure or of illness/death if exposed.

Assessment of confidence and likelihood based on evidence

Based on the evidence, there is **high confidence** that increasing frequency or intensity of extreme precipitation events will compromise recreational waters and sources of drinking water with pathogens, nutrients, and chemical contaminants from agricultural, wildlife, and urban sources.

There is consistent qualitative evidence that flooding associated with extreme precipitation events and storm surge results in loading of pathogens and nutrients to surface and groundwater (and drinking water distribution systems) through stormwater runoff and sewage overflows. However, other human and social factors modify risk, and there are no national-level studies upon which to draw conclusions regarding quantitative projections of increased exposure. Thus, the limited number of studies supports a **medium confidence** level that human exposure risk will increase due to changes in extreme events.

Water Infrastructure Failure

Key Finding 3: Increases in some extreme weather events and storm surges will increase the risk that infrastructure for drinking water, wastewater, and stormwater will fail due to either damage or exceedance of system capacity, especially in areas with aging infrastructure [*High Confidence*]. As a result, the risk of exposure to water-related pathogens, chemicals, and algal toxins will increase in recreational and shellfish harvesting waters and in drinking water where treatment barriers break down [*Medium Confidence*].

Description of evidence base

Water infrastructure in the United States is aging and may be inadequate or deteriorating. Combined sewers in many older cities were not designed to handle extreme precipitation events that are becoming more frequent with climate change. Multiple studies provide consistent, high-quality evidence that these systems are at risk of being overwhelmed during flood events or may be further damaged during other extreme weather events (e.g., storm surge), allowing contaminated surface water to run off into drinking water and recreational water sources.[10, 52, 59, 70, 76, 116] Drinking water source contamination may be exacerbated or insufficiently addressed by treatment processes at the plant or the distribution system. Drinking water treatment plants may be challenged by high pathogen loads and toxic cyanobacterial bloom events.[52, 55, 56] Multiple studies support a finding that climate change will place additional stresses on the capacity of drinking water treatment facilities and may increase the risk that water infrastructure, especially aging infrastructure, will fail through either damage or exceedance of system capacity.[6, 70, 74, 75]

Major uncertainties

The human health consequences of aging water infrastructure failure depend not only on the local and regional climate factors that contribute to damage or capacity challenges but also the nature of the system and the pressures on it, the population affected, and the timeliness and adequacy of the response—all of which are inherently local or regional factors. Due to the complicated local and regional specificity, there are no national projections of the human health impact of water infrastructure failure. Uncertainty remains regarding appropriate methods for projecting changes in illness rates, including how to integrate considerations of human behavior into modeling (current methods to assess exposure risk assume similar human behavior across time scales and geography). Methodological challenges are related to 1) baseline case reporting issues (underreporting and underdiagnosis), 2) accounting for the effects of potential adaptation strategies/public health interventions (for example, mitigating risk with improvements to current water and sewerage systems), and 3) accounting for changes in public healthcare infrastructure and access that can reduce the risk of exposure or of illness/death if exposed.

Assessment of confidence based on evidence

Based on the evidence found in the peer-reviewed literature, there is **high confidence** that the anticipated climate change related increases in some extreme weather events and in storm surge will increase the risk that water infrastructure for drinking water, wastewater, and stormwater will fail through either damage or exceedance of system capacity, with aging infrastructure being particularly vulnerable. Evidence shows contamination to or from these systems occurs with heavy precipitation and other extreme weather events. There is consistent qualitative evidence suggesting that projected climate change effects on extreme weather patterns— particularly extreme precipitation and storm surge—can adversely affect water infrastructure and lead to increased loading of pathogens, algal toxins, and contaminants. However, there are no national-level studies upon which to draw conclusions regarding quantitative projections of increased exposure. Thus, the limited number of studies supports a **medium confidence** level regarding risk of exposure.

DOCUMENTING UNCERTAINTY

This assessment relies on two metrics to communicate the degree of certainty in Key Findings. See Appendix 4: Documenting Uncertainty for more on assessments of likelihood and confidence.

Confidence Level
Very High
Strong evidence (established theory, multiple sources, consistent results, well documented and accepted methods, etc.), high consensus
High
Moderate evidence (several sources, some consistency, methods vary and/or documentation limited, etc.), medium consensus
Medium
Suggestive evidence (a few sources, limited consistency, models incomplete, methods emerging, etc.), competing schools of thought
Low
Inconclusive evidence (limited sources, extrapolations, inconsistent findings, poor documentation and/or methods not tested, etc.), disagreement or lack of opinions among experts

Likelihood
Very Likely
≥ 9 in 10
Likely
≥ 2 in 3
As Likely As Not
≈ 1 in 2
Unlikely
≤ 1 in 3
Very Unlikely
≤ 1 in 10

PHOTO CREDITS

Pg. 157–Hands cupping water: © iStockPhotos.com/ jacky9946

Pg. 158–Young women walking through floodwater: © Richard Ellis/Corbis

Pg. 163–Heavy rain: © iStockPhoto.com/AndreasWeber

Pg. 165–Family jumping in lake: © Juice Images/Corbis

Pg. 171–Razor clam dig: Courtesy of Vera Trainer/NOAA

References

1. Melillo, J.M., T.C. Richmond, and G.W. Yohe, Eds., 2014: *Climate Change Impacts in the United States: The Third National Climate Assessment*. U.S. Global Change Research Program, Washington, D.C., 842 pp. http://dx.doi.org/10.7930/J0Z31WJ2

2. Colford, J.M., Jr., S. Roy, M.J. Beach, A. Hightower, S.E. Shaw, and T.J. Wade, 2006: A review of household drinking water intervention trials and an approach to the estimation of endemic waterborne gastroenteritis in the United States. *Journal of Water and Health*, **4 Suppl 2,** 71-88. http://dx.doi.org/10.2166/wh.2006.018

3. Messner, M., S. Shaw, S. Regli, K. Rotert, V. Blank, and J. Soller, 2006: An approach for developing a national estimate of waterborne disease due to drinking water and a national estimate model application. *Journal of Water and Health*, **4 Suppl 2,** 201-240. http://dx.doi.org/10.2166/wh.2006.024

4. Reynolds, K.A., K.D. Mena, and C.P. Gerba, 2008: Risk of waterborne illness via drinking water in the United States. *Reviews of Environmental Contamination and Toxicology*, **192,** 117-158. http://dx.doi.org/10.1007/978-0-387-71724-1_4

5. Soller, J.A., T. Bartrand, N.J. Ashbolt, J. Ravenscroft, and T.J. Wade, 2010: Estimating the primary etiologic agents in recreational freshwaters impacted by human sources of faecal contamination. *Water Research*, **44,** 4736-4747. http://dx.doi.org/10.1016/j.watres.2010.07.064

6. Beach, M.J., S. Roy, J. Brunkard, J. Yoder, and M.C. Hlavsa, 2009: The changing epidemiology of waterborne disease outbreaks in the United States: Implications for system infrastructure and future planning. *Global Issues in Water, Sanitation, and Health: Workshop Summary*. Institute of Medicine. The National Academies Press, Washington, D.C., 156-168. http://dx.doi.org/10.17226/12658

7. Scallan, E., R.M. Hoekstra, F.J. Angulo, R.V. Tauxe, M.A. Widdowson, S.L. Roy, J.L. Jones, and P.M. Griffin, 2011: Foodborne illness acquired in the United States: Major pathogens. *Emerging Infectious Diseases*, **17,** 7-15. http://dx.doi.org/10.3201/eid1701.P11101

8. EPA, 2004: Report to Congress: Impacts and Control of CSOs and SSOs. EPA 833-R-04-001. U.S. Environmental Protection Agency, Office of Water, Washington, D.C. http://water.epa.gov/polwaste/npdes/cso/2004-Report-to-Congress.cfm

9. Montserrat, A., L. Bosch, M.A. Kiser, M. Poch, and L. Corominas, 2015: Using data from monitoring combined sewer overflows to assess, improve, and maintain combined sewer systems. *Science of the Total Environment*, **505,** 1053-1061. http://dx.doi.org/10.1016/j.scitotenv.2014.10.087

10. Patz, J.A., S.J. Vavrus, C.K. Uejio, and S.L. McLellan, 2008: Climate change and waterborne disease risk in the Great Lakes region of the US. *American Journal of Preventive Medicine*, **35,** 451-458. http://dx.doi.org/10.1016/j.amepre.2008.08.026

11. EPA, 2008: A Screening Assessment of the Potential Impacts of Climate Change on Combined Sewer Overflow Mitigation in the Great Lakes and New England Regions. EPA/600/R-07/033F, 50 pp. U.S. Environmental Protection Agency, Washington, D.C. http://cfpub.epa.gov/ncea/cfm/recordisplay.cfm?deid=188306

12. Cutter, S.L., W. Solecki, N. Bragado, J. Carmin, M. Fragkias, M. Ruth, and T. Wilbanks, 2014: Ch. 11: Urban systems, infrastructure, and vulnerability. *Climate Change Impacts in the United States: The Third National Climate Assessment*. Melillo, J.M., T.C. Richmond, and G.W. Yohe, Eds. U.S. Global Change Research Program, Washington, D.C., 282-296. http://dx.doi.org/10.7930/J0F769GR

13. Gargano, J.W., A.L. Freeland, M.A. Morrison, K. Stevens, L. Zajac, A. Wolkon, A. Hightower, M.D. Miller, and J.M. Brunkard, 2015: Acute gastrointestinal illness following a prolonged community-wide water emergency. *Epidemiology and Infection*, **143,** 2766-2776. http://dx.doi.org/10.1017/S0950268814003501

14. Converse, R.R., M.F. Piehler, and R.T. Noble, 2011: Contrasts in concentrations and loads of conventional and alternative indicators of fecal contamination in coastal stormwater. *Water Research*, **45,** 5229-5240. http://dx.doi.org/10.1016/j.watres.2011.07.029

15. Futch, J.C., D.W. Griffin, and E.K. Lipp, 2010: Human enteric viruses in groundwater indicate offshore transport of human sewage to coral reefs of the Upper Florida Keys. *Environmental Microbiology*, **12,** 964-974. http://dx.doi.org/10.1111/j.1462-2920.2010.02141.x

16. Futch, J.C., D.W. Griffin, K. Banks, and E.K. Lipp, 2011: Evaluation of sewage source and fate on southeast Florida coastal reefs. *Marine Pollution Bulletin*, **62,** 2308-2316. http://dx.doi.org/10.1016/j.marpolbul.2011.08.046

17. Levantesi, C., L. Bonadonna, R. Briancesco, E. Grohmann, S. Toze, and V. Tandoi, 2012: *Salmonella* in surface and drinking water: Occurrence and water-mediated transmission. *Food Research International*, **45,** 587-602. http://dx.doi.org/10.1016/j.foodres.2011.06.037

18. Molina, M., S. Hunter, M. Cyterski, L.A. Peed, C.A. Kelty, M. Sivaganesan, T. Mooney, L. Prieto, and O.C. Shanks, 2014: Factors affecting the presence of human-associated and fecal indicator real-time quantitative PCR genetic markers in urban-impacted recreational beaches. *Water Research*, **64,** 196-208. http://dx.doi.org/10.1016/j.watres.2014.06.036

19. Sauer, E.P., J.L. VandeWalle, M.J. Bootsma, and S.L. McLellan, 2011: Detection of the human specific *Bacteroides* genetic marker provides evidence of widespread sewage contamination of stormwater in the urban environment. *Water Research,* **45,** 4081-4091. http://dx.doi.org/10.1016/j.watres.2011.04.049

20. Sercu, B., L.C. Van De Werfhorst, J.L.S. Murray, and P.A. Holden, 2011: Sewage exfiltration as a source of storm drain contamination during dry weather in urban watersheds. *Environmental Science & Technology,* **45,** 7151-7157. http://dx.doi.org/10.1021/es200981k

21. Uejio, C.K., S.H. Yale, K. Malecki, M.A. Borchardt, H.A. Anderson, and J.A. Patz, 2014: Drinking water systems, hydrology, and childhood gastrointestinal illness in central and northern Wisconsin. *American Journal of Public Health,* **104,** 639-646. http://dx.doi.org/10.2105/ajph.2013.301659

22. Walthall, C., P. Backlund, J. Hatfield, L. Lengnick, E. Marshall, M. Walsh, S. Adkins, M. Aillery, E.A. Ainsworth, C. Amman, C.J. Anderson, I. Bartomeus, L.H. Baumgard, F. Booker, B. Bradley, D.M. Blumenthal, J. Bunce, K. Burkey, S.M. Dabney, J.A. Delgado, J. Dukes, A. Funk, K. Garrett, M. Glenn, D.A. Grantz, D. Goodrich, S. Hu, R.C. Izaurralde, R.A.C. Jones, S.-H. Kim, A.D.B. Leaky, K. Lewers, T.L. Mader, A. McClung, J. Morgan, D.J. Muth, M. Nearing, D.M. Oosterhuis, D. Ort, C. Parmesan, W.T. Pettigrew, W. Polley, R. Rader, C. Rice, M. Rivington, E. Rosskopf, W.A. Salas, L.E. Sollenberger, R. Srygley, C. Stöckle, E.S. Takle, D. Timlin, J.W. White, R. Winfree, L. Wright-Morton, and L.H. Ziska, 2012: Climate Change and Agriculture in the United States: Effects and Adaptation. USDA Technical Bulletin 1935, 186 pp. U.S. Department of Agriculture, Washington, D.C. http://www.usda.gov/oce/climate_change/effects_2012/CC and Agriculture Report (02-04-2013)b.pdf

23. GAO, 2008: Concentrated Animal Feeding Operations: EPA Needs More Information and a Clearly Defined Strategy to Protect Air and Water Quality from Pollutants of Concern. GAO-08-944, 79 pp. U.S. Government Accountability Office. http://www.gao.gov/new.items/d08944.pdf

24. Hribar, C., 2010: Understanding Concentrated Animal Feeding Operations and Their Impact on Communities. Schultz, M. (Ed.), 22 pp. National Association of Local Boards of Health, Bowling Green, OH. http://www.cdc.gov/nceh/ehs/docs/understanding_cafos_nalboh.pdf

25. NRC, 2010: *Toward Sustainable Agricultural Systems in the 21st Century.* National Research Council. The National Academies Press, Washington, D.C., 598 pp. http://dx.doi.org/10.17226/12832

26. Sun, L., K.E. Kunkel, L.E. Stevens, A. Buddenberg, J.G. Dobson, and D.R. Easterling, 2015: Regional Surface Climate Conditions in CMIP3 and CMIP5 for the United States: Differences, Similarities, and Implications for the U.S. National Climate Assessment. NOAA Technical Report NESDIS 144, 111 pp. National Oceanic and Atmospheric Administration, National Environmental Satellite, Data, and Information Service. http://www.nesdis.noaa.gov/technical_reports/NOAA_NESDIS_Technical_Report_144.pdf

27. USDA, 2014: 2012 Census of Agriculture. 695 pp. U.S. Department of Agriculture, National Agricultural Statistics Service, Washington, D.C. http://www.agcensus.usda.gov/Publications/2012/

28. Miller, W.A., D.J. Lewis, M. Lennox, M.G.C. Pereira, K.W. Tate, P.A. Conrad, and E.R. Atwill, 2007: Climate and on-farm risk factors associated with *Giardia duodenalis* cysts in storm runoff from California coastal dairies. *Applied and Environmental Microbiology,* **73,** 6972-6979. http://dx.doi.org/10.1128/aem.00100-07

29. Wilkes, G., J. Brassard, T.A. Edge, V. Gannon, C.C. Jokinen, T.H. Jones, R. Marti, N.F. Neumann, N.J. Ruecker, M. Sunohara, E. Topp, and D.R. Lapen, 2013: Coherence among different microbial source tracking markers in a small agricultural stream with or without livestock exclusion practices. *Applied and Environmental Microbiology,* **79,** 6207-6219. http://dx.doi.org/10.1128/aem.01626-13

30. Kilonzo, C., X. Li, E.J. Vivas, M.T. Jay-Russell, K.L. Fernandez, and E.R. Atwill, 2013: Fecal shedding of zoonotic food-borne pathogens by wild rodents in a major agricultural region of the central California coast. *Applied and Environmental Microbiology,* **79,** 6337-6344. http://dx.doi.org/10.1128/aem.01503-13

31. Fremaux, B., T. Boa, and C.K. Yost, 2010: Quantitative real-time PCR assays for sensitive detection of Canada goose-specific fecal pollution in water sources. *Applied and Environmental Microbiology,* **76,** 4886-4889. http://dx.doi.org/10.1128/aem.00110-10

32. Parmesan, C. and G. Yohe, 2003: A globally coherent fingerprint of climate change impacts across natural systems. *Nature,* **421,** 37-42. http://dx.doi.org/10.1038/nature01286

33. Craun, G.F., J.M. Brunkard, J.S. Yoder, V.A. Roberts, J. Carpenter, T. Wade, R.L. Calderon, J.M. Roberts, M.J. Beach, and S.L. Roy, 2010: Causes of outbreaks associated with drinking water in the United States from 1971 to 2006. *Clinical Microbiology Reviews,* **23,** 507-528. http://dx.doi.org/10.1128/cmr.00077-09

34. EPA, 2015: Public Drinking Water Systems Programs: Overview. U.S. Environmental Protection Agency, Washington, D.C. http://water.epa.gov/infrastructure/drinkingwater/pws/index.cfm

35. EPA, 2012: Private Drinking Water Wells. U.S. Environmental Protection Agency. http://water.epa.gov/drink/info/well/index.cfm

36. 42 USC. Sec 300f et seq., 1974: The Safe Drinking Water Act. http://www.gpo.gov/fdsys/pkg/USCODE-2011-title42/pdf/USCODE-2011-title42-chap6A-subchapXII-partA-sec300f.pdf

37. Hilborn, E.D., T.J. Wade, L. Hicks, L. Garrison, J. Carpenter, E. Adam, B. Mull, J.S. Yoder, V. Roberts, and J.W. Gargano, 2013: Surveillance for waterborne disease outbreaks associated with drinking water and other nonrecreational water — United States, 2009–2010. *MMWR. Morbidity and Mortality Weekly Report,* **62,** 714-720. http://www.cdc.gov/mmwr/preview/mmwrhtml/mm6235a3.htm

38. Curriero, F.C., J.A. Patz, J.B. Rose, and S. Lele, 2001: The association between extreme precipitation and waterborne disease outbreaks in the United States, 1948–1994. *American Journal of Public Health,* **91,** 1194-1199. http://dx.doi.org/10.2105/AJPH.91.8.1194

39. Brunkard, J.M., E. Ailes, V.A. Roberts, V. Hill, E.D. Hilborn, G.F. Craun, A. Rajasingham, A. Kahler, L. Garisson, L. Hicks, J. Carpenter, T.J. Wade, M.J. Beach, and J.S. Yoder, 2011: Surveillance for waterborne disease outbreaks associated with drinking water — United States, 2007–2008. *MMWR Surveillance Summaries,* **60(SS12),** 38-68. http://www.cdc.gov/mmwr/preview/mmwrhtml/ss6012a4.htm

40. Bradbury, K.R., M.A. Borchardt, M. Gotkowitz, S.K. Spencer, J. Zhu, and R.J. Hunt, 2013: Source and transport of human enteric viruses in deep municipal water supply wells. *Environmental Science & Technology,* **47,** 4096-4103. http://dx.doi.org/10.1021/es400509b

41. Drayna, P., S.L. McLellan, P. Simpson, S.-H. Li, and M.H. Gorelick, 2010: Association between rainfall and pediatric emergency department visits for acute gastrointestinal illness. *Environmental Health Perspectives,* **118,** 1439-1443. http://dx.doi.org/10.1289/ehp.0901671

42. Jofre, J., A.R. Blanch, and F. Lucena, 2010: Water-borne infectious disease outbreaks associated with water scarcity and rainfall events. *Water Scarcity in the Mediterranean: Perspectives under Global Change.* Sabater, S. and D. Barcelo, Eds. Springer, Berlin, 147-159. http://dx.doi.org/10.1007/698_2009_22

43. Auld, H., D. MacIver, and J. Klaassen, 2004: Heavy rainfall and waterborne disease outbreaks: The Walkerton example. *Journal of Toxicology and Environmental Health, Part A,* **67,** 1879-1887. http://dx.doi.org/10.1080/15287390490493475

44. Salvadori, M.I., J.M. Sontrop, A.X. Garg, L.M. Moist, R.S. Suri, and W.F. Clark, 2009: Factors that led to the Walkerton tragedy. *Kidney International,* **75,** S33-S34. http://dx.doi.org/10.1038/ki.2008.616

45. Fong, T.-T., L.S. Mansfield, D.L. Wilson, D.J. Schwab, S.L. Molloy, and J.B. Rose, 2007: Massive microbiological groundwater contamination associated with a waterborne outbreak in Lake Erie, South Bass Island, Ohio. *Environmental Health Perspectives,* **115,** 856-864. http://dx.doi.org/10.1289/ehp.9430

46. Hoxie, N.J., J.P. Davis, J.M. Vergeront, R.D. Nashold, and K.A. Blair, 1997: Cryptosporidiosis-associated mortality following a massive waterborne outbreak in Milwaukee, Wisconsin. *American Journal of Public Health,* **87,** 2032-2035. http://dx.doi.org/10.2105/ajph.87.12.2032

47. Mac Kenzie, W.R., N.J. Hoxie, M.E. Proctor, M.S. Gradus, K.A. Blair, D.E. Peterson, J.J. Kazmierczak, D.G. Addiss, K.R. Fox, J.B. Rose, and J.P. Davis, 1994: A massive outbreak in Milwaukee of Cryptosporidium infection transmitted through the public water supply. *The New England Journal of Medicine,* **331,** 161-167. http://dx.doi.org/10.1056/nejm199407213310304

48. Nichols, G., C. Lane, N. Asgari, N.Q. Verlander, and A. Charlett, 2009: Rainfall and outbreaks of drinking water related disease and in England and Wales. *Journal of Water Health,* **7,** 1-8. http://dx.doi.org/10.2166/wh.2009.143

49. Kozlica, J., A.L. Claudet, D. Solomon, J.R. Dunn, and L.R. Carpenter, 2010: Waterborne outbreak of Salmonella I 4,[5],12:i:-. *Foodborne Pathogens and Disease,* **7,** 1431-1433. http://dx.doi.org/10.1089/fpd.2010.0556

50. Clark, C.G., L. Price, R. Ahmed, D.L. Woodward, P.L. Melito, F.G. Rodgers, F. Jamieson, B. Ciebin, A. Li, and A. Ellis, 2003: Characterization of waterborne outbreak–associated *Campylobacter jejuni,* Walkerton, Ontario. *Emerging Infectious Diseases,* **9,** 1232-1241. http://dx.doi.org/10.3201/eid0910.020584

51. Walkerton Commission of Inquiry, 2002: Part One Report of the Walkerton Commission of Inquiry: The Events of May 2000 and Related Issues. 504 pp. Ontario Ministry of the Attorney General, Toronto, ONT. http://www.attorneygeneral.jus.gov.on.ca/english/about/pubs/walkerton/part1/

52. Delpla, I., A.-V. Jung, E. Baures, M. Clement, and O. Thomas, 2009: Impacts of climate change on surface water quality in relation to drinking water production. *Environment International,* **35,** 1225-1233. http://dx.doi.org/10.1016/j.envint.2009.07.001

53. EPA, 2015: 2015 Drinking Water Health Advisories for Two Cyanobacterial Toxins. EPA 820F15003. U.S. Environmental Protection Agency, Office of Water. http://www2.epa.gov/sites/production/files/2015-06/documents/cyanotoxins-fact_sheet-2015.pdf

54. EPA, 2015: Recommendations for Public Water Systems to Manage Cyanotoxins in Drinking Water. EPA 815-R-15-010. U.S. Environmental Protection Agency, Office of Water. http://www2.epa.gov/sites/production/files/2015-06/documents/cyanotoxin-management-drinking-water.pdf

55. Zamyadi, A., S.L. MacLeod, Y. Fan, N. McQuaid, S. Dorner, S. Sauvé, and M. Prévost, 2012: Toxic cyanobacterial breakthrough and accumulation in a drinking water plant: A monitoring and treatment challenge. *Water Research,* **46,** 1511-1523. http://dx.doi.org/10.1016/j.watres.2011.11.012

56. Zamyadi, A., S. Dorner, S. Sauve, D. Ellis, A. Bolduc, C. Bastien, and M. Prevost, 2013: Species-dependence of cyanobacteria removal efficiency by different drinking water treatment processes. *Water Research,* **47,** 2689-2700. http://dx.doi.org/10.1016/j.watres.2013.02.040

57. City of Toledo, 2014: Microcystin Event Preliminary Summary. 73 pp. City of Toledo Department of Public Utilities. http://toledo.oh.gov/media/132055/Microcystin-Test-Results.pdf

58. Georgakakos, A., P. Fleming, M. Dettinger, C. Peters-Lidard, T.C. Richmond, K. Reckhow, K. White, and D. Yates, 2014: Ch. 3: Water resources. *Climate Change Impacts in the United States: The Third National Climate Assessment.* Melillo, J.M., T.C. Richmond, and G.W. Yohe, Eds. U.S. Global Change Research Program, Washington, D.C., 69-112. http://dx.doi.org/10.7930/J0G44N6T

59. Cann, K.F., D.R. Thomas, R.L. Salmon, A.P. Wyn-Jones, and D. Kay, 2013: Extreme water-related weather events and waterborne disease. *Epidemiology and Infection,* **141,** 671-86. http://dx.doi.org/10.1017/s0950268812001653

60. Angelakis, A.N. and P. Gikas, 2014: Water reuse: Overview of current practices and trends in the world with emphasis on EU states. *Water Utility Journal,* **8,** 67-78. http://www.ewra.net/wuj/pdf/WUJ_2014_08_07.pdf

61. Jimenez, B. and T. Asano, 2008: Water reclamation and reuse around the world. *Water Reuse: An International Survey of Current Practice, Issues and Needs.* IWA Publishing, London, UK.

62. Wintgens, T., F. Salehi, R. Hochstrat, and T. Melin, 2008: Emerging contaminants and treatment options in water recycling for indirect potable use. *Water Science & Technology,* **57,** 99-107. http://dx.doi.org/10.2166/wst.2008.799

63. MacDonald, G.M., 2010: Water, climate change, and sustainability in the southwest. *Proceedings of the National Academy of Sciences,* **107,** 21256-21262. http://dx.doi.org/10.1073/pnas.0909651107

64. NRC, 2012: *Water Reuse: Potential for Expanding the Nation's Water Supply through Reuse of Municipal Wastewater.* National Academies Press, Washington, D.C.

65. Vo, P.T., H.H. Ngo, W. Guo, J.L. Zhou, P.D. Nguyen, A. Listowski, and X.C. Wang, 2014: A mini-review on the impacts of climate change on wastewater reclamation and reuse. *Science of the Total Environment,* **494-495,** 9-17. http://dx.doi.org/10.1016/j.scitotenv.2014.06.090

66. Bastian, R. and D. Murray, 2012: 2012 Guidelines for Water Reuse. EPA/600/R-12/618, 643 pp. U.S. EPA Office of Research and Development, Washington, D.C. http://nepis.epa.gov/Adobe/PDF/P100FS7K.pdf

67. Shuster, W.D., D. Lye, A. De La Cruz, L.K. Rhea, K. O'Connell, and A. Kelty, 2013: Assessment of residential rain barrel water quality and use in Cincinnati, Ohio. *Journal of the American Water Resources Association,* **49,** 753-765. http://dx.doi.org/10.1111/jawr.12036

68. Ahmed, W., A. Vieritz, A. Goonetilleke, and T. Gardner, 2010: Health risk from the use of roof-harvested rainwater in southeast Queensland, Australia, as potable or nonpotable water, determined using quantitative microbial risk assessment. *Applied and Environmental Microbiology,* **76,** 7382-7391. http://dx.doi.org/10.1128/AEM.00944-10

69. Lye, D.J., 2002: Health risks associated with consumption of untreated water from household roof catchment systems. *Journal of the American Water Resources Association,* **38,** 1301-1306.

70. Whitehead, P.G., R.L. Wilby, R.W. Battarbee, M. Kernan, and A.J. Wade, 2009: A review of the potential impacts of climate change on surface water quality. *Hydrological Sciences Journal,* **54,** 101-123. http://dx.doi.org/10.1623/hysj.54.1.101

71. Sterk, A., J. Schijven, T. de Nijs, and A.M. de Roda Husman, 2013: Direct and indirect effects of climate change on the risk of infection by water-transmitted pathogens. *Environmental Science & Technology,* **47,** 12648-12660. http://dx.doi.org/10.1021/es403549s

72. Schijven, J., M. Bouwknegt, A.M. de Roda Husman, S. Rutjes, B. Sudre, J.E. Suk, and J.C. Semenza, 2013: A decision support tool to compare waterborne and foodborne infection and/or illness risks associated with climate change. *Risk Analysis,* **33,** 2154-2167. http://dx.doi.org/10.1111/risa.12077

73. Smith, B.A., T. Ruthman, E. Sparling, H. Auld, N. Comer, I. Young, A.M. Lammerding, and A. Fazil, 2015: A risk modeling framework to evaluate the impacts of climate change and adaptation on food and water safety. *Food Research International,* **68,** 78-85. http://dx.doi.org/10.1016/j.foodres.2014.07.006

74. Levin, R.B., P.R. Epstein, T.E. Ford, W. Harrington, E. Olson, and E.G. Reichard, 2002: U.S. drinking water challenges in the twenty-first century. *Environmental Health Perspectives,* **110,** 43-52. PMC1241146

75. Rose, J.B., P.R. Epstein, E.K. Lipp, B.H. Sherman, S.M. Bernard, and J.A. Patz, 2001: Climate variability and change in the United States: Potential impacts on water- and foodborne diseases caused by microbiologic agents. *Environmental Health Perspectives,* **109 Suppl 2,** 211-221. PMC1240668

76. Perry, D., D. Bennett, U. Boudjou, M. Hahn, S. McLellan, and S. Elizabeth, 2012: Effect of climate change on sewer overflows in Milwaukee. *Proceedings of the Water Environment Federation, WEFTEC 2012: Session 30*, pp. 1857-1866. http://dx.doi.org/10.2175/193864712811725546

77. Hlavsa, M.C., V.A. Roberts, A. Kahler, E.D. Hilborn, T.J. Wade, L.C. Backer, and J.S. Yoder, 2014: Recreational water–associated disease outbreaks — United States, 2009–2010. *MMWR. Morbidity and Mortality Weekly Report, 63*, 6-10. http://www.cdc.gov/mmwr/preview/mmwrhtml/mm6301a2.htm

78. Stumpf, R.P., V. Fleming-Lehtinen, and E. Granéli, 2010: Integration of data for nowcasting of harmful algal blooms. *Proceedings of OceanObs'09: Sustained Ocean Observations and Information for Society (Volume 1)*, 21-25 September, Venice, Italy. Hall, J., D.E. Harrison, and D. Stammer, Eds. http://www.oceanobs09.net/proceedings/pp/pp36/index.php

79. Colford, J.M., Jr., K.C. Schiff, J.F. Griffith, V. Yau, B.F. Arnold, C.C. Wright, J.S. Gruber, T.J. Wade, S. Burns, J. Hayes, C. McGee, M. Gold, Y. Cao, R.T. Noble, R. Haugland, and S.B. Weisberg, 2012: Using rapid indicators for *Enterococcus* to assess the risk of illness after exposure to urban runoff contaminated marine water. *Water Research, 46*, 2176-2186. http://dx.doi.org/10.1016/j.watres.2012.01.033

80. Colford, J.M., Jr., T.J. Wade, K.C. Schiff, C.C. Wright, J.F. Griffith, S.K. Sandhu, S. Burns, M. Sobsey, G. Lovelace, and S.B. Weisberg, 2007: Water quality indicators and the risk of illness at beaches with nonpoint sources of fecal contamination. *Epidemiology, 18*, 27-35. http://dx.doi.org/10.1097/01.ede.0000249425.32990.b9

81. Wade, T.J., E. Sams, K.P. Brenner, R. Haugland, E. Chern, M. Beach, L. Wymer, C.C. Rankin, D. Love, Q. Li, R. Noble, and A.P. Dufour, 2010: Rapidly measured indicators of recreational water quality and swimming-associated illness at marine beaches: A prospective cohort study. *Environmental Health,* **Article 66**. http://dx.doi.org/10.1186/1476-069X-9-66

82. Haley, B.J., D.J. Cole, and E.K. Lipp, 2009: Distribution, diversity, and seasonality of waterborne salmonellae in a rural watershed. *Applied and Environmental Microbiology, 75*, 1248-1255. http://dx.doi.org/10.1128/aem.01648-08

83. Vereen, E., Jr., R.R. Lowrance, D.J. Cole, and E.K. Lipp, 2007: Distribution and ecology of campylobacters in coastal plain streams (Georgia, United States of America). *Applied and Environmental Microbiology, 73*, 1395-1403. http://dx.doi.org/10.1128/aem.01621-06

84. Vereen, E., Jr., R.R. Lowrance, M.B. Jenkins, P. Adams, S. Rajeev, and E.K. Lipp, 2013: Landscape and seasonal factors influence *Salmonella* and *Campylobacter* prevalence in a rural mixed use watershed. *Water Research, 47*, 6075-6085. http://dx.doi.org/10.1016/j.watres.2013.07.028

85. Bharti, A.R., J.E. Nally, J.N. Ricaldi, M.A. Matthias, M.M. Diaz, M.A. Lovett, P.N. Levett, R.H. Gilman, M.R. Willig, E. Gotuzzo, and J.M. Vinetz, 2003: Leptospirosis: A zoonotic disease of global importance. *The Lancet Infectious Diseases, 3*, 757-771. http://dx.doi.org/10.1016/S1473-3099(03)00830-2

86. Howell, D. and D. Cole, 2006: Leptospirosis: A waterborne zoonotic disease of global importance. *Georgia Epidemiology Report, 22*, 1-2. http://dph.georgia.gov/sites/dph.georgia.gov/files/related_files/site_page/ADES_Aug06GER.pdf

87. Lau, C.L., L.D. Smythe, S.B. Craig, and P. Weinstein, 2010: Climate change, flooding, urbanisation and leptospirosis: Fuelling the fire? *Transactions of the Royal Society of Tropical Medicine and Hygiene, 104*, 631-638. http://dx.doi.org/10.1016/j.trstmh.2010.07.002

88. Lau, C.L., A.J. Dobson, L.D. Smythe, E.J. Fearnley, C. Skelly, A.C.A. Clements, S.B. Craig, S.D. Fuimaono, and P. Weinstein, 2012: Leptospirosis in American Samoa 2010: Epidemiology, environmental drivers, and the management of emergence. *The American Journal of Tropical Medicine and Hygiene, 86*, 309-319. http://dx.doi.org/10.4269/ajtmh.2012.11-0398

89. Katz, A.R., A.E. Buchholz, K. Hinson, S.Y. Park, and P.V. Effler, 2011: Leptospirosis in Hawaii, USA, 1999–2008. *Emerging Infectious Diseases, 17*, 221-226. http://dx.doi.org/10.3201/eid1702.101109

90. Hartskeerl, R.A., M. Collares-Pereira, and W.A. Ellis, 2011: Emergence, control and re-emerging leptospirosis: Dynamics of infection in the changing world. *Clinical Microbiology and Infection, 17*, 494-501. http://dx.doi.org/10.1111/j.1469-0691.2011.03474.x

91. Desvars, A., S. Jégo, F. Chiroleu, P. Bourhy, E. Cardinale, and A. Michault, 2011: Seasonality of human leptospirosis in Reunion Island (Indian Ocean) and its association with meteorological data. *PLoS ONE, 6*, e20377. http://dx.doi.org/10.1371/journal.pone.0020377

92. Newton, A., M. Kendall, D.J. Vugia, O.L. Henao, and B.E. Mahon, 2012: Increasing rates of vibriosis in the United States, 1996–2010: Review of surveillance data from 2 systems. *Clinical Infectious Diseases, 54*, S391-S395. http://dx.doi.org/10.1093/cid/cis243

93. Dechet, A.M., P.A. Yu, N. Koram, and J. Painter, 2008: Nonfoodborne *Vibrio* infections: An important cause of morbidity and mortality in the United States, 1997–2006. *Clinical Infectious Diseases, 46*, 970-976. http://dx.doi.org/10.1086/529148

94. Yoder, J.S., M.C. Hlavsa, G.F. Craun, V. Hill, V. Roberts, P.A. Yu, L.A. Hicks, N.T. Alexander, R.L. Calderon, S.L. Roy, and M.J. Beach, 2008: Surveillance for waterborne disease and outbreaks associated with recreational water use and other aquatic facility-associated health events--United States, 2005-2006. *MMWR Surveillance Summaries,* **57(SS09),** 1-29. http://www.cdc.gov/mmWR/preview/mmwrhtml/ss5709a1.htm

95. Froelich, B., J. Bowen, R. Gonzalez, A. Snedeker, and R. Noble, 2013: Mechanistic and statistical models of total *Vibrio* abundance in the Neuse River Estuary. *Water Research,* **47,** 5783-5793. http://dx.doi.org/10.1016/j.watres.2013.06.050

96. Vezzulli, L., I. Brettar, E. Pezzati, P.C. Reid, R.R. Colwell, M.G. Höfle, and C. Pruzzo, 2012: Long-term effects of ocean warming on the prokaryotic community: Evidence from the vibrios. *The ISME Journal,* **6,** 21-30. http://dx.doi.org/10.1038/ismej.2011.89

97. Lipp, E.K., C. Rodriguez-Palacios, and J.B. Rose, 2001: Occurrence and distribution of the human pathogen *Vibrio vulnificus* in a subtropical Gulf of Mexico estuary. *The Ecology and Etiology of Newly Emerging Marine Diseases.* Porter, J.W., Ed. Springer, Dordrecht, 165-173. http://dx.doi.org/10.1007/978-94-017-3284-0_15

98. Louis, V.R., E. Russek-Cohen, N. Choopun, I.N.G. Rivera, B. Gangle, S.C. Jiang, A. Rubin, J.A. Patz, A. Huq, and R.R. Colwell, 2003: Predictability of Vibrio cholerae in Chesapeake Bay. *Applied and Environmental Microbiology,* **69,** 2773-2785. http://dx.doi.org/10.1128/aem.69.5.2773-2785.2003

99. Griffitt, K.J. and D.J. Grimes, 2013: Abundance and distribution of *Vibrio cholerae, V. parahaemolyticus,* and *V. vulnificus* following a major freshwater intrusion into the Mississippi Sound. *Microbial Ecology,* **65,** 578-583. http://dx.doi.org/10.1007/s00248-013-0203-6

100. Constantin de Magny, G., W. Long, C.W. Brown, R.R. Hood, A. Huq, R. Murtugudde, and R.R. Colwell, 2009: Predicting the distribution of *Vibrio* spp. in the Chesapeake Bay: A *Vibrio cholerae* case study. *EcoHealth,* **6,** 378-389. http://dx.doi.org/10.1007/s10393-009-0273-6

101. Hilborn, E.D., V.A. Roberts, L. Backer, E. DeConno, J.S. Egan, J.B. Hyde, D.C. Nicholas, E.J. Wiegert, L.M. Billing, M. DiOrio, M.C. Mohr, F.J. Hardy, T.J. Wade, J.S. Yoder, and M.C. Hlavsa, 2014: Algal bloom–associated disease outbreaks among users of freshwater lakes — United States, 2009–2010. *MMWR. Morbidity and Mortality Weekly Report,* **63,** 11-15. http://www.cdc.gov/mmwr/preview/mmwrhtml/mm6301a3.htm

102. Paerl, H.W. and J. Huisman, 2008: Blooms like it hot. *Science,* **320,** 57-58. http://dx.doi.org/10.1126/Science.1155398

103. Paerl, H.W., N.S. Hall, and E.S. Calandrino, 2011: Controlling harmful cyanobacterial blooms in a world experiencing anthropogenic and climatic-induced change. *Science of The Total Environment,* **409,** 1739-1745. http://dx.doi.org/10.1016/j.scitotenv.2011.02.001

104. Paerl, H.W. and V.J. Paul, 2012: Climate change: Links to global expansion of harmful cyanobacteria. *Water Research,* **46,** 1349-1363. http://dx.doi.org/10.1016/j.watres.2011.08.002

105. Paerl, H.W. and T.G. Otten, 2013: Blooms bite the hand that feeds them. *Science,* **342,** 433-434. http://dx.doi.org/10.1126/science.1245276

106. Carey, C.C., B.W. Ibelings, E.P. Hoffmann, D.P. Hamilton, and J.D. Brookes, 2012: Eco-physiological adaptations that favour freshwater cyanobacteria in a changing climate. *Water Research,* **46,** 1394-1407. http://dx.doi.org/10.1016/j.watres.2011.12.016

107. Kosten, S., V.L.M. Huszar, E. Bécares, L.S. Costa, E. van Donk, L.-A. Hansson, E. Jeppesen, C. Kruk, G. Lacerot, N. Mazzeo, L. De Meester, B. Moss, M. Lürling, T. Nões, S. Romo, and M. Scheffer, 2012: Warmer climates boost cyanobacterial dominance in shallow lakes. *Global Change Biology,* **18,** 118-126. http://dx.doi.org/10.1111/j.1365-2486.2011.02488.x

108. O'Neil, J.M., T.W. Davis, M.A. Burford, and C.J. Gobler, 2012: The rise of harmful cyanobacteria blooms: The potential roles of eutrophication and climate change. *Harmful Algae,* **14,** 313-334. http://dx.doi.org/10.1016/j.hal.2011.10.027

109. Elliott, J.A., 2010: The seasonal sensitivity of Cyanobacteria and other phytoplankton to changes in flushing rate and water temperature. *Global Change Biology,* **16,** 864-876. http://dx.doi.org/10.1111/j.1365-2486.2009.01998.x

110. Kirkpatrick, B., L.E. Fleming, L.C. Backer, J.A. Bean, R. Tamer, G. Kirkpatrick, T. Kane, A. Wanner, D. Dalpra, A. Reich, and D.G. Baden, 2006: Environmental exposures to Florida red tides: Effects on emergency room respiratory diagnoses admissions. *Harmful Algae,* **5,** 526-533. http://dx.doi.org/10.1016/j.hal.2005.09.004

111. Fleming, L.E., B. Kirkpatrick, L.C. Backer, C.J. Walsh, K. Nierenberg, J. Clark, A. Reich, J. Hollenbeck, J. Benson, Y.S. Cheng, J. Naar, R. Pierce, A.J. Bourdelais, W.M. Abraham, G. Kirkpatrick, J. Zaias, A. Wanner, E. Mendes, S. Shalat, P. Hoagland, W. Stephan, J. Bean, S. Watkins, T. Clarke, M. Byrne, and D.G. Baden, 2011: Review of Florida red tide and human health effects. *Harmful Algae,* **10,** 224-233. http://dx.doi.org/10.1016/j.hal.2010.08.006

112. Fleming, L.E., B. Kirkpatrick, L.C. Backer, J.A. Bean, A. Wanner, D. Dalpra, R. Tamer, J. Zaias, Y.-S. Cheng, R. Pierce, J. Naar, W. Abraham, R. Clark, Y. Zhou, M.S. Henry, D. Johnson, G. Van de Bogart, G.D. Bossart, M. Harrington, and D.G. Baden, 2005: Initial evaluation of the effects of aerosolized Florida red tide toxins (Brevetoxins) in persons with asthma. *Environmental Health Perspectives,* **113,** 650–657. http://dx.doi.org/10.1289/ehp.7500

113. Fleming, L.E., B. Kirkpatrick, L.C. Backer, J.A. Bean, A. Wanner, A. Reich, J. Zaias, Y.-S. Cheng, R. Pierce, J. Naar, W.M. Abraham, and D.G. Baden, 2007: Aerosolized red-tide toxins (Brevetoxins) and asthma. *Chest,* **131,** 187-194. http://dx.doi.org/10.1378/chest.06-1830

114. Thyng, K.M., R.D. Hetland, M.T. Ogle, X. Zhang, F. Chen, and L. Campbell, 2013: Origins of *Karenia brevis* harmful algal blooms along the Texas coast. *Limnology and Oceanography: Fluids and Environments,* **3,** 269-278. http://dx.doi.org/10.1215/21573689-2417719

115. Maier Brown, A.F., Q. Dortch, F.M. Van Dolah, T.A. Leighfield, W. Morrison, A.E. Thessen, K. Steidinger, B. Richardson, C.A. Moncreiff, and J.R. Pennock, 2006: Effect of salinity on the distribution, growth, and toxicity of Karenia spp. *Harmful Algae,* **5,** 199-212. http://dx.doi.org/10.1016/j.hal.2005.07.004

116. McLellan, S.L., E.J. Hollis, M.M. Depas, M. Van Dyke, J. Harris, and C.O. Scopel, 2007: Distribution and fate of *Escherichia coli* in Lake Michigan following contamination with urban stormwater and combined sewer overflows. *Journal of Great Lakes Research,* **33,** 566-580. http://dx.doi.org/10.3394/0380-1330(2007)33%5B566:dafoec%5D2.0.co;2

117. Corsi, S.R., M.A. Borchardt, S.K. Spencer, P.E. Hughes, and A.K. Baldwin, 2014: Human and bovine viruses in the Milwaukee River watershed: Hydrologically relevant representation and relations with environmental variables. *Science of The Total Environment,* **490,** 849-860. http://dx.doi.org/10.1016/j.scitotenv.2014.05.072

118. Duris, J.W., A.G. Reif, D.A. Krouse, and N.M. Isaacs, 2013: Factors related to occurrence and distribution of selected bacterial and protozoan pathogens in Pennsylvania streams. *Water Research,* **47,** 300-314. http://dx.doi.org/10.1016/j.watres.2012.10.006

119. Staley, C., K.H. Reckhow, J. Lukasik, and V.J. Harwood, 2012: Assessment of sources of human pathogens and fecal contamination in a Florida freshwater lake. *Water Research,* **46,** 5799-5812. http://dx.doi.org/10.1016/j.watres.2012.08.012

120. McBride, G.B., R. Stott, W. Miller, D. Bambic, and S. Wuertz, 2013: Discharge-based QMRA for estimation of public health risks from exposure to stormwater-borne pathogens in recreational waters in the United States. *Water Research,* **47,** 5282-5297. http://dx.doi.org/10.1016/j.watres.2013.06.001

121. NOAA, 2014: Sea Level Rise and Nuisance Flood Frequency Changes around the United States. NOAA Technical Report NOS CO-OPS 073, 58 pp. U.S. Department of Commerce, National Oceanic and Atmospheric Administration, National Ocean Service, Silver Spring, MD. http://tidesandcurrents.noaa.gov/publications/NOAA_Technical_Report_NOS_COOPS_073.pdf

122. Casman, E., B. Fischhoff, M. Small, H. Dowlatabadi, J. Rose, and M.G. Morgan, 2001: Climate change and cryptosporidiosis: A qualitative analysis. *Climatic Change,* **50,** 219-249. http://dx.doi.org/10.1023/a:1010623831501

123. Naumova, E.N., J.S. Jagai, B. Matyas, A. DeMaria, I.B. MacNeill, and J.K. Griffiths, 2007: Seasonality in six enterically transmitted diseases and ambient temperature. *Epidemiology and Infection,* **135,** 281-292. http://dx.doi.org/10.1017/S0950268806006698

124. Peeters, F., D. Straile, A. Lorke, and D.M. Livingstone, 2007: Earlier onset of the spring phytoplankton bloom in lakes of the temperate zone in a warmer climate. *Global Change Biology,* **13,** 1898-1909. http://dx.doi.org/10.1111/j.1365-2486.2007.01412.x

125. Suikkanen, S., M. Laamanen, and M. Huttunen, 2007: Long-term changes in summer phytoplankton communities of the open northern Baltic Sea. *Estuarine, Coastal and Shelf Science,* **71,** 580-592. http://dx.doi.org/10.1016/j.ecss.2006.09.004

126. Wiedner, C., J. Rücker, R. Brüggemann, and B. Nixdorf, 2007: Climate change affects timing and size of populations of an invasive cyanobacterium in temperate regions. *Oecologia,* **152,** 473-484. http://dx.doi.org/10.1007/s00442-007-0683-5

127. Wagner, C. and R. Adrian, 2009: Cyanobacteria dominance: Quantifying the effects of climate change. *Limnology and Oceanography,* **54,** 2460-2468. http://dx.doi.org/10.4319/lo.2009.54.6_part_2.2460

128. Vincent, W.F. and A. Quesada, 2012: Cyanobacteria in high latitude lakes, rivers and seas. *Ecology of Cyanobacteria II: Their Diversity in Space and Time.* Whitton, B.A., Ed. Springer, New York, 371-385. http://dx.doi.org/10.1007/978-94-007-3855-3

129. Padisak, J., 1997: Cylindrospermopsis raciborskii (Woloszynska) Seenayya et Subba Raju, an expanding, highly adaptive cyanobacterium: Worldwide distribution and review of its ecology. *Archiv Für Hydrobiologie Supplementband Monographische Beitrage* **107,** 563-593. http://real.mtak.hu/3229/1/1014071.pdf

130. Stüken, A., J. Rücker, T. Endrulat, K. Preussel, M. Hemm, B. Nixdorf, U. Karsten, and C. Wiedner, 2006: Distribution of three alien cyanobacterial species (Nostocales) in northeast Germany: *Cylindrospermopsis raciborskii, Anabaena bergii* and *Aphanizomenon aphanizomenoides. Phycologia,* **45,** 696-703. http://dx.doi.org/10.2216/05-58.1

131. Elliott, J.A., 2012: Is the future blue-green? A review of the current model predictions of how climate change could affect pelagic freshwater cyanobacteria. *Water Research,* **46,** 1364-1371. http://dx.doi.org/10.1016/j.watres.2011.12.018

132. Jacobs, J., S.K. Moore, K.E. Kunkel, and L. Sun, 2015: A framework for examining climate-driven changes to the seasonality and geographical range of coastal pathogens and harmful algae. *Climate Risk Management,* **8,** 16-27. http://dx.doi.org/10.1016/j.crm.2015.03.002

133. Urquhart, E.A., B.F. Zaitchik, D.W. Waugh, S.D. Guikema, and C.E. Del Castillo, 2014: Uncertainty in model predictions of *Vibrio vulnificus* response to climate variability and change: A Chesapeake Bay case study. *PLoS ONE,* **9,** e98256. http://dx.doi.org/10.1371/journal.pone.0098256

134. Froelich, B.A., T.C. Williams, R.T. Noble, and J.D. Oliver, 2012: Apparent loss of *Vibrio vulnificus* from North Carolina oysters coincides with a drought-induced increase in salinity. *Applied and Environmental Microbiology,* **78,** 3885-3889. http://dx.doi.org/10.1128/aem.07855-11

135. Copat, C., G. Arena, M. Fiore, C. Ledda, R. Fallico, S. Sciacca, and M. Ferrante, 2013: Heavy metals concentrations in fish and shellfish from eastern Mediterranean Sea: Consumption advisories. *Food and Chemical Toxicology,* **53,** 33-37. http://dx.doi.org/10.1016/j.fct.2012.11.038

136. Ho, K.K.Y. and K.M.Y. Leung, 2014: Organotin contamination in seafood and its implication for human health risk in Hong Kong. *Marine Pollution Bulletin,* **85,** 634-640. http://dx.doi.org/10.1016/j.marpolbul.2013.12.039

137. Shapiro, K., M. Silver, J. Largier, J. Mazet, W. Miller, M. Odagiri, and A. Schriewer, 2012: Pathogen aggregation: Understanding when, where, and why seafood contamination occurs. *Journal of Shellfish Research,* **31,** 345. http://dx.doi.org/10.2983/035.031.0124

138. FDA, 2005: Quantitative Risk Assessment on the Public Health Impact of Pathogenic *Vibrio parahaemolyticus* in Raw Oysters. 309 pp. U.S. Department of Health and Human Services, Food and Drug Administration, Center for Food Safety and Applied Nutrition. http://www.fda.gov/Food/FoodScienceResearch/RiskSafetyAssessment/ucm050421.htm

139. Bellou, M., P. Kokkinos, and A. Vantarakis, 2013: Shellfish-borne viral outbreaks: A systematic review. *Food and Environmental Virology,* **5,** 13-23. http://dx.doi.org/10.1007/s12560-012-9097-6

140. Iwamoto, M., T. Ayers, B.E. Mahon, and D.L. Swerdlow, 2010: Epidemiology of seafood-associated infections in the United States. *Clinical Microbiology Reviews,* **23,** 399-411. http://dx.doi.org/10.1128/Cmr.00059-09

141. Le Saux, J., O. Serais, J. Krol, S. Parnaudeau, P. Salvagnac, G. Delmas, V. Cicchelero, J. Claudet, P. Pothier, and K. Balay, 2009: Evidence of the presence of viral contamination in shellfish after short rainfall events. *6th International Conference on Molluscan Shellfish Safety,* Blenheim, New Zealand. Busby, P., Ed., pp. 256-252.

142. Wang, J. and Z. Deng, 2012: Detection and forecasting of oyster norovirus outbreaks: Recent advances and future perspectives. *Marine Environmental Research,* **80,** 62-69. http://dx.doi.org/10.1016/j.marenvres.2012.06.011

143. Riou, P., J.C. Le Saux, F. Dumas, M.P. Caprais, S.F. Le Guyader, and M. Pommepuy, 2007: Microbial impact of small tributaries on water and shellfish quality in shallow coastal areas. *Water Research,* **41,** 2774-2786. http://dx.doi.org/10.1016/j.watres.2007.03.003

144. Coulliette, A.D., E.S. Money, M.L. Serre, and R.T. Noble, 2009: Space/time analysis of fecal pollution and rainfall in an eastern North Carolina estuary. *Environmental Science & Technology,* **43,** 3728-3735. http://dx.doi.org/10.1021/es803183f

145. Lowther, J.A., K. Henshilwood, and D.N. Lees, 2008: Determination of norovirus contamination in oysters from two commercial harvesting areas over an extended period, using semiquantitative real-time reverse transcription PCR. *Journal of Food Protection,* **71,** 1427-1433.

146. Maalouf, H., M. Zakhour, J. Le Pendu, J.C. Le Saux, R.L. Atmar, and F.S. Le Guyader, 2010: Distribution in tissue and seasonal variation of norovirus genogroup I and II ligands in oysters. *Applied and Environmental Microbiology,* **76,** 5621-5630. http://dx.doi.org/10.1128/aem.00148-10

147. Woods, J.W. and W. Burkhardt, 2010: Occurrence of norovirus and hepatitis A virus in U.S. oysters. *Food and Environmental Virology,* **2,** 176-182. http://dx.doi.org/10.1007/s12560-010-9040-7

148. Crim, S.M., M. Iwamoto, J.Y. Huang, P.M. Griffin, D. Gilliss, A.B. Cronquist, M. Cartter, M. Tobin-D'Angelo, D. Blythe, K. Smith, S. Lathrop, S. Zansky, P.R. Cieslak, J. Dunn, K.G. Holt, S. Lance, R. Tauxe, and O.L. Henao, 2014: Incidence and trends of infections with pathogens transmitted commonly through food--Foodborne Diseases Active Surveillance Network, 10 U.S. sites, 2006-2013. *MMWR. Morbidity and Mortality Weekly Report,* **63,** 328-332. http://www.cdc.gov/mmwr/preview/mmwrhtml/mm6315a3.htm

149. Rippey, S.R., 1994: Infectious diseases associated with molluscan shellfish consumption. *Clinical Microbiology Reviews,* **7,** 419-425. http://dx.doi.org/10.1128/cmr.7.4.419

150. Lynch, M., J. Painter, R. Woodruff, and C. Braden, 2006: Surveillance for foodborne-disease outbreaks – United States, 1998-2002. *MMWR Surveillance Summaries,* **55(SS10),** 1-42. http://www.cdc.gov/mmwr/preview/mmwrhtml/ss5510a1.htm?_cid=ss

151. Vugia, D., A. Cronquist, J. Hadler, M. Tobin-D'Angelo, D. Blythe, K. Smith, K. Thornton, D. Morse, P. Cieslak, T. Jones, K. Holt, J. Guzewich, O. Henao, E. Scallan, F. Angulo, P. Griffin, R. Tauxe, and E. Barzilay, 2006: Preliminary FoodNet data on the incidence of infection with pathogens transmitted commonly through food – 10 states, United States, 2005. *MMWR. Morbidity and Mortality Weekly Report,* **55,** 392-395. http://www.cdc.gov/mmwr/preview/mmwrhtml/mm5514a2.htm

152. Martinez-Urtaza, J., J.C. Bowers, J. Trinanes, and A. DePaola, 2010: Climate anomalies and the increasing risk of *Vibrio parahaemolyticus* and *Vibrio vulnificus* illnesses. *Food Research International,* **43,** 1780-1790. http://dx.doi.org/10.1016/j.foodres.2010.04.001

153. Martinez-Urtaza, J., C. Baker-Austin, J.L. Jones, A.E. Newton, G.D. Gonzalez-Aviles, and A. DePaola, 2013: Spread of Pacific Northwest Vibrio parahaemolyticus strain. *The New England Journal of Medicine,* **369,** 1573-1574. http://dx.doi.org/10.1056/NEJMc1305535

154. McLaughlin, J.B., A. DePaola, C.A. Bopp, K.A. Martinek, N.P. Napolilli, C.G. Allison, S.L. Murray, E.C. Thompson, M.M. Bird, and J.P. Middaugh, 2005: Outbreak of *Vibrio parahaemolyticus* gastroenteritis associated with Alaskan oysters. *The New England Journal of Medicine,* **353,** 1463-1470. http://dx.doi.org/10.1056/NEJMoa051594

155. Newton, A.E., N. Garrett, S.G. Stroika, J.L. Halpin, M. Turnsek, and R.K. Mody, 2014: Notes from the field: Increase in Vibrio parahaemolyticus infections associated with consumption of Atlantic Coast shellfish — 2013. *MMWR. Morbidity and Mortality Weekly Report,* **63,** 335-336. http://origin.glb.cdc.gov/MMWR/preview/mmwrhtml/mm6315a6.htm?s_cid=mm6315a6_w

156. NSSP, 2011: National Shellfish Sanitation Program (NSSP) Guide for the Control of Molluscan Shellfish, 2011 Revision. 478 pp. U.S. Department of Health and Human Services, Public Health Service, Food and Drug Administration. http://www.fda.gov/downloads/Food/GuidanceRegulation/FederalStateFoodPrograms/UCM350344.pdf

157. Baker-Austin, C., J.A. Trinanes, N.G.H. Taylor, R. Hartnell, A. Siitonen, and J. Martinez-Urtaza, 2013: Emerging *Vibrio* risk at high latitudes in response to ocean warming. *Nature Climate Change,* **3,** 73-77. http://dx.doi.org/10.1038/nclimate1628

158. Ralston, E.P., H. Kite-Powell, and A. Beet, 2011: An estimate of the cost of acute health effects from food- and waterborne marine pathogens and toxins in the USA. *Journal of Water and Health,* **9,** 680-694. http://dx.doi.org/10.2166/wh.2011.157

159. Takemura, A.F., D.M. Chien, and M.F. Polz, 2014: Associations and dynamics of Vibrionaceae in the environment, from the genus to the population level. *Frontiers in Microbiology,* **5.** http://dx.doi.org/10.3389/fmicb.2014.00038

160. Hashizume, M., A.S.G. Faruque, T. Terao, M. Yunus, K. Streatfield, T. Yamamoto, and K. Moji, 2011: The Indian Ocean dipole and cholera incidence in Bangladesh: A time-series analysis. *Environmental Health Perspectives,* **119,** 239-244. http://dx.doi.org/10.1289/ehp.1002302

161. Lara, R.J., S.B. Neogi, M.S. Islam, Z.H. Mahmud, S. Yamasaki, and G.B. Nair, 2009: Influence of catastrophic climatic events and human waste on Vibrio distribution in the Karnaphuli Estuary, Bangladesh. *EcoHealth,* **6,** 279-286. http://dx.doi.org/10.1007/s10393-009-0257-6

162. Turner, J.W., B. Good, D. Cole, and E.K. Lipp, 2009: Plankton composition and environmental factors contribute to *Vibrio* seasonality. *The ISME Journal,* **3,** 1082-1092. http://dx.doi.org/10.1038/ismej.2009.50

163. Turner, J.W., L. Malayil, D. Guadagnoli, D. Cole, and E.K. Lipp, 2014: Detection of *Vibrio parahaemolyticus, Vibrio vulnificus* and *Vibrio cholerae* with respect to seasonal fluctuations in temperature and plankton abundance. *Environmental Microbiology,* **16,** 1019-1028. http://dx.doi.org/10.1111/1462-2920.12246

164. Campbell, L., R.J. Olson, H.M. Sosik, A. Abraham, D.W. Henrichs, C.J. Hyatt, and E.J. Buskey, 2010: First harmful *Dinophysis* (Dinophyceae, Dinophysiales) bloom in the U.S. is revealed by automated imaging flow cytometry. *Journal of Phycology,* **46,** 66-75. http://dx.doi.org/10.1111/j.1529-8817.2009.00791.x

165. Nishimura, T., S. Sato, W. Tawong, H. Sakanari, K. Uehara, M.M.R. Shah, S. Suda, T. Yasumoto, Y. Taira, H. Yamaguchi, and M. Adachi, 2013: Genetic diversity and distribution of the ciguatera-causing dinoflagellate *Gambierdiscus* spp. (Dinophyceae) in coastal areas of Japan. *PLoS ONE,* **8,** e60882. http://dx.doi.org/10.1371/journal.pone.0060882

166. Tester, P.A., R.P. Stumpf, F.M. Vukovich, P.K. Fowler, and J.T. Turner, 1991: An expatriate red tide bloom: Transport, distribution, and persistence. *Limnology and Oceanography,* **36,** 1053-1061. http://onlinelibrary.wiley.com/doi/10.4319/lo.1991.36.5.1053/pdf

167. Hallegraeff, G.M., 1993: A review of harmful algae blooms and their apparent global increase. *Phycologia,* **32,** 79-99. http://dx.doi.org/10.2216/i0031-8884-32-2-79.1

168. Van Dolah, F.M., 2000: Marine algal toxins: Origins, health effects, and their increased occurrence. *Environmental Health Perspectives,* **108,** 133-141. PMC1637787

169. Sellner, K.G., G.J. Doucette, and G.J. Kirkpatrick, 2003: Harmful algal blooms: Causes, impacts and detection. *Journal of Industrial Microbiology and Biotechnology,* **30,** 383-406. http://dx.doi.org/10.1007/s10295-003-0074-9

170. Chateau-Degat, M.-L., M. Chinain, N. Cerf, S. Gingras, B. Hubert, and É. Dewailly, 2005: Seawater temperature, Gambierdiscus spp. variability and incidence of ciguatera poisoning in French Polynesia. *Harmful Algae,* **4,** 1053-1062. http://dx.doi.org/10.1016/j.hal.2005.03.003

171. Gingold, D.B., M.J. Strickland, and J.J. Hess, 2014: Ciguatera fish poisoning and climate change: Analysis of National Poison Center Data in the United States, 2001–2011. *Environmental Health Perspectives, 122,* 580-586. http://dx.doi.org/10.1289/ehp.1307196

172. Tester, P.A., R.L. Feldman, A.W. Nau, S.R. Kibler, and R. Wayne Litaker, 2010: Ciguatera fish poisoning and sea surface temperatures in the Caribbean Sea and the West Indies. *Toxicon, 56,* 698-710. http://dx.doi.org/10.1016/j.toxicon.2010.02.026

173. Litaker, R.W., M.W. Vandersea, M.A. Faust, S.R. Kibler, M. Chinain, M.J. Holmes, W.C. Holland, and P.A. Tester, 2009: Taxonomy of Gambierdiscus including four new species, Gambierdiscus caribaeus, Gambierdiscus carolinianus, Gambierdiscus carpenteri and Gambierdiscus ruetzleri (Gonyaulacales, Dinophyceae). *Phycologia, 48,* 344-390. http://dx.doi.org/10.2216/07-15.1

174. Hales, S., P. Weinstein, and A. Woodward, 1999: Ciguatera (fish poisoning), El Niño, and Pacific sea surface temperatures. *Ecosystem Health, 5,* 20-25. http://dx.doi.org/10.1046/j.1526-0992.1999.09903.x

175. Erdner, D.L., J. Dyble, M.L. Parsons, R.C. Stevens, K.A. Hubbard, M.L. Wrabel, S.K. Moore, K.A. Lefebvre, D.M. Anderson, P. Bienfang, R.R. Bidigare, M.S. Parker, P. Moeller, L.E. Brand, and V.L. Trainer, 2008: Centers for oceans and human health: A unified approach to the challenge of harmful algal blooms. *Environmental Health, 7,* S2. http://dx.doi.org/10.1186/1476-069X-7-S2-S2

176. Moore, S.K., N.J. Mantua, B.M. Hickey, and V.L. Trainer, 2010: The relative influences of El Niño-Southern Oscillation and Pacific Decadal Oscillation on paralytic shellfish toxin accumulation in northwest Pacific shellfish. *Limnology and Oceanography, 55,* 2262-2274. http://dx.doi.org/10.4319/lo.2010.55.6.2262

177. Moore, S.K., N.J. Mantua, and E.P. Salathé, Jr., 2011: Past trends and future scenarios for environmental conditions favoring the accumulation of paralytic shellfish toxins in Puget Sound shellfish. *Harmful Algae, 10,* 521-529. http://dx.doi.org/10.1016/j.hal.2011.04.004

178. Trainer, V.L., B.-T.L. Eberhart, J.C. Wekell, N.G. Adams, L. Hanson, F. Cox, and J. Dowell, 2003: Paralytic shellfish toxins in Puget Sound, Washington state. *Journal of Shellfish Research, 22,* 213-223.

179. Hattenrath, T.K., D.M. Anderson, and C.J. Gobler, 2010: The influence of anthropogenic nitrogen loading and meteorological conditions on the dynamics and toxicity of *Alexandrium fundyense* blooms in a New York (USA) estuary. *Harmful Algae, 9,* 402-412. http://dx.doi.org/10.1016/j.hal.2010.02.003

180. Anderson, D.M., D.A. Couture, J.L. Kleindhinst, B.A. Keafer, D.J. McGillicuddy Jr., J.L. Martin, M.L. Richlen, J.M. Hickey, and A.R. Solow, 2014: Understanding interannual, decadal level variability in paralytic shellfish poisoning toxicity in the Gulf of Maine: The HAB Index. *Deep-Sea Research II, 103,* 264-276. http://dx.doi.org/10.1016/j.dsr2.2013.09.018

181. Nair, A., A.C. Thomas, and M.E. Borsuk, 2013: Interannual variability in the timing of New England shellfish toxicity and relationships to environmental forcing. *Science of The Total Environment, 447,* 255-266. http://dx.doi.org/10.1016/j.scitotenv.2013.01.023

182. Thomas, A.C., R. Weatherbee, H. Xue, and G. Liu, 2010: Interannual variability of shellfish toxicity in the Gulf of Maine: Time and space patterns and links to environmental variability. *Harmful Algae, 9,* 458-480. http://dx.doi.org/10.1016/j.hal.2010.03.002

183. McGillicuddy, D.J., Jr., D.W. Townsend, R. He, B.A. Keafer, J.L. Kleindhinst, Y. Li, J.P. Manning, D.G. Mountain, M.A. Thomas, and D.M. Anderson, 2011: Suppression of the 2010 *Alexandrium fundyense* bloom by changes in physical, biological, and chemical properties of the Gulf of Maine. *Limnology and Oceanography, 56,* 2411-2426. http://dx.doi.org/10.4319/lo.2011.56.6.2411

184. Parsons, M.L. and Q. Dortch, 2002: Sedimentological evidence of an increase in *Pseudo-nitzschia* (Bacillariophyceae) abundance in response to coastal eutrophication. *Limnology and Oceanography, 47,* 551-558. http://dx.doi.org/10.4319/lo.2002.47.2.0551

185. Fryxell, G.A., M.C. Villac, and L.P. Shapiro, 1997: The occurrence of the toxic diatom genus *Pseudo-nitzschia* (Bacillariophyceae) on the West Coast of the USA, 1920–1996: A review. *Phycologia, 36,* 419-437. http://dx.doi.org/10.2216/i0031-8884-36-6-419.1

186. Trainer, V.L., L. Moore, B.D. Bill, N.G. Adams, N. Harrington, J. Borchert, D.A.M. da Silva, and B.-T.L. Eberhart, 2013: Diarrhetic shellfish toxins and other lipophilic toxins of human health concern in Washington state. *Marine Drugs, 11,* 1815-1835. http://dx.doi.org/10.3390/md11061815

187. Belgrano, A., O. Lindahl, and B. Henroth, 1999: North Atlantic Oscillation primary productivity and toxic phytoplankton in the Gullmar Fjord, Sweden (1985-1996). *Proceedings of the Royal Society B: Biological Sciences, 266,* 425-430. http://dx.doi.org/10.1098/rspb.1999.0655

188. Anderson, D.M., A.D. Cembella, and G.M. Hallegraeff, 2012: Progress in understanding harmful algal blooms: Paradigm shifts and new technologies for research, monitoring, and management. *Annual Review of Marine Science, 4,* 143-76. http://dx.doi.org/10.1146/annurev-marine-120308-081121

189. Fu, F.X., A.O. Tatters, and D.A. Hutchins, 2012: Global change and the future of harmful algal blooms in the ocean. *Marine Ecology Progress Series,* **470,** 207-233. http://dx.doi.org/10.3354/meps10047

190. Moore, S.K., V.L. Trainer, N.J. Mantua, M.S. Parker, E.A. Laws, L.C. Backer, and L.E. Fleming, 2008: Impacts of climate variability and future climate change on harmful algal blooms and human health. *Environmental Health,* **7,** S4. http://dx.doi.org/10.1186/1476-069X-7-S2-S4

191. Hallegraeff, G.M., 2010: Ocean climate change, phytoplankton community responses, and harmful algal blooms: A formidable predictive challenge. *Journal of Phycology,* **46,** 220-235. http://dx.doi.org/10.1111/j.1529-8817.2010.00815.x

192. Backer, L.C. and S.K. Moore, 2012: Harmful algal blooms: Future threats in a warmer world. *Environmental Pollution and Its Relation to Climate Change.* El-Nemr, A., Ed. Nova Science Publishers, New York, 485-512.

193. Laws, E.A., 2007: Climate change, oceans, and human health. *Ocean Yearbook 21.* Chircop, A., S. Coffen-Smout, and M. McConnell, Eds. Bridge Street Books, 129-175. http://dx.doi.org/10.1163/221160007X00074

194. Hays, G.C., A.J. Richardson, and C. Robinson, 2005: Climate change and marine plankton. *TRENDS in Ecology and Evolution,* **20,** 337-344. http://dx.doi.org/10.1016/j.tree.2005.03.004

195. Berdalet, E., F. Peters, V.L. Koumandou, C. Roldán, Ó. Guadayol, and M. Estrada, 2007: Species-specific physiological response of dinoflagellates to quantified small-scale turbulence. *Journal of Phycology,* **43,** 965-977. http://dx.doi.org/10.1111/j.1529-8817.2007.00392.x

196. Margalef, R., M. Estrada, and D. Blasco, 1979: Functional morphology of organisms involved in red tides, as adapted to decaying turbulence. *Toxic Dinoflagellate Blooms.* Taylor, D.L. and H.H. Seliger, Eds. Elsevier North Holland, Amsterdam, 89-94.

197. Hinder, S.L., G.C. Hays, M. Edwards, E.C. Roberts, A.W. Walne, and M.B. Gravenor, 2012: Changes in marine dinoflagellate and diatom abundance under climate change. *Nature Climate Change,* **2,** 271-275. http://dx.doi.org/10.1038/nclimate1388

198. Moore, S.K., J.A. Johnstone, N.S. Banas, and E.P.S. Jr., 2015: Present-day and future climate pathways affecting *Alexandrium* blooms in Puget Sound, WA, USA. *Harmful Algae,* **48,** 1-11. http://dx.doi.org/10.1016/j.hal.2015.06.008

199. Kibler, S.R., P.A. Tester, K.E. Kunkel, S.K. Moore, and R.W. Litaker, 2015: Effects of ocean warming on growth and distribution of dinoflagellates associated with ciguatera fish poisoning in the Caribbean. *Ecological Modelling,* **316,** 194-210. http://dx.doi.org/10.1016/j.ecolmodel.2015.08.020

200. VanDerslice, J., 2011: Drinking water infrastructure and environmental disparities: Evidence and methodological considerations. *American Journal of Public Health,* **101,** S109-S114. http://dx.doi.org/10.2105/AJPH.2011.300189

201. Balazs, C.L. and I. Ray, 2014: The drinking water disparities framework: On the origins and persistence of inequities in exposure. *American Journal of Public Health,* **104,** 603-610. http://dx.doi.org/10.2105/AJPH.2013.301664

202. Heaney, C.D., S. Wing, S.M. Wilson, R.L. Campbell, D. Caldwell, B. Hopkins, S. O'Shea, and K. Yeatts, 2013: Public infrastructure disparities and the microbiological and chemical safety of drinking and surface water supplies in a community bordering a landfill. *Journal of Environmental Health,* **75,** 24-36. PMC4514614

203. Wilson, S.M., C.D. Heaney, and O. Wilson, 2010: Governance structures and the lack of basic amenities: Can community engagement be effectively used to address environmental in justice in underserved Black communities? *Environmental Justice,* **3,** 125-133. http://dx.doi.org/10.1089/env.2010.0014

204. Jepson, W., 2014: Measuring 'no-win' waterscapes: Experience-based scales and classification approaches to assess household water security in *colonias* on the US-Mexico border. *Geoforum,* **51,** 107-120. http://dx.doi.org/10.1016/j.geoforum.2013.10.002

205. Wescoat, J.L., Jr.,, L. Headington, and R. Theobald, 2007: Water and poverty in the United States. *Geoforum,* **38,** 801-814. http://dx.doi.org/10.1016/j.geoforum.2006.08.007

206. Hennessy, T.W., T. Ritter, R.C. Holman, D.L. Bruden, K.L. Yorita, L. Bulkow, J.E. Cheek, R.J. Singleton, and J. Smith, 2008: The relationship between in-home water service and the risk of respiratory tract, skin, and gastrointestinal tract infections among rural Alaska natives. *American Journal of Public Health,* **98,** 2072-2078. http://dx.doi.org/10.2105/ajph.2007.115618

207. Furth, D.P., 2010: What's in the water? Climate change, waterborne pathogens, and the safety of the rural Alaskan water supply. *Hastings West-Northwest Journal of Environmental Law and Policy,* **16,** 251-276.

208. Evengard, B., J. Berner, M. Brubaker, G. Mulvad, and B. Revich, 2011: Climate change and water security with a focus on the Arctic. *Global Health Action,* **4.** http://dx.doi.org/10.3402/gha.v4i0.8449

209. Lane, K., K. Charles-Guzman, K. Wheeler, Z. Abid, N. Graber, and T. Matte, 2013: Health effects of coastal storms and flooding in urban areas: A review and vulnerability assessment. *Journal of Environmental and Public Health,* **2013,** Article ID 913064. http://dx.doi.org/10.1155/2013/913064

210. Xu, Z., P.E. Sheffield, W. Hu, H. Su, W. Yu, X. Qi, and S. Tong, 2012: Climate change and children's health—A call for research on what works to protect children. *International Journal of Environmental Research and Pulic Health*, **9**, 3298-3316. http://dx.doi.org/10.3390/ijerph9093298

211. Bernstein, A.S. and S.S. Myers, 2011: Climate change and children's health. *Current Opinion in Pediatrics*, **23**, 221-226. http://dx.doi.org/10.1097/MOP.0b013e3283444c89

212. Kistin, E.J., J. Fogarty, R.S. Pokrasso, M. McCally, and P.G. McCornick, 2010: Climate change, water resources and child health. *Archives of Disease in Childhood*, **95**, 545-549. http://dx.doi.org/10.1136/adc.2009.175307

213. Lopman, B.A., A.J. Hall, A.T. Curns, and U.D. Parashar, 2011: Increasing rates of gastroenteritis hospital discharges in US adults and the contribution of norovirus, 1996-2007. *Clinical Infectious Diseases*, **52**, 466-474. http://dx.doi.org/10.1093/cid/ciq163

214. Rylander, C., J.O. Odland, and T.M. Sandanger, 2013: Climate change and the potential effects on maternal and pregnancy outcomes: An assessment of the most vulnerable--the mother, fetus, and newborn child. *Global Health Action*, **6**, 19538. http://dx.doi.org/10.3402/gha.v6i0.19538

215. CDC, 2012: Cryptosporidiosis Surveillance —United States, 2009–2010 and Giardiasis Surveillance —United States, 2009–2010. *MMWR Surveillance Summaries*, **61(5)**, 1-23. http://www.cdc.gov/mmwr/pdf/ss/ss6105.pdf

216. Judd, N.L., C.H. Drew, C. Acharya, Marine Resources for Future Generations, T.A. Mitchell, J.L. Donatuto, G.W. Burns, T.M. Burbacher, and E.M. Faustman, 2005: Framing scientific analyses for risk management of environmental hazards by communities: Case studies with seafood safety issues. *Environmental Health Perspectives*, **113**, 1502-1508. http://dx.doi.org/10.1289/ehp.7655

217. Donatuto, J.L., T.A. Satterfield, and R. Gregory, 2011: Poisoning the body to nourish the soul: Prioritising health risks and impacts in a Native American community. *Health, Risk & Society*, **13**, 103-127. http://dx.doi.org/10.1080/136985 75.2011.556186

218. Lefebvre, K.A. and A. Robertson, 2010: Domoic acid and human exposure risks: A review. *Toxicon*, **56**, 218-230. http://dx.doi.org/10.1016/j.toxicon.2009.05.034

219. Lewitus, A.J., R.A. Horner, D.A. Caron, E. Garcia-Mendoza, B.M. Hickey, M. Hunter, D.D. Huppert, R.M. Kudela, G.W. Langlois, J.L. Largier, E.J. Lessard, R. RaLonde, J.E.J. Rensel, P.G. Strutton, V.L. Trainer, and J.F. Tweddle, 2012: Harmful algal blooms along the North American west coast region: History, trends, causes, and impacts. *Harmful Algae*, **19**, 133-159. http://dx.doi.org/10.1016/j.hal.2012.06.009

220. Yoder, J.S., S. Straif-Bourgeois, S.L. Roy, T.A. Moore, G.S. Visvesvara, R.C. Ratard, V.R. Hill, J.D. Wilson, A.J. Linscott, R. Crager, N.A. Kozak, R. Sriram, J. Narayanan, B. Mull, A.M. Kahler, C. Schneeberger, A.J. da Silva, M. Poudel, K.L. Baumgarten, L. Xiao, and M.J. Beach, 2012: Primary amebic meningoencephalitis deaths associated with sinus irrigation using contaminated tap water. *Clinical Infectious Diseases*, **55**, e79-e85. http://dx.doi.org/10.1093/cid/cis626

221. Kemble, S.K., R. Lynfield, A.S. DeVries, D.M. Drehner, W.F. Pomputius, M.J. Beach, G.S. Visvesvara, A.J. da Silva, V.R. Hill, J.S. Yoder, L. Xiao, K.E. Smith, and R. Danila, 2012: Fatal *Naegleria fowleri* infection acquired in Minnesota: Possible expanded range of a deadly thermophilic organism. *Clinical Infectious Diseases*, **54**, 805-809. http://dx.doi.org/10.1093/cid/cir961

222. Goudot, S., P. Herbelin, L. Mathieu, S. Soreau, S. Banas, and F. Jorand, 2012: Growth dynamic of Naegleria fowleri in a microbial freshwater biofilm. *Water Research*, **46**, 3958-3966. http://dx.doi.org/10.1016/j.watres.2012.05.030

223. Puzon, G.J., J.A. Lancaster, J.T. Wylie, and J.J. Plumb, 2009: Rapid detection of Naegleria fowleri in water distribution pipeline biofilms and drinking water supplies. *Environmental Science & Technology*, **43**, 6691-6696. http://dx.doi.org/10.1021/es900432m

224. Cope, J.R., R.C. Ratard, V.R. Hill, T. Sokol, J.J. Causey, J.S. Yoder, G. Mirani, B. Mull, K.A. Mukerjee, J. Narayanan, M. Doucet, Y. Qvarnstrom, C.N. Poole, O.A. Akingbola, J. Ritter, Z. Xiong, A. da Silva, D. Roellig, R. Van Dyke, H. Stern, L. Xiao, and M.J. Beach, 2015: First association of a primary amebic meningoencephalitis death with culturable Naegleria fowleri in tap water from a US treated public drinking water system. *Clinical Infectious Diseases*, **60**, e36-e42. http://dx.doi.org/10.1093/cid/civ017

225. Carrie, J., F. Wang, H. Sanei, R.W. Macdonald, P.M. Outridge, and G.A. Stern, 2010: Increasing contaminant burdens in an arctic fish, Burbot (*Lota lota*), in a warming climate. *Environmental Science & Technology*, **44**, 316-322. http://dx.doi.org/10.1021/es902582y

226. Balbus, J.M., A.B. Boxall, R.A. Fenske, T.E. McKone, and L. Zeise, 2013: Implications of global climate change for the assessment and management of human health risks of chemicals in the natural environment. *Environmental Toxicology and Chemistry*, **32**, 62-78. http://dx.doi.org/10.1002/etc.2046

7 FOOD SAFETY, NUTRITION, AND DISTRIBUTION

Lead Authors

Lewis Ziska
U.S. Department of Agriculture

Allison Crimmins*
U.S. Environmental Protection Agency

Contributing Authors

Allan Auclair
U.S. Department of Agriculture

Stacey DeGrasse
U.S. Food and Drug Administration

Jada F. Garofalo
Centers for Disease Control and Prevention

Ali S. Khan
University of Nebraska Medical Center

Irakli Loladze
Bryan College of Health Sciences

Adalberto A. Pérez de León
U.S. Department of Agriculture

Allan Showler
U.S. Department of Agriculture

Jeanette Thurston
U.S. Department of Agriculture

Isabel Walls
U.S. Department of Agriculture

Acknowledgements: Steve Gendel, Formerly of the U.S. Food and Drug Administration

Recommended Citation: Ziska, L., A. Crimmins, A. Auclair, S. DeGrasse, J.F. Garofalo, A.S. Khan, I. Loladze, A.A. Pérez de León, A. Showler, J. Thurston, and I. Walls, 2016: Ch. 7: Food Safety, Nutrition, and Distribution. *The Impacts of Climate Change on Human Health in the United States: A Scientific Assessment.* U.S. Global Change Research Program, Washington, DC, 189–216. http://dx.doi.org/10.7930/J0ZP4417

On the web: health2016.globalchange.gov

*Chapter Coordinators

7 FOOD SAFETY, NUTRITION, AND DISTRIBUTION

Key Findings

Increased Risk of Foodborne Illness
Key Finding 1: Climate change, including rising temperatures and changes in weather extremes, is expected to increase the exposure of food to certain pathogens and toxins *[Likely, High Confidence]*. This will increase the risk of negative health impacts *[Likely, Medium Confidence]*, but actual incidence of foodborne illness will depend on the efficacy of practices that safeguard food in the United States *[High Confidence]*.

Chemical Contaminants in the Food Chain
Key Finding 2: Climate change will increase human exposure to chemical contaminants in food through several pathways *[Likely, Medium Confidence]*. Elevated sea surface temperatures will lead to greater accumulation of mercury in seafood *[Likely, Medium Confidence]*, while increases in extreme weather events will introduce contaminants into the food chain *[Likely, Medium Confidence]*. Rising carbon dioxide concentrations and climate change will alter incidence and distribution of pests, parasites, and microbes *[Very Likely, High Confidence]*, leading to increases in the use of pesticides and veterinary drugs *[Likely, Medium Confidence]*.

Rising Carbon Dioxide Lowers Nutritional Value of Food
Key Finding 3: The nutritional value of agriculturally important food crops, such as wheat and rice, will decrease as rising levels of atmospheric carbon dioxide continue to reduce the concentrations of protein and essential minerals in most plant species *[Very Likely, High Confidence]*.

Extreme Weather Limits Access to Safe Foods
Key Finding 4: Increases in the frequency or intensity of some extreme weather events associated with climate change will increase disruptions of food distribution by damaging existing infrastructure or slowing food shipments *[Likely, High Confidence]*. These impediments lead to increased risk for food damage, spoilage, or contamination, which will limit availability of and access to safe and nutritious food depending on the extent of disruption and the resilience of food distribution infrastructure *[Medium Confidence]*.

7.1 Introduction

A safe and nutritious food supply is a vital component of food security. Food security, in a public health context, can be summarized as permanent access to a sufficient, safe, and nutritious food supply needed to maintain an active and healthy lifestyle.[1]

The impacts of climate change on food production, prices, and trade for the United States and globally have been widely examined, including in the U.S. Global Change Research Program (USGCRP) report, "Climate Change, Global Food Security, and the U.S. Food System," in the most recent Intergovernmental Panel on Climate Change report, and elsewhere.[1, 2, 3, 4, 5, 6, 7] An overall finding of the USGCRP report was that "climate change is very likely to affect global, regional, and local food security by disrupting food availability, decreasing access to food, and making utilization more difficult."[1]

This chapter focuses on some of the less reported aspects of food security, specifically, the impacts of climate change on food safety, nutrition, and distribution in the context of human health in the United States. While ingestion of contaminated seafood is discussed in this chapter, details on the exposure pathways of water related pathogens (for example, through recreational or drinking water) are discussed in Chapter 6: Water-Related Illness.

Systems and processes related to food safety, nutrition, and production are inextricably linked to their physical and biological environment.[5, 8] Although production is important, for most developed countries such as the United States, food shortages are uncommon; rather, nutritional quality and food safety are the primary health concerns.[5, 9] Certain populations, such as the poor, children, and Indigenous populations, may be more vulnerable to climate impacts on food safety, nutrition, and distribution (see also Ch. 9: Populations of Concern).

Farm to Table
The Potential Interactions of Rising CO₂ and Climate Change on Food Safety and Nutrition

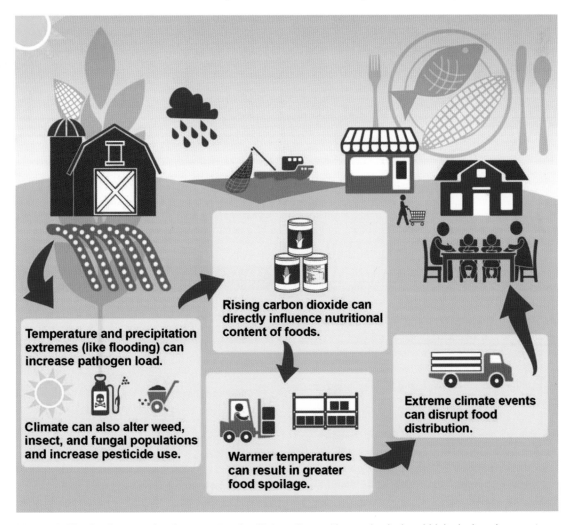

Figure 1: The food system involves a network of interactions with our physical and biological environments as food moves from production to consumption, or from "farm to table." Rising CO₂ and climate change will affect the quality and distribution of food, with subsequent effects on food safety and nutrition.

Climate Change and Health—*Salmonella*

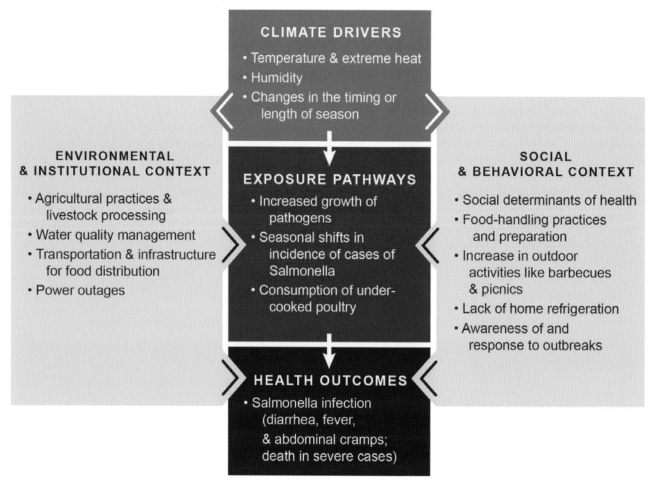

Figure 2: This conceptual diagram for a *Salmonella* example illustrates the key pathways by which humans are exposed to health threats from climate drivers, and potential resulting health outcomes (center boxes). These exposure pathways exist within the context of other factors that positively or negatively influence health outcomes (gray side boxes). Key factors that influence vulnerability for individuals are shown in the right box, and include social determinants of health and behavioral choices. Key factors that influence vulnerability at larger scales, such as natural and built environments, governance and management, and institutions, are shown in the left box. All of these influencing factors can affect an individual's or a community's vulnerability through changes in exposure, sensitivity, and adaptive capacity and may also be affected by climate change. See Ch. 1: Introduction for more information.

There are two overarching means by which increasing carbon dioxide (CO_2) and climate change alter safety, nutrition, and distribution of food. The first is associated with rising global temperatures and the subsequent changes in weather patterns and extreme climate events.[13, 14, 15] Current and anticipated changes in climate and the physical environment have consequences for contamination, spoilage, and the disruption of food distribution.

The second pathway is through the direct CO_2 "fertilization" effect on plant photosynthesis. Higher concentrations of CO_2 stimulate growth and carbohydrate production in some plants, but can lower the levels of protein and essential minerals in a number of widely consumed crops, including wheat, rice, and potatoes, with potentially negative implications for human nutrition.[16]

Terminology

Food Safety – Those conditions and measures necessary for food production, processing, storage, and distribution in order to ensure a safe, sound, wholesome product that is fit for human consumption.[10]

Foodborne Illness or Disease – Foodborne illness (sometimes called "food poisoning") is a common public health problem. Each year, one in six Americans reports getting sick by consuming contaminated foods or beverages.[11] Foodborne disease is caused by ingestion of contaminated food. Many different disease-causing microbes, or pathogens, can contaminate foods, so there are many different foodborne infections. In addition, food contaminated by toxins or chemicals can also result in foodborne illness.[12]

7.2 Food Safety

Although the United States has one of the safest food supplies in the world,[17] food safety remains an important public health issue. In the United States, the Centers for Disease Control and Prevention (CDC) estimate that there are 48 million cases of foodborne illnesses per year, with approximately 3,000 deaths.[12] As climate change drives changes in environmental variables such as ambient temperature, precipitation, and weather extremes (particularly flooding and drought), increases in foodborne illnesses are expected.[18, 19]

Most acute illnesses are caused by foodborne viruses (specifically *noroviruses*), followed by bacterial pathogens (such as *Salmonella*; see Table 1). Of the common foodborne illnesses in the United States, most deaths are caused by *Salmonella*, followed by the parasite *Toxoplasma gondii*.[20, 21, 22, 23] In addition, climate change impacts on the transport of chemical contaminants or accumulation of pesticides or heavy metals (such as mercury) in food, can also represent significant health threats in the food chain.[22, 24, 25, 26, 27, 28]

How Climate Affects Food Safety

Climate already influences food safety within an agricultural system—prior to, during, and after the harvest, and during transport, storage, preparation, and consumption. Changes in climate factors, such as temperature, precipitation, and extreme weather are key drivers of pathogen introduction, food contamination, and foodborne disease, as well as changes in the level of exposure to specific contaminants and chemical residues for crops and livestock.[29, 30, 31]

The impact of climate on food safety occurs through multiple pathways. Changes in air and water temperatures, weather-related changes, and extreme events can shift the seasonal and geographic occurrence of bacteria, viruses, pests, parasites, fungi, and other chemical contaminants.[23, 30, 31, 32, 33] For example:

- Higher temperatures can increase the number of pathogens already present on produce[34] and seafood.[35, 36]

Seasonality of Human Illnesses Associated With Foodborne Pathogens

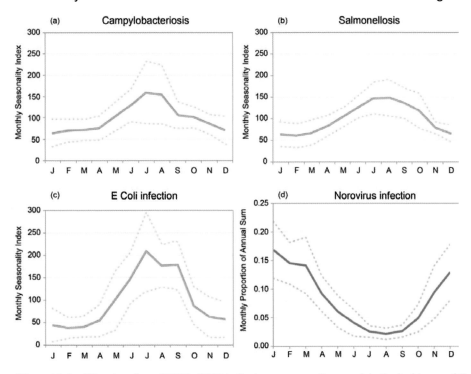

Figure 3: A review of the published literature from 1960 to 2010 indicates a summertime peak in the incidence of illnesses associated with infection from a) *Campylobacter*, b) *Salmonella*, and c) *Escherichia coli* (*E. coli*). For these three pathogens, the monthly seasonality index shown here on the y-axis indicates the global disease incidence above or below the yearly average, which is denoted as 100. For example, a value of 145 for the month of July for Salmonellosis would mean that the proportion of cases for that month was 45% higher than the 12 month average. Unlike these three pathogens, incidence of norovirus, which can be attained through food, has a wintertime peak. The y-axis of the norovirus incidence graph (d) uses a different metric than (a–c): the monthly proportion of the annual sum of norovirus cases in the northern hemisphere between 1997 and 2011. For example, a value of 0.12 for March would indicate that 12% of the annual cases occurred during that month). Solid line represents the average; confidence intervals (dashed lines) are plus and minus one standard deviation. (Figure sources: a, b, and c: adapted from Lal et al. 2012; d: Ahmed et al. 2013)[49, 183]

Table 1. Foodborne Illness and Climate Change

Foodborne Hazard	Symptoms	Estimated Annual Illness and Disease	Other Climate Drivers	Temperature/ Humidity Relationship
Norovirus	Vomiting, non-bloody diarrhea with abdominal pain, nausea, aches, low grade fever	• 5,500,000 illnesses • 15,000 hospitalizations • 150 deaths	Extreme weather events (such as heavy precipitation and flooding)	**Pathogens Favoring Colder/ Dryer Conditions**
Listeria monocytogene	Fever, muscle aches, and rarely diarrhea. Intensive infection can lead to miscarriage, stillbirth, premature delivery, or life-threatening infections (meningitis).	• 1,600 illnesses • 1,500 hospitalizations • 260 deaths		
Toxoplasma	Minimal to mild illness with fever, serious illness in rare cases. Inflammation of the brain and infection of other organs, birth defects.	• 87,000 illnesses • 4,400 hospitalizations • 330 deaths		
Campylobacter	Diarrhea, cramping, abdominal pain, nausea, and vomiting. In serious cases can be life-threatening.	• 850,000 illnesses • 8,500 hospitalizations • 76 deaths	Changes in the timing or length of seasons, precipitation and flooding	
Salmonella spp. (non typhoidal)	Diarrhea, fever, and abdominal cramps; in severe cases death.	• 1,000,000 illnesses • 19,000 hospitalizations • 380 deaths	Extreme weather events, changes in the timing or length of seasons	
Vibrio vulnificus and parahaemolyticus	When ingested: watery diarrhea often with abdominal cramping, nausea, vomiting, fever and chills. Can cause liver disease. When exposed to an open wound: infection of the skin.	• 35,000 illnesses • 190 hospitalizations • 40 deaths	Sea surface temperature, extreme weather events	
Escherichia coli (E coli)	E. coli usually causes mild diarrhea. More severe pathogenic types, such as enterohemorrhagic E. Coli (EHEC), are associated with hemolytic uremic syndrome (a toxin causing destruction of red blood cells, leading to kidney failure).	• 200,000 illnesses • 2,400 hospitalizations • 20 deaths	Extreme weather events, changes in the timing or length of seasons	**Pathogens Favoring Warmer/ Wetter Conditions**

Estimated annual number of foodborne illnesses and deaths in the United States. (Adapted from Scallan et al. 2011; Akil et al. 2014; Kim et al. 2015; Lal et al. 2012)[20, 48, 49, 80]

- Bacterial populations can increase during food storage which, depending on time and temperature, can also increase food spoilage rates.[37]

- Sea surface temperature is directly related to seafood exposure to pathogens (see Ch. 6: Water-Related Illness).[38, 39, 40]

- Precipitation has been identified as a factor in the contamination of irrigation water and produce,[30, 31, 33, 41] which has been linked to foodborne illness outbreaks.[42, 43]

- Extreme weather events like dust storms or flooding can introduce toxins to crops during development (see Ch. 4: Extreme Events).[44]

- Changing environmental conditions and soil properties may result in increases in the incidence of heavy metals in the food supply.[45, 46, 47]

Climate Impacts on Pathogen Prevalence

While climate change affects the prevalence of pathogens harmful to human health, the extent of exposure and resulting illness will depend on individual and institutional sensitivity and adaptive capacity, including human behavior and the effectiveness of food safety regulatory, surveillance, monitoring, and communication systems.

Rising Temperature and Humidity

Climate change will influence the fate, transport, transmission, viability, and multiplication rate of pathogens in the food chain. For example, increases in average global temperatures and humidity will lead to changes in the geographic range, seasonal occurrence, and survivability of certain pathogens.[9, 48, 49, 50]

Ongoing changes in temperature and humidity will not affect all foodborne pathogens equally (Table 1). The occurrence of some pathogens, such as *Salmonella, Escherichia coli (E. coli)*, and *Campylobacter*, could increase with climate change because these pathogens thrive in warm, humid conditions. For example, *Salmonella* on raw chicken will double in number approximately every hour at 70°F, every 30 minutes at 80°F, and every 22 minutes at 90°F.[51, 52]

There is a summertime peak in the incidence of illnesses associated with these specific pathogens (see Figure 3).[18, 48, 53, 54] This peak may be related not only to warmer temperatures favoring pathogen growth but also to an increase in outdoor activities, such as barbecues and picnics. Risk for foodborne illness is higher when food is prepared outdoors where the safety controls that a kitchen provides—thermostat-controlled cooking, refrigeration, and washing facilities—are usually not available.[5, 18, 19, 48, 55, 56]

Norovirus, the most common cause of stomach flu, can be transmitted by consumption of contaminated food. Although norovirus generally has a winter seasonal peak (see Figure 3), changing climate parameters, particularly temperature and rainfall, may influence its incidence and spread. Overall, localized climate impacts could improve health outcomes (fewer cases during warmer winters) or worsen them (elevated transmission during floods), such that projected trends in overall health outcomes for norovirus remain unclear.[48, 57]

Rising ocean temperatures can increase the risk of pathogen exposure from ingestion of contaminated seafood. For example, significantly warmer coastal waters in Alaska from 1997 to 2004 were associated with an outbreak in 2004 of *Vibrio parahaemolyticus*, a bacterium that causes gastrointestinal illnesses when contaminated seafood is ingested.[58] *Vibrio parahaemolyticus* is one of the leading causes of seafood-related gastroenteritis in the United States and is associated with the consumption of raw oysters harvested from warm-water estuaries.[59] Similarly, the emergence of a related bacterium, *Vibrio vulnificus*, may also be associated with high water temperatures.[40] While increasing average water temperatures were implicated in a 2004 outbreak,[58] ambient air temperature also affects pathogen levels of multiple species of *Vibrio* in shellfish.[35, 36] For example, *Vibrio vulnificus* may increase 10- to 100-fold when oysters are stored at ambient temperatures for ten hours before refrigeration.[60] Increases in ambient ocean water and air temperatures would accelerate *Vibrio* growth in shellfish, potentially necessitating changes in post-harvest controls to minimize the increased risk of exposure. (For more information on *Vibrio* and other water-related pathogens, including contamination of recreational and drinking water, see Ch. 6: Water-Related Illness).

Finally, climate change is projected to result in warmer winters, earlier springs, and an increase in the overall growing season in many regions.[61, 62] While there are potential food production benefits from such changes, warmer and longer growing seasons could also alter the timing and occurrence of pathogen transmissions in food and the chance of human exposure.[63, 64, 65]

Extreme Events

In addition to the effects of increasing average temperature and humidity on pathogen survival and growth, increases in temperature and precipitation extremes can contribute to changes in pathogen transmission, multiplication, and survivability. More frequent and severe heavy rainfall events can in-

> *Climate change will influence the fate, transport, transmission, viability, and multiplication rate of pathogens in the food chain.*

Crops Susceptible to Mycotoxin Infections

Climate change will expand the geographical range where mold growth and mycotoxin production occur.[9, 32, 37, 75] Corn, a major U.S. crop, is especially susceptible to mold growth and mycotoxin production.[81] Human dietary exposure to these toxins has resulted in illness and death in tropical regions, or where their presence remains unregulated.[82] In the United States, regulations are designed to prevent mycotoxins entering the food supply.

Aflatoxins (naturally occurring mycotoxins found in corn) are known carcinogens and can also cause impaired development in children, immune suppression, and, with severe exposure, death.[82, 83, 84] Recent models show that aflatoxin contamination in corn may increase with climate change in Europe.[85] Other commodities susceptible to contamination by mycotoxins include peanuts, cereal grains, and fruit.[37]

crease infection risk from most pathogens, particularly when it leads to flooding.[66] Flooding, and other weather extremes, can increase the incidence and levels of pathogens in food production, harvesting, and processing environments. Groundwater and surface water used for irrigation, harvesting, and washing can be contaminated with runoff or flood waters that carry partially or untreated sewage, manure, or other wastes containing foodborne contaminants.[55, 67, 68, 69, 70, 71] The level of *Salmonella* in water is elevated during times of monthly maximum precipitation in the summer and fall months;[56, 72] consequently the likelihood of *Salmonella* in water may increase in regions experiencing increased total or heavy precipitation events.

Water is also an important factor in food processing. Climate and weather extremes, such as flooding or drought, can reduce water quality and increase the risk of pathogen transfer during the handling and storage of food following harvest.[9]

The direct effect of drought on food safety is less clear. Dry conditions can pose a risk for pathogen transmission due to reduced water quality, increased risk of runoff when rains

do occur, and increased pathogen concentration in reduced water supplies if such water is used for irrigation, food processing, or livestock management.[29, 31, 55, 73] Increasing drought generally leads to an elevated risk of exposure to pathogens such as norovirus and *Cryptosporidium*.[66] However, drought and extreme heat events could also decrease the survivability of certain foodborne pathogens, affecting establishment and transmission, and thus reducing human exposure.[66, 74]

Mycotoxins and Phycotoxins

Mycotoxins are toxic chemicals produced by molds that grow on crops prior to harvest and during storage. Prior to harvest, increasing temperatures and drought can stress plants, making them more susceptible to mold growth.[75] Warm and moist conditions favor mold growth directly and affect the biology of insect vectors that transmit molds to crops. Post-harvest contamination is also affected by environmental parameters, including extreme temperatures and moisture. If crops are not dried and stored at low humidity, mold growth and mycotoxin production can increase to very high levels.[76, 77]

Phycotoxins are toxic chemicals produced by certain harmful freshwater and marine algae that may affect the safety of drinking water and shellfish or other seafood. For example, the alga responsible for producing ciguatoxin (the toxin that causes the illness known as ciguatera fish poisoning) thrives in warm water (see also Ch. 6: Water-Related Illness). Projected increases in sea surface temperatures may expand the endemic range of ciguatoxin-producing algae and increase ciguatera fish poisoning incidence following ingestion.[78] Predicted increases in sea surface temperature of 4.5° to 6.3°F (2.5° to 3.5°C) could yield increases in ciguatera fish poisoning cases of 200% to 400%.[79]

Crop dusting of a corn field in Iowa.

Once introduced into the food chain, these poisonous toxins can result in adverse health outcomes, with both acute and chronic effects. Current regulatory laws and management strategies safeguard the food supply from mycotoxins and phycotoxins; however, increases in frequency and range of their prevalence may increase the vulnerability of the food safety system.

Climate Impacts on Chemical Contaminants

Climate change will affect human exposure to metals, pesticides, pesticide residues, and other chemical contaminants. However, resulting incidence of illness will depend on the genetic predisposition of the person exposed, type of contaminant, and extent of exposure over time.[86]

Metals and Other Chemical Contaminants

There are a number of environmental contaminants, such as polychlorinated biphenyls, persistent organic pollutants, dioxins, pesticides, and heavy metals, which pose a human health risk when they enter the food chain. Extreme events may facilitate the entry of such contaminants into the food chain, particularly during heavy precipitation and flooding.[45, 46, 47] For example, chemical contaminants in floodwater following Hurricane Katrina included spilled oil, pesticides, heavy metals, and hazardous waste.[47, 87]

Methylmercury is a form of mercury that can be absorbed into the bodies of animals, including humans, where it can have adverse neurological effects. Elevated water temperatures may lead to higher concentrations of methylmercury in fish and mammals.[88, 89] This is related to an increase in metabolic rates and increased mercury uptake at higher water temperatures.[28, 90, 91] Human exposure to dietary mercury is influenced by the amount of mercury ingested, which can vary with the species, age, and size of the fish. If future fish consumption patterns are unaltered, increasing ocean temperature would likely increase mercury exposure in human diets. Methylmercury exposure can affect the development of children, particularly if exposed in utero.[92]

Pesticides

Climate change is likely to exhibit a wide range of effects on the biology of plant and livestock pests (weeds, insects, and microbes). Rising minimum winter temperatures and longer growing seasons are very likely to alter pest distribution and populations.[93, 94, 95] In addition, rising average temperature and CO_2 concentration are also likely to increase the range and distribution of pests, their impact, and the vulnerability of host plants and animals.[3, 96, 97]

Pesticides are chemicals generally regulated for use in agriculture to protect plants and animals from pests; chemical management is the primary means for agricultural pest control in the United States and most developed countries. Because climate and CO_2 will intensify pest distribution and populations,[98, 99] increases in pesticide use are expected.[100, 101] In addition, the efficacy of chemical management may be reduced in the context of climate change. This decline in efficacy can reflect CO_2-induced increases in the herbicide tolerance of certain weeds or climate-induced shifts in invasive weed, insect,

Impacts of Rising CO_2 on the Nutritional Value of Crops

Protein. Protein content of major food crops is very likely to decline significantly as atmospheric CO_2 concentrations increase to between 540 and 960 parts per million (ppm),[129, 134, 135, 137] the range projected by the end of this century (see description of Representative Concentration Pathways in Appendix 1: Technical Support Document).[14] Current atmospheric concentrations of CO_2 are approximately 400 ppm.[138]

Minerals and trace elements. Rising CO_2 levels are very likely to lower the concentrations of essential micro- and macroelements such as iron, zinc, calcium, magnesium, copper, sulfur, phosphorus, and nitrogen in most plants (including major cereals and staple crops).[16, 128, 132, 133, 139, 140]

Wheat grown in southeast Washington state, August, 2008.

Ratio of major macronutrients (carbohydrates to protein). It is very likely that rising CO_2 will alter the relative proportions of major macronutrients in many crops by increasing carbohydrate content (starch and sugars) while at the same time decreasing protein content.[16] An increase in dietary carbohydrates-to-protein ratio can have unhealthy effects on human metabolism and body mass.[136, 141, 142, 143]

and plant pathogen populations[100, 102, 103, 104, 105, 106, 107, 108] as well as climate-induced changes that enhance pesticide degradation or affect coverage.[108, 109]

Increased pest pressures and reductions in the efficacy of pesticides are likely to lead to increased pesticide use, contamination in the field, and exposure within the food chain.[110] Increased exposure to pesticides could have implications for human health.[5, 29, 44] However, the extent of pesticide use and potential exposure may also reflect climate change induced choices for crop selection and land use.

Pesticide Residues

Climate change, especially increases in temperature, may be important in altering the transmission of vector-borne diseases in livestock by influencing the life cycle, range, and reproductive success of disease vectors.[8, 65] Potential changes in veterinary practices, including an increase in the use of parasiticides and other animal health treatments, are likely to be adopted to maintain livestock health in response to climate-induced changes in pests, parasites, and microbes.[5, 23, 110] This could increase the risk of pesticides entering the food chain or lead to evolution of pesticide resistance, with subsequent implications for the safety, distribution, and consumption of livestock and aquaculture products.[111, 112, 113]

Climate change may affect aquatic animal health through temperature-driven increases in disease.[114] The occurrence of increased infections in aquaculture with rising temperature has been observed for some diseases (such as *Ichthyophthirius multifiliis* and *Flavobacterium columnare*)[115] and is likely to result in greater use of aquaculture drugs.[76]

7.3 Nutrition

While sufficient *quantity* of food is an obvious requirement for food security, food *quality* is essential to fulfill basic nutritional needs. Globally, chronic dietary deficiencies of micronutrients such as vitamin A, iron, iodine, and zinc contribute to "hidden hunger," in which the consequences of the micronutrient insufficiency may not be immediately visible or easily observed. This type of micronutrient deficiency constitutes one of the world's leading health risk factors and adversely affects metabolism, the immune system, cognitive development and maturation—particularly in children. In addition, micronutrient deficiency can exacerbate the effects of diseases and can be a factor in prevalence of obesity.[116, 117, 118, 119, 120, 121]

In developed countries with abundant food supplies, like the United States, the health burden of malnutrition may not be intuitive and is often underappreciated. In the United States, although a number of foods are supplemented with nutrients, it is estimated that the diets of 38% and 45% of the population fall below the estimated average requirements for calcium and magnesium, respectively.[122] Approximately 12% of the population is at risk for zinc deficiency, including perhaps as much as 40% of the elderly.[123] In addition, nutritional deficiencies of magnesium, iron, selenium, and other essential micronutrients can occur in overweight and obese individuals, whose diets might reflect excessive intake of calories and refined carbohydrates but insufficient intake of vitamins and essential minerals.[119, 124, 125, 126]

Effects of Carbon Dioxide on Protein and Minerals

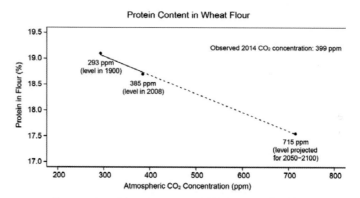

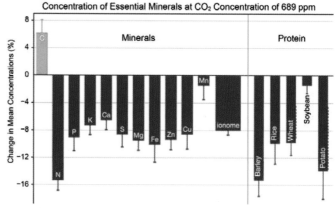

Figure 4: Direct effect of rising atmospheric carbon dioxide (CO_2) on the concentrations of protein and minerals in crops. The top figure shows that the rise in CO_2 concentration from 293 ppm (at the beginning of the last century) to 385 ppm (global average in 2008) to 715 ppm (projected to occur by 2100 under the RCP8.5 and RCP6.0 pathways),[184] progressively lowers protein concentrations in wheat flour (the average of four varieties of spring wheat). The lower figure—the average effect on 125 plant species and cultivars—shows that a doubling of CO_2 concentration from preindustrial levels diminishes the concentration of essential minerals in wild and crop plants, including ionome (all the inorganic ions present in an organism) levels, and also lowers protein concentrations in barley, rice, wheat and potato. (Figure source: Experimental data from Ziska et al. 2004 (top figure), Taub et al. 2008, and Loladze 2014 (bottom figure)).[16, 129, 134]

How Rising CO₂ Affects Nutrition

Though rising CO_2 stimulates plant growth and carbohydrate production, it reduces the nutritional value (protein and minerals) of most food crops (Figure 4).[16, 127, 128, 129, 130, 131, 132, 133] This direct effect of rising CO_2 on the nutritional value of crops represents a potential threat to human health.[16, 133, 134, 135, 136]

Protein

As CO_2 increases, plants need less protein for photosynthesis, resulting in an overall decline in protein concentration in plant tissues.[134, 135] This trend for declining protein levels is evident for wheat flour derived from multiple wheat varieties when grown under laboratory conditions simulating the observed increase in global atmospheric CO_2 concentration since 1900.[129] When grown at the CO_2 levels projected for 2100 (540–958 ppm), major food crops, such as barley, wheat, rice, and potato, exhibit 6% to 15% lower protein concentrations relative to ambient levels (315–400 ppm).[16, 134, 135] In contrast, protein content is not anticipated to decline significantly for corn or sorghum.[135]

While protein is an essential aspect of human dietary needs, the projected human health impacts of a diet including plants with reduced protein concentration from increasing CO_2 are not well understood and may not be of considerable threat in the United States, where dietary protein deficiencies are uncommon.

Micronutrients

The ongoing increase in atmospheric CO_2 is also very likely to deplete other elements essential to human health (such as calcium, copper, iron, magnesium, and zinc) by 5% to 10% in most plants.[16] The projected decline in mineral concentrations in crops has been attributed to at least two distinct effects of elevated CO_2 on plant biology. First, rising CO_2 increases carbohydrate accumulation in plant tissues, which can, in turn, dilute the content of other nutrients, including minerals. Second, high CO_2 concentrations reduce plant demands for water, resulting in fewer nutrients being drawn into plant roots.[133, 144, 145]

The ongoing increase in CO_2 concentrations reduces the amount of essential minerals per calorie in most crops, thus reducing nutrient density. Such a reduction in crop quality may aggravate existing nutritional deficiencies, particularly for populations with pre-existing health conditions (see Ch. 9: Populations of Concern).

Carbohydrate-to-Protein Ratio

Elevated CO_2 tends to increase the concentrations of carbohydrates (starch and sugars) and reduce the concentrations of protein.[134] The overall effect is a significant increase in the ratio of carbohydrates to protein in plants exposed to increasing CO_2.[16] There is growing evidence that a dietary increase in this ratio can adversely affect human metabolism[143] and body composition.[141]

7.4 Distribution and Access

A reliable and resilient food distribution system is essential for access to a safe and nutritious food supply. Access to food is characterized by transportation and availability, which are defined by infrastructure, trade management, storage requirements, government regulation, and other socioeconomic factors.[146]

The shift in recent decades to a more global food market has resulted in a greater dependency on food transport and distribution, particularly for growing urban populations. Consequently, any climate-related disturbance to food distribution and transport may have significant impacts not only on safety and quality but also on food access. The effects of climate change on each of these interfaces will differ based on geographic, social, and economic factors.[4] Ultimately, the outcome of climate-related disruptions and damages to the food transportation system will be strongly influenced by the resilience of the system, as well as the adaptive capacity of individuals, populations, and institutions.

How Extreme Events Affect Food Distribution and Access

Projected increases in the frequency or severity of some extreme events will interrupt food delivery, particularly for vulnerable transport routes.[13, 15, 147, 148] The degree of disruption is related to three factors: a) popularity of the transport pathway, b) availability of alternate routes, and c) timing or seasonality of the extreme event.[149] As an example, the food transportation system in the United States frequently moves large volumes of grain by water. In the case of an extreme weather event affecting a waterway, there are few, if any, alternate pathways for transport.[150] This presents an especially relevant risk to food access if an extreme event, like flooding or drought, coincides with times of agricultural distribution, such as the fall harvest.

Immediately following an extreme event, food supply and safety can be compromised.[150, 151, 152] Hurricanes or other storms can disrupt food distribution infrastructure, damage food supplies,[7] and limit access to safe and nutritious food, even in areas not directly affected by such events (see also Ch. 4: Extreme Events).[153] For example, the Gulf Coast transportation network is vulnerable to storm surges of 23 feet.[154] Following Hurricane Katrina in 2005, where storm surges of 25 to 28 feet were recorded along parts of the Gulf Coast, grain transportation by rail or barge was severely slowed due to physical damage to infrastructure and the displacement of employees.[151, 155] Barriers to food transport may also affect food markets, reaching consumers in the form of increased food costs.[156]

The risk for food spoilage and contamination in storage facilities, supermarkets, and homes is likely to increase due to the impacts of extreme weather events, particularly those that result in power outages, which may expose food to ambient tem-

Case Study: Extreme Drought and the Mississippi River, 2012

Low water conditions on Mississippi River near St. Louis, MO, on December 5, 2012. Photo source: St. Louis District, U.S. Army Corps of Engineers.

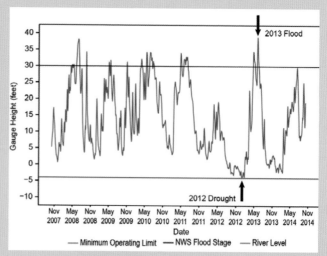

Figure 5: Mississippi River gauge height at St. Louis, MO, from October 2007 through October 2014 showing low water conditions during the 2012 drought and water levels above flood stage in 2013. (Figure source: adapted from USGS 2015)[185]

The summer (June through August) of 2012 was the second hottest on record for the contiguous United States.[159] High temperatures and a shortage of rain led to one of the most severe summer droughts the nation has seen and posed serious impacts to the Mississippi River watershed, a major transcontinental shipping route for Midwestern agriculture.[160, 161] This drought resulted in significant food and economic losses due to reductions in barge traffic, the volume of goods carried, and the number of Americans employed by the tugboat industry.[162] The 2012 drought was immediately followed by flooding throughout the Mississippi in the spring of 2013, which also resulted in disruptions of barge traffic and food transport. These swings in precipitation, from drought to flooding, are consistent with projected increases in the frequency or severity of some types of extreme weather under continued climate change.[7, 62, 152]

peratures inadequate for safe storage.[152] Storm-related power grid disruptions have steadily increased since 2000.[157] Between 2002 and 2012, extreme weather caused 58% of power outage events, 87% of which affected 50,000 or more customers.[157] Power outages are often linked to an increase in illness. For example, in August of 2003, a sudden power outage affected over 60 million people in the northeastern United States and Canada. New York City's Department of Health and Mental Hygiene detected a statistically significant citywide increase in diarrheal illness resulting from consumption of spoiled foods due to lost refrigeration capabilities.[158]

7.5 Populations of Concern

Climate change, combined with other social, economic, and political conditions, may increase the vulnerability of many different populations to food insecurity or food-related illness.[163] However, not all populations are equally vulnerable.[7, 62] Infants and young children, pregnant women, the elderly, low-income populations, agricultural workers, and those with weakened immune systems or who have underlying medical conditions are more susceptible to the effects of climate change on food safety, nutrition, and access.

Children may be especially vulnerable because they eat more food by body weight than adults, and do so during important stages of physical and mental growth and development. Children are also more susceptible to severe infection or complications from *E. coli* infections, such as hemolytic uremic syndrome.[164, 165, 166] Agricultural field workers, especially pesticide applicators, may experience increased exposure as pesticide applications increase with rising pest loads, which could also lead to higher pesticide levels in the children of these field workers.[167, 168] People living in low-income urban areas, those with limited access to supermarkets,[169, 170] and the elderly may have difficulty accessing safe and nutritious food after disruptions associated with extreme weather events. Climate change will also affect U.S. Indigenous peoples' access to both wild and cultivated traditional foods associated with their nutrition, cultural practices, local economies, and community health[171] (see also Ch. 6 Water-Related Illness and Ch. 9: Populations of Concern). All of the health impacts described in this chapter can have significant consequences on mental health and well-being (see Ch. 8 Mental Health).

7.6 Emerging Issues

Climate and food allergies. Food allergies in the United States currently affect between 1% and 9% of the population,[172] but have increased significantly among children under age 18 since 1997.[173] Rising CO_2 levels can reduce protein content and alter protein composition in certain plants, which has the potential to alter allergenic sensitivity. For example, rising CO_2 has been shown to increase the concentration of the *Amb a 1* protein—the allergenic protein most associated with ragweed pollen.[174] However, at present, the question of how rising levels of CO_2 and climate change affect allergenic properties of food is uncertain and requires more research.[175]

Heavy metals. Arsenic and other heavy metals occur naturally in some groundwater sources.[176] Climate change can exacerbate drought and competition for water, resulting in the use of poorer-quality water sources.[177, 178] Because climate and rising CO_2 levels can also influence the extent of water loss through the crop canopy, poorer water quality could lead to changes in the concentrations of arsenic and potentially other heavy metals (like cadmium and selenium) in plant tissues. Additional information is needed to determine how rising levels of CO_2 and climate change affect heavy metal accumulation in food and the consequences for human exposure.

Zoonosis and livestock. Zoonotic diseases, which are spread from animals to humans, can be transmitted through direct contact with an infected animal or through the consumption of contaminated food or water. Climate change could potentially increase the rate of zoonoses, through environmental change that alters the biology or evolutionary rate of disease vectors or the health of animal hosts. The impact of rising levels of CO_2 and climate change on the transmission of disease through zoonosis remains a fundamental issue of potential global consequence.

Foodborne pathogen contamination of fresh produce by insect vectors. Climate change will alter the range and distribution of insects and other microorganisms that can transmit bacterial pathogens such as *Salmonella* to fresh produce.[179, 180, 181] Additional information is needed regarding the role of climate change on the transmission to and development of food pathogens through insect vectors.

7.7 Research Needs

In addition to the emerging issues identified above, the authors highlight the following potential areas for additional scientific and research activity on food safety, nutrition and distribution, based on their review of the literature. Understanding climate change impacts in the context of the current food safety infrastructure will be improved by enhanced surveillance of foodborne diseases and contaminant levels, improved understanding of CO_2 impacts on nutritional quality of food, and more accurate models of the impacts of extreme events on food access and delivery.

Future assessments can benefit from research activities that:

- synthesize and assess efforts to identify and respond to current and projected food safety concerns and their impacts on human health within the existing and future food safety infrastructure;

- develop, test, and expand integrated assessment models to enhance understanding of climate and weather variability, particularly extreme events, and the role of human responses, including changes in farming technology and management, on health risks within the food chain; and

- examine the impacts of rising CO_2 and climate change on human and livestock nutritional needs, as well as the impacts of changing nutritional sources on disease vulnerability.[1]

Supporting Evidence

The chapter was developed through technical discussions of relevant evidence and expert deliberation by the report authors at several workshops, teleconferences, and email exchanges. The authors considered inputs and comments submitted by the public, the National Academies of Sciences, and Federal agencies. For additional information on the overall report process, see Appendices 2 and 3. The author team also engaged in targeted consultations during multiple exchanges with contributing authors, who provided additional expertise on subsets of the Traceable Accounts associated with each Key Finding.

Because the impacts of climate change on food production, prices, and trade for the United States and globally have been widely examined elsewhere, including in the most recent report from the Intergovernmental Panel on Climate Change,[2, 3, 4, 5, 6, 7] this chapter focuses only on the impacts of climate change on food safety, nutrition, and distribution in the context of human health in the United States. Many nutritional deficiencies and food-related illnesses are of critical importance globally, particularly those causing diarrheal epidemics or mycotoxin poisoning, and affect U.S. interests abroad; but the primary focus of this chapter is to address climate impacts on the food safety concerns most important in the United States. Thus, the literature cited in this chapter is specific to the United States or of demonstrated relevance to developed countries. The placement of health threats from seafood was determined based on pre- and post-ingestion risks: while ingestion of contaminated seafood is discussed in this chapter, details on the exposure pathways of water-related pathogens (for example, through recreational or drinking water) are discussed in Chapter 6: Water-Related Illness.

KEY FINDING TRACEABLE ACCOUNTS

Increased Risk of Foodborne Illness

Key Finding 1: Climate change, including rising temperatures and changes in weather extremes, is expected to increase the exposure of food to certain pathogens and toxins *[Likely, High Confidence]*. This will increase the risk of negative health impacts *[Likely, Medium Confidence]*, but actual incidence of foodborne illness will depend on the efficacy of practices that safeguard food in the United States *[High Confidence]*.

Description of evidence base

Multiple lines of research have shown that changes in weather extremes, such as increased extreme precipitation (leading to flooding and runoff events), can result in increased microbial and chemical contamination of crops and water in agricultural environments, with increases in human exposure.[55, 56, 72] During times of drought, plants become weaker and more susceptible to stress, which can result in mold growth and mycotoxin production if plants are held in warm, moist environments.[32, 75]

While studies that link climate change to specific outbreaks of foodborne illness are limited, numerous studies have documented that many microbial foodborne illnesses increase with increasing ambient temperature.[18, 19] There is very strong evidence that certain bacteria grow more rapidly at higher temperatures and can increase the prevalence of pathogens and toxins in food.[32, 34, 54] Case studies have demonstrated that lack of refrigerated storage, particularly during very warm weather, leads to increases in microbial growth and higher exposure to pathogens.[5, 18, 19, 48, 60]

Major uncertainties

Concentrations of pathogens and toxins in food are expected to increase, resulting in an increase in the risk of human exposure to infectious foodborne pathogens and toxins. However, the number or severity of foodborne illnesses due to climate change is uncertain. Much of this uncertainty is due to having controls in place to protect public health. For example, contaminated crops are likely to be destroyed before consumption, and certain pathogens in food, like mycotoxins, are highly regulated in the United States. Consequently, the extent of exposure and foodborne illness will depend on regulatory, surveillance, monitoring, and communication systems, and on how, and to what extent, climate change alters these adaptive capacities. Furthermore, for certain pathogens, it is not yet clear whether the impact of climate change on a pathogen will be positive or negative. For example, climate change could lead to fewer cases of norovirus infection in the winter, but worsening health outcomes are also possible due to elevated transmission of norovirus during floods. Similarly drought can reduce water quality, increase runoff, and increase pathogen concentration, but can also decrease the survivability of certain foodborne pathogens.

Assessment of confidence and likelihood based on evidence

There is *high confidence* that rising temperature and increases in flooding, runoff events, and drought will *likely* lead to increases in the occurrence and transport of pathogens in agricultural environments, which will increase the risk of food contamination and human exposure to pathogens and toxins. However, the actual prevalence of disease will depend on the response of regulatory systems and, for certain pathogens, the relative importance of multiple climate drivers with opposing impacts on exposure. Thus there is *medium confidence* that these impacts of climate change on exposure to pathogens and toxins will *likely* lead to negative health outcomes. There is a *high confidence* that the actual incidence of foodborne illness will depend on the efficacy of practices that safeguard food in the United States.

Chemical Contaminants in the Food Chain

Key Finding 2: Climate change will increase human exposure to chemical contaminants in food through several pathways *[Likely, Medium Confidence]*. Elevated sea surface temperatures will lead to greater accumulation of mercury in seafood *[Likely, Medium Confidence]*, while increases in extreme weather events will introduce contaminants into the food chain *[Likely, Medium Confidence]*. Rising carbon dioxide concentrations and climate change will alter incidence and distribution of pests, parasites, and microbes *[Very Likely, High Confidence]*, leading to increases in the use of pesticides and veterinary drugs *[Likely, Medium Confidence]*.

Description of evidence base

There are a number of established pathways by which climate change will intensify chemical contaminants within the food chain. Multiple studies have shown that increases in ocean temperatures are likely to increase the potential for mercury exposure, likely due to the increased uptake and concentration of mercury in fish and mammals at higher metabolic rates associated with warmer ambient temperatures.[28, 88, 89, 90] Another pathway includes extreme weather events, which can move chemical contaminants such as lead into agricultural fields and pastures (as well as into drinking or recreational water sources—see Chapter 6: Water-Related Illness).[45, 46, 87] A final pathway is through rising minimum winter temperatures and longer growing seasons, which will very likely alter pest distribution and populations. A large body of literature shows that temperature, carbon dioxide (CO_2) concentrations, and water availability are also likely to affect pest development, number of pest generations per year, changes in pest range, rate of infestation, and host plant and animal susceptibility.[3, 50, 76, 96, 97] Empirical models and an analysis of long-term in situ data indicate that rising temperatures will result in increased pest pressures.[100, 101, 105] These changes are expected to result in increased use of pesticides,[100, 102] which can lead to increased human exposure.[86]

Major uncertainties

Each of the pathways described in the evidence base has variable levels of uncertainty associated with each step of the exposure pathway.[110] For all these pathways, projecting the specific consequences on human health in the Unites States is challenging, due to the variability in type of pathogen or contaminant, time and duration of exposures, individual sensitivity (for example, genetic predisposition) and individual or institutional adaptive capacity. While increasing exposure to chemicals will exacerbate potential health risks, the nature of those risks will depend on the specific epidemiological links between exposure and human health as well as availability and access to health services. Resulting incidence of illness will depend on the genetic predisposition of the person exposed, type of contaminant, and extent of exposure over time.[86]

Assessment of confidence and likelihood based on evidence

Although it is *likely* that climate change will increase human exposure to chemical contaminants, the specific pathway(s) of exposure have varying levels of uncertainty associated with them and hence there is *medium confidence* regarding the overall extent of exposure. This chapter focuses on three such pathways. First, it is *likely* that elevated sea surface temperatures will result in increased bioaccumulation of mercury in seafood, but there is *medium confidence* regarding human illness because rates of accumulation and exposure vary according to the type of seafood ingested, and because of the role of varying individual sensitivity and individual or institutional adaptive capacity (particularly behavioral choices). Similarly, it is *likely* that extreme events will increase contaminants into agricultural soil and the food chain. However, there is *medium confidence* regarding exposure because the specific nature of the contaminant and the food source will vary, and because the extent of exposure will depend on risk management, communication of public health threats, and the effectiveness of regulatory, surveillance, and monitoring systems within the current food safety network. There is *high confidence* that it is *very likely* that rising CO_2 and climate change will alter pest incidence and distribution. There is *medium confidence* that such changes in incidence and distribution are *likely* to increase chemical management and the use of veterinary drugs in livestock. However, in all these pathways, the specific consequences on human health in the Unites States are uncertain, due primarily to the variability in type of pathogen or contaminant, time and duration of exposures, individual sensitivity (for example, genetic predisposition), and individual or institutional adaptive capacity.

Rising Carbon Dioxide Lowers Nutritional Value of Food

Key Finding 3: The nutritional value of agriculturally important food crops, such as wheat and rice, will decrease as rising levels of atmospheric carbon dioxide continue to reduce the concentrations of protein and essential minerals in most plant species *[Very Likely, High Confidence]*.

Description of evidence base

The nutritional response of crops to rising carbon dioxide is well documented, particularly among C_3 cereals such as rice and wheat, which make up the bulk of human caloric input. C_3 species are about 95% of all plant species and represent those species most likely to respond to an increase in atmospheric CO_2 concentrations.

There is strong evidence and consensus that protein concentrations in plants strongly correlate with nitrogen concentrations. CO_2-induced declines in nitrogen concentrations have been observed in nearly a hundred individual studies and several meta-analyses.[16, 133, 137, 139, 140] A meta-analysis of the effect of CO_2 on protein by crop covers

228 observations on wheat, rice, soybeans, barley and potato, [134] and was recently repeated for the United States, Japan, and Australia,[135] covering 138 mean observations on nitrogen/protein in wheat, rice, peas, maize, and sorghum. There is very strong evidence that rising CO_2 reduces protein content in non-leguminous C_3 crops, including wheat, rice, potato, and barley. There is also good agreement across studies that the ongoing increase in CO_2 elevates the overall carbohydrate content in C_3 plants.[16]

Another meta-analysis quantifies the role of increasing CO_2 in altering the ionome (the mineral nutrient and trace element composition of an organism) of plants, including major crops.[16] This meta-analysis of 7,761 observations indicates that increasing CO_2 also significantly reduces the mineral concentrations (calcium, magnesium, iron, zinc, copper, sulfur, potassium, and phosphorus) in C_3 plants, including grains and edible parts of other crops, while also substantially increasing the ratio of total non-structural carbohydrates (starch and sugars) to minerals and to protein.

Furthermore, these studies show the quality of current crops to be lower relative to the crops raised in the past with respect to protein and minerals.[16, 134] Direct experimental evidence shows that protein concentrations in wheat flour progressively declined with rising CO_2 concentrations representing levels in 1900 (approximately 290 ppm), 2008 (approximately 385 ppm), and the CO_2 concentrations projected to occur later in this century (approximately 715 ppm).[129]

Major uncertainties

While the general response and the direction in the change of crop quality is evident; there is uncertainty in the extent of variation in both protein and ionome among different crop varieties. There is little evidence regarding the CO_2 effects on complex micronutrients such as carotenoids (vitamin A, lutein, and zeaxanthin). Although protein, micronutrients, and ratio of carbohydrates to protein are all essential aspects of human dietary needs, the projected human health impacts of nutritional changes with increasing CO_2 are still being evaluated. There remains a high level of uncertainty regarding how reductions in crop quality affect human nutrition by contributing to or aggravating existing chronic dietary deficiencies and obesity risks, particularly in the United States where dietary protein deficiencies are uncommon.

Assessment of confidence and likelihood based on evidence

Based on the evidence, there is **high confidence** that the rapid increase in atmospheric CO_2 has resulted in a reduction in the level of protein and minerals relative to the amount of carbohydrates present for a number of important crop species (including a number of globally important cereals such as wheat, barley and rice), and will **very likely** continue to do so as atmospheric CO_2 concentration continues to rise.

Extreme Weather Limits Access to Safe Foods

Key Finding 4: Increases in the frequency or intensity of some extreme weather events associated with climate change will increase disruptions of food distribution by damaging existing infrastructure or slowing food shipments *[Likely, High Confidence]*. These impediments lead to increased risk for food damage, spoilage, or contamination, which will limit availability of and access to safe and nutritious food, depending on the extent of disruption and the resilience of food distribution infrastructure *[Medium Confidence]*.

Description of evidence base

It is well documented in assessment literature that climate models project an increase in the frequency and intensity of some extreme weather events.[14, 15] Because the food transportation system moves large volumes at a time, has limited alternative routes, and is dependent on the timing of the growing and harvest seasons, it is likely that the projected increase in the frequency and intensity of extreme weather events[13, 14] will also increase the frequency of food supply chain disruptions (including risks to food availability and access)[147, 148, 149, 150, 151, 152, 156] and the risk for food spoilage and contamination.[152, 163] Recent extreme events have demonstrated a clear linkage to the disruption of food distribution and access.[151, 161] Case studies show that such events, particularly those that result in power outages, may also expose food to temperatures inadequate for safe storage,[152] with increased risk of illness. For example, New York City's Department of Health and Mental Hygiene detected a statistically significant citywide increase in diarrheal illness resulting from consumption of spoiled foods due to lost refrigeration capabilities after a 2003 power outage.[158]

Major uncertainties

The extent to which climate-related disruptions to the food distribution system will affect food supply, safety, and human health, including incidences of illnesses, remains uncertain. This is because the impacts of any one extreme weather event are determined by the type, severity, and intensity of the event, the geographic location in which it occurs, infrastructure resiliency, and the social vulnerabilities or adaptive capacity of the populations at risk.

Assessment of confidence and likelihood based on evidence

Given the evidence base and current uncertainties, there is **high confidence** that projected increases in the frequency and severity of extreme events will **likely** lead to damage of existing food supplies and disruptions to food distribution infrastructure. There is **medium confidence** that these damages and disruptions will increase risk for food damage, spoilage, or contamination, which will limit availability and access to safe and nutritious foods because of uncertainties surrounding the extent of the disruptions and individual, community, or institutional sensitivity to impacts. There are further uncertainties surrounding how the specific dynamics of the extreme event, such as the geographic location in which it occurs, as well as the social vulnerabilities or adaptive capacity of the populations at risk, will impact human health.

DOCUMENTING UNCERTAINTY

See Appendix 4: Documenting Uncertainty for more information on assessments of confidence and likelihood used in this report.

Confidence Level	Likelihood
Very High	**Very Likely**
Strong evidence (established theory, multiple sources, consistent results, well documented and accepted methods, etc.), high consensus	≥ 9 in 10
High	**Likely**
Moderate evidence (several sources, some consistency, methods vary and/or documentation limited, etc.), medium consensus	≥ 2 in 3
	As Likely As Not
	≈ 1 in 2
Medium	**Unlikely**
Suggestive evidence (a few sources, limited consistency, models incomplete, methods emerging, etc.), competing schools of thought	≤ 1 in 3
	Very Unlikely
	≤ 1 in 10
Low	
Inconclusive evidence (limited sources, extrapolations, inconsistent findings, poor documentation and/or methods not tested, etc.), disagreement or lack of opinions among experts	

PHOTO CREDITS

Pg. 189–Farmer holding wheat: © Dan Lamont/Corbis

Pg. 190–Family enjoying outdoor grilling party: © Hill Street Studios/Blend Images/Corbis

Pg. 196–Helicopter crop dusting: © Lucas Payne/AgStock Images/Corbis

Pg. 197–Farmer holding wheat: © Dan Lamont/Corbis

References

1. Brown, M.E., J.M. Antle, P. Backlund, E.R. Carr, W.E. Easterling, M.K. Walsh, C. Ammann, W. Attavanich, C.B. Barrett, M.F. Bellemare, V. Dancheck, C. Funk, K. Grace, J.S.I. Ingram, H. Jiang, H. Maletta, T. Mata, A. Murray, M. Ngugi, D. Ojima, B. O'Neill, and C. Tebaldi, 2015: Climate Change, Global Food Security and the U.S. Food System. 146 pp. U.S. Global Change Research Program. http://www.usda.gov/oce/climate_change/FoodSecurity2015Assessment/FullAssessment.pdf

2. Porter, J.R., L. Xie, A.J. Challinor, K. Cochrane, S.M. Howden, M.M. Iqbal, D.B. Lobell, and M.I. Travasso, 2014: Food security and food production systems. *Climate Change 2014: Impacts, Adaptation, and Vulnerability. Part A: Global and Sectoral Aspects. Contribution of Working Group II to the Fifth Assessment Report of the Intergovernmental Panel of Climate Change.* Field, C.B., V.R. Barros, D.J. Dokken, K.J. Mach, M.D. Mastrandrea, T.E. Bilir, M. Chatterjee, K.L. Ebi, Y.O. Estrada, R.C. Genova, B. Girma, E.S. Kissel, A.N. Levy, S. MacCracken, P.R. Mastrandrea, and L.L. White, Eds. Cambridge University Press, Cambridge, United Kingdom and New York, NY, USA, 485-533. http://www.ipcc.ch/pdf/assessment-report/ar5/wg2/WGIIAR5-Chap7_FINAL.pdf

3. Chakraborty, S. and A.C. Newton, 2011: Climate change, plant diseases and food security: An overview. *Plant Pathology,* **60,** 2-14. http://dx.doi.org/10.1111/j.1365-3059.2010.02411.x

4. Gregory, P.J., J.S.I. Ingram, and M. Brklacich, 2005: Climate change and food security. *Philosophical Transactions of the Royal Society B: Biological Sciences,* **360,** 2139-2148. http://dx.doi.org/10.1098/rstb.2005.1745

5. Lake, I.R., L. Hooper, A. Abdelhamid, G. Bentham, A.B.A. Boxall, A. Draper, S. Fairweather-Tait, M. Hulme, P.R. Hunter, G. Nichols, and K.W. Waldron, 2012: Climate change and food security: Health impacts in developed countries. *Environmental Health Perspectives,* **120,** 1520-1526. http://dx.doi.org/10.1289/ehp.1104424

6. Walthall, C., P. Backlund, J. Hatfield, L. Lengnick, E. Marshall, M. Walsh, S. Adkins, M. Aillery, E.A. Ainsworth, C. Amman, C.J. Anderson, I. Bartomeus, L.H. Baumgard, F. Booker, B. Bradley, D.M. Blumenthal, J. Bunce, K. Burkey, S.M. Dabney, J.A. Delgado, J. Dukes, A. Funk, K. Garrett, M. Glenn, D.A. Grantz, D. Goodrich, S. Hu, R.C. Izaurralde, R.A.C. Jones, S.-H. Kim, A.D.B. Leaky, K. Lewers, T.L. Mader, A. McClung, J. Morgan, D.J. Muth, M. Nearing, D.M. Oosterhuis, D. Ort, C. Parmesan, W.T. Pettigrew, W. Polley, R. Rader, C. Rice, M. Rivington, E. Rosskopf, W.A. Salas, L.E. Sollenberger, R. Srygley, C. Stöckle, E.S. Takle, D. Timlin, J.W. White, R. Winfree, L. Wright-Morton, and L.H. Ziska, 2012: Climate Change and Agriculture in the United States: Effects and Adaptation. USDA Technical Bulletin 1935, 186 pp. U.S. Department of Agriculture, Washington, D.C. http://www.usda.gov/oce/climate_change/effects_2012/CC%20and%20Agriculture%20Report%20(02-04-2013)b.pdf

7. IPCC, 2014: Climate Change 2014: Impacts, Adaptation, and Vulnerability. Part A: Global and Sectoral Aspects. Contribution of Working Group II to the Fifth Assessment Report of the Intergovernmental Panel on Climate Change. Field, C.B., V.R. Barros, D.J. Dokken, K.J. Mach, M.D. Mastrandrea, T.E. Bilir, M. Chatterjee, K.L. Ebi, Y.O. Estrada, R.C. Genova, B. Girma, E.S. Kissel, A.N. Levy, S. MacCracken, P.R. Mastrandrea, and L.L. White (Eds.), 1132 pp. Cambridge University Press, Cambridge, UK and New York, NY. http://www.ipcc.ch/report/ar5/wg2/

8. IOM, 2012: *Improving Food Safety Through a One Health Approach: Workshop Summary.* Choffnes, E.R., D.A. Relman, L. Olsen, R. Hutton, and A. Mack, Eds. Institute of Medicine. The National Academies Press, Washington, D.C., 418 pp. http://dx.doi.org/10.17226/13423

9. Jaykus, L., M. Woolridge, J.M. Frank, M. Miraglia, A. McQuatters-Gollop, C. Tirado, R. Clarke, and M. Friel, 2008: Climate Change: Implications for Food Safety. 49 pp. Food and Agriculture Organisation of the United Nations. http://www.fao.org/docrep/010/i0195e/i0195e00.HTM

10. FAO, 2014: Definitions of Purposes of the Codex Alimentarius. Food and Agriculture Organization of the United Nations. http://www.fao.org/docrep/005/y2200e/y2200e07.htm

11. CDC, 2014: CDC Estimates of Foodborne Illness in the United States. Centers for Disease Control and Prevention, Atlanta, GA. http://www.cdc.gov/foodborneburden/2011-foodborne-estimates.html

12. CDC, 2015: Food Safety: Foodborne Germs and Illnesses. Centers for Disease Control and Prevention, Atlanta, GA. http://www.cdc.gov/foodsafety/foodborne-germs.html

13. IPCC, 2012: Managing the Risks of Extreme Events and Disasters to Advance Climate Change Adaptation. A Special Report of Working Groups I and II of the Intergovernmental Panel on Climate Change. Field, C.B., V. Barros, T.F. Stocker, D. Qin, D.J. Dokken, K.L. Ebi, M.D. Mastrandrea, K.J. Mach, G.-K. Plattner, S.K. Allen, M. Tignor, and P.M. Midgley (Eds.), 582 pp. Cambridge University Press, Cambridge, UK and New York, NY. http://ipcc-wg2.gov/SREX/images/uploads/SREX-All_FINAL.pdf

14. IPCC, 2013: Summary for policymakers. *Climate Change 2013: The Physical Science Basis. Contribution of Working Group I to the Fifth Assessment Report of the Intergovernmental Panel on Climate Change.* Stocker, T.F., D. Qin, G.-K. Plattner, M. Tignor, S.K. Allen, J. Boschung, A. Nauels, Y. Xia, V. Bex, and P.M. Midgley, Eds. Cambridge University Press, Cambridge, United Kingdom and New York, NY, USA, 1–30. http://dx.doi.org/10.1017/CBO9781107415324.004

15. Walsh, J., D. Wuebbles, K. Hayhoe, J. Kossin, K. Kunkel, G. Stephens, P. Thorne, R. Vose, M. Wehner, J. Willis, D. Anderson, S. Doney, R. Feely, P. Hennon, V. Kharin, T. Knutson, F. Landerer, T. Lenton, J. Kennedy, and R. Somerville, 2014: Ch. 2: Our changing climate. *Climate Change Impacts in the United States: The Third National Climate Assessment.* Melillo, J.M., T.C. Richmond, and G.W. Yohe, Eds. U.S. Global Change Research Program, Washington, D.C., 19-67. http://dx.doi.org/10.7930/J0KW5CXT

16. Loladze, I., 2014: Hidden shift of the ionome of plants exposed to elevated CO2 depletes minerals at the base of human nutrition. *eLife,* **3,** e02245. http://dx.doi.org/10.7554/eLife.02245

17. Curtis, D., A. Hill, A. Wilcock, and S. Charlebois, 2014: Foodborne and waterborne pathogenic bacteria in selected Organisation for Economic Cooperation and Development (OECD) countries. *Journal of Food Science,* **79,** R1871-R1876. http://dx.doi.org/10.1111/1750-3841.12646

18. Kovats, R.S., S.J. Edwards, S. Hajat, B.G. Armstrong, K.L. Ebi, and B. Menne, 2004: The effect of temperature on food poisoning: A time-series analysis of salmonellosis in ten European countries. *Epidemiology and Infection,* **132,** 443-453. http://dx.doi.org/10.1017/s0950268804001992

19. USDA, 2013: Foodborne Illness Peaks in Summer-Why? United States Department of Agriculture, Food Safety and Inspection Service. http://www.fsis.usda.gov/wps/portal/fsis/topics/food-safety-education/get-answers/food-safety-fact-sheets/foodborne-illness-and-disease/foodborne-illness-peaks-in-summer/ct_index

20. Scallan, E., P.M. Griffin, F.J. Angulo, R.V. Tauxe, and R.M. Hoekstra, 2011: Foodborne illness acquired in the United States: Unspecified agents. *Emerging Infectious Diseases,* **17,** 16-22. http://dx.doi.org/10.3201/eid1701.P21101

21. Scallan, E., R.M. Hoekstra, F.J. Angulo, R.V. Tauxe, M.A. Widdowson, S.L. Roy, J.L. Jones, and P.M. Griffin, 2011: Foodborne illness acquired in the United States: Major pathogens. *Emerging Infectious Diseases,* **17,** 7-15. http://dx.doi.org/10.3201/eid1701.P11101

22. Painter, J.A., R.M. Hoekstra, T. Ayers, R.V. Tauxe, C.R. Braden, F.J. Angulo, and P.M. Griffin, 2013: Attribution of foodborne illnesses, hospitalizations, and deaths to food commodities by using outbreak data, United States, 1998–2008. *Emerging Infectious Diseases,* **19,** 407-415. http://dx.doi.org/10.3201/eid1903.111866

23. Utaaker, K.S. and L.J. Robertson, 2015: Climate change and foodborne transmission of parasites: A consideration of possible interactions and impacts for selected parasites. *Food Research International,* **68,** 16-23. http://dx.doi.org/10.1016/j.foodres.2014.06.051

24. Pennotti, R., E. Scallan, L. Backer, J. Thomas, and F.J. Angulo, 2013: Ciguatera and scombroid fish poisoning in the United States. *Foodborne Pathogens and Disease,* **10,** 1059-1066. http://dx.doi.org/10.1089/fpd.2013.1514

25. Tchounwou, P.B., W.K. Ayensu, N. Ninashvili, and D. Sutton, 2003: Review: Environmental exposure to mercury and its toxicopathologic implications for public health. *Environmental Toxicology,* **18,** 149-175. http://dx.doi.org/10.1002/tox.10116

26. Marques, A., M.L. Nunes, S.K. Moore, and M.S. Strom, 2010: Climate change and seafood safety: Human health implications. *Food Research International,* **43,** 1766-1779. http://dx.doi.org/10.1016/j.foodres.2010.02.010

27. Downs, S.G., C.L. MacLeod, and J.N. Lester, 1998: Mercury in precipitation and its relation to bioaccumulation in fish: A literature review. *Water, Air, and Soil Pollution,* **108,** 149-187. http://dx.doi.org/10.1023/a:1005023916816

28. Booth, S. and D. Zeller, 2005: Mercury, food webs, and marine mammals: Implications of diet and climate change for human health. *Environmental Health Perspectives,* **113,** 521-526. http://dx.doi.org/10.1289/ehp.7603

29. Boxall, A.B.A., A. Hardy, S. Beulke, T. Boucard, L. Burgin, P.D. Falloon, P.M. Haygarth, T. Hutchinson, R.S. Kovats, G. Leonardi, L.S. Levy, G. Nichols, S.A. Parsons, L. Potts, D. Stone, E. Topp, D.B. Turley, K. Walsh, E.M.H. Wellington, and R.J. Williams, 2009: Impacts of climate change on indirect human exposure to pathogens and chemicals from agriculture. *Environmental Health Perspectives,* **117,** 508-514. http://dx.doi.org/10.1289/ehp.0800084

30. Strawn, L.K., E.D. Fortes, E.A. Bihn, K.K. Nightingale, Y.T. Gröhn, R.W. Worobo, M. Wiedmann, and P.W. Bergholz, 2013: Landscape and meteorological factors affecting prevalence of three food-borne pathogens in fruit and vegetable farms. *Applied and Environmental Microbiology,* **79,** 588-600. http://dx.doi.org/10.1128/aem.02491-12

31. Liu, C., N. Hofstra, and E. Franz, 2013: Impacts of climate change on the microbial safety of pre-harvest leafy green vegetables as indicated by Escherichia coli O157 and Salmonella spp. *International Journal of Food Microbiology,* **163,** 119-128. http://dx.doi.org/10.1016/j.ijfoodmicro.2013.02.026

32. Wu, F., D. Bhatnagar, T. Bui-Klimke, I. Carbone, R. Hellmich, G. Munkvold, P. Paul, G. Payne, and E. Takle, 2011: Climate change impacts on mycotoxin risks in US maize. *World Mycotoxin Journal,* **4,** 79-93. http://dx.doi.org/10.3920/WMJ2010.1246

33. Castro-Ibáñez, I., M.I. Gil, J.A. Tudela, and A. Allende, 2015: Microbial safety considerations of flooding in primary production of leafy greens: A case study. *Food Research International,* **68,** 62-69. http://dx.doi.org/10.1016/j.foodres.2014.05.065

34. Lake, I.R., I.A. Gillespie, G. Bentham, G.L. Nichols, C. Lane, G.K. Adak, and E.J. Threlfall, 2009: A re-evaluation of the impact of temperature and climate change on food-borne illness. *Epidemiology and Infection,* **137,** 1538-1547. http://dx.doi.org/10.1017/s0950268809002477

35. Cook, D.W., 1994: Effect of time and temperature on multiplication of *Vibrio vulnificus* in postharvest Gulf Coast shellstock oysters. *Applied and Environmental Microbiology,* **60,** 3483-3484. PMC201838

36. Gooch, J., A. DePaola, J. Bowers, and D. Marshall, 2002: Growth and survival of *Vibrio parahaemolyticus* in postharvest American oysters. *Journal of Food Protection,* **65,** 911-1053.

37. Paterson, R.R.M. and N. Lima, 2010: How will climate change affect mycotoxins in food? *Food Research International,* **43,** 1902-1914. http://dx.doi.org/10.1016/j.foodres.2009.07.010

38. Lipp, E.K., R. Kurz, R. Vincent, C. Rodriguez-Palacios, S.R. Farrah, and J.B. Rose, 2001: The effects of seasonal variability and weather on microbial fecal pollution and enteric pathogens in a subtropical estuary. *Estuaries,* **24,** 266-276. http://dx.doi.org/10.2307/1352950

39. Kovats, R.S., M.J. Bouma, S. Hajat, E. Worrall, and A. Haines, 2003: El Niño and health. *The Lancet,* **362,** 1481-1489. http://dx.doi.org/10.1016/s0140-6736(03)14695-8

40. Paz, S., N. Bisharat, E. Paz, O. Kidar, and D. Cohen, 2007: Climate change and the emergence of *Vibrio vulnificus* disease in Israel. *Environmental Research,* **103,** 390-396. http://dx.doi.org/10.1016/j.envres.2006.07.002

41. Monge, R. and M. Chinchilla, 1996: Presence of *Cryptosporidium* oocysts in fresh vegetables. *Journal of Food Protection,* **59,** 202-203.

42. Solomon, E.B., S. Yaron, and K.R. Matthews, 2002: Transmission of *Escherichia coli* O157:H7 from contaminated manure and irrigation water to lettuce plant tissue and its subsequent internalization. *Applied and Environmental Microbiology,* **68,** 397-400. http://dx.doi.org/10.1128/aem.68.1.397-400.2002

43. Sales-Ortells, H., X. Fernandez-Cassi, N. Timoneda, W. Dürig, R. Girones, and G. Medema, 2015: Health risks derived from consumption of lettuces irrigated with tertiary effluent containing norovirus. *Food Research International,* **68,** 70-77. http://dx.doi.org/10.1016/j.foodres.2014.08.018

44. Miraglia, M., H.J.P. Marvin, G.A. Kleter, P. Battilani, C. Brera, E. Coni, F. Cubadda, L. Croci, B. De Santis, S. Dekkers, L. Filippi, R.W.A. Hutjes, M.Y. Noordam, M. Pisante, G. Piva, A. Prandini, L. Toti, G.J. van den Born, and A. Vespermann, 2009: Climate change and food safety: An emerging issue with special focus on Europe. *Food and Chemical Toxicology,* **47,** 1009-1021. http://dx.doi.org/10.1016/j.fct.2009.02.005

45. Foulds, S.A., P.A. Brewer, M.G. Macklin, W. Haresign, R.E. Betson, and S.M.E. Rassner, 2014: Flood-related contamination in catchments affected by historical metal mining: An unexpected and emerging hazard of climate change. *Science of the Total Environment,* **476-477,** 165-180. http://dx.doi.org/10.1016/j.scitotenv.2013.12.079

46. Umlauf, G., G. Bidoglio, E.H. Christoph, J. Kampheus, F. Kruger, D. Landmann, A.J. Schulz, R. Schwartz, K. Severin, B. Stachel, and D. Stehr, 2005: The situation of PCDD/Fs and dioxin-like PCBs after the flooding of River Elbe and Mulde in 2002. *Acta Hydrochimica et Hydrobiologica,* **33,** 543-554. http://dx.doi.org/10.1002/aheh.200400597

47. Rotkin-Ellman, M., G. Solomon, C.R. Gonzales, L. Agwaramgbo, and H.W. Mielke, 2010: Arsenic contamination in New Orleans soil: Temporal changes associated with flooding. *Environmental Research,* **110,** 19-25. http://dx.doi.org/10.1016/j.envres.2009.09.004

48. Kim, Y.S., K.H. Park, H.S. Chun, C. Choi, and G.J. Bahk, 2015: Correlations between climatic conditions and foodborne disease. *Food Research International,* **68,** 24-30. http://dx.doi.org/10.1016/j.foodres.2014.03.023

49. Lal, A., S. Hales, N. French, and M.G. Baker, 2012: Seasonality in human zoonotic enteric diseases: A systematic review. *PLoS ONE,* **7,** e31883. http://dx.doi.org/10.1371/journal.pone.0031883

50. Pérez de León, A.A., P.D. Teel, A.N. Auclair, M.T. Messenger, F.D. Guerrero, G. Schuster, and R.J. Miller, 2012: Integrated strategy for sustainable cattle fever tick eradication in USA is required to mitigate the impact of global change. *Frontiers in Physiology,* **3,** Article 195. http://dx.doi.org/10.3389/fphys.2012.00195

51. Baranyi, J. and M.L. Tamplin, 2004: ComBase: A common database on microbial responses to food environments. *Journal of Food Protection,* **67,** 1967-1971.

52. Oscar, T.P., 2009: Predictive model for survival and growth of Salmonella typhimurium DT104 on chicken skin during temperature abuse. *Journal of Food Protection,* **72,** 304-314.

53. CDC, 2014: *Salmonella*: Multistate Outbreak of Multi-drug-Resistant Salmonella Heidelberg Infections Linked to Foster Farms Brand Chicken (Final Update), July 31, 2014. Centers for Disease Control and Prevention, Atlanta, GA. http://www.cdc.gov/salmonella/heidelberg-10-13/index.html

54. Ravel, A., E. Smolina, J.M. Sargeant, A. Cook, B. Marshall, M.D. Fleury, and F. Pollari, 2010: Seasonality in human salmonellosis: Assessment of human activities and chicken contamination as driving factors. *Foodborne Pathogens and Disease,* **7,** 785-794. http://dx.doi.org/10.1089/fpd.2009.0460

55. Rose, J.B., P.R. Epstein, E.K. Lipp, B.H. Sherman, S.M. Bernard, and J.A. Patz, 2001: Climate variability and change in the United States: Potential impacts on water- and foodborne diseases caused by microbiologic agents. *Environmental Health Perspectives,* **109 Suppl 2,** 211-221. PMC1240668

56. Semenza, J.C., S. Herbst, A. Rechenburg, J.E. Suk, C. Höser, C. Schreiber, and T. Kistemann, 2012: Climate change impact assessment of food- and waterborne diseases. *Critical Reviews in Environmental Science and Technology,* **42,** 857-890. http://dx.doi.org/10.1080/10643389.2010.534706

57. Rohayem, J., 2009: Norovirus seasonality and the potential impact of climate change. *Clinical Microbiology and Infection,* **15,** 524-527. http://dx.doi.org/10.1111/j.1469-0691.2009.02846.x

58. McLaughlin, J.B., A. DePaola, C.A. Bopp, K.A. Martinek, N.P. Napolilli, C.G. Allison, S.L. Murray, E.C. Thompson, M.M. Bird, and J.P. Middaugh, 2005: Outbreak of *Vibrio parahaemolyticus* gastroenteritis associated with Alaskan oysters. *The New England Journal of Medicine,* **353,** 1463-1470. http://dx.doi.org/10.1056/NEJMoa051594

59. Martinez-Urtaza, J., J.C. Bowers, J. Trinanes, and A. DePaola, 2010: Climate anomalies and the increasing risk of *Vibrio parahaemolyticus* and *Vibrio vulnificus* illnesses. *Food Research International,* **43,** 1780-1790. http://dx.doi.org/10.1016/j.foodres.2010.04.001

60. Cook, D.W., 1997: Refrigeration of oyster shellstock: Conditions which minimize the outgrowth of *Vibrio vulnificus*. *Journal of Food Protection,* **60,** 349-352.

61. Kunkel, K.E., L.E. Stevens, S.E. Stevens, L. Sun, E. Janssen, D. Wuebbles, and J.G. Dobson, 2013: Regional Climate Trends and Scenarios for the U.S. National Climate Assessment: Part 9. Climate of the Contiguous United States. NOAA Technical Report NESDIS 142-9, 85 pp. U.S. Department of Commerce, National Oceanic and Atmospheric Administration, National Environmental Satellite, Data, and Information Service, Washington, D.C. http://www.nesdis.noaa.gov/technical_reports/NOAA_NESDIS_Tech_Report_142-9-Climate_of_the_Contiguous_United_States.pdf

62. Melillo, J.M., T.C. Richmond, and G.W. Yohe, Eds., 2014: *Climate Change Impacts in the United States: The Third National Climate Assessment.* U.S. Global Change Research Program, Washington, D.C., 842 pp. http://dx.doi.org/10.7930/J0Z31WJ2

63. Polley, L. and R.C.A. Thompson, 2009: Parasite zoonoses and climate change: Molecular tools for tracking shifting boundaries. *Trends in Parasitology,* **25,** 285-291. http://dx.doi.org/10.1016/j.pt.2009.03.007

64. Mills, J.N., K.L. Gage, and A.S. Khan, 2010: Potential influence of climate change on vector-borne and zoonotic diseases: A review and proposed research plan. *Environmental Health Perspectives,* **118,** 1507-1514. http://dx.doi.org/10.1289/ehp.0901389

65. Esteve-Gassent, M.D., A.A. Pérez de León, D. Romero-Salas, T.P. Feria-Arroyo, R. Patino, I. Castro-Arellano, G. Gordillo-Pérez, A. Auclair, J. Goolsby, R.I. Rodriguez-Vivas, and J.G. Estrada-Franco, 2014: Pathogenic landscape of transboundary zoonotic diseases in the Mexico–US border along the Rio Grande. *Frontiers in Public Health,* **2,** 177. http://dx.doi.org/10.3389/fpubh.2014.00177

66. Schijven, J., M. Bouwknegt, A.M. de Roda Husman, S. Rutjes, B. Sudre, J.E. Suk, and J.C. Semenza, 2013: A decision support tool to compare waterborne and foodborne infection and/or illness risks associated with climate change. *Risk Analysis,* **33,** 2154-2167. http://dx.doi.org/10.1111/risa.12077

67. Patz, J.A., S.J. Vavrus, C.K. Uejio, and S.L. McLellan, 2008: Climate change and waterborne disease risk in the Great Lakes region of the US. *American Journal of Preventive Medicine,* **35,** 451-458. http://dx.doi.org/10.1016/j.amepre.2008.08.026

68. Tornevi, A., O. Bergstedt, and B. Forsberg, 2014: Precipitation effects on microbial pollution in a river: Lag structures and seasonal effect modification. *PloS ONE,* **9,** e98546. http://dx.doi.org/10.1371/journal.pone.0098546

69. Cho, K.H., Y.A. Pachepsky, J.H. Kim, A.K. Guber, D.R. Shelton, and R. Rowland, 2010: Release of *Escherichia coli* from the bottom sediment in a first-order creek: Experiment and reach-specific modeling. *Journal of Hydrology,* **391,** 322-332. http://dx.doi.org/10.1016/j.jhydrol.2010.07.033

70. Jamieson, R., D.M. Joy, H. Lee, R. Kostaschuk, and R. Gordon, 2005: Transport and deposition of sediment-associated *Escherichia coli* in natural streams. *Water Research,* **39,** 2665-2675. http://dx.doi.org/10.1016/j.watres.2005.04.040

71. Jamieson, R.C., D.M. Joy, H. Lee, R. Kostaschuk, and R.J. Gordon, 2005: Resuspension of sediment-associated *Escherichia coli* in a natural stream. *Journal of Environment Quality,* **34,** 581-589. http://dx.doi.org/10.2134/jeq2005.0581

72. Fleury, M., D.F. Charron, J.D. Holt, O.B. Allen, and A.R. Maarouf, 2006: A time series analysis of the relationship of ambient temperature and common bacterial enteric infections in two Canadian provinces. *International Journal of Biometeorology,* **50,** 385-391. http://dx.doi.org/10.1007/s00484-006-0028-9

73. Senhorst, H.A. and J.J. Zwolsman, 2005: Climate change and effects on water quality: A first impression. *Water Science & Technology,* **51,** 53-59.

74. Zhang, G., L. Ma, L.R. Beuchat, M.C. Erickson, V.H. Phelan, and M.P. Doyle, 2009: Heat and drought stress during growth of lettuce (*Lactuca sativa* L.) does not promote internalization of *Escherichia coli* O157: H7. *Journal of Food Protection,* **72,** 2471-2475.

75. Cotty, P.J. and R. Jaime-Garcia, 2007: Influences of climate on aflatoxin producing fungi and aflatoxin contamination. *International Journal of Food Microbiology,* **119,** 109-115. http://dx.doi.org/10.1016/j.ijfoodmicro.2007.07.060

76. Tirado, M.C., R. Clarke, L.A. Jaykus, A. McQuatters-Gollop, and J.M. Frank, 2010: Climate change and food safety: A review. *Food Research International,* **43,** 1745-1765. http://dx.doi.org/10.1016/j.foodres.2010.07.003

77. Medina, A., A. Rodriguez, and N. Magan, 2014: Effect of climate change on *Aspergillus flavus* and aflatoxin B1 production. *Frontiers in Microbiology,* **5.** http://dx.doi.org/10.3389/fmicb.2014.00348

78. Gould, L.H., K.A. Walsh, A.R. Vieira, K. Herman, I.T. Williams, A.J. Hall, and D. Cole, 2013: Surveillance for foodborne disease Outbreaks — United States, 1998–2008. *MMWR. Morbidity and Mortality Weekly Report,* **62,** 1-34. http://www.cdc.gov/mmwr/preview/mmwrhtml/ss6202a1.htm

79. Gingold, D.B., M.J. Strickland, and J.J. Hess, 2014: Ciguatera fish poisoning and climate change: Analysis of National Poison Center Data in the United States, 2001–2011. *Environmental Health Perspectives,* **122,** 580-586. http://dx.doi.org/10.1289/ehp.1307196

80. Akil, L., H.A. Ahmad, and R.S. Reddy, 2014: Effects of climate change on Salmonella infections. *Foodborne Pathogens and Disease,* **11,** 974-980. http://dx.doi.org/10.1089/fpd.2014.1802

81. Strosnider, H., E. Azziz-Baumgartner, M. Banziger, R. Bhat, R. Breiman, M. Brune, K. DeCock, A. Dilley, J. Groopman, K. Hell, S. Henry, D. Jeffers, C. Jolly, J. P, G. Kibata, L. Lewis, X. Liu, G. Luber, L. McCoy, P. Mensah, M. Miraglia, A. Misore, H. Njapau, C. Ong, M. Onsongo, S. Page, D. Park, M. Patel, T. Phillips, M. Pineiro, J. Pronczuk, H. Rogers, C. Rubin, M. Sabino, A. Schaafsma, G. Shephard, J. Stroka, C. Wild, J. Williams, and D. Wilson, 2006: Workgroup report: Public health strategies for reducing aflatoxin exposure in developing countries. *Environmental Health Perspectives,* **114,** 1898-1903. http://dx.doi.org/10.1289/ehp.9302 http://www.jstor.org/stable/4119604

82. Lewis, L., M. Onsongo, H. Njapau, H. Schurz-Rogers, G. Luber, S. Kieszak, J. Nyamongo, L. Backer, A.M. Dahiye, A. Misore, K. DeCock, and C. Rubin, 2005: Aflatoxin contamination of commercial maize products during an outbreak of acute aflatoxicosis in eastern and central Kenya. *Environmental Health Perspectives,* **113,** 1763-1767. http://dx.doi.org/10.1289/ehp.7998

83. Wild, C.P. and Y.Y. Gong, 2010: Mycotoxins and human disease: A largely ignored global health issue. *Carcinogenesis,* **31,** 71-82. http://dx.doi.org/10.1093/carcin/bgp264

84. Liu, Y. and F. Wu, 2010: Global burden of aflatoxin-induced hepatocellular carcinoma: A risk assessment. *Environmental Health Perspectives,* **118,** 818-824. http://dx.doi.org/10.1289/ehp.0901388

85. Battilani, P., V. Rossi, P. Giorni, A. Pietri, A. Gualla, H.J. Van der Fels-Klerx, C.J.H. Booij, A. Moretti, A. Logrieco, F. Miglietta, P. Toscano, M. Miraglia, B. De Santis, and C. Brera, 2012: Modelling, Predicting and Mapping the Emergence of Aflatoxins in Cereals in the EU due to Climate Change. Question No. EFSA-Q-2009-00812. European Food Safety Authority, Parma, Italy. http://www.efsa.europa.eu/sites/default/files/scientific_output/files/main_documents/223e.pdf

86. Alavanja, M.C.R., M.K. Ross, and M.R. Bonner, 2013: Increased cancer burden among pesticide applicators and others due to pesticide exposure. *CA: A Cancer Journal for Clinicians,* **63,** 120-142. http://dx.doi.org/10.3322/caac.21170

87. Presley, S.M., T.R. Rainwater, G.P. Austin, S.G. Platt, J.C. Zak, G.P. Cobb, E.J. Marsland, K. Tian, B. Zhang, T.A. Anderson, S.B. Cox, M.T. Abel, B.D. Leftwich, J.R. Huddleston, R.M. Jeter, and R.J. Kendall, 2006: Assessment of pathogens and toxicants in New Orleans, LA following Hurricane Katrina. *Environmental Science & Technology,* **40,** 468-474. http://dx.doi.org/10.1021/es052219p

88. Carrie, J., F. Wang, H. Sanei, R.W. Macdonald, P.M. Outridge, and G.A. Stern, 2010: Increasing contaminant burdens in an arctic fish, Burbot (*Lota lota*), in a warming climate. *Environmental Science & Technology,* **44,** 316-322. http://dx.doi.org/10.1021/es902582y

89. Bodaly, R.A., J.W.M. Rudd, R.J.P. Fudge, and C.A. Kelly, 1993: Mercury concentrations in fish related to size of remote Canadian shield lakes. *Canadian Journal of Fisheries and Aquatic Sciences,* **50,** 980-987. http://dx.doi.org/10.1139/f93-113

90. Dijkstra, J.A., K.L. Buckman, D. Ward, D.W. Evans, M. Dionne, and C.Y. Chen, 2013: Experimental and natural warming elevates mercury concentrations in estuarine fish. *PLoS ONE,* **8,** e58401. http://dx.doi.org/10.1371/journal.pone.0058401

91. Pack, E.C., C.H. Kim, S.H. Lee, C.H. Lim, D.G. Sung, M.H. Kim, K.H. Park, S.-S. Hong, K.M. Lim, D.W. Choi, and S.W. Kim, 2014: Effects of environmental temperature change on mercury absorption in aquatic organisms with respect to climate warming. *Journal of Toxicology and Environmental Health, Part A,* **77,** 1477-1490. http://dx.doi.org/10.1080/15287394.2014.955892

92. González-Estecha, M., A. Bodas-Pinedo, M.A. Rubio-Herrera, N. Martell-Claros, E.M. Trasobares-Iglesias, J.M. Ordóñez-Iriarte, J.J. Guillén-Pérez, M.A. Herráiz-Martínez, J.A. García-Donaire, R. Farré-Rovira, E. Calvo-Manuel, J.R. Martínez-Álvarez, M.T. Llorente-Ballesteros, M. Sáinz-Martín, T. Martínez-Astorquiza, M.J. Martínez-García, I. Bretón Lesmes, M.A. Cuadrado-Cenzual, S. Prieto-Menchero, C. Gallardo-Pino, R. Moreno-Rojas, P. Bermejo-Barrera, M. Torres-Moreno, M. Arroyo-Fernández, and A. Calle-Pascual, 2014: Effectos sobre la salud del metilmercurio en ninos y adultos: Estudios nacionales e internacionales [The effects of methylmercury on health in children and adults: National and international studies]. *Nutricion Hospitalaria,* **30,** 989-1007. http://dx.doi.org/10.3305/nh.2014.30.5.7728

93. Bebber, D.P., 2015: Range-expanding pests and pathogens in a warming world. *Annual Review of Phytopathology,* **53,** 335-356. http://dx.doi.org/10.1146/annurev-phyto-080614-120207

94. Sanchez-Guillen, R.A., A. Cordoba-Aguilar, B. Hansson, J. Ott, and M. Wellenreuther, 2015: Evolutionary consequences of climate-induced range shifts in insects. *Biological Reviews,* **July 6**. http://dx.doi.org/10.1111/brv.12204

95. Bale, J.S. and S.A.L. Hayward, 2010: Insect overwintering in a changing climate. *Journal of Experimental Biology,* **213,** 980-994. http://dx.doi.org/10.1242/jeb.037911

96. Rosenzweig, C., A. Iglesias, X.B. Yang, P.R. Epstein, and E. Chivian, 2001: Climate change and extreme weather events: Implications for food production, plant diseases, and pests. *Global Change and Human Health,* **2,** 90-104. http://dx.doi.org/10.1023/a:1015086831467

97. Ziska, L.H. and G.B. Runion, 2007: Future weed, pest, and disease problems for plants. *Agroecosystems in a Changing Climate.* Newton, P.C.D., R.A. Carran, G.R. Edwards, and P.A. Niklaus, Eds. CRC Press, Boca Raton, FL, 261-287. http://www.ars.usda.gov/SP2UserFiles/Place/60100500/csr/ResearchPubs/runion/runion_07a.pdf

98. Zavala, J.A., C.L. Casteel, E.H. DeLucia, and M.R. Berenbaum, 2008: Anthropogenic increase in carbon dioxide compromises plant defense against invasive insects. *PNAS,* **105,** 5129-5133. http://dx.doi.org/10.1073/pnas.0800568105

99. O'Neill, B.F., A.R. Zangerl, E.H. DeLucia, C. Casteel, J.A. Zavala, and M.R. Berenbaum, 2011: Leaf temperature of soybean grown under elevated CO2 increases *Aphis glycines* (Hemiptera: Aphididae) population growth. *Insect Science,* **18,** 419-425. http://dx.doi.org/10.1111/j.1744-7917.2011.01420.x

100. Chen, C.-C. and B.A. McCarl, 2001: An investigation of the relationship between pesticide usage and climate change. *Climatic Change,* **50,** 475-487. http://dx.doi.org/10.1023/a:1010655503471

101. Ziska, L.H., 2014: Increasing minimum daily temperatures are associated with enhanced pesticide use in cultivated soybean along a latitudinal gradient in the Mid-Western United States. *PLoS ONE,* **9,** e98516. http://dx.doi.org/10.1371/journal.pone.0098516

102. Ziska, L.H., J.R. Teasdale, and J.A. Bunce, 1999: Future atmospheric carbon dioxide may increase tolerance to glyphosate. *Weed Science,* **47,** 608-615.

103. Ziska, L.H., S. Faulkner, and J. Lydon, 2004: Changes in biomass and root:shoot ratio of field-grown Canada thistle (*Cirsium arvense*), a noxious, invasive weed, with elevated CO_2: Implications for control with glyphosate. *Weed Science,* **52,** 584-588. http://dx.doi.org/10.1614/WS-03-161R

104. McDonald, A., S. Riha, A. DiTommaso, and A. DeGaetano, 2009: Climate change and the geography of weed damage: Analysis of U.S. maize systems suggests the potential for significant range transformations. *Agriculture, Ecosystems & Environment,* **130,** 131-140. http://dx.doi.org/10.1016/j.agee.2008.12.007

105. Manea, A., M.R. Leishman, and P.O. Downey, 2011: Exotic C4 grasses have increased tolerance to glyphosate under elevated carbon dioxide. *Weed Science,* **59,** 28-36. http://dx.doi.org/10.1614/ws-d-10-00080.1

106. Ziska, L.H., 2011: Climate change, carbon dioxide and global crop production: Food security and uncertainty. *Handbook on Climate Change and Agriculture.* Dinar, A. and R. Mendelsohn, Eds. Edward Elgar Publishing, Cheltenham, United Kingdom, 9-31.

107. DiTommaso, A., Q. Zhong, and D.R. Clements, 2014: Identifying climate change as a factor in the establishment and persistence of invasive weeds in agricultural crops. *Invasive Species and Global Climate Change.* Ziska, L.H. and J.S. Dukes, Eds. CABI Books, Wallingford, Oxfordshire, UK, 253-270.

108. Ziska, L.H. and L.L. McConnell, 2015: Climate change, carbon dioxide, and pest biology: Monitor, mitigate, manage. *Journal of Agricultural and Food Chemistry.* http://dx.doi.org/10.1021/jf506101h

109. Bailey, S.W., 2004: Climate change and decreasing herbicide persistence. *Pest Management Science,* **60,** 158-162. http://dx.doi.org/10.1002/ps.785

110. Delcour, I., P. Spanoghe, and M. Uyttendaele, 2015: Literature review: Impact of climate change on pesticide use. *Food Research International,* **68,** 7-15. http://dx.doi.org/10.1016/j.foodres.2014.09.030

111. WHO, 2014: Antimicrobial Resistance: Global Report on Surveillance. 257 pp. World Health Organization, Geneva. http://www.who.int/drugresistance/documents/surveillancereport/en/

112. Cooper, K.M., C. McMahon, I. Fairweather, and C.T. Elliott, 2015: Potential impacts of climate change on veterinary medicinal residues in livestock produce: An island of Ireland perspective. *Trends in Food Science & Technology,* **44,** 21-35. http://dx.doi.org/10.1016/j.tifs.2014.03.007

113. Gormaz, J.G., J.P. Fry, M. Erazo, and D.C. Love, 2014: Public health perspectives on aquaculture. *Current Environmental Health Reports,* **1,** 227-238. http://dx.doi.org/10.1007/s40572-014-0018-8

114. Lafferty, K.D., J.W. Porter, and S.E. Ford, 2004: Are diseases increasing in the ocean? *Annual Review of Ecology, Evolution, and Systematics, 35,* 31-54. http://dx.doi.org/10.1146/annurev.ecolsys.35.021103.105704

115. Karvonen, A., P. Rintamäki, J. Jokela, and E.T. Valtonen, 2010: Increasing water temperature and disease risks in aquatic systems: Climate change increases the risk of some, but not all, diseases. *International Journal for Parasitology, 40,* 1483-1488. http://dx.doi.org/10.1016/j.ijpara.2010.04.015

116. Bhaskaram, P., 2002: Micronutrient malnutrition, infection, and immunity: An overview. *Nutrition Reviews, 60,* S40-S45. http://dx.doi.org/10.1301/00296640260130722

117. Ezzati, M., A.D. Lopez, A. Rodgers, S. Vander Hoorn, and C.J.L. Murray, 2002: Selected major risk factors and global and regional burden of disease. *The Lancet, 360,* 1347-1360. http://dx.doi.org/10.1016/s0140-6736(02)11403-6

118. Kennedy, G., G. Nantel, and P. Shetty, 2003: The scourge of "hidden hunger": Global dimensions of micronutrient deficiencies. *Food Nutrition and Agriculture 32,* 8-16. ftp://193.43.36.93/docrep/fao/005/y8346m/y8346m01.pdf

119. Kaidar-Person, O., B. Person, S. Szomstein, and R.J. Rosenthal, 2008: Nutritional deficiencies in morbidly obese patients: A new form of malnutrition? *Obesity Surgery, 18,* 1028-1034. http://dx.doi.org/10.1007/s11695-007-9350-5

120. Stein, A.J., 2010: Global impacts of human mineral malnutrition. *Plant and Soil, 335,* 133-154. http://dx.doi.org/10.1007/s11104-009-0228-2

121. Gibson, R.S., 2012: Zinc deficiency and human health: Etiology, health consequences, and future solutions. *Plant and Soil, 361,* 291-299. http://dx.doi.org/10.1007/s11104-012-1209-4

122. Fulgoni, V.L., D.R. Keast, R.L. Bailey, and J. Dwyer, 2011: Foods, fortificants, and supplements: Where do Americans get their nutrients? *Journal of Nutrition, 141,* 1847-1854. http://dx.doi.org/10.3945/jn.111.142257

123. Song, Y., S.W. Leonard, M.G. Traber, and E. Ho, 2009: Zinc deficiency affects DNA damage, oxidative stress, antioxidant defenses, and DNA repair in rats. *Journal of Nutrition, 139,* 1626-1631. http://dx.doi.org/10.3945/jn.109.106369

124. Flancbaum, L., S. Belsley, V. Drake, T. Colarusso, and E. Tayler, 2006: Preoperative nutritional status of patients undergoing Roux-en-Y gastric bypass for morbid obesity. *Journal of Gastrointestinal Surgery, 10,* 1033-1037. http://dx.doi.org/10.1016/j.gassur.2006.03.004

125. Kimmons, J.E., H.M. Blanck, B.C. Tohill, J. Zhang, and L.K. Khan, 2006: Associations between body mass index and the prevalence of low micronutrient levels among US adults. *Medscape General Medicine, 8,* 59. PMC1868363

126. Lopez-Ridaura, R., W.C. Willett, E.B. Rimm, S. Liu, M.J. Stampfer, J.E. Manson, and F.B. Hu, 2004: Magnesium intake and risk of type 2 diabetes in men and women. *Diabetes Care, 27,* 134-140. http://dx.doi.org/10.2337/diacare.27.1.134

127. Conroy, J.P., 1992: Influence of elevated atmospheric CO_2 concentrations on plant nutrition. *Australian Journal of Botany, 40,* 445-456. http://dx.doi.org/10.1071/bt9920445

128. Manderscheid, R., J. Bender, H.J. Jäger, and H.J. Weigel, 1995: Effects of season long CO_2 enrichment on cereals. II. Nutrient concentrations and grain quality. *Agriculture, Ecosystems & Environment, 54,* 175-185. http://dx.doi.org/10.1016/0167-8809(95)00602-o

129. Ziska, L.H., C.F. Morris, and E.W. Goins, 2004: Quantitative and qualitative evaluation of selected wheat varieties released since 1903 to increasing atmospheric carbon dioxide: Can yield sensitivity to carbon dioxide be a factor in wheat performance? *Global Change Biology, 10,* 1810-1819. http://dx.doi.org/10.1111/j.1365-2486.2004.00840.x

130. Högy, P. and A. Fangmeier, 2008: Effects of elevated atmospheric CO_2 on grain quality of wheat. *Journal of Cereal Science, 48,* 580-591. http://dx.doi.org/10.1016/j.jcs.2008.01.006

131. Högy, P. and A. Fangmeier, 2009: Atmospheric CO_2 enrichment affects potatoes: 2. Tuber quality traits. *European Journal of Agronomy, 30,* 85-94. http://dx.doi.org/10.1016/j.eja.2008.07.006

132. Fernando, N., J. Panozzo, M. Tausz, R.M. Norton, N. Neumann, G.J. Fitzgerald, and S. Seneweera, 2014: Elevated CO_2 alters grain quality of two bread wheat cultivars grown under different environmental conditions. *Agriculture, Ecosystems & Environment, 185,* 24-33. http://dx.doi.org/10.1016/j.agee.2013.11.023

133. Loladze, I., 2002: Rising atmospheric CO₂ and human nutrition: Toward globally imbalanced plant stoichiometry? *Trends in Ecology & Evolution,* **17,** 457-461. http://dx.doi.org/10.1016/s0169-5347(02)02587-9

134. Taub, D.R., B. Miller, and H. Allen, 2008: Effects of elevated CO₂ on the protein concentration of food crops: A meta-analysis. *Global Change Biology,* **14,** 565-575. http://dx.doi.org/10.1111/j.1365-2486.2007.01511.x

135. Myers, S.S., A. Zanobetti, I. Kloog, P. Huybers, A.D.B. Leakey, A.J. Bloom, E. Carlisle, L.H. Dietterich, G. Fitzgerald, T. Hasegawa, N.M. Holbrook, R.L. Nelson, M.J. Ottman, V. Raboy, H. Sakai, K.A. Sartor, J. Schwartz, S. Seneweera, M. Tausz, and Y. Usui, 2014: Increasing CO₂ threatens human nutrition. *Nature,* **510,** 139-142. http://dx.doi.org/10.1038/nature13179

136. Raubenheimer, D., G.E. Machovsky-Capuska, A.K. Gosby, and S. Simpson, 2014: Nutritional ecology of obesity: From humans to companion animals. *British Journal of Nutrition,* **113,** S26-S39. http://dx.doi.org/10.1017/s0007114514002323

137. Jablonski, L.M., X. Wang, and P.S. Curtis, 2002: Plant reproduction under elevated CO₂ conditions: A meta-analysis of reports on 79 crop and wild species. *New Phytologist,* **156,** 9-26. http://dx.doi.org/10.1046/j.1469-8137.2002.00494.x

138. Tans, P. and R. Keeling, 2012: Trends in Atmospheric Carbon Dioxide, Full Mauna Loa CO₂ Record. NOAA Earth System Research Laboratory. http://www.esrl.noaa.gov/gmd/ccgg/trends/

139. Cotrufo, M.F., P. Ineson, and A. Scott, 1998: Elevated CO2 reduces the nitrogen concentration of plant tissues. *Global Change Biology,* **4,** 43-54. http://dx.doi.org/10.1046/j.1365-2486.1998.00101.x

140. Taub, D.R. and X. Wang, 2008: Why are nitrogen concentrations in plant tissues lower under elevated CO₂? A critical examination of the hypotheses. *Journal of Integrative Plant Biology,* **50,** 1365-1374. http://dx.doi.org/10.1111/j.1744-7909.2008.00754.x

141. Layman, D.K., R.A. Boileau, D.J. Erickson, J.E. Painter, H. Shiue, C. Sather, and D.D. Christou, 2003: A reduced ratio of dietary carbohydrate to protein improves body composition and blood lipid profiles during weight loss in adult women. *Journal of Nutrition,* **133,** 411-417. http://jn.nutrition.org/content/133/2/411.full.pdf+html

142. Krieger, J.W., H.S. Sitren, M.J. Daniels, and B. Langkamp-Henken, 2006: Effects of variation in protein and carbohydrate intake on body mass and composition during energy restriction: A meta-regression. *The American Journal of Clinical Nutrition,* **83,** 260-274. http://ajcn.nutrition.org/content/83/2/260.long

143. Ebbeling, C.B., J.F. Swain, H.A. Feldman, W.W. Wong, D.L. Hachey, E. Garcia-Lago, and D.S. Ludwig, 2012: Effects of dietary composition on energy expenditure during weight-loss maintenance. *JAMA - Journal of the American Medical Association,* **307,** 2627-2634. http://dx.doi.org/10.1001/jama.2012.6607

144. Grifferty, A. and S. Barrington, 2000: Zinc uptake by young wheat plants under two transpiration regimes. *Journal of Environment Quality,* **29,** 443-446. http://dx.doi.org/10.2134/jeq2000.00472425002900020011x

145. McGrath, J.M. and D.B. Lobell, 2013: Reduction of transpiration and altered nutrient allocation contribute to nutrient decline of crops grown in elevated CO₂ concentrations. *Plant, Cell & Environment,* **36,** 697-705. http://dx.doi.org/10.1111/pce.12007

146. Ericksen, P.J., 2008: Conceptualizing food systems for global environmental change research. *Global Environmental Change,* **18,** 234-245. http://dx.doi.org/10.1016/j.gloenvcha.2007.09.002

147. Koetse, M.J. and P. Rietveld, 2009: The impact of climate change and weather on transport: An overview of empirical findings. *Transportation Research Part D: Transport and Environment,* **14,** 205-221. http://dx.doi.org/10.1016/j.trd.2008.12.004

148. Koetse, M.J. and P. Rietveld, 2012: Adaptation to climate change in the transport sector. *Transport Reviews,* **32,** 267-286. http://dx.doi.org/10.1080/01441647.2012.657716

149. Zhang, Y. and A. Erera, 2012: Consequence assessment for complex food transportation systems facing catastrophic disruptions. *Homeland Security Affairs,* **4.** http://hdl.handle.net/10945/25011

150. McGuirk, M., S. Shuford, T.C. Peterson, and P. Pisano, 2009: Weather and climate change implications for surface transportation in the USA. *WMO Bulletin,* **58,** 84-93. http://citeseerx.ist.psu.edu/viewdoc/download?doi=10.1.1.459.696&rep=rep1&type=pdf

151. USDA, 2005: Grain Transportation Report, November 17, 2005. 20 pp. U.S. Department of Agriculture, Transportation Services Branch. http://apps.ams.usda.gov/SearchReports/Documents/steldev3100982.pdf

152. IPCC, 2007: Climate Change 2007: Impacts, Adaptation and Vulnerability. Contribution of Working Group II to the Fourth Assessment Report of the Intergovernmental Panel on Climate Change. Parry, M.L., O.F. Canziani, J.P. Palutikof, P.J. van der Linden, and C.E. Hanson (Eds.), 976 pp. Cambridge University Press, Cambridge, UK. https://www.ipcc.ch/pdf/assessment-report/ar4/wg2/ar4_wg2_full_report.pdf

153. Blake, E.S., T.B. Kimberlain, R.J. Berg, J.P. Cangialosi, and J.L. Beven, II, 2013: Tropical Cyclone Report: Hurricane Sandy (AL182012), 22-29 October 2012. 157 pp. National Hurricane Center. http://www.nhc.noaa.gov/data/tcr/AL182012_Sandy.pdf

154. CCSP, 2008: Impacts of Climate Change and Variability on Transportation Systems and Infrastructure: Gulf Study, Phase I. A Report by the U.S. Climate Change Science Program and the Subcommittee on Global Change Research. Savonis, M.J., V.R. Burkett, and J.R. Potter (Eds.), 445 pp. U.S. Department of Transportation, Washington, D.C. http://downloads.globalchange.gov/sap/sap4-7/sap4-7-final-all.pdf

155. NOAA, 2014: Storm Surge Overview. National Oceanic and Atmospheric Administration, National Weather Service, National Hurricane Center, Miami, FL. http://www.nhc.noaa.gov/surge/

156. Lal, R., J.A. Delgado, J. Gulliford, D. Nielsen, C.W. Rice, and R.S. Van Pelt, 2012: Adapting agriculture to drought and extreme events. *Journal of Soil and Water Conservation,* **67,** 162A-166A. http://dx.doi.org/10.2489/jswc.67.6.162A

157. DOE, 2012: Electric Disturbance Events (OE-417). U.S. Department of Energy, Office of Electricity Delivery & Energy Reliability, Washington, D.C. http://www.oe.netl.doe.gov/oe417.aspx

158. Marx, M.A., C.V. Rodriguez, J. Greenko, D. Das, R. Heffernan, A.M. Karpati, F. Mostashari, S. Balter, M. Layton, and D. Weiss, 2006: Diarrheal illness detected through syndromic surveillance after a massive power outage: New York City, August 2003. *American Journal of Public Health,* **96,** 547-553. http://dx.doi.org/10.2105/ajph.2004.061358

159. NOAA, 2012: NCDC Announces Warmest Year on Record for Contiguous U.S. National Oceanic and Atmospheric Administration, National Centers for Environmental Information. http://www.ncdc.noaa.gov/news/ncdc-announces-warmest-year-record-contiguous-us

160. USDA, 2013: U.S. Drought 2012: Farm and Food Impacts. United States Department of Agriculture, Economic Research Service. http://www.ers.usda.gov/topics/in-the-news/us-drought-2012-farm-and-food-impacts.aspx#.UgzjpG3W5AA

161. Olson, K.R. and L.W. Morton, 2014: Dredging of the fractured bedrock-lined Mississippi River Channel at Thebes, Illinois. *Journal of Soil and Water Conservation,* **69,** 31A-35A. http://dx.doi.org/10.2489/jswc.69.2.31A

162. AWO, 2015: Jobs and Economy: Industry Facts. American Waterways Operators, Arlington, VA. http://americanwaterways.com/initiatives/jobs-economy/industry-facts

163. Schmidhuber, J. and F.N. Tubiello, 2007: Global food security under climate change. *Proceedings of the National Academy of Sciences,* **104,** 19703-19708. http://dx.doi.org/10.1073/pnas.0701976104

164. CDC, 2014: *E. coli* Infection and Food Safety. Centers for Disease Control and Prevention, Atlanta, GA. http://www.cdc.gov/features/ecoliinfection/

165. Tarr, P.I., C.A. Gordon, and W.L. Chandler, 2005: Shiga-toxin-producing *Escherichia coli* and haemolytic uraemic syndrome. *The Lancet,* **365,** 19-25. http://dx.doi.org/10.1016/S0140-6736(05)71144-2

166. Wong, C.S., J.C. Mooney, J.R. Brandt, A.O. Staples, S. Jelacic, D.R. Boster, S.L. Watkins, and P.I. Tarr, 2012: Risk factors for the hemolytic uremic syndrome in children infected with *Escherichia coli* O157:H7: A multivariate analysis. *Clinical Infectious Diseases,* **55,** 33-41. http://dx.doi.org/10.1093/cid/cis299

167. Forman, J., J. Silverstein, Committee on Nutrition, and Council on Environmental Health, 2012: Organic foods: Health and environmental advantages and disadvantages. *Pediatrics,* **130,** e1406-e1415. http://dx.doi.org/10.1542/peds.2012-2579

168. Eskenazi, B., L.G. Rosas, A.R. Marks, A. Bradman, K. Harley, N. Holland, C. Johnson, L. Fenster, and D.B. Barr, 2008: Pesticide toxicity and the developing brain. *Basic & Clinical Pharmacology & Toxicology,* **102,** 228-236. http://dx.doi.org/10.1111/j.1742-7843.2007.00171.x

169. Hendrickson, D., C. Smith, and N. Eikenberry, 2006: Fruit and vegetable access in four low-income food deserts communities in Minnesota. *Agriculture and Human Values,* **23,** 371-383. http://dx.doi.org/10.1007/s10460-006-9002-8

170. Furey, S., C. Strugnell, and H. McIlveen, 2001: An investigation of the potential existence of "food deserts" in rural and urban areas of Northern Ireland. *Agriculture and Human Values,* **18,** 447-457. http://dx.doi.org/10.1023/a:1015218502547

171. Bennett, T.M.B., N.G. Maynard, P. Cochran, R. Gough, K. Lynn, J. Maldonado, G. Voggesser, S. Wotkyns, and K. Cozzetto, 2014: Ch. 12: Indigenous Peoples, Lands, and Resources. *Climate Change Impacts in the United States: The Third National Climate Assessment.* Melillo, J.M., T.C. Richmond, and G.W. Yohe, Eds. U.S. Global Change Research Program, Washington, DC, 297-317. http://dx.doi.org/10.7930/J09G5JR1

172. Schneider Chafen, J.J., S.J. Newberry, M.A. Riedl, D.M. Bravata, M. Maglione, M.J. Suttorp, V. Sundaram, N.M. Paige, A. Towfigh, B.J. Hulley, and P.G. Shekelle, 2010: Diagnosing and managing common food allergies: A systematic review. *JAMA - Journal of the American Medical Association,* **303,** 1848-1856. http://dx.doi.org/10.1001/jama.2010.582

173. Branum, A.M. and S.L. Lukacs, 2008: Food Allergy Among U.S. Children: Trends in Prevalence and Hospitalizations. NCHS Data Brief No. 10, October 2008, 8 pp. National Center for Health Statistics, Hyattsville, MD. http://www.cdc.gov/nchs/data/databriefs/db10.pdf

174. Singer, B.D., L.H. Ziska, D.A. Frenz, D.E. Gebhard, and J.G. Straka, 2005: Increasing Amb a 1 content in common ragweed (*Ambrosia artemisiifolia*) pollen as a function of rising atmospheric CO_2 concentration. *Functional Plant Biology,* **32,** 667-670. http://dx.doi.org/10.1071/fp05039

175. Beggs, P.J., 2004: Impacts of climate change on aeroallergens: Past and future. *Clinical & Experimental Allergy,* **34,** 1507-1513. http://dx.doi.org/10.1111/j.1365-2222.2004.02061.x

176. Welch, A.H., D.B. Westjohn, D.R. Helsel, and R.B. Wanty, 2000: Arsenic in ground water of the United States: Occurrence and geochemistry. *Ground Water,* **38,** 589-604. http://dx.doi.org/10.1111/j.1745-6584.2000.tb00251.x

177. Appleyard, S.J., J. Angeloni, and R. Watkins, 2006: Arsenic-rich groundwater in an urban area experiencing drought and increasing population density, Perth, Australia. *Applied Geochemistry,* **21,** 83-97. http://dx.doi.org/10.1016/j.apgeochem.2005.09.008

178. Levin, R.B., P.R. Epstein, T.E. Ford, W. Harrington, E. Olson, and E.G. Reichard, 2002: U.S. drinking water challenges in the twenty-first century. *Environmental Health Perspectives,* **110,** 43-52. PMC1241146

179. Soto-Arias, J.P., R. Groves, and J.D. Barak, 2013: Interaction of phytophagous insects with *Salmonella enterica* on plants and enhanced persistence of the pathogen with *Macrosteles quadrilineatus* infestation or *Frankliniella occidentalis* feeding. *PLoS ONE,* **8,** e79404. http://dx.doi.org/10.1371/journal.pone.0079404

180. Kopanic, R.J., B.W. Sheldon, and C. Wright, 1994: Cockroaches as vectors of Salmonella: Laboratory and field trials. *Journal of Food Protection,* **57,** 125-131.

181. Soto-Arias, J.P., R.L. Groves, and J.D. Barak, 2014: Transmission and retention of *Salmonella enterica* by phytophagous hemipteran insects. *Applied and Environmental Microbiology,* **80,** 5447-5456. http://dx.doi.org/10.1128/aem.01444-14

182. Rose, J.B. and F. Wu, 2015: Waterborne and foodborne diseases. *Climate Change and Public Health.* Levy, B. and J. Patz, Eds. Oxford University Press, Oxford, UK, 157-172. http://dx.doi.org/10.1093/med/9780190202453.003.0008

183. Ahmed, S.M., B.A. Lopman, and K. Levy, 2013: A systematic review and meta-analysis of the global seasonality of norovirus. *PLoS ONE,* **8,** e75922. http://dx.doi.org/10.1371/journal.pone.0075922

184. Meinshausen, M., S.J. Smith, K. Calvin, J.S. Daniel, M.L.T. Kainuma, J.-F. Lamarque, K. Matsumoto, S.A. Montzka, S.C.B. Raper, K. Riahi, A. Thomson, G.J.M. Velders, and D.P.P. van Vuuren, 2011: The RCP greenhouse gas concentrations and their extensions from 1765 to 2300. *Climatic Change,* **109,** 213-241. http://dx.doi.org/10.1007/s10584-011-0156-z

185. USGS, 2015: USGS Water Resources: 07010000 Mississippi River at St. Louis, MO. National Water Information System, United States Geological Survey. http://waterdata.usgs.gov/nwis/inventory?agency_code=USGS&site_no=07010000

8 MENTAL HEALTH AND WELL-BEING

Lead Author

Daniel Dodgen
U.S. Department of Health and Human Services, Office of the
Assistant Secretary for Preparedness and Response

Contributing Authors

Darrin Donato
U.S. Department of Health and Human Services, Office of the
Assistant Secretary for Preparedness and Response

Nancy Kelly
U.S. Department of Health and Human Services, Substance Abuse
and Mental Health Services Administration

Annette La Greca
University of Miami

Joshua Morganstein
Uniformed Services University of the Health Sciences

Joseph Reser
Griffith University

Josef Ruzek
U.S. Department of Veterans Affairs

Shulamit Schweitzer
U.S. Department of Health and Human Services, Office of the
Assistant Secretary for Preparedness and Response

Mark M. Shimamoto*
U.S. Global Change Research Program, National Coordination Office

Kimberly Thigpen Tart
National Institutes of Health

Robert Ursano
Uniformed Services University of the Health Sciences

Acknowledgements: Anthony Barone, U.S. Department of Health and Human Services, Office of the Assistant Secretary for Preparedness and Response; **Kathleen Danskin,** U.S. Department of Health and Human Services, Office of the Assistant Secretary for Preparedness and Response; **Trina Dutta,** Formerly of the U.S. Department of Health and Human Services, Substance Abuse and Mental Health Services Administration; **Ilya Fischhoff,** U.S. Global Change Research Program, National Coordination Office; **Lizna Makhani,** Formerly of the U.S. Department of Health and Human Services, Substance Abuse and Mental Health Services Administration

Recommended Citation: Dodgen, D., D. Donato, N. Kelly, A. La Greca, J. Morganstein, J. Reser, J. Ruzek, S. Schweitzer, M.M. Shimamoto, K. Thigpen Tart, and R. Ursano, 2016: Ch. 8: Mental Health and Well-Being. *The Impacts of Climate Change on Human Health in the United States: A Scientific Assessment.* U.S. Global Change Research Program, Washington, DC, 217–246. http://dx.doi. org/10.7930/J0TX3C9H

8 MENTAL HEALTH AND WELL-BEING

Key Findings

Exposure to Disasters Results in Mental Health Consequences

Key Finding 1: Many people exposed to climate-related or weather-related disasters experience stress and serious mental health consequences. Depending on the type of the disaster, these consequences include post-traumatic stress disorder (PTSD), depression, and general anxiety, which often occur at the same time *[Very High Confidence]*. The majority of affected people recover over time, although a significant proportion of exposed individuals develop chronic psychological dysfunction *[High Confidence]*.

Specific Groups of People Are at Higher Risk

Key Finding 2: Specific groups of people are at higher risk for distress and other adverse mental health consequences from exposure to climate-related or weather-related disasters. These groups include children, the elderly, women (especially pregnant and post-partum women), people with preexisting mental illness, the economically disadvantaged, the homeless, and first responders *[High Confidence]*. Communities that rely on the natural environment for sustenance and livelihood, as well as populations living in areas most susceptible to specific climate change events, are at increased risk for adverse mental health outcomes *[High Confidence]*.

Climate Change Threats Result in Mental Health Consequences and Social Impacts

Key Finding 3: Many people will experience adverse mental health outcomes and social impacts from the threat of climate change, the perceived direct experience of climate change, and changes to one's local environment *[High Confidence]*. Media and popular culture representations of climate change influence stress responses and mental health and well-being *[Medium Confidence]*.

Extreme Heat Increases Risks for People with Mental Illness

Key Finding 4: People with mental illness are at higher risk for poor physical and mental health due to extreme heat *[High Confidence]*. Increases in extreme heat will increase the risk of disease and death for people with mental illness, including elderly populations and those taking prescription medications that impair the body's ability to regulate temperature *[High Confidence]*.

Introduction

The effects of global climate change on mental health and well-being are integral parts of the overall climate-related human health impacts. Mental health consequences of climate change range from minimal stress and distress symptoms to clinical disorders, such as anxiety, depression, post-traumatic stress, and suicidal thoughts.[1, 2, 3, 4, 5] Other consequences include effects on the everyday life, perceptions, and experiences of individuals and communities attempting to understand and respond appropriately to climate change and its implications.[3, 6, 7]

The social and mental health consequences of extreme weather events have been the focus of research for more than three decades.[3, 4, 5, 8, 9, 10] The mental health and well-being consequences of extreme events, particularly natural disasters, are common and form a significant part of the overall effects on health. These consequences of climate change related impacts rarely occur in isolation, but often interact with other social and environmental stressors.

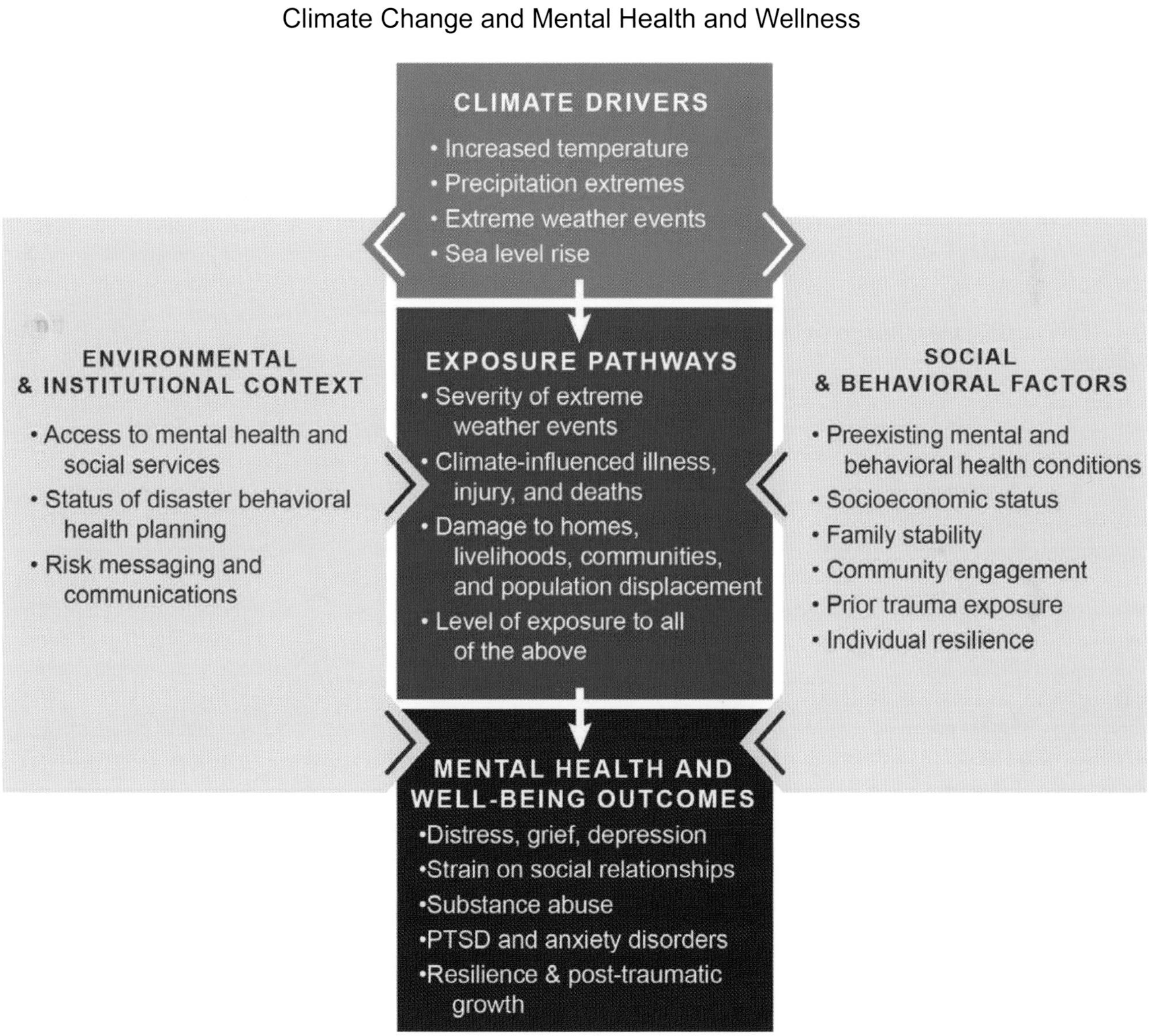

Climate Change and Mental Health and Wellness

Figure 1: This conceptual diagram illustrates the key pathways by which humans are exposed to health threats from climate drivers, and potential resulting mental health and well-being outcomes (center boxes). These exposure pathways exist within the context of other factors that positively or negatively influence health outcomes (gray side boxes). Key factors that influence health outcomes and vulnerability for individuals are shown in the right box, and include social determinants of health and behavioral choices. Key factors that influence health outcomes and vulnerability at larger community or societal scales, such as natural and built environments, governance and management, and institutions, are shown in the left box. All of these influencing factors may also be affected by climate change. See Chapter 1: Introduction for more information.

The threat of climate change is a key psychological and emotional stressor. Individuals and communities are affected both by direct experience of local events attributed to climate change and by exposure to information regarding climate change and its effects.[10, 11, 12, 13, 14, 15] For example, public communication and media messages about climate change and its projected consequences can affect perceptions of physical and societal risks and consequently affect mental health and well-being. The interactive and cumulative nature of climate change effects on health, mental health, and well-being are critical factors in understanding the overall consequences of climate change on human health.[16]

People have inherent capabilities to adjust to new information and experiences and adopt new behaviors to cope with change. There is also an array of interventions and treatments that mental health practitioners use to address mental health conditions and stress reactions. These interventions occur within the context of health systems that have finite resources to deliver these services. These considerations are not discussed in detail, as this chapter focuses on the state of the science regarding the effects of climate change on mental health and well-being, rather than potential actions that could be taken in response to the impacts and risks associated with climate change.

8.1 Effects of Climate Change on Mental Health and Well-being

The cumulative and interactive effects of climate change, as well as the threat and perception of climate change, adversely impact individual and societal health, mental health, and well-being. Figure 2 illustrates how climate change impacts create cascading and inter-related mental, physical, and community health effects. These impacts include exposures to higher temperatures and extreme weather events as well as vector-borne disease transmission, degraded air and water quality, and diminished food safety and security.

Extreme Weather Events

In the United States, the mental health impacts of extreme weather mainly have been studied in response to hurricanes and floods[17, 18, 19, 20, 21, 22, 23, 24] and, to a lesser extent, wildfires.[25, 26, 27, 28] Though many studies discuss the mental health impacts of specific historical events, they are demonstrative of the types of mental health issues that could arise as climate change leads to further increases in the frequency, severity, or duration of some types of extreme weather (see Ch. 1: Introduction and Ch. 4: Extreme Events). The mental health impacts of these events, such as hurricanes, floods, and drought, can be expected to increase as more people experience the stress—and often trauma—of these disasters.

Residents and volunteers in Queens, New York City, filter through clothes and food supplies from donors following Superstorm Sandy on November 3, 2012. A majority of individuals psychologically affected by a traumatic event recover over time, and some experience a set of positive changes that known as post-traumatic growth as a result of coping with or experiencing a traumatic event.

Many people exposed to climate- or weather-related natural disasters experience stress reactions and serious mental health consequences, including symptoms of post-traumatic stress disorder (PTSD), depression, and general anxiety, which often occur simultaneously.[29, 30, 31, 32, 33, 34] Mental health effects include grief/bereavement, increased substance use or misuse, and suicidal thoughts.[19, 35, 36, 37, 38] All of these reactions have the potential to interfere with the individual's functioning and well-being, and are especially problematic for certain groups (see "8.2 Populations of Concern" on page 223).

> *The mental health impacts of hurricanes, floods, and drought can be expected to increase as more people experience the stress—and often trauma—of these disasters.*

Exposure to life threatening events, like highly destructive hurricanes such as Hurricane Katrina in 2005, have been associated with acute stress, PTSD, and higher rates of depression and suicide in affected communities.[18, 20, 23, 30, 39, 40, 41, 42, 43, 44, 45, 46, 47] These mental health consequences are of particular concern for people facing recurring disasters, posing a cumulative psychological toll. Following exposure to Hurricane Katrina, veterans with preexisting mental illness had a 6.8 times greater risk for developing any additional mental illness, compared to those veterans without a preexisting mental illness.[48] Following hurricanes, increased levels of PTSD have been experienced by individuals who perceive members of their community as being less supportive or helpful to one another.[49]

Depression and general anxiety are also common consequences of extreme events (such as hurricanes and floods) that involve a loss of life, resources, or social support and social networks or events that involve extensive relocation and life disruption.[20, 21, 23, 29, 30, 31, 33, 37, 41, 46, 50, 51, 52, 53, 54] For example, long-term anxiety and depression, PTSD, and increased aggression (in children) have been found to be associated with floods.[55] First responders following a disaster also experience increased rates of anxiety and depression.[37]

Increases from pre-disaster rates have been observed in interpersonal and domestic violence, including intimate partner violence,[5, 56] particularly toward women, in the wake of climate- or weather-related disasters.[37, 57, 58] High-risk coping behaviors, such as alcohol abuse, can also increase following extreme weather events.[37, 38, 59, 60, 61, 62] Individuals who use alcohol to cope with stress and those with preexisting alcohol use disorders are most vulnerable to increased alcohol use following extreme weather events.[62]

Persons directly affected by a climate- or weather-related disaster are at increased incidence of suicidal thoughts and behaviors. Increases in both suicidal thoughts (from 2.8% to 6.4%) and actual suicidal plans (from 1.0% to 2.5%) were observed in residents 18 months after Hurricane Katrina.[19] Following Hurricanes Katrina and Rita, a study of internally displaced women living in temporary housing found reported rates of suicide attempt and completion to be 78.6 times and 14.7 times the regional average, respectively.[63] In the six months following 1992's Hurricane Andrew, the rate of homicide-suicides doubled to two per month in Miami-Dade County, where the hurricane hit, compared to an average of

one per month during the prior five-year period that did not include hurricane activity of the same scale.[64]

Climate- or weather-related disasters can strain the resources available to provide adequate mental (or even immediate physical) health care, due to the increased number of individuals who experience severe stress and mental health reactions. Communities adversely affected by these events also have diminished interpersonal and social networks available to support mental health needs and recovery due to the destruction and disruption caused by the event.[65]

Drought

Many regions in the United States have experienced drought (see Ch 1: Introduction and Ch. 4: Extreme Events).[66] Long-term drought, unlike sudden extreme weather events, has a slow onset and long duration.[66, 67] Long-term drought interacts over time with multiple environmental and social stressors to disrupt lives and livelihoods and the functioning of individuals, households, and communities.[68, 69, 70] Prolonged drought can have visible and long-term impacts on landscapes, on rural agricultural industries and communities, and on individual and community resilience.[71, 72, 73]

Cascading and interacting economic, social, and daily life circumstances have accompanied prolonged drought in rural regions. Drought-related worry and psychological distress

Impact of Climate Change on Physical, Mental, and Community Health

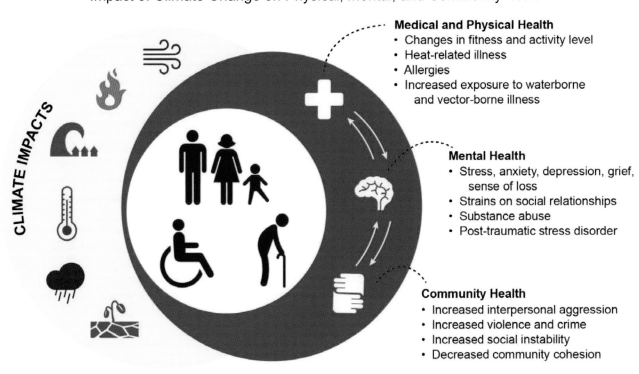

Figure 2: At the center of the diagram are human figures representing adults, children, older adults, and people with disabilities. The left circle depicts climate impacts including air quality, wildfire, sea level rise and storm surge, heat, storms, and drought. The right circle shows the three interconnected health domains that will be affected by climate impacts—Medical and Physical Health, Mental Health, and Community Health. (Figure source: adapted from Clayton et al. 2014).[5]

An elderly couple walk to the Superdome days after Hurricane Katrina made landfall. New Orleans, Louisiana, September 1, 2005.

increased in drought-declared Australian regions, particularly for those experiencing loss of livelihood and industry.[2, 72, 74, 75, 76] Long-term drought has been linked to increased incidence of suicide among male farmers in Australia.[2, 77]

Extreme Heat

The majority (80.7%) of the U.S. population lives in cities and urban areas[78] and urbanization is expected to increase in the future.[79] People in cities may experience greater exposure to heat-related health effects during heat waves (see Ch. 2: Temperature-Related Death and Illness). The impact of extreme heat on mental health is associated with increased incidence of disease and death, aggressive behavior, violence, and suicide and increases in hospital and emergency room admissions for those with mental health or psychiatric conditions.[80, 81, 82, 83, 84, 85, 86, 87]

Individuals with mental illness are especially vulnerable to extreme heat or heat waves. In six case-control studies involving 1,065 heat wave-related deaths, preexisting mental illness was found to triple the risk of death due to heat wave exposure.[88] The risk of death also increases during hot weather for patients with psychosis, dementia, and substance misuse.[84] Hospital admissions have been shown to increase for those with mental illness as a result of extreme heat, increasing ambient temperatures, and humidity.[81, 86, 87] An increased death rate has also been observed in those with mental illness among cases admitted to the emergency department with a diagnosis of heat-related pathology.[82]

People who are isolated and have difficulty caring for themselves—often characteristics of the elderly or those with a mental illness—are also at higher risk for heat-related incidence of disease and death.[86, 88] Fewer opportunities for social interaction and increased isolation[89, 90, 91] put people at elevated risk for not only heat-related illness and death but also decline in mental health and, in some cases, increases in aggression and violence.[5] Hotter temperatures and poorer air quality limit people's outdoor activities. For many, reductions in outdoor exercise and stress-reducing activities lead to diminished physical health, increased stress, and poor mental health.[5]

There may be a link between extreme heat (climate change related or otherwise) and increasing violence, aggressive motives, and/or aggressive behavior.[80, 92, 93, 94] The frequency of interpersonal violence and intergroup conflict may increase with more extreme precipitation and hotter temperatures.[83] These impacts can include heightened aggression, which may result in increased interpersonal violence and violent crime, negatively impacting individual and societal mental health and well-being.[85] Given projections of increasing temperatures (see Ch. 2: Temperature-Related Death and Illness), there is potential for increases in human conflict, but the causal linkages between climate change and conflict are complex and the evidence is still emerging.[83, 95, 96]

Threat of Climate Change as a Stressor

Many people are routinely exposed to images, headlines, and risk messages about the threat of current and projected climate change. Forty percent of Americans report hearing about climate change in the media at least once a month.[97]

Noteworthy environmental changes associated with climate change constitute a powerful environmental stressor—an ongoing and stress-inducing condition or aspect of an individual's everyday environment.[69, 98, 99] Equally concerning are adverse impacts relating to people's connections to place and identity, and consequent sense of loss and disconnection.[11]

About half of Americans reported being worried about climate change in a 2015 survey. However, these people tended to see climate change as a relatively distant threat: 36% said global warming would harm them personally, while more expected harm to come to people in other countries and to future generations.[97] Public risk perceptions of the phenomenon and threat of climate change is associated with stigma, dread risk (such as a heightened fear of low-probability, high-consequence events), and uncertainty about the future.[3, 7, 10, 70, 100, 101, 102, 103, 104, 105, 106, 107]

Many individuals experience a range of adverse psychological responses to the hybrid risk of climate change impacts. A hybrid risk is an ongoing threat or event, which is perceived or understood as reflecting both natural and human causes and processes. These responses include heightened risk perceptions, preoccupation, general anxiety, pessimism, helplessness, eroded sense of self and collective control, stress, distress, sadness, loss, and guilt.[1, 4, 5, 16, 56, 108, 109, 110, 111, 112]

Media representations of serious environmental risks, such as climate change, are thought to elicit strong emotional responses,[7, 113] in part dependent on how climate change information is presented.[114] People experience the threat of climate change through frequent media coverage describing events and future risks attributed to climate change. They also are directly exposed to increasingly visible changes in local environments and seasonal patterns, and in the frequency, magnitude, and intensity of extreme weather events.[6, 115] Furthermore, between 2012 and 2013, roughly a third of U.S. survey respondents report that they have personally experienced the effects of global warming.[12, 13] Exposure to climate change through the media could cause undue stress if the media coverage is scientifically inaccurate or discouraging. However, effective risk communication promotes adaptive and preventive individual or collective action.[4, 5, 116, 117, 118, 119]

Resilience and Recovery

A majority of individuals psychologically affected by a traumatic event (such as a climate-related disaster) will recover over time.[120] A set of positive changes that can occur in a person as a result of coping with or experiencing a traumatic event is called post-traumatic growth.[121, 122, 123, 124] An array of intervention approaches used by mental health practitioners also may reduce the adverse consequence of traumatic events. While most people who are exposed to a traumatic event can be expected to recover over time, a significant proportion (up to 20%) of individuals directly exposed develop chronic levels of psychological dysfunction, which may not get better or be resolved.[21, 35, 47, 53, 125, 126, 127, 128] Multiple risk factors contribute to these adverse psychological effects, including disaster-related factors such as physical injury, death, or loss of a loved one;[18, 23, 51, 129] loss of resources such as possessions or property;[20, 30, 44, 46, 47] and displacement.[32, 130, 131, 132, 133, 134] Life events and stressors secondary to extreme events also affect mental health, including loss of jobs and social connections, financial worries, loss of social support, and family distress or dysfunction.[18, 20, 46, 47, 129, 135]

People experience the threat of climate change through frequent media coverage.

Disaster-related stress reactions and accompanying psychological impacts occur in many individuals directly exposed to the event and can continue over extended time periods (up to a year or more). For example, three months after Hurricane Andrew, 38% of children (age 8 to 12 years) living in affected areas of south Florida reported symptom levels consistent with a "probable diagnosis" of PTSD. At 10 months post-disaster, this proportion declined to about 18%,[21, 44] representing a substantial decrease but still indicating a significant number of individuals with serious mental health issues resulting from the disaster event.

Emerging evidence shows that individuals who are actively involved in climate change adaptation or mitigation actions experience appreciable health and well-being benefit from such engagement.[110, 136] These multiple psychological and environmental benefits do not necessarily minimize distress. However, when people do have distress related to relevant media exposure or to thinking about or discussing climate change, taking action to address the issue can buffer against distress.[110, 136] Such engagement both addresses the threat and helps manage the emotional responses as people come to terms with—and adjust their understandings and lives in the context of—climate change.

8.2 Populations of Concern

Populations of concern will be at higher risk for poor mental health outcomes as the negative effects of climate change progress.[10, 137] In addition to the populations described below, farmers, those with limited mobility, immigrants, those living in coastal areas, those from Indigenous communities or tribes,[138, 139] and veterans are also expected to experience higher risk of poor mental health outcomes (see also Ch. 9: Populations of Concern).[1, 10, 140, 141, 142, 143, 144, 145]

Children are at particular risk for distress, anxiety, and other adverse mental health effects in the aftermath of an extreme event.

Children

Children are at particular risk for distress, anxiety, and other adverse mental health effects in the aftermath of an extreme event. As children are constantly developing, their reactions will vary by age and developmental level. Children have been shown to possess an innate resilience to adverse events,[146, 147, 148, 149] but despite this resilience, children can and do exhibit various stress symptoms when exposed to a traumatic event. These symptoms will depend on the developmental stage of the child, the level and type of exposure, the amount of destruction seen, and that particular child's risk factors and protective factors.[150]

Children are dependent on others for care and a significant predictor of mental health and well-being in a child is the mental health status of the primary caregiver.[5, 151] If the primary caregiver's mental health needs are being addressed, then a child will fare better after experiencing a disaster or other trauma.[5, 150, 151, 152, 153]

The potential exists for an array of difficult emotional and behavioral responses in children shortly after a disaster, such as depression, clinginess, aggressiveness, and social withdrawal, some of which are normal and expected and will resolve over time with proper support. However, children may be at a higher risk than adults of having symptoms persist in the long-term. Significantly more children than adults have shown continued PTSD symptoms more than two years post-disaster, and, in general, children are more likely to be impaired by a disaster.[141] Chronic stress from the acute and ongoing impacts of climate change may alter biological stress response systems and make growing children more at risk for developing mental health conditions later in life, such as anxiety, depression, and other clinically diagnosable disorders.[151]

Women, Pregnant Women, and Post-partum Mothers

Post-disaster stress symptoms are often reported more frequently by women than men.[154, 155] Women have higher prevalence of PTSD and other mental health disorders after disasters than do men,[156] and are prone to greater worry and feelings of vulnerability,[157] anxiety disorders, and other adverse mental health outcomes.[141, 158] Increases in domestic violence towards women are also common after a disaster.[5, 56]

Pregnant and postpartum women can be quite resilient, but their resilience diminishes when social supports are reduced, when they have experienced injury, illness, or danger due to the disaster, and when they have lived through multiple disaster experiences.[39, 57, 159] Estimates indicated that there were 56,100 pregnant women and 74,900 infants directly affected by Hurricane Katrina[160] and that pregnant women with high hurricane exposure and severe hurricane experiences were at a significantly increased risk for PTSD and depression.[156] The increases in PTSD and depression found in pregnant women exposed to Hurricane Katrina were likely due to the severity of the event and the intensity of the disaster experience rather than a general exposure to the event.[42, 156]

The many consequences of natural disasters, such as destruction of homes, and of gradual climate change impacts, such as rising temperatures, incidence of vector-borne illness, water-borne illness, and even compromised food,[160] can all contribute to the emotional stress that women have while pregnant, nursing, or responsible for young children. Nutrition is essential to women's health and well-being, especially if pregnant or nursing. Access to clean water and food is critical, and the lack of either may affect women's ability to cope with the impacts of climate change. Poor nutrition can lead to difficult pregnancies, delivery problems, low birth weight, and even death of a newborn, all of which can be immensely stressful to the mother.[161]

Elderly

In the United States, the number of individuals 65 years of age and older is expected to climb from 47.8 million by the end of 2015 to 98 million in 2060, an increase from 14.9% of the population to 23.6%.[162] The aging population may have difficulty responding to the challenges of climate change, as they tend to have higher rates of untreated depression and physical ailments that contribute to their overall vulnerability, such as increased susceptibility to heat and accompanying physical and mental health and well-being impacts.

Physical health problems are associated with the development of mental health problems,[163, 164] particularly among older adults.[137, 165] Long-term exposure to air pollution is linked with poorer cognitive function and an increased rate of cognitive decline among the elderly.[166, 167, 168, 169, 170] Greater flood exposure, lack of social support, higher stoicism, and the use of maladaptive coping are all associated with greater deterioration in mental health after floods for seniors.[17] The mental health consequences experienced by the elderly in response to a disaster may ultimately be due to challenges they face with physical health, mobility, and difficulty managing trauma in response to the disaster.[142]

Economically Disadvantaged

People living in poverty and with fewer socioeconomic resources have less capacity to adapt to the challenges brought by climate change. They are less able to evacuate should there be a natural disaster, and are more exposed to harmful conditions created by heat waves and poor air quality. Low-income people disproportionately experience the most negative impacts and weather-related mental distress due to more fragile overall health, reduced mobility, reduced access to health care, and economic limitations that reduce the ability to buy goods and services that could provide basic comfort and mitigate the effects of disasters.[140, 143]

A home owner reacts after firefighters arrive to take over the protection of his home and two of his neighbors' homes in Rim Forest, California, October 3, 2003.

Many low-income people in the United States are employed in climate-dependent sectors, such as agriculture and fishing, or live in weather- and temperature-vulnerable areas, such as cities, flood zones, and drought-prone areas (see Ch. 9: Populations of Concern). As observed internationally, such individuals also have higher levels of distress and are more vulnerable to experiencing poor mental health due to extreme weather events or other climate change impacts.[137, 171] Farming or rural communities may be particularly vulnerable to the negative mental health outcomes associated with drought. For example, older farmers in Australia reported experiencing an overwhelming sense of loss as a result of chronic drought and its economic consequences.[172]

Emergency Workers and First Responders

Emergency workers and first responders, including healthcare workers and public safety workers, are exposed to deaths, injuries, diseases, and mental stress caused by climate and weather-related disasters. As some extreme weather events increase in frequency and severity (see Ch. 4: Extreme Events), there will be an increased need for emergency response workers involved in rescue and cleanup.[173] Firefighters, emergency medical service providers, healthcare workers, those recovering human remains, and non-traditional first responders who may be involved with supporting the community after a natural disaster are all at increased risk for mental health consequences, including substance use, both in the short term and long term.[174, 175]

The very nature of the work, which involves being exposed to a traumatic event and helping others in crisis, frequently working long hours in difficult environments and away from loved ones, increases the susceptibility of first responders and emergency workers to experiencing negative mental health consequences. The level of stress and distress in responders increases when the injured are children or people they know.[176] Vicarious trauma or identifying with the victim's suffering, and being overwhelmed by the number and scope of injuries, can also adversely impact the general mental health and well-being of all responders.[176, 177]

Rates of PTSD among first responders have ranged from 13% to 18% up to four years following large-scale response events.[174] Among Australian firefighters with PTSD, a large proportion (77%) also presented with simultaneously occurring mental health conditions, such as depression, panic disorder, or phobic disorders.[174] In a study of Coast Guard responders to Hurricanes Katrina and Rita, local responders were three times more likely to report depression than those who were not local.[178]

Extreme weather events and natural disasters can cause damage to infrastructure (such as power grids, roads, and transportation) and buildings and put response workers at increased risk of traumatic injury and death (see Ch. 4: Extreme

Events).[179] The impacts of more frequent and intense weather events result in increased stress for responders and threaten their overall mental health and well-being.[37, 177, 180]

People Who Are Homeless

About 30% of people who are chronically homeless suffer from some form of mental illness.[181] The majority of homeless populations live in urban and suburban areas, where they are more vulnerable to health risks from exposure to heat waves due to the urban heat island effect.[182] The combination of risk factors, including high rates of mental illness and the geographical location of the homeless, make the homeless very vulnerable to the effects of extreme heat.

Some extreme weather events are projected to become more frequent and severe, and those who become homeless due to these disasters are at increased risk for post-traumatic stress symptoms. People experiencing homelessness are also vulnerable to acquiring a vector-borne illness. Increases in human–mosquito exposure have been observed after hurricanes, such as after Hurricane Katrina.[183] For the homeless population, Lyme disease and West Nile virus have the potential to compound already high rates of mental illness with additional cognitive, neurological, and mental health complications that can result from these vector-borne illnesses.[184, 185]

Individuals with Prior or Preexisting Mental Illness

As of 2013, there were an estimated 43.8 million adults aged 18 or older in the United States who had any mental illness in the past year, representing 18.5% of all adults in the United States.[186] An estimated 2.6 million youth age 12–17 had a major depressive episode during the past year.[186] People with mental illness and those using medications to treat a variety of mental health disorders such as depression, anxiety, and other mood disorders are particularly vulnerable to extreme weather events and extreme heat.[137] Between 2005 and 2010, approximately 6% of the U.S. adolescents aged 12–19 reported using medications to treat a mental illness.[187] As the U.S. population and average age increases, the total number of U.S. adults with depressive disorders is projected to increase from 33.9 million to 45.8 million from 2005 to 2050—a 35% increase, with those over 65 years old having the largest increase (117%) in depressive disorders.[188] As the number of people with mental health disorders increases, so will the number taking medications for these disorders, giving rise to a larger population vulnerable to the effects of extreme heat and extreme weather events.

Extreme weather events carry threats of psychological trauma and disruption to behavioral health services systems. Individuals with mental health and stress-related disorders, such as PTSD, depression, anxiety, sleep difficulties, and sometimes those who abuse drugs or alcohol, can experience an exacerbation of symptoms following a traumatic event. When infrastructure is damaged and communication lines are weakened, mental health services and personal support networks are also disrupt-

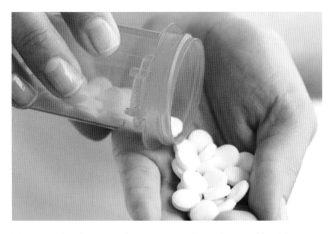

Many medications used to treat a variety of mental health disorders interfere with temperature regulation and heat elimination and may directly induce hyperthermia.

ed, leaving those with a mental illness vulnerable to experiencing additional negative mental health consequences (see Ch. 4: Extreme Events).

Many medications used to treat a variety of mental health disorders interfere with temperature regulation and heat elimination and may directly induce hyperthermia. Being dehydrated can also influence the way some medications such as lithium (used to stabilize mood)[82, 189] or anti-epileptics work in the body.[190] One of the major underlying risks for death due to extreme heat is the use of medications that affect the body's ability to regulate heat or that have neurological effects, increasing susceptibility to the effect of heat.[191]

After the 2012 heat wave in Wisconsin, nearly 52% of the heat-related deaths studied occurred among people with at least one mental illness, and half of those were taking a medication that treats mental illness and sensitizes people to heat.[192] Certain drugs prescribed for depression, sleep disorders, psychosis, and anxiety-related disorders were found to be independent risk factors for heat-related hospitalization cases at an emergency department studied after the 2003 heat wave in France.[82] Many studies have found increased susceptibility to heat for people taking certain classes of medication typically used to treat mental health disorders and other conditions, as well as for alcohol- and drug-dependent people.[81, 89, 189, 193, 194, 195, 196, 197]

Several other factors, besides the effects or side effects of medication use, might explain why people with mental illness are vulnerable to heat-related death.[196, 198] Isolation and deficits in care, common to those with severe mental illness, are critical characteristics of those with the highest rates of heat-related illness and death, as these factors lower the likelihood of utilizing preventive strategies such as showers and cooling shelters during times of extreme heat.[192] Those with mental illness often experience poorer overall health and have fewer social supports. Persons with a combination of mental and physical disorders and who are taking more than one kind of medication are also at greater risk of heat-related death.

8.3 Emerging Issues

Multiple issues warrant further attention regarding the impact of climate change on individuals' and communities' mental health and well-being. Broadly, these include: 1) the impacts of mass evacuation and relocation before, during, and after extreme weather events; 2) the influence of individuals' understandings and attitudes toward climate change and associated risk perceptions on their disaster-related psychological reactions; and 3) the cumulative effects of media presentations of extreme events on mental health and well-being.

A more specific emerging issue is the effect of extreme temperatures on mental health, in particular suicide. Some studies report a connection between higher temperature and suicide;[199] with some indicating increased risk of suicide.[200, 201, 202] The association between hotter temperatures and suicide appears to be stronger for violent suicide methods than for non-violent suicide methods,[203] and there is emergent evidence that deaths by suicide may increase above certain temperatures, suggesting hot weather may trigger impulsive and aggressive behaviors.[201, 204] More studies are needed to better understand the relationship, as negative correlations have been found,[205, 206] as well as no correlation at all.[207, 208, 209]

Children who use methylphenidate (for example, to treat attention deficit disorder) and are engaging in physical activity in hot and humid environments may also be at heightened risk for heat-related illness.[210] More studies are needed to assess what the impact will be on children who use behavior modification medications during extreme heat. In addition, more frequent and prolonged heat waves may increase the amount of time spent indoors, which could have an effect on mental health, particularly for children and those who use the outdoors for exercise and stress management.

As more is learned about the relationship between climate change and vector-borne illnesses, it will be important to further understand the scope of mental health consequences for those who become infected. Chapter 5 (Vector-Borne Diseases) addresses the complex relationship between climate change and vector-borne illnesses, focusing primarily on West Nile Virus (WNV) and Lyme disease. Individuals infected with either WNV or Lyme disease may experience a range of mental health consequences following infection that can include reduced cognitive function as well depression associated with other symptoms, such as fatigue, pain, and muscle and joint aches.[184, 185] These mental health symptoms can last for months but usually resolve over time.

Clinical depression has been observed in patients who are infected with WNV.[211, 212] In a long-term observational study, 35% of participants were found to have new-onset depression. Those with the more severe neuroinvasive forms of WNV are at greater risk for depression between 13 to 18 months post-infection.[212] People who are left with limited mobility as a result of WNV infection can experience long-term mental health impacts.[212] Patient experiences, such as undergoing an extended treatment process or experiencing stress in family or work life due to a lingering illness, can result in mental health consequences.

Poor air quality may have an effect on depression and suicide.[213, 214, 215] While the current literature is not robust enough to imply causation, studies have found significant associations between short-term exposure to air pollution (sulfur dioxide [SO_2], particulate matter smaller than 2.5 microns [$PM_{2.5}$], nitrogen dioxide [NO_2], and carbon monoxide [CO]) and emergency department admissions for depressive episodes in Canada.[213, 214] Recent studies conducted outside of the United States also found associations between air pollution, including aeroallergens, and risk of suicide and emergency department admissions for suicide attempts.[215, 216, 217] These emerging issues may prove to be a significant impact if air quality conditions worsen in the United States.

The severity of risks to mental health and well-being for Indigenous populations that have a close connection to the environment, and in some cases lower economic resources, is also a concern.[144, 145, 218, 219] All of these areas will require further study.

With regard to the impact of climate change related food safety risks, increased CO_2 levels could decrease the nutritional value of some foods (see Ch. 7: Food Safety). Malnutrition (specifically, iron deficiencies) can cause fatigue and depression in children and adolescents.[220] More needs to be learned regarding the mental health and well-being impacts that will result from changes to food composition, quality, and safety due to climate change.

Climate change and rising CO_2 levels may increase the incidence of food allergies.[221] Such an increase in food allergies would have an impact on mental health status, where those with food allergies have higher rates of stress and anxiety.[222] Food allergy in children and adolescents has been connected to psychological distress, including anxiety and depression. Parents of children with food allergies have been found to have higher rates of stress and anxiety than parents of children without food allergies.[223] Those with food allergies have higher rates of major depression, bipolar disorder, panic disorder, and social phobia than those with no food allergy.[224]

8.4 Research Needs

In addition to the emerging issues identified above, the authors highlight the following potential areas for additional scientific and research activity on mental health and well-being, based on their review of the literature. Studies of the broad range of health effects of climate change should incorporate mental health effects and consequences, since many mental health impacts are secondary to other health problems. In addition, the U.S does not currently have sustained psychological and social impact assessments or monitoring programs and measures necessary to identify important changes in mental health and well-being associated with climate change. National psychosocial impact assessment and monitoring programs could enhance the development of standardized methodologies and measures of psychological and social pathways needed to better predict mental health and well-being outcomes.

Future assessments can benefit from research activities that:

- better understand how other health risks from gradual climate change affect mental health, including exposures to extreme heat, poor air quality, diminished food safety and security, and increased vector-borne risks;

- explore the associations between extreme temperatures and violent behavior, including violent suicide;

- develop efficient questionnaires and other methods of collecting data on mental health, psychological, and social impacts for use in epidemiological studies of other health impacts of climate change;

- identify predictors or risk factors for adverse psychological outcomes following weather-related or climate-related disasters;

- further improve evidence-based practices to facilitate recovery and post-traumatic growth following extreme events;

- identify the best practices for adaptation and prevention strategies to reduce the impacts of extreme heat on people with mental illness, including patients taking medications that increase their vulnerability to heat stress;

- improve understanding of the effects of secondary exposure, including cumulative media representations of climate change, as well as how an individual's understanding of the threat of climate change affects their psychological well-being and resilience; and

- enhance understanding of the mental health and psychosocial impacts of long-term displacement, relocation, or loss of culturally significant geographic features, particularly for Indigenous populations.

Supporting Evidence

The chapter was developed through technical discussions of relevant evidence and expert deliberation by the report authors at several workshops, teleconferences, and email exchanges. The authors considered inputs and comments submitted by the public, the National Academies of Sciences, and Federal agencies. For additional information on the overall report process, see Appendices 2 and 3.

Areas of focus for the Mental Health and Well-Being chapter were determined based on the most relevant available scientific literature relating to mental health, wellness, and climate change, as well as the mental health impacts of events associated with climate change. Much of the evidence on these impacts has been compiled in countries outside the United States; however, the scenarios are similar and the evidence directly relevant to the situation in the United States, and thus this literature has been considered in the chapter. The evidence-base on mental health and wellness following extreme weather disasters is both well-established and relevant to climate change. The existence of highly relevant scientific literature on specific concerns directly influenced by climate change—such as the effects of extreme heat, stress associated with the threat and perception of climate change, and special population risks—resulted in the inclusion of these more targeted topics. Although significant scientific literature for resilience exists, in-depth discussions of adaptation, coping, and treatment approaches are outside the scope of this chapter, but are discussed in brief in the Resilience and Recovery section.

KEY FINDING TRACEABLE ACCOUNTS

Exposure to Weather-Related Disasters Results in Mental Health Consequences

Key Finding 1: Many people exposed to climate-related or weather-related disasters experience stress and serious mental health consequences. Depending on the type of the disaster, these consequences include post-traumatic stress disorder (PTSD), depression, and general anxiety, which often occur at the same time *[Very High Confidence]*. The majority of affected people recover over time, although a significant proportion of exposed individuals develop chronic psychological dysfunction *[High Confidence]*.

Description of evidence base

Very strong evidence from multiple studies shows a consensus that many people exposed to climate- or weather-related natural disasters experience stress reactions and serious psychological harm, which often occur simultaneously.[30, 31, 32, 33, 34] Though many of these studies describe the mental health impacts of specific historical events, they demonstrate the

types of mental health issues that will continue to arise as climate change leads to increases in the frequency, severity, and duration of extreme climate- and weather-related events such as floods, hurricanes, droughts, and wildfires.[17, 18, 19, 20, 21, 22, 23, 24, 25, 29, 41, 50, 51, 68, 70] Strong support is found in a number of recent studies for the potential for climate change-related psychological effects, including grief/bereavement, increased substance use or misuse, and thoughts of suicide.[19, 35, 36, 37, 38, 60]

Research on individual resilience and recovery shows that a majority of individuals psychologically affected by a traumatic event will recover over time. However, a convincing body of recent research shows that a significant proportion (typically up to 20%) of individuals directly exposed to the event will develop chronic levels of psychological dysfunction, which may not get better or be resolved.[21, 35, 47, 53, 125, 126, 127, 128]

Major uncertainties

There remains some uncertainty about the degree to which future extreme weather and climate events will impact mental health and wellness. An increase in the scope, frequency, or severity of these events will increase the number of people impacted and the degree to which they are affected. However, efforts that effectively increase preparation for both the physical and psychological consequences of extreme weather- and climate-related events could decrease the impact on mental health and well-being.

Assessment of confidence and likelihood based on evidence

Numerous and recent studies have examined the mental health and wellness impacts of climate- and weather-related events among a variety of populations. Taken as a whole, the strength of this scientific evidence provides **very high confidence** regarding the adverse impacts of environmental changes and events associated with global climate change on individual and societal mental health and well-being, and **high confidence** that these impacts will be long-lasting for a significant portion of the impacted population.

Specific Groups of People Are at Higher Risk

Key Finding 2: Specific groups of people are at higher risk for distress and other adverse mental health consequences from exposure to climate-related or weather-related disasters. These groups include children, the elderly, women (especially pregnant and post-partum women), people with preexisting mental illness, the economically disadvantaged, the homeless, and first-responders *[High Confidence]*. Communities that rely on the natural environment for sustenance and livelihood, as well as populations living in areas most susceptible to specific climate change events, are at increased risk for adverse mental health outcomes *[High Confidence]*.

Description of evidence base

Multiple studies have identified specific populations within the United States that are particularly vulnerable to the mental health impacts of climate change events.[1, 10, 137, 140, 142, 143] Some evidence suggests that children are at particular risk for distress, anxiety, and PTSD.[141, 150, 151] Highly cited studies of the elderly show that high rates of physical and mental health disorders leave them more vulnerable to the impacts of climate change.[17, 142, 163, 164] A large body of post-disaster studies shows that women often have a higher prevalence of PTSD[156] and other adverse psychological outcomes. [5, 39, 56, 57, 141, 157, 158, 159, 160] Research strongly suggests that people who currently suffer from psychological disorders will face additional challenges from climate change impacts.[81, 82, 84, 86, 87] Strong evidence suggests that people living in poverty disproportionately experience the most negative impacts,[140, 143] in part because they have less capacity to evacuate to avoid natural disasters, and because they are more frequently exposed to harmful environmental conditions such as heat waves and poor air quality.[162] Similarly, the majority (91%) of homeless populations live in urban and suburban areas, where they are more vulnerable to certain weather- and climate-related health risks.[182]

A number of studies of disaster responders point to an increased risk of mental and physical health problems following climate-related disasters.[174, 175, 178, 179] More frequent and intense weather events will increase the likelihood of this threat.[37, 177, 180]

Several studies show that those living in drought-prone areas are vulnerable to high levels of distress.[137, 171, 172] In addition, evidence suggests those living in Arctic or other coastal areas, such as Indigenous communities or tribes, tend to be more reliant on natural resources that could be diminished by climate change, which can lead to an increased risk of poor mental health outcomes.[138, 139, 144, 145, 218, 219]

Major uncertainties

While there is uncertainty around the magnitude of effect, there is general agreement that climate-related disasters cause emotional and behavioral responses that will increase the likelihood of a mental illness or effect. Understanding how exposure, sensitivity, and adaptive capacity change over time and location for specific populations of concern is challenging. Uncertainties remain with respect to the underlying social determinants of health, public health interventions or outreach, adaptation options, and climate impacts at fine local scales.

Assessment of confidence and likelihood based on evidence

The combined breadth and strength of the scientific literature supports **high confidence** that certain vulnerable populations will face psychological tolls in the aftermath of climate-related disasters. An increase in adverse climate-related events will result in increased exposure of such populations of concern and an increased likelihood of elevated risk for mental health consequences. There is also **high confidence** that natural-resource-dependent communities and populations living in areas most susceptible to specific climate change events are at increased risk for adverse mental health outcomes.

Climate Change Threats Result in Mental Health Consequences and Social Impacts

Key Finding 3: Many people will experience adverse mental health outcomes and social impacts from the threat of climate change, the perceived direct experience of climate change, and changes to one's local environment *[High Confidence]*. Media and popular culture representations of climate change influence stress responses and mental health and well-being *[Medium Confidence]*.

Description of evidence base

A strong combination of mental health epidemiological research, social science-based national survey research, social and clinical psychology, environmental risk perception research, and disaster mental health research supports the finding that the threat of climate change and perceptions of its related physical environment changes and extreme events together constitute a significant environmental stressor.[3, 7, 10, 11, 69, 70, 98, 99, 100, 101, 102, 103, 104, 105, 106, 107]

A large number of recent studies that have evaluated responses to the hybrid risk (risk that is part natural and part human-caused) of climate change impacts specifically reveal that many individuals experience a range of adverse psychological responses.[1, 4, 5, 16, 56, 108, 109, 110, 111, 112]

Major uncertainties

Major uncertainties derive from the distinction between people's objective and subjective exposure and experience of environmental threats. The multimedia information environment to which individuals are exposed and its coverage of climate change and related events can contribute to complicated public perceptions and strong emotional responses related to climate change as a social, environmental, and political issue.[7, 113, 114] If media exposure is inaccurate or discouraging, that could cause undue stress.

However, accurate risk information dissemination can result in adaptive and preventive individual and collective action.[4, 5, 116, 117, 118, 119] The relative dearth of long term impact assessment and monitoring programs relating to the psychosocial impacts of climate change necessitates reliance on smaller-scale, typically cross-sectional studies and research surveys that are often limited by their use of single-item indicators rather than standardized, climate change-specific, multi-item psychometric measures.

Assessment of confidence and likelihood based on evidence
The large body of well-documented scientific evidence provides **high confidence** that adverse mental health outcomes and social impacts can result from the threat of climate change, the perceived experience of climate change, and changes to one's local environment. Emerging evidence suggests there is **medium confidence** that media representations of climate change influence stress responses and mental health and well-being.

Extreme Heat Increases Risks for People with Mental Illness

Key Finding 4: People with mental illness are at higher risk for poor physical and mental health due to extreme heat *[High Confidence]*. Increases in extreme heat will increase the risk of disease and death for people with mental illness, including elderly populations and those taking prescription medications that impair the body's ability to regulate temperature *[High Confidence]*.

Description of evidence base
Mental, behavioral, and cognitive disorders can be triggered or exacerbated by heat waves. An increased susceptibility to heat due to medication use for psychiatric and other mental health disorders, as well as for alcohol- and drug-dependent people, is supported by numerous studies,[81, 189, 193, 194, 195, 196, 197] and the influence of dehydration on the effects of psychotropic medications is well-documented.[82, 189, 190]

A significant body of evidence shows that the combination of mental illness and extreme heat can result in increases in hospitalizations and even death.[81, 82, 84, 86, 87] Furthermore, six case-control studies, involving 1,065 heat wave-related deaths, have found that preexisting mental illness tripled the risk of death.[88] In a more recent heat wave study, close to 52% of the heat-related fatalities were of people with at least one mental illness and half of those were taking a psychotropic medication.[192]

Major uncertainties
Uncertainties include whether pharmaceutical companies will develop new medications to treat mental illness and other health conditions that make individuals less susceptible to heat, whether strategies for prevention of heat-related

illness and death are implemented, and whether individuals begin to adapt over time to increases in heat. Prevention, detection, and treatment of mental illness without the use of medications that negatively impact the body's ability to regulate heat could moderate the magnitude of extreme heat's impact on those predicted to have psychiatric and stress related disorders.

Assessment of confidence and likelihood based on evidence
A large body of established scientific evidence shows there is **high confidence** that people with mental illness are at greater risk for poor physical and mental health outcomes from climate change. Similarly, there is **high confidence** that exposure to extreme heat will exacerbate such outcomes, particularly for the elderly and those who take certain prescription medications to treat their mental illnesses. Given predictions of growth in the subgroup of the population who have mental health conditions and who take pharmaceuticals that sensitize them to heat, increases in the number of people experiencing related negative health outcomes due to climate change is expected to occur.

DOCUMENTING UNCERTAINTY

This assessment relies on two metrics to communicate the degree of certainty in Key Findings. See Appendix 4: Documenting Uncertainty for more on assessments of likelihood and confidence.

Confidence Level	Likelihood
Very High Strong evidence (established theory, multiple sources, consistent results, well documented and accepted methods, etc.), high consensus	**Very Likely** ≥ 9 in 10
High Moderate evidence (several sources, some consistency, methods vary and/or documentation limited, etc.), medium consensus	**Likely** ≥ 2 in 3
	As Likely As Not ≈ 1 in 2
Medium Suggestive evidence (a few sources, limited consistency, models incomplete, methods emerging, etc.), competing schools of thought	**Unlikely** ≤ 1 in 3
	Very Unlikely ≤ 1 in 10
Low Inconclusive evidence (limited sources, extrapolations, inconsistent findings, poor documentation and/or methods not tested, etc.), disagreement or lack of opinions among experts	

PHOTO CREDITS

Pg. 217–Man in smoke: © Carlos Avila Gonzalez/San
  Francisco Chronicle/Corbis

Pg. 218–Rescuer and resident hugging: © U.S. Air Force/Staff
  Sgt. James L. Harper, Jr.

Pg. 223–Remote control: ©flickr/flash.pro

Pg. 220–Clothing drive: © Alec McClure/Demotix/Corbis

Pg. 222–Elderly couple in flooded street: © Michael
  Ainsworth Dallas Morning News/Corbis

Pg. 224–Children looking through window: © Aurora/Aurora
  Photo/Corbis

Pg. 225–Man in smoke: © Carlos Avila Gonzalez/San
  Francisco Chronicle/Corbis

Pg. 226–Woman with medication: © JGI/Jamie Grill/Blend
  Images/Corbis

References

1. Doherty, T.J. and S. Clayton, 2011: The psychological impacts of global climate change. *American Psychologist,* **66,** 265-276. http://dx.doi.org/10.1037/a0023141

2. Hanigan, I.C., C.D. Butler, P.N. Kokic, and M.F. Hutchinson, 2012: Suicide and drought in New South Wales, Australia, 1970–2007. *Proceedings of the National Academy of Sciences,* **109,** 13950-13955. http://dx.doi.org/10.1073/pnas.1112965109

3. ISSC and UNESCO, 2013: World Social Science Report 2013: Changing Global Environments. 609 pp. OECD Publishing and UNESCO Publishing, Paris, France. http://www.worldsocialscience.org/documents/wss-report-2013-full-text.pdf

4. Doherty, T.J., 2015: Mental health impacts. *Climate Change and Public Health.* Patz, J. and B.S. Levy, Eds. Oxford University Press, New York.

5. Clayton, S., C.M. Manning, and C. Hodge, 2014: Beyond Storms & Droughts: The Psychological Impacts of Climate Change. 51 pp. American Psychological Association and ecoAmerica, Washington, D.C. http://ecoamerica.org/wp-content/uploads/2014/06/eA_Beyond_Storms_and_Droughts_Psych_Impacts_of_Climate_Change.pdf

6. Schmidt, A., A. Ivanova, and M.S. Schäfer, 2013: Media attention for climate change around the world: A comparative analysis of newspaper coverage in 27 countries. *Global Environmental Change,* **23,** 1233-1248. http://dx.doi.org/10.1016/j.gloenvcha.2013.07.020

7. Smith, N. and H. Joffe, 2013: How the public engages with global warming: A social representations approach. *Public Understanding of Science,* **22,** 16-32. http://dx.doi.org/10.1177/0963662512440913

8. Chen, R.S., E. Boulding, and S.H. Schneider, Eds., 1983: *Social Science Research and Climate Change: An Interdisciplinary Appraisal.* Springer, Dordrecht, The Netherlands, 255 pp.

9. Stern, P.C., 1992: Psychological dimensions of global environmental change. *Annual Review of Psychology,* **43,** 269-302. http://dx.doi.org/10.1146/annurev.ps.43.020192.001413

10. Swim, J.K., P.C. Stern, T.J. Doherty, S. Clayton, J.P. Reser, E.U. Weber, R. Gifford, and G.S. Howard, 2011: Psychology's contributions to understanding and addressing global climate change. *American Psychologist,* **66,** 241-250. http://dx.doi.org/10.1037/a0023220

11. Devine-Wright, P., 2013: Think global, act local? The relevance of place attachments and place identities in a climate changed world. *Global Environmental Change,* **23,** 61-69. http://dx.doi.org/10.1016/j.gloenvcha.2012.08.003

12. Leiserowitz, A., E. Maibach, C. Roser-Renouf, G. Feinberg, and P. Howe, 2012: Climate Change in The American Mind: Americans' Global Warming Beliefs and Attitudes in September 2012. 31 pp. Yale Project on Climate Change Communication, Yale University and George Mason University, New Haven, CT. http://environment.yale.edu/climate-communication/files/Climate-Beliefs-September-2012.pdf

13. Leiserowitz, A., E. Maibach, C. Roser-Renouf, G. Feinberg, and P. Howe, 2013: Climate Change in the American Mind: Americans' Global Warming Beliefs and Attitudes In April 2013. 29 pp. Yale Project on Climate Change Communication, Yale University and George Mason University, New Haven, CT. http://environment.yale.edu/climate-communication/files/Climate-Beliefs-April-2013.pdf

14. Fresque-Baxter, J.A. and D. Armitage, 2012: Place identity and climate change adaptation: A synthesis and framework for understanding. *Wiley Interdisciplinary Reviews: Climate Change,* **3,** 251-256. http://dx.doi.org/10.1002/wcc.164

15. Reser, J.P., G.L. Bradley, and M.C. Ellul, 2014: Encountering climate change: 'Seeing' is more than 'believing'. *Wiley Interdisciplinary Reviews: Climate Change,* **5,** 521-537. http://dx.doi.org/10.1002/wcc.286

16. Reser, J.P. and J.K. Swim, 2011: Adapting to and coping with the threat and impacts of climate change. *American Psychologist,* **66,** 277-289. http://dx.doi.org/10.1037/a0023412

17. Bei, B., C. Bryant, K.M. Gilson, J. Koh, P. Gibson, A. Komiti, H. Jackson, and F. Judd, 2013: A prospective study of the impact of floods on the mental and physical health of older adults. *Aging & Mental Health,* **17,** 992-1002. http://dx.doi.org/10.1080/13607863.2013.799119

18. Galea, S., M. Tracy, F. Norris, and S.F. Coffey, 2008: Financial and social circumstances and the incidence and course of PTSD in Mississippi during the first two years after Hurricane Katrina. *Journal of Traumatic Stress,* **21,** 357-368. http://dx.doi.org/10.1002/jts.20355

19. Kessler, R.C., S. Galea, M.J. Gruber, N.A. Sampson, R.J. Ursano, and S. Wessely, 2008: Trends in mental illness and suicidality after Hurricane Katrina. *Molecular Psychiatry,* **13,** 374-384. http://dx.doi.org/10.1038/sj.mp.4002119

20. La Greca, A.M., W.K. Silverman, B. Lai, and J. Jaccard, 2010: Hurricane-related exposure experiences and stressors, other life events, and social support: Concurrent and prospective impact on children's persistent posttraumatic stress symptoms. *Journal of Consulting and Clinical Psychology,* **78,** 794-805. http://dx.doi.org/10.1037/a0020775

21. La Greca, A.M., B.S. Lai, M.M. Llabre, W.K. Silverman, E.M. Vernberg, and M.J. Prinstein, 2013: Children's post-disaster trajectories of PTS symptoms: Predicting chronic distress. *Child & Youth Care Forum,* **42,** 351-369. http://dx.doi.org/10.1007/s10566-013-9206-1

22. Lowe, S.R., M. Tracy, M. Cerda, F.H. Norris, and S. Galea, 2013: Immediate and longer-term stressors and the mental health of Hurricane Ike survivors. *Journal of Traumatic Stress,* **26,** 753-761. http://dx.doi.org/10.1002/jts.21872

23. Norris, F.H., K. Sherrieb, and S. Galea, 2010: Prevalence and consequences of disaster-related illness and injury from Hurricane Ike. *Rehabilitation Psychology,* **55,** 221-230. http://dx.doi.org/10.1037/a0020195

24. Pietrzak, R.H., M. Tracy, S. Galea, D.G. Kilpatrick, K.J. Ruggiero, J.L. Hamblen, S.M. Southwick, and F.H. Norris, 2012: Resilience in the face of disaster: Prevalence and longitudinal course of mental disorders following Hurricane Ike. *PLoS ONE,* **7,** e38964. http://dx.doi.org/10.1371/journal.pone.0038964

25. Maida, C.A., N.S. Gordon, A. Steinberg, and G. Gordon, 1989: Psychosocial impact of disasters: Victims of the Baldwin Hills fire. *Journal of Traumatic Stress,* **2,** 37-48. http://dx.doi.org/10.1007/BF00975765

26. Marshall, G.N., T.L. Schell, M.N. Elliott, N.R. Rayburn, and L.H. Jaycox, 2007: Psychiatric disorders among adults seeking emergency disaster assistance after a wildland-urban interface fire. *Psychiatric Services,* **58,** 509-514. http://dx.doi.org/10.1176/appi.ps.58.4.509

27. Jones, R.T., D.P. Ribbe, P.B. Cunningham, J.D. Weddle, and A.K. Langley, 2002: Psychological impact of fire disaster on children and their parents. *Behavior Modification,* **26,** 163-186. http://dx.doi.org/10.1177/0145445502026002003

28. Langley, A.K. and R.T. Jones, 2005: Coping efforts and efficacy, acculturation, and post-traumatic symptomatology in adolescents following wildfire. *Fire Technology,* **41,** 125-143. http://dx.doi.org/10.1007/s10694-005-6387-7

29. Felton, J.W., D.A. Cole, and N.C. Martin, 2013: Effects of rumination on child and adolescent depressive reactions to a natural disaster: The 2010 Nashville flood. *Journal of Abnormal Psychology,* **122,** 64-73. http://dx.doi.org/10.1037/a0029303

30. Lai, B.S., A.M. La Greca, B.A. Auslander, and M.B. Short, 2013: Children's symptoms of posttraumatic stress and depression after a natural disaster: Comorbidity and risk factors. *Journal of Affective Disorders,* **146,** 71-78. http://dx.doi.org/10.1016/j.jad.2012.08.041

31. Ruggiero, K.J., K. Gros, J.L. McCauley, H.S. Resnick, M. Morgan, D.G. Kilpatrick, W. Muzzy, and R. Acierno, 2012: Mental health outcomes among adults in Galveston and Chambers counties after Hurricane Ike. *Disaster Medicine and Public Health Preparedness,* **6,** 26-32. http://dx.doi.org/10.1001/dmp.2012.7

32. Sastry, N. and M. VanLandingham, 2009: One year later: Mental illness prevalence and disparities among New Orleans residents displaced by Hurricane Katrina. *American Journal of Public Health,* **99,** S725-S731. http://dx.doi.org/10.2105/ajph.2009.174854

33. Scheeringa, M.S. and C.H. Zeanah, 2008: Reconsideration of harm's way: Onsets and comorbidity patterns of disorders in preschool children and their caregivers following Hurricane Katrina. *Journal of Clinical Child & Adolescent Psychology,* **37,** 508-518. http://dx.doi.org/10.1080/15374410802148178

34. Tracy, M., F.H. Norris, and S. Galea, 2011: Differences in the determinants of posttraumatic stress disorder and depression after a mass traumatic event. *Depression and Anxiety,* **28,** 666-675. http://dx.doi.org/10.1002/da.20838

35. Lowe, S.R. and J.E. Rhodes, 2013: Trajectories of psychological distress among low-income, female survivors of Hurricane Katrina. *American Journal of Orthopsychiatry,* **83,** 398-412. http://dx.doi.org/10.1111/ajop.12019

36. North, C.S., C.L. Ringwalt, D. Downs, J. Derzon, and D. Galvin, 2011: Postdisaster course of alcohol use disorders in systematically studied survivors of 10 disasters. *Archives of General Psychiatry,* **68,** 173-180. http://dx.doi.org/10.1001/archgenpsychiatry.2010.131

37. Osofsky, H.J., J.D. Osofsky, J. Arey, M.E. Kronenberg, T. Hansel, and M. Many, 2011: Hurricane Katrina's first responders: The struggle to protect and serve in the aftermath of the disaster. *Disaster Medicine and Public Health Preparedness,* **5,** S214-S219. http://dx.doi.org/10.1001/dmp.2011.53

38. Rohrbach, L.A., R. Grana, E. Vernberg, S. Sussman, and P. Sun, 2009: Impact of Hurricane Rita on adolescent substance use. *Psychiatry: Interpersonal and Biological Processes,* **72,** 222-237. http://dx.doi.org/10.1521/psyc.2009.72.3.222

39. Harville, E.W., X. Xiong, G. Pridjian, K. Elkind-Hirsch, and P. Buekens, 2009: Postpartum mental health after Hurricane Katrina: A cohort study. *BMC Pregnancy Childbirth,* **9,** 21. http://dx.doi.org/10.1186/1471-2393-9-21

40. Roberts, M.C. and J.F. Kelly, Eds., 2008: *Psychologists Responding to Hurricane Katrina. Professional Psychology: Research and Practice, Volume 39.* American Psychological Association, Washington, D.C., 112 pp.

41. Boscarino, J.A., S.N. Hoffman, R.E. Adams, C.R. Figley, and R. Solhkhah, 2014: Mental health outcomes among vulnerable residents after Hurricane Sandy: Implications for disaster research and planning. *American Journal of Disaster Medicine,* **9,** 107-120. http://dx.doi.org/10.5055/ajdm.2014.0147

42. Ehrlich, M., E. Harville, X. Xiong, P. Buekens, G. Pridjian, and K. Elkind-Hirsch, 2010: Loss of resources and hurricane experience as predictors of postpartum depression among women in southern Louisiana. *Journal of Women's Health,* **19,** 877-884. http://dx.doi.org/10.1089/jwh.2009.1693

43. Garrison, C.Z., E.S. Bryant, C.L. Addy, P.G. Spurrier, J.R. Freedy, and D.G. Kilpatrick, 1995: Posttraumatic stress disorder in adolescents after Hurricane Andrew. *Journal of the American Academy of Child and Adolescent Psychiatry,* **34,** 1193-1201. http://dx.doi.org/10.1097/00004583-199509000-00017

44. La Greca, A., W.K. Silverman, E.M. Vernberg, and M.J. Prinstein, 1996: Symptoms of posttraumatic stress in children after Hurricane Andrew: A prospective study. *Journal of Consulting and Clinical Psychology,* **64,** 712-723. http://dx.doi.org/10.1037/0022-006X.64.4.712

45. Osofsky, H.J., J.D. Osofsky, M. Kronenberg, A. Brennan, and T.C. Hansel, 2009: Posttraumatic stress symptoms in children after Hurricane Katrina: Predicting the need for mental health services. *American Journal of Orthopsychiatry,* **79,** 212-220. http://dx.doi.org/10.1037/a0016179

46. Paul, L.A., M. Price, D.F. Gros, K.S. Gros, J.L. McCauley, H.S. Resnick, R. Acierno, and K.J. Ruggiero, 2014: The associations between loss and posttraumatic stress and depressive symptoms following Hurricane Ike. *The Journal of Clinical Psychology,* **70,** 322-332. http://dx.doi.org/10.1002/jclp.22026

47. Pietrzak, R.H., P.H. Van Ness, T.R. Fried, S. Galea, and F.H. Norris, 2013: Trajectories of posttraumatic stress symptomatology in older persons affected by a large-magnitude disaster. *Journal of Psychiatric Research,* **47,** 520-526. http://dx.doi.org/10.1016/j.jpsychires.2012.12.005

48. Sullivan, G., J.J. Vasterling, X. Han, A.T. Tharp, T. Davis, E.A. Deitch, and J.I. Constans, 2013: Preexisting mental illness and risk for developing a new disorder after hurricane Katrina. *The Journal of Nervous and Mental Disease,* **201,** 161-166. http://dx.doi.org/10.1097/NMD.0b013e31827f636d

49. Ursano, R.J., J.B.A. McKibben, D.B. Reissman, X. Liu, L. Wang, R.J. Sampson, and C.S. Fullerton, 2014: Posttraumatic stress disorder and community collective efficacy following the 2004 Florida hurricanes. *PLoS ONE,* **9,** e88467. http://dx.doi.org/10.1371/journal.pone.0088467

50. Boscarino, J.A., S.N. Hoffman, H.L. Kirchner, P.M. Erlich, R.E. Adams, C.R. Figley, and R. Solhkhah, 2013: Mental health outcomes at the Jersey Shore after Hurricane Sandy. *International Journal of Emergency Mental Health and Human Resilience,* **15,** 147-158. http://www.omicsonline.com/open-access/previousissue-international-journal-of-emergency-mental-health-and-human-resilience.pdf

51. Galea, S., C.R. Brewin, M. Gruber, R.T. Jones, D.W. King, L.A. King, R.J. McNally, R.J. Ursano, M. Petukhova, and R.C. Kessler, 2007: Exposure to hurricane-related stressors and mental illness after Hurricane Katrina. *Archives of General Psychiatry,* **64,** 1427-1434. http://dx.doi.org/10.1001/archpsyc.64.12.1427

52. Pina, A.A., I.K. Villalta, C.D. Ortiz, A.C. Gottschall, N.M. Costa, and C.F. Weems, 2008: Social support, discrimination, and coping as predictors of posttraumatic stress reactions in youth survivors of Hurricane Katrina. *Journal of Clinical Child & Adolescent Psychology,* **37,** 564-574. http://dx.doi.org/10.1080/15374410802148228

53. Weems, C.F. and R.A. Graham, 2014: Resilience and trajectories of posttraumatic stress among youth exposed to disaster. *Journal of Child and Adolescent Psychopharmacology,* **24,** 2-8. http://dx.doi.org/10.1089/cap.2013.0042

54. Silove, D. and Z. Steel, 2006: Understanding community psychosocial needs after disasters: Implications for mental health services. *Journal of Postgraduate Medicine,* **52,** 121-125. http://www.jpgmonline.com/text.asp?2006/52/2/121/25157

55. Ahern, M., R.S. Kovats, P. Wilkinson, R. Few, and F. Matthies, 2005: Global health impacts of floods: Epidemiologic evidence. *Epidemiologic Reviews,* **27,** 36-46. http://dx.doi.org/10.1093/epirev/mxi004

56. Fritze, J.G., G.A. Blashki, S. Burke, and J. Wiseman, 2008: Hope, despair and transformation: Climate change and the promotion of mental health and wellbeing. *International Journal of Mental Health Systems,* **2,** 13. http://dx.doi.org/10.1186/1752-4458-2-13

57. Harville, E.W., C.A. Taylor, H. Tesfai, X. Xiong, and P. Buekens, 2011: Experience of Hurricane Katrina and reported intimate partner violence. *Journal of Interpersonal Violence,* **26,** 833-845. http://dx.doi.org/10.1177/0886260510365861

58. Wenden, A.L., 2011: Women and climate change: Vulnerabilities and challenges. *Climate Change and Human Well-being: Global challenges and opportunities.* Weissbecker, I., Ed. Springer-Verlag, New York, 119-133. http://dx.doi.org/10.1007/978-1-4419-9742-5

59. Flory, K., B.L. Hankin, B. Kloos, C. Cheely, and G. Turecki, 2009: Alcohol and cigarette use and misuse among Hurricane Katrina survivors: Psychosocial risk and protective factors. *Substance Use & Misuse,* **44,** 1711-1724. http://dx.doi.org/10.3109/10826080902962128

60. Williams, A.R., B. Tofighi, J. Rotrosen, J.D. Lee, and E. Grossman, 2014: Psychiatric comorbidity, red flag behaviors, and associated outcomes among office-based buprenorphine patients following Hurricane Sandy. *Journal of Urban Health,* **91,** 366-375. http://dx.doi.org/10.1007/s11524-014-9866-7

61. Beaudoin, C.E., 2011: Hurricane Katrina: Addictive behavior trends and predictors. *Public Health Reports,* **126,** 400-409. PMC3072862

62. North, C.S., A. Kawasaki, E.L. Spitznagel, and B.A. Hong, 2004: The course of PTSD, major depression, substance abuse, and somatization after a natural disaster. *Journal of Nervous & Mental Disease,* **192,** 823-829.

63. Larrance, R., M. Anastario, and L. Lawry, 2007: Health status among internally displaced persons in Louisiana and Mississippi travel trailer parks. *Annals of Emergency Medicine,* **49,** 590-601.e12. http://dx.doi.org/doi:10.1016/j.annemergmed.2006.12.004

64. Lew, E.O. and C.V. Wetli, 1996: Mortality from Hurricane Andrew. *Journal of Forensic Sciences,* **41,** 449-452.

65. Kaniasty, K., 2012: Predicting social psychological well-being following trauma: The role of postdisaster social support. *Psychological Trauma: Theory, Research, Practice, and Policy,* **4,** 22-33. http://dx.doi.org/10.1037/a0021412

66. Melillo, J.M., T.C. Richmond, and G.W. Yohe, eds. *Climate Change Impacts in the United States: The Third National Climate Assessment.* 2014, U.S. Global Change Research Program: Washington, DC. 842. http://dx.doi.org/10.7930/J0Z31WJ2. http://nca2014.globalchange.gov

67. NOAA, 2014: State of the Climate: Drought for Annual 2014. National Oceanic and Atmospheric Administration, National Climatic Data Center. http://www.ncdc.noaa.gov/sotc/drought/201413

68. Dai, A., 2011: Drought under global warming: A review. *Wiley Interdisciplinary Reviews: Climate Change,* **2,** 45-65. http://dx.doi.org/10.1002/wcc.81

69. Evans, G.W. and S.J. Lepore, 2008: Psychosocial processes linking the environment and mental health. *The Impact of the Environment On Psychiatric Disorder.* Freeman, S. and S. Stansfeld, Eds. Routledge, London, UK, 127-157.

70. Weissbecker, I., Ed., 2011: *Climate Change and Human Well-being: Global Challenges and Opportunities.* Springer-Verlag, New York, 220 pp. http://dx.doi.org/10.1007/978-1-4419-9742-5

71. Albrecht, G., G.M. Sartore, L. Connor, N. Higginbotham, S. Freeman, B. Kelly, H. Stain, A. Tonna, and G. Pollard, 2007: Solastalgia: The distress caused by environmental change. *Australasian Psychiatry,* **15,** S95-S98. http://dx.doi.org/10.1080/10398560701701288

72. Sartore, G.M., B. Kelly, H. Stain, G. Albrecht, and N. Higginbotham, 2008: Control, uncertainty, and expectations for the future: A qualitative study of the impact of drought on a rural Australian community. *Rural and Remote Health,* **8,** Article 950. http://www.rrh.org.au/articles/subviewnew.asp?ArticleID=950

73. Seery, M.D., E.A. Holman, and R.C. Silver, 2010: Whatever does not kill us: Cumulative lifetime adversity, vulnerability, and resilience. *Journal of Personality and Social Psychology,* **99,** 1025-1041. http://dx.doi.org/10.1037/a0021344

74. Dean, J.G. and H.J. Stain, 2010: Mental health impact for adolescents living with prolonged drought. *Australian Journal of Rural Health,* **18,** 32-37. http://dx.doi.org/10.1111/j.1440-1584.2009.01107.x

75. Obrien, L.V., H.L. Berry, C. Coleman, and I.C. Hanigan, 2014: Drought as a mental health exposure. *Environmental Research,* **131,** 181-187. http://dx.doi.org/10.1016/j.envres.2014.03.014

76. Stain, H.J., B. Kelly, V.J. Carr, T.J. Lewin, M. Fitzgerald, and L. Fragar, 2011: The psychological impact of chronic environmental adversity: Responding to prolonged drought. *Social Science & Medicine,* **73,** 1593-1599. http://dx.doi.org/10.1016/j.socscimed.2011.09.016

77. Guiney, R., 2012: Farming suicides during the Victorian drought: 2001–2007. *Australian Journal of Rural Health,* **20,** 11-15. http://dx.doi.org/10.1111/j.1440-1584.2011.01244.x

78. U.S. Census Bureau, 2010: 2010 Census Urban and Rural Classification and Urban Area Criteria: Urban, Urbanized Area, Urban Cluster, and Rural Population, 2010 and 2000, United States. https://www.census.gov/geo/reference/ua/urban-rural-2010.html

79. Cutter, S.L., W. Solecki, N. Bragado, J. Carmin, M. Fragkias, M. Ruth, and T. Wilbanks, 2014: Ch. 11: Urban systems, infrastructure, and vulnerability. *Climate Change Impacts in the United States: The Third National Climate Assessment.* Melillo, J.M., T.C. Richmond, and G.W. Yohe, Eds. U.S. Global Change Research Program, Washington, D.C., 282-296. http://dx.doi.org/10.7930/J0F769GR

80. Anderson, C.A., 2012: Climate change and violence. *The Encyclopedia of Peace Psychology.* Christie, D.J., Ed. Wiley-Blackwell, Hoboken, NJ, 128-132. http://dx.doi.org/10.1002/9780470672532.wbepp032

81. Hansen, A., P. Bi, M. Nitschke, P. Ryan, D. Pisaniello, and G. Tucker, 2008: The effect of heat waves on mental health in a temperate Australian city. *Environmental Health Perspectives,* **116,** 1369-1375. http://dx.doi.org/10.1289/ehp.11339

82. Martin-Latry, K., M.P. Goumy, P. Latry, C. Gabinski, B. Bégaud, I. Faure, and H. Verdoux, 2007: Psychotropic drugs use and risk of heat-related hospitalisation. *European Psychiatry,* **22,** 335-338. http://dx.doi.org/10.1016/j.eurpsy.2007.03.007

83. Hsiang, S.M., M. Burke, and E. Miguel, 2013: Quantifying the influence of climate on human conflict. *Science,* **341,** 1235367. http://dx.doi.org/10.1126/science.1235367

84. Page, L.A., S. Hajat, R.S. Kovats, and L.M. Howard, 2012: Temperature-related deaths in people with psychosis, dementia and substance misuse. *The British Journal of Psychiatry,* **200,** 485-490. http://dx.doi.org/10.1192/bjp.bp.111.100404

85. Ranson, M., 2014: Crime, weather, and climate change. *Journal of Environmental Economics and Management,* **67,** 274-302. http://dx.doi.org/10.1016/j.jeem.2013.11.008

86. Vida, S., M. Durocher, T.B.M.J. Ouarda, and P. Gosselin, 2012: Relationship between ambient temperature and humidity and visits to mental health emergency departments in Québec. *Psychiatric Services,* **63,** 1150-1153. http://dx.doi.org/10.1176/appi.ps.201100485

87. Wang, X., E. Lavigne, H. Ouellette-kuntz, and B.E. Chen, 2014: Acute impacts of extreme temperature exposure on emergency room admissions related to mental and behavior disorders in Toronto, Canada. *Journal of Affective Disorders,* **155,** 154-161. http://dx.doi.org/10.1016/j.jad.2013.10.042

88. Bouchama, A., M. Dehbi, G. Mohamed, F. Matthies, M. Shoukri, and B. Menne, 2007: Prognostic factors in heat wave-related deaths: A meta-analysis. *Archives of Internal Medicine,* **167,** 2170-2176. http://dx.doi.org/10.1001/archinte.167.20.ira70009

89. Kaiser, R., C.H. Rubin, A.K. Henderson, M.I. Wolfe, S. Kieszak, C.L. Parrott, and M. Adcock, 2001: Heat-related death and mental illness during the 1999 Cincinnati heat wave. *The American Journal of Forensic Medicine and Pathology,* **22,** 303-307.

90. Naughton, M.B., A. Henderson, M.C. Mirabelli, R. Kaiser, J.L. Wilhelm, S.M. Kieszak, C.H. Rubin, and M.A. McGeehin, 2002: Heat-related mortality during a 1999 heat wave in Chicago. *American Journal of Preventive Medicine,* **22,** 221-227. http://dx.doi.org/10.1016/S0749-3797(02)00421-X

91. Semenza, J.C., C.H. Rubin, K.H. Falter, J.D. Selanikio, W.D. Flanders, H.L. Howe, and J.L. Wilhelm, 1996: Heat-related deaths during the July 1995 heat wave in Chicago. *New England Journal of Medicine,* **335,** 84-90. http://dx.doi.org/10.1056/NEJM199607113350203

92. Anderson, C.A., B.J. Bushman, and R.W. Groom, 1997: Hot years and serious and deadly assault: Empirical tests of the heat hypothesis. *Journal of Personality and Social Psychology,* **73,** 1213-1223. http://dx.doi.org/10.1037/0022-3514.73.6.1213

93. Larrick, R.P., T.A. Timmerman, A.M. Carton, and J. Abrevaya, 2011: Temper, temperature, and temptation: Heat-related retaliation in baseball. *Psychological Science,* **22,** 423-428. http://dx.doi.org/10.1177/0956797611399292

94. Mares, D., 2013: Climate change and levels of violence in socially disadvantaged neighborhood groups. *Journal of Urban Health,* **90,** 768-783. http://dx.doi.org/10.1007/s11524-013-9791-1

95. Hsiang, S.M. and M. Burke, 2014: Climate, conflict, and social stability: What does the evidence say? *Climatic Change,* **123,** 39-55. http://dx.doi.org/10.1007/s10584-013-0868-3

96. Buhaug, H., J. Nordkvelle, T. Bernauer, T. Bohmelt, M. Brzoska, J.W. Busby, A. Ciccone, H. Fjelde, E. Gartzke, N.P. Gleditsch, J.A. Goldstone, H. Hegre, H. Holtermann, V. Koubi, J.S.A. Link, P.M. Link, P. Lujala, J. O'Loughlin, C. Raleigh, J. Scheffran, J. Schilling, T.G. Smith, O.M. Theisen, R.S.J. Tol, H. Urdal, and N. von Uexkull, 2014: One effect to rule them all? A comment on climate and conflict. *Climatic Change,* **127,** 391-397. http://dx.doi.org/10.1007/s10584-014-1266-1

97. Leiserowitz, A., E. Maibach, C. Roser-Renouf, G. Feinberg, and S. Rosenthal, 2015: Climate Change in the American Mind: March 2015. Yale University and George Mason University, New Haven, CT: Yale Project on Climate Change Communication. http://environment.yale.edu/climate-communication/files/Global-Warming-CCAM-March-2015.pdf

98. Aldwin, C. and D. Stokols, 1988: The effects of environmental change on individuals and groups: Some neglected issues in stress research. *Journal of Environmental Psychology,* **8,** 57-75. http://dx.doi.org/10.1016/S0272-4944(88)80023-9

99. Bell, P.A., T.C. Greene, J.D. Fisher, and A. Baum, 2001: *Environmental Psychology,* 5th ed. Harcourt College Publishers, Fort Worth, TX, 634 pp.

100. Leiserowitz, A., 2006: Climate change risk perception and policy preferences: The role of affect, imagery, and values. *Climatic Change,* **77,** 45-72. http://dx.doi.org/10.1007/s10584-006-9059-9

101. Lorenzoni, I., A. Leiserowitz, M. De Franca Doria, W. Poortinga, and N.F. Pidgeon, 2006: Cross-national comparisons of image associations with "global warming" and "climate change" among laypeople in the United States of America and Great Britain. *Journal of Risk Research,* **9,** 265-281. http://dx.doi.org/10.1080/13669870600613658

102. Loewenstein, G.F., E.U. Weber, C.K. Hsee, and N. Welch, 2001: Risk as feelings. *Psychological Bulletin,* **127,** 267-286. http://dx.doi.org/10.1037/0033-2909.127.2.267

103. Slovic, P., 2000: *The Perception of Risk.* Earthscan, London, 473 pp.

104. Slovic, P., M. Layman, N. Kraus , J. Flynn, J. Chalmers, and G. Gesell, 2001: Perceived risk, stigma, and potential economic impacts of a high-level nuclear waste repository in Nevada. *Risk, Media and Stigma: Understanding Public Challenges to Modern Science and Technology.* Flynn, J., P. Slovic, and H. Kunreuther, Eds. Earthscan, London, UK, 87-105.

105. Slovic, P., 2010: *The Feeling of Risk: New Perspectives on Risk Perception.* Earthscan, New York, 456 pp.

106. Smith, E., J. Wasiak, A. Sen, F. Archer, and F.M. Burkle, Jr., 2009: Three decades of disasters: A review of disaster-specific literature from 1977-2009. *Prehospital and Disaster Medicine,* **24,** 306-311. http://dx.doi.org/10.1017/S1049023X00007020

107. Lever-Tracey, C., Ed., 2010: *Routledge International Handbook of Climate Change and Society.* Routledge Press, New York.

108. Agho, K., G. Stevens, M. Taylor, M. Barr, and B. Raphael, 2010: Population risk perceptions of global warming in Australia. *Environmental Research,* **110,** 756-763. http://dx.doi.org/10.1016/j.envres.2010.09.007

109. Reser, J.P., G.L. Bradley, A.I. Glendon, M.C. Ellul, and R. Callaghan, 2012: Public Risk Perceptions, Understandings and Responses To Climate Change and Natural Disasters in Australia, 2010 and 2011. 245 pp. National Climate Change Adaptation Research Facility, Gold Coast, Australia. https://www.nccarf.edu.au/sites/default/files/attached_files_publications/Reser_2012_Public_risk_perceptions_Second_survey_report.pdf

110. Reser, J.P., G.L. Bradley, and M.C. Ellul, 2012: Coping with climate change: Bringing psychological adaptation in from the cold. *Handbook of the Psychology of Coping: Psychology of Emotions, Motivations and Actions.* Molinelli, B. and V. Grimaldo, Eds. Nova Science Publishers, New York, 1-34.

111. Searle, K. and K. Gow, 2010: Do concerns about climate change lead to distress? *International Journal of Climate Change Strategies and Management,* **2,** 362-379. http://dx.doi.org/10.1108/17568691011089891

112. Ferguson, M.A. and N.R. Branscombe, 2010: Collective guilt mediates the effect of beliefs about global warming on willingness to engage in mitigation behavior. *Journal of Environmental Psychology,* **30,** 135-142. http://dx.doi.org/10.1016/j.jenvp.2009.11.010

113. Nabi, R.L. and W. Wirth, 2008: Exploring the role of emotion in media effects: An introduction to the special issue. *Media Psychology,* **11,** 1-6. http://dx.doi.org/10.1080/15213260701852940

114. Myers, T.A., M.C. Nisbet, E.W. Maibach, and A.A. Leiserowitz, 2012: A public health frame arouses hopeful emotions about climate change. *Climatic Change,* **113,** 1105-1112. http://dx.doi.org/10.1007/s10584-012-0513-6

115. Boykoff, M.T. and J.T. Roberts, 2007: Media Coverage of Climate Change: Current Trends, Strengths, Weaknesses. Human Development Report Office Occasional Paper. United National Development Programme Human Development Report 2007/2008. http://www.researchgate.net/profile/Max_Boykoff2/publication/228637999_Media_coverage_of_climate_change_Current_trends_strengths_weaknesses/links/02e7e528bf129aba0b000000.pdf

116. Feldman, L., P.S. Hart, and T. Milosevic, 2015: Polarizing news? Representations of threat and efficacy in leading US newspapers' coverage of climate change. *Public Understanding of Science,* Published online 30 July 2015. http://dx.doi.org/10.1177/0963662515595348

117. Gow, K., Ed., 2009: *Meltdown: Climate Change, Natural Disasters, and Other Catastrophes--Fears and Concerns of the Future.* Nova Science Publishers, New York, 430 pp.

118. Marshall, J.P., Ed., 2009: *Depth Psychology, Disorder and Climate Change.* Jung Downunder Books, Sydney, NSW, 475 pp.

119. Smith, N.W. and H. Joffe, 2009: Climate change in the British press: The role of the visual. *Journal of Risk Research,* **12,** 647-663. http://dx.doi.org/10.1080/13669870802586512

120. Bonanno, G.A., 2004: Loss, trauma, and human resilience: Have we underestimated the human capacity to thrive after extremely aversive events? *American Psychologist,* **59,** 20-28. http://dx.doi.org/10.1037/0003-066x.59.1.20

121. Shakespeare-Finch, J.E., S.G. Smith, K.M. Gow, G. Embelton, and L. Baird, 2003: The prevalence of post-traumatic growth in emergency ambulance personnel. *Traumatology,* **9,** 58-71. http://dx.doi.org/10.1177/153476560300900104

122. Tedeschi, R.G. and L.G. Calhoun, 1996: The posttraumatic growth inventory: Measuring the positive legacy of trauma. *Journal of Traumatic Stress,* **9,** 455-471. http://dx.doi.org/10.1007/BF02103658

123. Lowe, S.R., E.E. Manove, and J.E. Rhodes, 2013: Posttraumatic stress and posttraumatic growth among low-income mothers who survived Hurricane Katrina. *Journal of Consulting and Clinical Psychology,* **81,** 877-889. http://dx.doi.org/10.1037/a0033252

124. Taku, K., A. Cann, L.G. Calhoun, and R.G. Tedeschi, 2008: The factor structure of the posttraumatic growth inventory: A comparison of five models using confirmatory factor analysis. *Journal of Traumatic Stress,* **21,** 158-164. http://dx.doi.org/10.1002/jts.20305

125. Kronenberg, M.E., T.C. Hansel, A.M. Brennan, H.J. Osofsky, J.D. Osofsky, and B. Lawrason, 2010: Children of Katrina: Lessons learned about postdisaster symptoms and recovery patterns. *Child Development,* **81,** 1241-1259. http://dx.doi.org/10.1111/j.1467-8624.2010.01465.x

126. Mills, L.D., T.J. Mills, M. Macht, R. Levitan, A. De Wulf, and N.S. Afonso, 2012: Post-traumatic stress disorder in an emergency department population one year after Hurricane Katrina. *The Journal of Emergency Medicine,* **43,** 76-82. http://dx.doi.org/10.1016/j.jemermed.2011.06.124

127. Self-Brown, S., B.S. Lai, J.E. Thompson, T. McGill, and M.L. Kelley, 2013: Posttraumatic stress disorder symptom trajectories in Hurricane Katrina affected youth. *Journal of Affective Disorders,* **147,** 198-204. http://dx.doi.org/10.1016/j.jad.2012.11.002

128. Wadsworth, M.E., C.D. Santiago, and L. Einhorn, 2009: Coping with displacement from Hurricane Katrina: Predictors of one-year post-traumatic stress and depression symptom trajectories. *Anxiety, Stress & Coping,* **22,** 413-432. http://dx.doi.org/10.1080/10615800902855781

129. McLaughlin, K.A., J.A. Fairbank, M.J. Gruber, R.T. Jones, M.D. Lakoma, B. Pfefferbaum, N.A. Sampson, and R.C. Kessler, 2009: Serious emotional disturbance among youths exposed to Hurricane Katrina 2 years postdisaster. *Journal of the American Academy of Child and Adolescent Psychiatry,* **48,** 1069-1078. http://dx.doi.org/10.1097/CHI.0b013e3181b76697

130. Abramson, D., T. Stehling-Ariza, R. Garfield, and I. Redlener, 2008: Prevalence and predictors of mental health distress post-Katrina: Findings from the Gulf Coast Child and Family Health Study. *Disaster Medicine and Public Health Preparedness,* **2,** 77-86. http://dx.doi.org/10.1097/DMP.0b013e318173a8e7

131. Hansel, T.C., J.D. Osofsky, H.J. Osofsky, and P. Friedrich, 2013: The effect of long-term relocation on child and adolescent survivors of Hurricane Katrina. *Journal of Traumatic Stress,* **26,** 613-620. http://dx.doi.org/10.1002/jts.21837

132. Jacob, B., A.R. Mawson, M. Payton, and J.C. Guignard, 2008: Disaster mythology and fact: Hurricane Katrina and social attachment. *Public Health Reports,* **123,** 555-566. PMC2496928

133. Davydov, D.M., R. Stewart, K. Ritchie, and I. Chaudieu, 2010: Resilience and mental health. *Clinical Psychology Review,* **30,** 479-495. http://dx.doi.org/10.1016/j.cpr.2010.03.003

134. Freedy, J.R. and W.M. Simpson, Jr., 2007: Disaster-related physical and mental health: A role for the family physician. *American Family Physician,* **75,** 841-846. http://www.aafp.org/afp/2007/0315/p841.pdf

135. Banks, D.M. and C.F. Weems, 2014: Family and peer social support and their links to psychological distress among hurricane-exposed minority youth. *American Journal of Orthopsychiatry,* **84,** 341-352. http://dx.doi.org/10.1037/ort0000006

136. Bradley, G.L., J.P. Reser, A.I. Glendon, and M.C. Ellul, 2014: Distress and coping in response to climate change. *Stress and Anxiety: Applications to Social and Environmental Threats, Psychological Wellbeing, Occupational Challenges, and Developmental Psychology.* Kaniasty, K., P. Buchwald, S. Howard, and K.A. Moore, Eds. Logos Verlag, Berlin, 33-42.

137. Berry, H.L., K. Bowen, and T. Kjellstrom, 2010: Climate change and mental health: A causal pathways framework. *International Journal of Public Health,* **55,** 123-132. http://dx.doi.org/10.1007/s00038-009-0112-0

138. Berner, J., C. Furgal, P. Bjerregaard, M. Bradley, T. Curtis, E.D. Fabo, J. Hassi, W. Keatinge, S. Kvernmo, S. Nayha, H. Rintamaki, and J. Warren, 2005: Ch. 15: Human Health. *Arctic Climate Impact Assessment.* Cambridge University Press, Cambridge, UK, 863-906. http://www.acia.uaf.edu/pages/scientific.html

139. Maldonado, J.K., C. Shearer, R. Bronen, K. Peterson, and H. Lazrus, 2013: The impact of climate change on tribal communities in the US: Displacement, relocation, and human rights. *Climatic Change,* **120,** 601-614. http://dx.doi.org/10.1007/s10584-013-0746-z

140. Swim, J., S. Clayton, T. Doherty, R. Gifford, G. Howard, J. Reser, P. Stern, and E. Weber, 2010: Psychology & Global Climate Change: Addressing a Multi-Faceted Phenomenon and Set of Challenges. 108 pp. The American Psychological Association Task Force on the Interface between Psychology and Global Climate Change. http://www.apa.org/science/about/publications/climate-change-booklet.pdf

141. Norris, F.H., M.J. Friedman, P.J. Watson, C.M. Byrne, E. Diaz, and K. Kaniasty, 2002: 60,000 disaster victims speak: Part I. An empirical review of the empirical literature, 1981–2001. *Psychiatry: Interpersonal and Biological Processes,* **65,** 207-39. http://dx.doi.org/10.1521/psyc.65.3.207.20173

142. Somasundaram, D.J. and W.A.C.M. van de Put, 2006: Management of trauma in special populations after a disaster. *The Journal of Clinical Psychiatry,* **67,** 64-73. http://www.ucalgary.ca/psychiatry/files/psychiatry/j-clin-monograph-supplement-feb-06.pdf - page=66

143. Bourque, F. and A. Cunsolo Willox, 2014: Climate change: The next challenge for public mental health? *International Review of Psychiatry,* **26,** 415-422. http://dx.doi.org/10.3109/09540261.2014.925851

144. Turner, N.J. and H. Clifton, 2009: "It's so different today": Climate change and Indigenous lifeways in British Columbia, Canada. *Global Environmental Change,* **19,** 180-190. http://dx.doi.org/10.1016/j.gloenvcha.2009.01.005

145. Cunsolo Willox, A., S.L. Harper, J.D. Ford, K. Landman, K. Houle, V.L. Edge, and Rigolet Inuit Community Government, 2012: "From this place and of this place:" Climate change, sense of place, and health in Nunatsiavut, Canada. *Social Science & Medicine,* **75,** 538-547. http://dx.doi.org/10.1016/j.socscimed.2012.03.043

146. Masten, A.S., 2014: Global perspectives on resilience in children and youth. *Child Development,* **85,** 6-20. http://dx.doi.org/10.1111/cdev.12205

147. Masten, A.S. and A.J. Narayan, 2012: Child development in the context of disaster, war, and terrorism: Pathways of risk and resilience. *Annual Review of Psychology,* **63,** 227-257. http://dx.doi.org/10.1146/annurev-psych-120710-100356

148. Sapienza, J.K. and A.S. Masten, 2011: Understanding and promoting resilience in children and youth. *Current Opinion in Psychiatry,* **24,** 267-273. http://dx.doi.org/10.1097/YCO.0b013e32834776a8

149. Southwick, S.M., G.A. Bonanno, A.S. Masten, C. Panter-Brick, and R. Yehuda, 2014: Resilience definitions, theory, and challenges: Interdisciplinary perspectives. *European Journal of Psychotraumatology,* **5.** http://dx.doi.org/10.3402/ejpt.v5.25338

150. Joshi, P.T. and S.M. Lewin, 2004: Disaster, terrorism and children: Addressing the effects of traumatic events on children and their families is critical to long-term recovery and resilience. *Psychiatric Annals,* **34,** 710-716.

151. Simpson, D.M., I. Weissbecker, and S.E. Sephton, 2011: Extreme weather-related events: Implications for mental health and well-being. *Climate Change and Human Well-being: Global Challenges and Opportunities.* Weissbecker, I., Ed. Springer-Verlag, New York, 57-78. http://dx.doi.org/10.1007/978-1-4419-9742-5

152. Tees, M.T., E.W. Harville, X. Xiong, P. Buekens, G. Pridjian, and K. Elkind-Hirsch, 2010: Hurricane Katrina-related maternal stress, maternal mental health, and early infant temperament. *Maternal and Child Health Journal,* **14,** 511-518. http://dx.doi.org/10.1007/s10995-009-0486-x

153. Enarson, E., A. Fothergill, and L. Peek, 2007: Gender and disaster: Foundations and directions. *Handbook of Disaster Research.* Rodriguez, H., E.L. Quarantelli, and R.R. Dynes, Eds. Springer, New York, 130-146. http://dx.doi.org/10.1007/978-0-387-32353-4_8

154. Dasgupta, S., I. Siriner, and P. Sarathi De, Eds., 2010: *Women's Encounter With Disaster.* Frontpage, London, United Kingdom, 281 pp.

155. Rahman, M.S., 2013: Climate change, disaster and gender vulnerability: A study on two divisions of Bangladesh. *American Journal of Human Ecology,* **2,** 72-82. http://dx.doi.org/10.11634/216796221302315

156. Xiong, X., E.W. Harville, D.R. Mattison, K. Elkind-Hirsch, G. Pridjian, and P. Buekens, 2010: Hurricane Katrina experience and the risk of post-traumatic stress disorder and depression among pregnant women. *American Journal of Disaster Medicine,* **5,** 181-187. PMC3501144

157. Trumbo, C., M. Lueck, H. Marlatt, and L. Peek, 2011: The effect of proximity to Hurricanes Katrina and Rita on subsequent hurricane outlook and optimistic bias. *Risk Analysis,* **31,** 1907-1918. http://dx.doi.org/10.1111/j.1539-6924.2011.01633.x

158. Corrarino, J.E., 2008: Disaster-related mental health needs of women and children. *MCN, The American Journal of Maternal/Child Nursing,* **33,** 242-248. http://dx.doi.org/10.1097/01.NMC.0000326079.26870.e3

159. Harville, E., X. Xiong, and P. Buekens, 2010: Disasters and perinatal health: A systematic review. *Obstetrical & Gynecological Survey,* **65,** 713-728. http://dx.doi.org/10.1097/OGX.0b013e31820eddbe

160. Callaghan, W.M., S.A. Rasmussen, D.J. Jamieson, S.J. Ventura, S.L. Farr, P.D. Sutton, T.J. Mathews, B.E. Hamilton, K.R. Shealy, D. Brantley, and S.F. Posner, 2007: Health concerns of women and infants in times of natural disasters: Lessons learned from Hurricane Katrina. *Maternal and Child Health Journal,* **11,** 307-311. http://dx.doi.org/10.1007/s10995-007-0177-4

161. Triunfo, S. and A. Lanzone, 2015: Impact of maternal under nutrition on obstetric outcomes. *Journal of Endocrinological Investigation,* **38,** 31-38. http://dx.doi.org/10.1007/s40618-014-0168-4

162. U.S. Census Bureau, 2014: 2014 National Population Projections: Summary Tables. Table 4: Projections of the Native-Born Population by Sex and Selected Age Groups for the United States: 2015 to 2060 (NP2014-T4). U.S. Department of Commerce, Washington, D.C. http://www.census.gov/population/projections/data/national/2014/summarytables.html

163. Miller, G., E. Chen, and S.W. Cole, 2009: Health psychology: Developing biologically plausible models linking the social world and physical health. *Annual Review of Psychology,* **60,** 501-524. http://dx.doi.org/10.1146/annurev.psych.60.110707.163551

164. Prince, M., V. Patel, S. Saxena, M. Maj, J. Maselko, M.R. Phillips, and A. Rahman, 2007: No health without mental health. *The Lancet,* **370,** 859-877. http://dx.doi.org/10.1016/S0140-6736(07)61238-0

165. Katz, I.R., 1996: On the inseparability of mental and physical health in aged persons: Lessons from depression and medical comorbidity. *The American Journal of Geriatric Psychiatry,* **4,** 1-16. http://dx.doi.org/10.1097/00019442-199624410-00001

166. Power, M.C., M.G. Weisskopf, S.E. Alexeeff, B.A. Coull, A. Spiro, III,, and J. Schwartz, 2011: Traffic-related air pollution and cognitive function in a cohort of older men. *Environmental Health Perspectives,* **119,** 682-687. http://dx.doi.org/10.1289/ehp.1002767

167. Ranft, U., T. Schikowski, D. Sugiri, J. Krutmann, and U. Kramer, 2009: Long-term exposure to traffic-related particulate matter impairs cognitive function in the elderly. *Environmental Research,* **109,** 1004-1011. http://dx.doi.org/10.1016/j.envres.2009.08.003

168. Wellenius, G.A., L.D. Boyle, B.A. Coull, W.P. Milberg, A. Gryparis, J. Schwartz, M.A. Mittleman, and L.A. Lipsitz, 2012: Residential proximity to nearest major roadway and cognitive function in community-dwelling seniors: Results from the MOBILIZE Boston Study. *Journal of the American Geriatrics Society,* **60,** 2075-2080. http://dx.doi.org/10.1111/j.1532-5415.2012.04195.x

169. Weuve, J., R.C. Puett, J. Schwartz, J.D. Yanosky, F. Laden, and F. Grodstein, 2012: Exposure to particulate air pollution and cognitive decline in older women. *Archives of Internal Medicine,* **172,** 219-227. http://dx.doi.org/10.1001/archinternmed.2011.683

170. Tonne, C., A. Elbaz, S. Beevers, and A. Singh-Manoux, 2014: Traffic-related air pollution in relation to cognitive function in older adults. *Epidemiology,* **25,** 674-681. http://dx.doi.org/10.1097/ede.0000000000000144

171. Coêlho, A.E., J.G. Adair, and J.S.P. Mocellin, 2004: Psychological responses to drought in northeastern Brazil. *Revista Interamericana de Psicología/Interamerican Journal of Psychology,* **38,** 95-103. http://www.psicorip.org/Resumos/PerP/RIP/RIP036a0/RIP03811.pdf

172. Polain, J.D., H.L. Berry, and J.O. Hoskin, 2011: Rapid change, climate adversity and the next 'big dry': Older farmers' mental health. *Australian Journal of Rural Health,* **19,** 239-243. http://dx.doi.org/10.1111/j.1440-1584.2011.01219.x

173. Keim, M.E., 2008: Building human resilience: The role of public health preparedness and response as an adaptation to climate change. *American Journal of Preventive Medicine,* **35,** 508-516. http://dx.doi.org/10.1016/j.amepre.2008.08.022

174. Benedek, D.M., C. Fullerton, and R.J. Ursano, 2007: First responders: Mental health consequences of natural and human-made disasters for public health and public safety workers. *Annual Review of Public Health,* **28,** 55-68. http://dx.doi.org/10.1146/annurev.publhealth.28.021406.144037

175. Laugharne, J., G. Van de Watt, and A. Janca, 2011: After the fire: The mental health consequences of fire disasters. *Current Opinion in Psychiatry,* **24,** 72-77. http://dx.doi.org/10.1097/YCO.0b013e32833f5e4e

176. Alexander, D.A. and S. Klein, 2009: First responders after disasters: A review of stress reactions, at-risk, vulnerability, and resilience factors. *Prehospital and Disaster Medicine,* **24,** 87-94. http://dx.doi.org/10.1017/s1049023x00006610

177. Fullerton, C.S., J.E. McCarroll, R.J. Ursano, and K.M. Wright, 1992: Psychological responses of rescue workers: Fire fighters and trauma. *American Journal of Orthopsychiatry,* **62,** 371-378. http://dx.doi.org/10.1037/h0079363

178. Rusiecki, J.A., D.L. Thomas, L. Chen, R. Funk, J. McKibben, and M.R. Dayton, 2014: Disaster-related exposures and health effects among US Coast Guard responders to Hurricanes Katrina and Rita. *Journal of Occupational and Environmental Medicine,* **56,** 820-833. http://dx.doi.org/10.1097/jom.0000000000000188

179. Fayard, G.M., 2009: Fatal work injuries involving natural disasters, 1992–2006. *Disaster Medicine and Public Health Preparedness,* **3,** 201-209. http://dx.doi.org/10.1097/DMP.0b013e3181b65895

180. Kiefer, M., J. Lincoln, P. Schulte, and B. Jacklitsch, 2014: Climate Change and Occupational Safety and Health (NIOSH Science Blog, September 22, 2014). Centers for Disease Control and Prevention, Atlanta, GA. http://blogs.cdc.gov/niosh-science-blog/2014/09/22/climate-change/

181. SAMHSA, 2011: Current Statistics on the Prevalence and Characteristics of People Experience Homelessness in the United States. 23 pp. Substance Abuse and Mental Health Services Administration. http://homeless.samhsa.gov/ResourceFiles/hrc_factsheet.pdf

182. Ramin, B. and T. Svoboda, 2009: Health of the homeless and climate change. *Journal of Urban Health,* **86,** 654-664. http://dx.doi.org/10.1007/s11524-009-9354-7

183. Caillouet, K.A., S.R. Michaels, X. Xiong, I. Foppa, and D.M. Wesson, 2008: Increase in West Nile neuroinvasive disease after Hurricane Katrina. *Emerging Infectious Diseases,* **14,** 804-807. http://dx.doi.org/10.3201/eid1405.071066

184. Isaac, M.L. and E.B. Larson, 2014: Medical conditions with neuropsychiatric manifestations. *Medical Clinics of North America,* **98,** 1193-1208. http://dx.doi.org/10.1016/j.mcna.2014.06.012

185. Murray, K.O., M. Resnick, and V. Miller, 2007: Depression after infection with West Nile virus. *Emerging Infectious Diseases,* **13,** 479-481. http://dx.doi.org/10.3201/eid1303.060602

186. SAMHSA, 2014: Results from the 2013 National Survey on Drug Use and Health: Mental Health Findings. NSDUH Series H-49, HHS Publication No. (SMA) 14-4887. U.S. Department of Health and Human Services, Substance Abuse and Mental Health Services Administration, Rockville, MD. http://www.samhsa.gov/data/sites/default/files/NSDUHmhfr2013/NSDUHmhfr2013.pdf

187. Jonas, B.S., Q. Gu, and J.R. Albertorio-Diaz, 2013: Psychotropic Medication Use Among Adolescents: United States, 2005-2010. NCHS Data Brief No. 135, December 2013, 8 pp. National Center for Health Statistics, Hyattsville, MD. http://www.cdc.gov/nchs/data/databriefs/db135.pdf

188. Heo, M., C.F. Murphy, K.R. Fontaine, M.L. Bruce, and G.S. Alexopoulos, 2008: Population projection of US adults with lifetime experience of depressive disorder by age and sex from year 2005 to 2050. *International Journal of Geriatric Psychiatry,* **23,** 1266-1270. http://dx.doi.org/10.1002/gps.2061

189. Cusack, L., C. de Crespigny, and P. Athanasos, 2011: Heatwaves and their impact on people with alcohol, drug and mental health conditions: A discussion paper on clinical practice considerations. *Journal of Advanced Nursing,* **67,** 915-922. http://dx.doi.org/10.1111/j.1365-2648.2010.05551.x

190. Stöllberger, C., W. Lutz, and J. Finsterer, 2009: Heat-related side-effects of neurological and non-neurological medication may increase heatwave fatalities. *European Journal of Neurology,* **16,** 879-882. http://dx.doi.org/10.1111/j.1468-1331.2009.02581.x

191. Berko, J., D.D. Ingram, S. Saha, and J.D. Parker, 2014: Deaths Attributed to Heat, Cold, and Other Weather Events in the United States, 2006–2010. National Health Statistics Reports No. 76, July 30, 2014, 15 pp. National Center for Health Statistics, Hyattsville, MD. http://www.cdc.gov/nchs/data/nhsr/nhsr076.pdf

192. Christenson, M.L., S.D. Geiger, and H.A. Anderson, 2013: Heat-related fatalities in Wisconsin during the summer of 2012. *WMJ,* **112,** 219-23. https://wmstest.ancillapartners.com/_WMS/publications/wmj/pdf/112/5/219.pdf

193. Kwok, J.S.S. and T.Y.K. Chan, 2005: Recurrent heat-related illnesses during antipsychotic treatment. *The Annals of Pharmacotherapy,* **39,** 1940-1942. http://dx.doi.org/10.1345/aph.1G130

194. Faunt, J.D., T.J. Wilkinson, P. Aplin, P. Henschke, M. Webb, and R.K. Penhall, 1995: The effete in the heat: Heat-related hospital presentations during a ten day heat wave. *Australian and New Zealand Journal of Medicine,* **25,** 117-121. http://dx.doi.org/10.1111/j.1445-5994.1995.tb02822.x

195. Weir, E., 2002: Heat wave: First, protect the vulnerable. *Canadian Medical Association Journal,* **167,** 169. PMC117098

196. Kovats, R.S., H. Johnson, and C. Griffiths, 2006: Mortality in southern England during the 2003 heat wave by place of death. *Health Statistics Quarterly,* **29,** 6-8. http://www.ons.gov.uk/ons/rel/hsq/health-statistics-quarterly/no--29--spring-2006/mortality-in-southern-england-during-the-2003-heat-wave-by-place-of-death.pdf

197. Nitschke, M., G.R. Tucker, and P. Bi, 2007: Morbidity and mortality during heatwaves in metropolitan Adelaide. *The Medical Journal of Australia,* **187,** 662-665. https://www.mja.com.au/system/files/issues/187_11_031207/nit10385_fm.pdf

198. Page, L.A. and L.M. Howard, 2010: The impact of climate change on mental health (but will mental health be discussed at Copenhagen?). *Psychological Medicine,* **40,** 177-180. http://dx.doi.org/10.1017/S0033291709992169

199. Deisenhammer, E.A., G. Kemmler, and P. Parson, 2003: Association of meteorological factors with suicide. *Acta Psychiatrica Scandinavica,* **108,** 455-459. http://dx.doi.org/10.1046/j.0001-690X.2003.00219.x

200. Lee, H.-C., H.-C. Lin, S.-Y. Tsai, C.-Y. Li, C.-C. Chen, and C.-C. Huang, 2006: Suicide rates and the association with climate: A population-based study. *Journal of Affective Disorders,* **92,** 221-226. http://dx.doi.org/10.1016/j.jad.2006.01.026

201. Page, L.A., S. Hajat, and R.S. Kovats, 2007: Relationship between daily suicide counts and temperature in England and Wales. *The British Journal of Psychiatry,* **191,** 106-112. http://dx.doi.org/10.1192/bjp.bp.106.031948

202. Preti, A., G. Lentini, and M. Maugeri, 2007: Global warming possibly linked to an enhanced risk of suicide: Data from Italy, 1974–2003. *Journal of Affective Disorders,* **102,** 19-25. http://dx.doi.org/10.1016/j.jad.2006.12.003

203. Maes, M., F. De Meyer, P. Thompson, D. Peeters, and P. Cosyns, 1994: Synchronized annual rhythms in violent suicide rate, ambient temperature and the light-dark span. *Acta Psychiatrica Scandinavica,* **90,** 391-396. http://dx.doi.org/10.1111/j.1600-0447.1994.tb01612.x

204. Qi, X., S. Tong, and W. Hu, 2009: Preliminary spatiotemporal analysis of the association between socio-environmental factors and suicide. *Environmental Health,* **8,** Article 46. http://dx.doi.org/10.1186/1476-069X-8-46

205. Souêtre, E., E. Salvati, J.L. Belugou, P. Douillet, T. Braccini, and G. Darcourt, 1987: Seasonality of suicides: Environmental, sociological and biological covariations. *Journal of Affective Disorders,* **13,** 215-225. http://dx.doi.org/10.1016/0165-0327(87)90040-1

206. Souêtre, E., T.A. Wehr, P. Douillet, and G. Darcourt, 1990: Influence of environmental factors on suicidal behavior. *Psychiatry Research,* **32,** 253-263. http://dx.doi.org/10.1016/0165-1781(90)90030-9

207. Partonen, T., J. Haukka, H. Nevanlinna, and J. Lönnqvist, 2004: Analysis of the seasonal pattern in suicide. *Journal of Affective Disorders,* **81,** 133-139. http://dx.doi.org/10.1016/s0165-0327(03)00137-x

208. Dixon, P.G., A.N. McDonald, K.N. Scheitlin, J.E. Stapleton, J.S. Allen, W.M. Carter, M.R. Holley, D.D. Inman, and J.B. Roberts, 2007: Effects of temperature variation on suicide in five U.S. counties, 1991-2001. *International Journal of Biometeorology,* **51,** 395-403. http://dx.doi.org/10.1007/s00484-006-0081-4

209. Ruuhela, R., L. Hiltunen, A. Venäläinen, P. Pirinen, and T. Partonen, 2009: Climate impact on suicide rates in Finland from 1971 to 2003. *International Journal of Biometeorology,* **53,** 167-175. http://dx.doi.org/10.1007/s00484-008-0200-5

210. Thoenes, M.M., 2011: Heat-related illness risk with methylphenidate use. *Journal of Pediatric Health Care,* **25,** 127-132. http://dx.doi.org/10.1016/j.pedhc.2010.07.006

211. Berg, P.J., S. Smallfield, and L. Svien, 2010: An investigation of depression and fatigue post West Nile virus infection. *South Dakota Medicine,* **63,** 127-129, 131-133.

212. Nolan, M.S., A.M. Hause, and K.O. Murray, 2012: Findings of long-term depression up to 8 years post infection from West Nile virus. *Journal of Clinical Psychology,* **68,** 801-808. http://dx.doi.org/10.1002/jclp.21871

213. Szyszkowicz, M., 2007: Air pollution and emergency department visits for depression in Edmonton, Canada. *International Journal of Occupational Medicine and Environmental Health,* **20,** 241-245. http://dx.doi.org/10.2478/v10001-007-0024-2

214. Szyszkowicz, M., B. Rowe, and I. Colman, 2009: Air pollution and daily emergency department visits for depression. *International Journal of Occupational Medicine and Environmental Health,* **22,** 355-362. http://dx.doi.org/10.2478/v10001-009-0031-6

215. Szyszkowicz, M., J.B. Willey, E. Grafstein, B.H. Rowe, and I. Colman, 2010: Air pollution and emergency department visits for suicide attempts in Vancouver, Canada. *Environmental Health Insights,* **4,** 79-86. http://dx.doi.org/10.4137/ehi.s5662

216. Kim, C., S.H. Jung, D.R. Kang, H.C. Kim, K.T. Moon, N.W. Hur, D.C. Shin, and I. Suh, 2010: Ambient particulate matter as a risk factor for suicide. *American Journal of Psychiatry,* **167,** 1100-1107. http://dx.doi.org/10.1176/appi.ajp.2010.09050706

217. Qin, P., B.L. Waltoft, P.B. Mortensen, and T.T. Postolache, 2013: Suicide risk in relation to air pollen counts: A study based on data from Danish registers. *BMJ Open,* **3,** e002462. http://dx.doi.org/10.1136/bmjopen-2012-002462

218. Maldonado, J.K., B. Colombi, and R. Pandya, Eds., 2014: *Climate Change and Indigenous Peoples in the United States: Impacts, Experiences and Actions.* Springer, New York, 174 pp. http://dx.doi.org/10.1007/978-3-319-05266-3

219. Ford, J.D., 2012: Indigenous health and climate change. *American Journal of Public Health,* **102,** 1260-1266. http://dx.doi.org/10.2105/ajph.2012.300752

220. NCTSN, 2003: Review of Child and Adolescent Refugee Mental Health: White Paper from the National Child Traumatic Stress Network Refugee Trauma Task Force. 49 pp. National Child Traumatic Stress Network http://www.nctsnet.org/nctsn_assets/pdfs/reports/refugeereview.pdf

221. Beggs, P.J. and N.E. Walczyk, 2008: Impacts of climate change on plant food allergens: A previously unrecognized threat to human health. *Air Quality, Atmosphere & Health,* **1,** 119-123. http://dx.doi.org/10.1007/s11869-008-0013-z

222. Teufel, M., T. Biedermann, N. Rapps, C. Hausteiner, P. Henningsen, P. Enck, and S. Zipfel, 2007: Psychological burden of food allergy. *World Journal of Gastroenterology,* **13,** 3456-3465. http://dx.doi.org/10.3748/wjg.v13.i25.3456

223. Cummings, A.J., R.C. Knibb, R.M. King, and J.S. Lucas, 2010: The psychosocial impact of food allergy and food hypersensitivity in children, adolescents and their families: A review. *Allergy,* **65,** 933-945. http://dx.doi.org/10.1111/j.1398-9995.2010.02342.x

224. Patten, S.B. and J.V.A. Williams, 2007: Self-reported allergies and their relationship to several Axis I disorders in a community sample. *International Journal of Psychiatry in Medicine,* **37,** 11-22. http://dx.doi.org/10.2190/L811-0738-10NG-7157

9 POPULATIONS OF CONCERN

Lead Authors

Janet L. Gamble
U.S. Environmental Protection Agency

John Balbus
National Institutes of Health

Contributing Authors

Martha Berger
U.S. Environmental Protection Agency

Karen Bouye
Centers for Disease Control and Prevention

Vince Campbell
Centers for Disease Control and Prevention

Karletta Chief
The University of Arizona

Kathryn Conlon
Centers for Disease Control and Prevention

Allison Crimmins*
U.S. Environmental Protection Agency

Barry Flanagan
Centers for Disease Control and Prevention

Cristina Gonzalez-Maddux
formerly of the Institute for Tribal Environmental Professionals

Elaine Hallisey
Centers for Disease Control and Prevention

Sonja Hutchins
Centers for Disease Control and Prevention

Lesley Jantarasami*
U.S. Environmental Protection Agency

Samar Khoury
Association of Schools and Programs of Public Health

Max Kiefer
Centers for Disease Control and Prevention, National Institute for Occupational Safety and Health

Jessica Kolling
Centers for Disease Control and Prevention

Kathy Lynn
University of Oregon

Arie Manangan
Centers for Disease Control and Prevention

Marian McDonald
Centers for Disease Control and Prevention

Rachel Morello-Frosch
University of California, Berkeley

Margaret Hiza Redsteer
U.S. Geological Survey

Perry Sheffield
Icahn School of Medicine at Mount Sinai, New York

Kimberly Thigpen Tart
National Institutes of Health

Joanna Watson
Centers for Disease Control and Prevention, National Institute for Occupational Safety and Health

Kyle Powys Whyte
Michigan State University

Amy Funk Wolkin
Centers for Disease Control and Prevention

Acknowledgements: Larry Campbell, Swinomish Indian Tribal Community; **Jean Paul Chretien**, U.S. Department of Defense; **Patricia A. Cochran**, Alaska Native Science Commission; **Jamie Donatuto**, Swinomish Indian Tribal Community; **James Persson**, U.S. Department of Defense

Recommended Citation: Gamble, J.L., J. Balbus, M. Berger, K. Bouye, V. Campbell, K. Chief, K. Conlon, A. Crimmins, B. Flanagan, C. Gonzalez-Maddux, E. Hallisey, S. Hutchins, L. Jantarasami, S. Khoury, M. Kiefer, J. Kolling, K. Lynn, A. Manangan, M. McDonald, R. Morello-Frosch, M.H. Redsteer, P. Sheffield, K. Thigpen Tart, J. Watson, K.P. Whyte, and A.F. Wolkin, 2016: Ch. 9: Populations of Concern. *The Impacts of Climate Change on Human Health in the United States: A Scientific Assessment.* U.S. Global Change Research Program, Washington, DC, 247–286. http://dx.doi.org/10.7930/J0Q81B0T

On the web: health2016.globalchange.gov

*Chapter Coordinator

9 POPULATIONS OF CONCERN

Key Findings

Vulnerability Varies Over Time and Is Place-Specific
Key Finding 1: Across the United States, people and communities differ in their exposures, their inherent sensitivity, and their adaptive capacity to respond to and cope with climate change related health threats *[Very High Confidence]*. Vulnerability to climate change varies across time and location, across communities, and among individuals within communities *[Very High Confidence]*.

Health Impacts Vary with Age and Life Stage
Key Finding 2: People experience different inherent sensitivities to the impacts of climate change at different ages and life stages *[High Confidence]*. For example, the very young and the very old are particularly sensitive to climate-related health impacts.

Social Determinants of Health Interact with Climate Factors to Affect Health Risks
Key Finding 3: Climate change threatens the health of people and communities by affecting exposure, sensitivity, and adaptive capacity *[High Confidence]*. Social determinants of health, such as those related to socioeconomic factors and health disparities, may amplify, moderate, or otherwise influence climate-related health effects, particularly when these factors occur simultaneously or close in time or space *[High Confidence]*.

Mapping Tools and Vulnerability Indices Identify Climate Health Risks
Key Finding 4: The use of geographic data and tools allows for more sophisticated mapping of risk factors and social vulnerabilities to identify and protect specific locations and groups of people *[High Confidence]*.

9.1 Introduction

Climate change is already causing, and is expected to continue to cause, a range of health impacts that vary across different population groups in the United States. The vulnerability of any given group is a function of its sensitivity to climate change related health risks, its exposure to those risks, and its capacity for responding to or coping with climate variability and change. Vulnerable groups of people, described here as *populations of concern*, include those with low income, some communities of color, immigrant groups (including those with limited English proficiency), Indigenous peoples, children and pregnant women, older adults, vulnerable occupational groups, persons with disabilities, and persons with preexisting or chronic medical conditions. Planners and public health officials, politicians and physicians, scientists and social service providers are tasked with understanding and responding to the health impacts of climate change. Collectively, their characterization of vulnerability should consider how populations of concern experience disproportionate, multiple, and complex risks to their health and well-being in response to climate change.

Some groups face a number of stressors related to both climate and non-climate factors. For example, people living in impoverished urban or isolated rural areas, floodplains, coastlines, and other at-risk locations are more vulnerable not only to extreme weather and persistent climate change but also to social and economic stressors. Many of these stressors can occur simultaneously or consecutively. Over time, this "accumulation" of multiple, complex stressors is expected to become more evident[1] as climate impacts interact with stressors associated with existing mental and physical health conditions and with other socioeconomic and demographic factors.

9.2 A Framework for Understanding Vulnerability

Some populations of concern demonstrate relatively greater vulnerability to the health impacts of climate change. The definitions of the following key concepts are important to understand how some people or communities are disproportionately affected by climate-related health risks (Figure 1). Definitions are adapted from the Intergovernmental Panel on Climate Change (IPCC) and the National Research Council (NRC).[2,3]

- **Vulnerability** is the tendency or predisposition to be adversely affected by climate-related health effects, and encompasses three elements: exposure, sensitivity or susceptibility to harm, and the capacity to adapt to or to cope with change. *Exposure* is contact between a person and one or more biological, chemical, or physical stressors, including stressors affected by climate change. Contact may occur in a single instance or repeatedly over time, and may occur in one location or over a wider geographic area. *Sensitivity* is the degree to which people or communities are affected,

Food is distributed to people in need at Catholic Community Service in Wheaton, MD, November 23, 2010. Populations of concern experience disproportionate, multiple, and complex risks to their health and well-being in reponse to climate change.

either adversely or beneficially, by climate variability and change. *Adaptive capacity* is the ability of communities, institutions, or people to adjust to potential hazards, to take advantage of opportunities, or to respond to consequences. A related term, *resilience,* is the ability to prepare and plan for, absorb, recover from, and more successfully adapt to adverse events. People and communities with strong adaptive capacity have greater resilience.

- **Risk** is the potential for consequences to develop where something of value (such as human health) is at stake and where the outcome is uncertain. Risk is often represented as the probability of the occurrence of a hazardous event multiplied by the expected severity of the impacts of that event.

- **Stressors** are events or trends, whether related to climate change or other factors, that increase vulnerability to health effects.

People or communities can have greater or lesser vulnerability to health risks depending on social, political, and economic factors that are collectively known as social determinants of health.[5] Some groups are disproportionately disadvantaged by social determinants of health that limit resources and opportunities for health-promoting behaviors and conditions of daily life, such as living/working circumstances and access to healthcare services.[5] In disadvantaged groups, social determinants of health interact with the three elements of vulnerability by contributing to increased exposure, increased sensitivity, and reduced adaptive capacity (Figure 2). Health risks and vulnerability may increase in locations or instances where combinations of social determinants of health that amplify health threats occur simultaneously or close in time or space.[6,7] For example, people with limited economic resources living in areas with deteriorating infrastructure are more likely to experience disproportionate impacts and are less

Determinants of Vulnerability

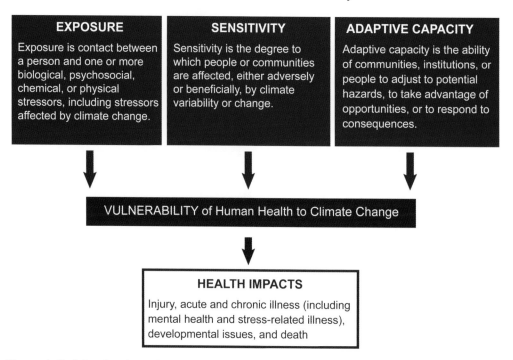

Figure 1: Defining the determinants of vulnerability to health impacts associated with climate change, including exposure, sensitivity, and adaptive capacity. (Figure source: adapted from Turner et al. 2003)[4]

able to recover following extreme events,[8, 9] increasing their vulnerability to climate-related health effects. Understanding the role of social determinants of health can help characterize climate change impacts and identify public health interventions or actions to reduce or prevent exposures in populations of concern.[6, 7, 10]

Factors that Contribute to Exposure

Exposures to climate-related variability and change are determined by a range of factors that individually and collectively shape the nature and extent of exposures. These factors include:

• *Occupation:* Certain occupations have a greater risk of exposure to climate impacts. People working outdoors or performing duties that expose them to extreme weather, such as emergency responders, utility repair crews, farm workers, construction workers, and other outdoor laborers, are at particular risk.[11]

• *Time spent in risk-prone locations:* Where a person lives, goes to school, works, or spends leisure time will contribute to exposure. Locations with greater health threats include urban areas (due to, for example, the "heat island" effect or air quality concerns), areas where airborne allergens and other air pollutants occur at levels that aggravate respiratory illnesses, communities experiencing depleted water supplies or vulnerable energy and transportation infrastructure, coastal and other flood-prone areas, and locations affected

by drought and wildfire.[12, 13, 14]

• *Responses to extreme events:* A person's ability or, in some cases, their choice whether to evacuate or shelter-in-place in response to an extreme event such as a hurricane, flood, or wildfire affects their exposure to health threats. Low-income populations are generally less likely to evacuate in response to a warning (see Ch. 4: Extreme Events).[8]

• *Socioeconomic status:* Persons living in poverty are more likely to be exposed to extreme heat and air pollution.[15, 16] Poverty also determines, at least in part, how people perceive the risks to which they are exposed, how they respond to evacuation orders and other emergency warnings, and their ability to evacuate or relocate to a less risk-prone location (see Ch. 8: Mental Health).[8]

• *Infrastructure condition and access:* Older buildings may expose occupants to increased indoor air pollutants and mold, stagnant airflow, or high indoor temperatures (see Ch. 3: Air Quality Impacts). Persons preparing for or responding to flooding, wildfires, or other weather-related emergencies may be hampered by disruption to transportation, utilities, medical, or communication infrastructure. Lack of access to these resources, in either urban or rural settings, can increase a person's vulnerability (see Ch. 4: Extreme Events).[17,18]

• **Compromised mobility, cognitive function, and other mental or behavioral factors:** These factors can lead to increased exposure to climate-related health impacts if people are not aware of health threats or are unable to take actions to avoid, limit, or respond to risks.[19] People with access and functional needs may be particularly at risk if these factors interfere with their ability to access or receive medical care before, during, or after a disaster or emergency.

Characterizing Biological Sensitivity

The sensitivity of human communities and individuals to climate change stressors is determined, at least in part, by biological traits. Among those traits are the overall health status, age, and life stage. From fetus, to infant, to toddler, to child, to adolescent, to adult, to the elderly, persons at every life stage have varying sensitivity to climate change impacts.[12, 20, 21] For instance, the relatively immature immune systems of very young children make them more sensitive to aeroallergen exposure (such as airborne pollens). In addition to life stage,

people experiencing long-term chronic medical and/or psychological conditions are more sensitive to climate stressors. Persons with asthma or chronic obstructive pulmonary disease (COPD) are more sensitive to exposures to wildfire smoke and other respiratory irritants. Social and economic factors also affect disparities in the prevalence of chronic medical conditions that aggravate biological sensitivity.[22, 23]

Adaptive Capacity and Response to Climate Change

Many of the same factors that contribute to exposure or sensitivity also influence the ability of both individuals and communities to adapt to climate variability and change. Socioeconomic status, the condition and accessibility of infrastructure, the accessibility of health care, certain demographic characteristics, human and social capital (the skills, knowledge, experience, and social cohesion of a community), and other institutional resources all contribute to the timeliness and effectiveness of adaptive capacity (see Ch. 1: Introduction and Ch. 4: Extreme Events).

Intersection of Social Determinants of Health and Vulnerability

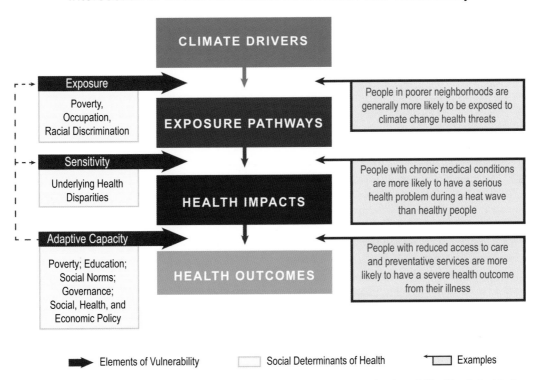

Figure 2: Social determinants of health interact with the three elements of vulnerability. The left side boxes provide examples of social determinants of health associated with each of the elements of vulnerability. Increased exposure, increased sensitivity and reduced adaptive capacity all affect vulnerability at different points in the causal chain from climate drivers to health outcomes (middle boxes). Adaptive capacity can influence exposure and sensitivity and also can influence the resilience of individuals or populations experiencing health impacts by influencing access to care and preventive services. The right side boxes provide illustrative examples of the implications of social determinants on increased exposure, increased sensitivity, and reduced adaptive capacity.

Nursing students and faculty at Emory University School of Nursing in Atlanta, Georgia, volunteering to give checkups in migrant workers' camps, June 12, 2006.

9.3 Populations of Concern

Communities of Color, Low Income, Immigrants, and Limited English Proficiency Groups

In the United States, some communities of color, low-income groups, people with limited English proficiency (LEP), and certain immigrant groups (especially those who are undocumented) live with many of the factors that contribute to their vulnerability to the health impacts of climate change (see Section 9.2). These populations are at increased risk of exposure given their higher likelihood of living in risk-prone areas (such as urban heat islands, isolated rural areas, or coastal and other flood-prone areas), areas with older or poorly maintained infrastructure, or areas with an increased burden of air pollution.[24, 25, 26, 27] These groups of people also experience relatively greater incidence of chronic medical conditions, such as cardiovascular and kidney disease, diabetes, asthma, and COPD,[28, 29, 30] which can be exacerbated by climate-related health impacts.[24, 31, 32, 33, 34] Socioeconomic and educational factors, limited transportation, limited access to health education, and social isolation related to language deficiencies collectively impede their ability to prepare for, respond to, and cope with climate-related health risks.[24, 26, 34, 35, 36, 37, 38, 39, 40, 41] These populations also may have limited access to medical care and may not be able to afford medications or other treatments.[30, 38] For LEP and undocumented persons, high poverty rates, language and cultural barriers, and citizenship status limit access to and use of health care and other social services and make these groups more hesitant to seek out help that might compromise their immigration status in the United States.[39, 42, 43, 44, 45, 46]

The number of people of color in the United States who may be affected by heightened vulnerability to climate-related health risks will continue to grow. Currently, Hispanics or Latinos, Blacks or African Americans, American Indians and Alaska Natives, Asian Americans, and Native Hawaiians and Pacific Islanders represent 37% of the total U.S. population.[47, 48] By

2042, they are projected to become the majority.[49] People of color already constitute the majority in four states (California, Hawaii, New Mexico, and Texas) and in many cities.[48] Numbers of LEP and undocumented immigrant populations have also increased. In 2011, LEP groups comprised approximately 9% (25.3 million individuals) of the U.S. population aged five and older.[50] In 2010, approximately 11.2 million people in the United States were undocumented.[51]

Vulnerability to Climate-Related Health Stressors

Key climate impacts for some communities of color and low-income, LEP, and immigrant populations include heat waves, other extreme weather events, poor air quality, food safety, infectious diseases, and psychological stressors.

Race is an important factor in vulnerability to climate-related stress, but it can be difficult to isolate the role of race from other related socioeconomic and geographic factors. Some racial minorities are also members of low-income groups, immigrants, and people with limited English proficiency, and it is their socioeconomic status (SES) that contributes most directly to their vulnerability to climate change-related stressors. SES is a measure of a person's economic and social status, often defined by income, education, and occupation. Additional factors such as age, gender, preexisting medical conditions, psychosocial factors, and physical and mental stress are also associated with vulnerability to climate change. Because many of these variables are highly related to one another, statistical models must account for these factors in order to accurately measure the relative importance of various risk factors.[52, 53] For instance, minority race and low SES are jointly linked to increased prevalence of underlying health conditions that may affect sensitivity to climate change. When adjusted for age, gender, and level of education, the number of potential life-years lost from all causes of death was found to be 35% greater for Blacks than for Whites in the United States,[54] indicating an independent effect of race.

Extreme heat events. Some communities of color and some low-income, homeless, and immigrant populations are more exposed to heat waves,[55, 56] as these groups often reside in urban areas affected by heat island effects.[13, 15, 24, 57] In addition, these populations are likely to have limited adaptive capacity due to a lack of adequately insulated housing, inability to afford or to use air conditioning, inadequate access to public shelters such as cooling centers, and inadequate access to both routine and emergency health care.[24, 26, 29, 34, 35, 38] These social, economic, and health risk factors give rise to the observed increase in deaths and disease from extreme heat in some immigrant and impoverished communities.[24, 32, 33] Elevated risks for mortality associated with exposures to high ambient temperatures are also reported for Blacks as compared to Whites,[32, 40, 58, 59] a finding that persists once air conditioning use is accounted for (see also Ch. 2: Temperature-Related Death and Illness).[60]

Other weather extremes. As observed during and after Hurricane Katrina and Hurricane/Post-Tropical Cyclone Sandy, some communities of color and low-income people experienced increased illness or injury, death, or displacement due to poor-quality housing, lack of access to emergency communications, lack of access to transportation, inadequate access to health care services and medications, limited post-disaster employment, and limited or no health and property insurance.[61, 62, 63, 64, 65, 66] Following a 2006 flood in El Paso, Texas, Hispanic ethnicity was identified as a significant risk factor for adverse health effects after controlling for other important socioeconomic factors (for example, age and housing quality).[67] Adaptation measures to address these risk factors—such as providing transportation during evacuations or targeted employment assistance during the recovery phase—may help reduce or eliminate these health impact disparities, but may not be readily available or affordable (see also Ch. 4: Extreme Events).[61, 62, 63, 65, 66]

Degraded air quality. Climate change impacts on outdoor air quality will increase exposure in urban areas where large proportions of minority, low-income, homeless, and immigrant populations reside. Fine particulate matter and ozone levels already exceed National Ambient Air Quality Standards in many urban areas.[26, 27, 68, 69] Given the relatively higher rates of cardiovascular and respiratory diseases in low-income urban populations,[26, 28, 30] these populations are more sensitive to degraded air quality, resulting in increases in illness, hospitalization, and premature death.[70, 71, 72, 73, 74, 75, 76, 77, 78] In addition, climate change can contribute to increases in aeroallergens, which exacerbate asthma, an illness that is relatively more common among some communities of color and low-income groups. People of color are especially impacted by air pollution due to both disproportionate exposures for persons living in urban areas as well as higher prevalence of underlying diseases, such as asthma and COPD, which increase their inherent sensitivity. In 2000, the prevalence of asthma was 122 per 1,000 Black persons and 104 per 1,000 White persons in the United States. At that time, asthma mortality was approximately three times higher among Blacks as compared to Whites (see also Ch. 1: Introduction; Ch. 3: Air Quality Impacts).[59]

Waterborne and vector-borne diseases. Climate change is expected to increase exposure to waterborne pathogens that cause a variety of illnesses—most commonly gastrointestinal illness and diarrhea (see also Ch. 6: Water-Related Illness). Health risks increase in crowded shelter conditions following floods or hurricanes,[79] which suggests that some low-income groups living in crowded housing (particularly prevalent among foreign-born or Hispanic populations)[80] may face increased exposure risk. Substandard or deteriorating water infrastructure (including sewerage, drainage, and storm water systems, and drinking water systems) in both urban and rural low-income areas also contribute to increased risk of expo-

sure to waterborne pathogens.[81, 82] Low-income populations in some regions may also be more vulnerable to the changes in the distribution of some vector-borne diseases that are expected to result from climate change. For example, higher incidence of West Nile virus disease has been linked to poverty and to urban location in the southeastern and northeastern United States, respectively (see also Ch. 5: Vector-Borne Diseases).[83, 84]

Food safety and security. Climate change affects food safety and is projected to reduce the nutrient and protein content of some crops, like wheat and rice. Some communities of color and low-income populations are more likely to be affected because they spend a relatively larger portion of their household income on food compared to more affluent households. These groups often suffer from poor-quality diets and limited access to full-service grocery stores that offer healthy and affordable dietary choices (see also Ch. 7: Food Safety).[36, 37, 85, 86]

Psychological stress. Some communities of color, low-income populations, immigrants, and LEP groups are more likely to experience stress-related mental health impacts, particularly during and after extreme events. Other contributing factors include barriers in accessing and affording mental health care, such as counseling in native languages, and the availability and affordability of appropriate medications (see also Ch. 8: Mental Health).[87, 88]

Indigenous Peoples in the United States

A number of health risks are higher among Indigenous populations, such as poor mental health related to historical or personal trauma, alcohol abuse, suicide, infant/child mortality, environmental exposures from pollutants or toxic substances, and diabetes caused by inadequate or improper diets.[89, 90, 91, 92, 93, 94, 95, 96] Because of existing vulnerabilities, Indigenous people, especially those who are dependent on the environment for sustenance or who live in geographically isolated or impoverished communities, are likely to experience greater exposure and lower resilience to climate-related health effects. Indigenous Arctic communities have already experienced difficulty adapting to climate change effects such as reductions in sea ice thickness, thawing permafrost, increases in coastal erosion[97, 98, 99, 100] and landslide frequency,[101] alterations in the ranges of some fish,[102] increased weather unpredictability,[103] and northward advance of the tree line.[104] These climate changes have disrupted traditional hunting and subsistence practices and may threaten infrastructure such as the condition of housing, transportation, and pipelines,[103] which ultimately may force relocation of villages.[105]

Food safety and security. Examples of how climate changes can affect the health of Indigenous peoples include changes in the abundance and nutrient content of certain foodstuffs, such as berries for Alaska Native communities;[106] declining moose populations in Minnesota, which are significant to many Ojibwe peoples and an important source of dietary protein;[107, 108] rising temperatures and lack of available water for farming among Navajo people;[109] and declines in tradition-al rice harvests among the Ojibwe in the Upper Great Lakes region.[110] Traditional foods and livelihoods are embedded in Indigenous cultural beliefs and subsistence practices.[111, 112, 113, 114, 115, 116, 117] Climate impacts on traditional foods may result in poor nutrition and increased obesity and diabetes.[118]

Changes in aquatic habitats and species also affect subsistence fishing.[119] Rising temperatures affect water quality and avail-ability. Lower oxygen levels in freshwater and seawater de-grade water quality and promote the growth of disease-caus-ing bacteria, viruses, and parasites.[120] Warming can exacerbate shellfish disease and make mercury more readily absorbed in fish tissue. Elevated sea surface temperatures, consistent with projected trends in climate warming, have been associated with increased accumulation of methylmercury in fish and increased human exposure.[121] Mercury is a neurotoxin that adversely affects people at all life stages, particularly during the prenatal stage (see also Ch. 6: Water-Related Illness; Ch. 7: Food Safety).[121, 122, 123] In addition, oceans are becoming more acidic as they absorb some of the carbon dioxide (CO_2) added to the atmosphere by fossil fuel burning and other sources, and this change in acidity can lower shellfish survival.[120] This affects Indigenous peoples on the West and Gulf Coasts and Alaska Natives whose livelihoods depend on shellfish har-vests.[124] Rising sea levels will also destroy fresh and saltwater habitats that some Indigenous peoples located along the Gulf Coast rely upon for subsistence food.[125]

Water security. Indigenous peo-ples may lack access to water resources and to adequate in-frastructure for water treatment and supply. A significant number of Indigenous persons living on remote reservations lack indoor plumbing and rely on unregu-lated water supplies that are vulnerable to drought, changes in water quality, and contamination of water in local systems.[109, 126] Existing infrastructure may be poorly maintained or in need of significant and costly upgrades.[127] Heavy rainfall events and warm temperatures have been linked to diarrheal outbreaks and bacterial contamination of drinking water sources (see Ch. 6: Water-Related Illness). Acute diarrheal disease has been shown to disproportionately affect children on the Fort Apache reservation in Arizona,[128] and result in higher over-all hospitalization rates for American Indian/Alaska Native infants.[129] Increased extreme precipitation and potential increases in cyanobacterial blooms (see Ch. 6: Water-Related

Indigenous deckhand pulls in net of geoducks near Suquamish, Washington, January 17, 2007. Traditional foods and livelihoods are embedded in Indigenous cultural beliefs and subsistence practices.

Illness) are also expected to stress existing water infrastruc-ture on tribal lands and increase exposure to waterborne pathogens.[122, 130]

Loss of cultural identity. Climate change threatens sacred ceremonial and cultural practices through changing the availability of culturally relevant plant and animal species.[95, 130] Climate-related threats may compound historical impacts associated with colonialism, as well as current effects on tribal culture as more young people leave reservations for educa-tion and employment opportunities. Loss of tribal territory and disruption of cultural resources and traditional ways of life[131, 132] lead to loss of cultural identity.[133, 134, 135] The loss of medicinal plants due to climate change may leave ceremonial and traditional practitioners without the resources they need to practice traditional healing.[114, 136] The relocation of young people may reduce interactions across generations and under-mine the sharing of traditional knowledge, tribal lore, and oral history.[137, 138]

Degraded infrastructure and other impacts. Rising tempera-tures may damage transportation infrastructure on tribal lands. Changing ice or thawing permafrost, flooding, and drought-related dust storms may block roads and cut off communities from access to evacuation routes and emergency medical care or social services.[139] Poor air quality from blowing dust affects southwestern Indigenous communities, particu-larly in Arizona and New Mexico, and is likely to worsen with drought conditions.[140] Exposure to impaired air quality also affects Indigenous communities, especially those downwind from urban areas or industrial complexes.

> *Because of existing vulnerabilities, Indigenous people, especially those who are dependent on the environment for sustenance or who live in geographically isolated or impoverished communities, are likely to experience greater exposure and lower resilience to climate-related health effects.*

Children and Pregnant Women

Children are vulnerable to adverse health effects associated with environmental exposures due to factors related to their immature physiology and metabolism, their unique exposure pathways, their biological sensitivities, and limits to their adaptive capacity. Children pass through a series of windows of vulnerability that begin in the womb and continue through their second decade of life. Children have a proportionately higher intake of air, food, and water relative to their body weight compared to adults.[20] They also share unique behaviors and interactions with their environment that may increase their exposure to environmental contaminants. For example, small children often play indoors on the floor or outdoors on the ground and place hands and other objects in their mouths, increasing their exposure to dust and other contaminants, such as pesticides, mold spores, and allergens.[141] There is, however, large variation in vulnerability among children at different life stages due to differing physiology and behaviors (Figure 3). Climate change—interacting with factors such as economic status, diet, living situation, and stage of development—will increase children's exposure to health threats.[12, 21, 142, 143, 144] The impact of poverty on children's health is a critical factor to consider in ascertaining how climate change will be manifest in children. Poor and low-income households have difficulty accessing health care and meeting the basic needs that are crucial for healthy child development. In addition, children in poverty are less likely to have access to air conditioning to mitigate the effects of extreme heat. Children living in poverty are also less likely to be able to respond to or escape from extreme weather events.[12, 21, 142, 143, 144]

Vulnerability to Climate-Related Health Stressors

Extreme heat events. An increase in the frequency and intensity of extreme heat events (see Ch. 2: Temperature-Related Death and Illness) will affect children who spend time outdoors or in non-climate-controlled indoor settings. Student athletes and other children who are susceptible to heat-related illnesses when they exercise or play outdoors in hot and humid weather may be poorly acclimated to physical exertion in the heat. Some 9,000 high school athletes in the United States are treated for exertional heat illness (such as heat stroke and muscle cramps) each year, with the greatest risk among high school football players.[145, 146] This appears to be a worsening trend. Between 1997 and 2006, emergency department visits for all heat-related illness increased 133% and youth made up almost 50% of those cases.[147] From 2000 through 2013, the number of deaths due to heat stroke doubled among U.S. high school and college football players.[148] Other data show effects of extreme heat on children of all ages, including increases in heat illness, fluid and electrolyte imbalances, and asthma. Children in homes or schools without air conditioning are also more vulnerable during heat events.

Other weather extremes. Climate change is likely to affect the mental health and well-being of children, primarily by increasing exposure to traumatic weather events that result in injury, death, or displacement. In 2003, more than 10% of U.S. children from infancy to 18 years of age reported experiencing a disaster (fire, tornado, flood, hurricane, earthquake, etc.) during their lifetimes.[149] Exposures to traumatic events can impact children's capacity to regulate emotions, undermine cognitive development and academic performance, and contribute to post-traumatic stress disorder (PTSD) and other psychiatric disorders (such as depression, anxiety, phobia, and panic).[150] Children's ability to cope with disasters is affected by factors such as socioeconomic status, available support systems, and timeliness of treatment. Negative mental health effects in children, if untreated, can extend into adulthood.[150] (See Ch. 4: Extreme Events; Ch. 8: Mental Health).

Vulnerability to the Health Impacts of Climate Change at Different Life Stages.

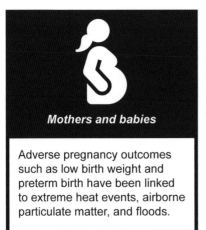

Mothers and babies

Adverse pregnancy outcomes such as low birth weight and preterm birth have been linked to extreme heat events, airborne particulate matter, and floods.

Infants and toddlers

Young children's biological sensitivity places them at greater risk from asthma, diarrheal illness, and heat-related illness.

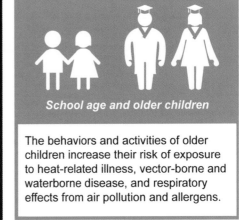

School age and older children

The behaviors and activities of older children increase their risk of exposure to heat-related illness, vector-borne and waterborne disease, and respiratory effects from air pollution and allergens.

Figure 3: Children's vulnerability to climate change results from distinct exposures, biological sensitivities (developing bodies and immune systems), and limitations to adaptive capacity (dependency on caregivers) at different life stages.

Climate-related exposures may lead to adverse pregnancy and newborn health outcomes.

Degraded air quality. Several factors make children more sensitive to the effects of respiratory hazards, including lung development that continues through adolescence, the size of the child's airways, their level of physical activity, and body weight. Climate change has the potential to affect future ground-level ozone concentrations, particulate matter concentrations, and levels of some aeroallergens. Ground-level ozone and particulate matter are associated with increases in asthma episodes and other adverse respiratory effects in children.[151, 152, 153] Nearly seven million, or about 9%, of children in the United States, suffer from asthma.[154] Asthma accounts for 10 million missed school days each year.[155] Particulate matter such as dust and emissions from coal-fired electricity generation plants is also associated with decreases in lung maturation in children.[156]

Changes in climate also contribute to longer, more severe pollen seasons that may be associated with increases in childhood asthma episodes and other allergic illnesses. Children may also be exposed to indoor air pollutants, including both particulate matter originating outdoors and indoor sources such as tobacco smoke and mold. In addition, high outdoor temperatures may increase the amount of time children spend indoors. Homes, childcare centers, and schools—places where children spend large amounts of their time—are all settings where indoor air quality issues may affect children's health. In communities where these buildings are insufficiently supplied with screens, air conditioning, humidity controls, or pest control, children's health may be at risk.[157] (See Ch. 3: Air Quality Impacts).

Waterborne illnesses. Climate change induced increases in heavy rainfall, flooding, and coastal storm events are expected to increase children's risk of gastrointestinal illness from ingestion of or contact with contaminated water.[61, 142, 143, 158] An increased association between heavy rainfall and increased acute gastrointestinal illness has already been observed in children in the United States.[159] Children may be especially vulnerable to recreational exposures to waterborne pathogens, in part because they swallow roughly twice as much water as adults while swimming.[160] In addition, children comprised

40% of swimming-related eye and ear infections from the waterborne bacteria *Vibrio alginolyticus* during the period 1997–2006[161] and 66% (ages 1–19) of those seeking treatment for illness associated with harmful algal bloom toxins in 2009–2010.[162] (See Ch. 6. Water-Related Illness).

Vector-Borne and other infectious diseases. The changes in the distribution of infectious diseases that are expected to result from climate change may introduce new exposures to children (see Ch. 5: Vector-Borne Disease). Due to physiological vulnerability or changes in their body's immune system, fetuses, pregnant women, and children are at increased risk of acquiring or having complications from certain infectious diseases such as listeriosis,[163] dengue fever,[164] and influenza.[165] Children spend more time outdoors than adults, increasing their exposure to mosquito and tick bites that can cause vector-borne diseases that disproportionately affect children such as La Crosse encephalitis or Lyme disease.[21, 143, 166] Lyme disease is most frequently reported among male children aged 5 to 9 years, and a disproportionate increasing trend was observed in all children from 1992 to 2006.[167, 168]

Food safety and security. Climate change, including rising levels of atmospheric CO_2, significantly reduces food quality and threatens availability and access for children. Because of the importance of nutrition during certain stages of physical and mental growth and development, the direct effect of the continued rise of CO_2 on reducing food quality will be an increasingly significant issue for children globally.[169, 170, 171] For the United States, disruptions in food production or distribution due to extreme events such as drought can increase costs and limit availability or access,[172, 173] particularly for food-insecure households, which include nearly 16% of households with children in the United States.[174] Children are also more susceptible to severe infection or complications from *Escherichia coli* infections, such as hemolytic uremic syndrome.[175] (See Ch. 7: Food Safety).

Vulnerability Related to Life Stage
Prenatal and pregnancy outcomes for mothers and babies.
Climate-related exposures may lead to adverse pregnancy and newborn health outcomes, including spontaneous abortion, low birth weight (less than 5.5 pounds), preterm birth (birth before 37 weeks of pregnancy), increased neonatal death, dehydration and associated renal failure, malnutrition, diarrhea, and respiratory disease.[21, 176] Other risk factors that may influence maternal and newborn health include water scarcity, poverty, and population displacement.[21, 176] The rate of preterm births is relatively high in the United States (1 of every 9 infants born),[177] where they contribute substantially to neonatal death and illness. Of the 1.2 million preterm births estimated to occur annually in high-income countries, more than 500 thousand (42% of the total) occur in the United States.[178] Extreme heat events have been associated with adverse birth outcomes such as low birth weight, preterm birth,

and infant mortality,[179, 180, 181] as well as congenital cataracts.[182] Newborns are especially sensitive to ambient temperatures that are too high or too low because their capacity for regulating body temperature is limited.[183]

In addition, exposure of pregnant women to inhaled particulate matter is associated with negative birth outcomes.[184, 185, 186, 187, 138, 189] Incidences of diarrheal diseases and dehydration may increase in extent and severity, which can be associated with adverse effects on pregnancy outcomes and the health of newborns.[176] Floods are associated with an increased risk of maternal exposure to environmental toxins and mold, reduced access to safe food and water, psychological stress, and disrupted health care. Other flood-related health outcomes for mothers and babies include maternal risk of anemia (a condition associated with low red blood cell counts sometimes caused by low iron intake), eclampsia (a condition that can cause seizures in pregnant women), and spontaneous abortion.[190, 191, 192, 193]

Infants and toddlers. Infants and toddlers are particularly sensitive to air pollutants, extreme heat, and microbial water contamination, which are all affected by climate change. Ozone exposure in young children and exposure to air pollutants and toxins in wildfire smoke are associated with increased asthma risk and other respiratory illnesses.[78, 142] Young children and infants are particularly vulnerable to heat-related illness and death, as their bodies are less able to adapt to heat than are adults.[32, 40, 58, 143, 194] Children under four years of age experience higher hospital admissions for respiratory illnesses during heat waves.[195] Rates of diarrheal illness have been shown to be higher in children under age five in the United States,[196] and climate change is expected to increase children's risk of gastrointestinal illness from ingestion or contact with contaminated water (see also Ch. 6: Water-Related Illness).[61, 142, 143, 158]

Older Adults

Older adults (generally defined as persons aged 65 and older) are vulnerable to the health impacts associated with climate change and weather extremes.[12, 197, 198, 199] The number of older adults in the United States is projected to grow substantially in the coming decades. The nation's older adult population (ages 65 and older) will nearly double in number from 2015 through 2050, from approximately 48 million to 88 million.[200] Of those 88 million older adults, a little under 19 million will be 85 years of age and older.[201] This projected population growth is largely due to the aging of the Baby Boomer generation (an estimated 76 million people born in the United States between 1946 and 1964), along with increases in lifespan and survivorship.[19] Older adults in the United States are not uniform with regard to their climate-related vulnerabilities, but are a diverse group with distinct subpopulations that can be identified not only by age but also by race, educational attainment, socioeconomic status, social support networks, overall physical and mental health, and disability status.[198, 202]

Vulnerability to Climate-Related Health Stressors

The potential climate change related health impacts for older adults include rising temperatures and heat waves; increased risk of more intense hurricanes (Categories IV and V), floods, droughts, and wildfires; degraded air quality; exposure to infectious diseases; and other climate-related hazards.[120]

Extreme heat events. Older adults exposed to extreme heat can experience multiple adverse effects.[203] In the coming decades, extreme heat events are projected to become more frequent, more intense, and of longer duration, especially in higher latitudes and large metropolitan areas.[24, 204] Between 1979 and 2004, 5,279 deaths were reported in the United States related to heat exposure, with those deaths reported most commonly among adults aged 65 and older.[205] Disease incidence among older adults is expected to increase even in regions with relatively modest temperature changes (as demonstrated by case studies of a 2006 California heat wave).[40] In New York City, extreme high temperatures were associated with an increase in hospital admissions for cardiovascular and respiratory disorders, with the elderly among the most affected. Hospital admissions for respiratory illness were greatest for the elderly, with a 4.7% increase per degree Centigrade increase.[33] Future climate-related increases in summertime temperatures may increase the risk of death in older people with chronic conditions, particularly those suffering from congestive heart failure and diabetes.[206] The percentage of older adults with diabetes, which puts individuals at higher risk for heat-related illness and death, has increased from 9.1% in 1980 to 19.9% in 2009.[207]

Other weather extremes. Hurricanes and other severe weather events lead to physical, mental, or emotional trauma before, during, and after the event.[208] The need to evacuate an area can pose increased health and safety risks for older adults, especially those who are poor or reside in nursing or assisted-living facilities.[209, 210] Moving patients to a sheltering facility is complicated, costly, and time-consuming and requires concurrent transfer of medical records, medications, and medical equipment (see also Ch. 4: Extreme Events).[210, 211]

Degraded air quality. Climate change can affect air quality by increasing ground-level ozone, fine particulate matter, aeroallergens, wildfire smoke, and dust (see Ch. 3: Air Quality Impacts).[212, 213] Exposure to ground-level ozone varies with age and can affect lung function and increase emergency department visits and hospital admissions, even for healthy adults. Air pollution can also exacerbate asthma and COPD and can increase the risk of heart attack in older adults, especially those who are also diabetic or obese.[214]

Vector-Borne and waterborne diseases. The changes in the distribution of disease vectors like ticks and mosquitoes that are expected to result from climate change may increase exposures to pathogens in older adult populations (see Ch. 5:

Vector-Borne Diseases). Some vector-borne diseases, notably mosquito-borne West Nile and St. Louis encephalitis viruses,[215, 216] pose a greater health risk among sensitive older adults with already compromised immune systems. Climate change is also expected to increase exposure risk to waterborne pathogens in sources of drinking water and recreational water. Older adults have a higher risk of contracting gastrointestinal illnesses from contaminated drinking and recreational water and suffering severe health outcomes and death (see Ch. 6: Water-Related Illness).[217, 218, 219, 220]

Interactions with Non-Climate Stressors

Vulnerability related to locations and condition of the built environment. Older adults are particularly vulnerable to climate change related health effects depending on their geographic location and characteristics of their homes, such as the quality of construction and amenities. More than half of the elderly U.S. adult population is concentrated in 170 counties (5% of all U.S. counties), and approximately 20% of older Americans live in a county in which a hurricane or tropical storm made landfall over the last decade.[221] For example, Florida is a traditional retirement destination with an older adult population accounting for 16.8% of the total in 2010, nearly four percentage points higher than the national average.[222] The increasing severity of tropical storms may pose particular risks for older adults in coastal zones.[223] Other geographic risk factors common to older adults are the urban heat island effect, urban sprawl (which affects mobility), characteristics of the built environment, and perceptions of neighborhood safety.[224, 225]

In neighborhoods where safety and crime are a concern, older residents may fear venturing out of their homes, thus increasing their social isolation and risk of health impacts during events such as heat waves.[224] Degraded infrastructure, including the condition of housing and public transportation, is associated with higher numbers of heat-related deaths in older adults. In multi-story residential buildings in which residents rely on elevators, electricity loss makes it difficult, if not impossible, for elderly residents and those with disabilities to leave the building to obtain food, medicine, and other needed services.[226] Also, older adults who own air-conditioning units may not utilize them during heat waves due to high operating costs.[12, 227, 228, 229]

Vulnerability related to physiological factors. Older adults are more sensitive to weather-related events due to age-related physiological factors. Elevated risks for cardiovascular deaths related to exposure to extreme heat have been observed in older adults.[32, 230] Generally poorer physical health conditions, such as long-term chronic illnesses, are exacerbated by climate change.[227, 228, 231, 232] In addition, aging can impair the mechanisms that regulate body temperature, particularly for those taking medications that interfere with regulation of body temperature, including psychotropic medications used to treat a variety of mental illnesses such as depression, anxiety, and psy-

chosis.[233] Respiratory impairments already experienced by older adults will be exacerbated by increased exposure to outdoor air pollutants (especially ozone and fine particulate matter), aeroallergens, and wildfire smoke—all of which may be exacerbated by climate change.[199, 213]

Vulnerability related to disabilities. Some functional limitations and mobility impairments increase older adults' sensitivity to climate change, particularly extreme events. In 2010, 49.8% of older adults (over 65) were reported to have a disability, compared to 16.6% of people aged 21–64.[234] Dementia occurs at a rate of 5% of the U.S. population aged 71 to 79 years, with an increase to more than 37% at age 90 and older.[235] Older adults with mobility or cognitive impairments are likely to experience greater vulnerability to health risks due to difficulty responding to, evacuating, and recovering from extreme events.[12, 231]

Occupational Groups

Climate change may increase the prevalence and severity of known occupational hazards and exposures, as well as the emergence of new ones. Outdoor workers are often among the first to be exposed to the effects of climate change. Climate change is expected to affect the health of outdoor workers through increases in ambient temperature, degraded air quality, extreme weather, vector-borne diseases, industrial exposures, and changes in the built environment.[11] Workers affected by climate change include farmers, ranchers, and other agricultural workers; commercial fishermen; construction workers; paramedics, firefighters and other first responders; and transportation workers. Also, laborers exposed to hot indoor work environments (such as steel mills, dry cleaners, manufacturing facilities, warehouses, and other areas that lack air conditioning) are at risk for extreme heat exposure.[236, 237, 238]

For some groups, such as migrant workers and day laborers, the health effects of climate change can be cumulative, with occupational exposures exacerbated by exposures associated with poorly insulated housing and lack of air conditioning. Workers may also be exposed to adverse occupational and climate-related conditions that the general public may altogether avoid, such as direct exposure to wildfires.

Extreme heat events. Higher temperatures or longer, more frequent periods of heat may result in more cases of heat-related illnesses (for example, heat stroke and heat exhaustion) and fatigue among workers,[237, 238, 239, 240, 241] especially among more physically demanding occupations. Heat stress and fatigue can also result in reduced vigilance, safety lapses, reduced work capacity, and increased risk of injury. Elevated temperatures can increase levels of air pollution, including ground-level ozone, resulting in increased worker exposure and subsequent risk of respiratory illness (see also Ch. 2: Temperature-Related Death and Illness).[11, 236, 237, 242]

Other weather extremes. Some extreme weather events and natural disasters, such as floods, storms, droughts, and wildfires, are becoming more frequent and intense (see also Ch. 4: Extreme Events).[120] An increased need for complex emergency responses will expose rescue and recovery workers to physical and psychological hazards.[205, 243] The safety of workers and their ability to recognize and avoid workplace hazards may be impaired by damage to infrastructure and disrupted communication.

From 2000 to 2013, almost 300 U.S. wildfire firefighters were killed while on duty.[244] With the frequency and severity of wildfires projected to increase, more firefighters will be exposed. Common workplace hazards faced on the fire line include being overrun by fire (as happened during the Yarnell Hill Fire in Arizona in 2013 that killed 19 firefighters);[245]

heat-related illnesses and injuries; smoke inhalation; vehicle-related injuries (including aircraft); slips, trips, and falls; and exposure to particulate matter and other air pollutants in wildfire smoke. In addition, wildland fire fighters are at risk of rhabdomyolysis (a breakdown of muscle tissue) that is associated with prolonged and intense physical exertion.[246]

Other workplace exposures to outdoor health hazards. Other climate-related health threats for outdoor workers include increased waterborne and foodborne pathogens, increased duration of aeroallergen exposure with longer pollen seasons,[247, 248] and expanded habitat ranges of disease-carrying vectors that may influence the risk of human exposure to diseases such as West Nile virus or Lyme disease (see also Ch. 5: Vector-Borne Diseases).[249]

Vulnerability of the U.S. Armed Forces

Another emerging area of interest, but one where research is limited and key research questions remain, is the relationship between climate change and occupational safety and health hazards posed to members of the U.S. Armed Forces. The U.S. Department of Defense (DoD) recognizes that climate change will affect its operating environment, roles, and missions both within the United States and abroad.[250, 251, 252] The DoD faces unique challenges in protecting the health of its personnel from climate change impacts.

Military personnel who train and conduct operations in hot environments are at risk for heat-related illness. The incidence of heat illness among active duty U.S. military personnel is several-fold higher than the summertime incidence in the general U.S. population (147 per

Soldiers race for first place during an annual physical training competition in Fort Riley, Kansas.

100,000 among the military versus 21.5 per 100,000 in the general population per year).[253, 254] A large proportion of military heat illness cases occur in training settings in the southern United States,[253] where climate change may increase future risk.

Exposure to some climate-sensitive infectious diseases also may be increased among military personnel who work extensively in field settings. For example, Lyme disease is the most commonly reported vector-borne disease in the list of Armed Forces Reportable Medical events, which covers diseases that may represent significant threats to public health and military operations. Lyme disease incidence is highest in military units in the Northeast United States, and in some cases is substantially higher than the Centers for Disease Control and Prevention (CDC) estimates for the population of the state in which the unit is located.[255] Coccidioidomycosis, or "valley fever," caused by inhalation of fungal spores, is an occupational hazard for military personnel training in the southwestern United States or other endemic areas (see also Ch 4: Extreme Events).[256, 257] Military personnel are stationed and deployed globally, and may face higher risk of climate-sensitive infections that are rare in the United States. Recent examples include chikungunya,[258, 259] dengue fever,[260] leishmaniasis,[261] and malaria.[262, 263, 264, 265]

The DoD's climate change adaptation plan includes several health-related initiatives to understand and mitigate such threats, including assessment of projected climate change on health risks to DoD personnel, health surveillance demands, and distribution of disease vectors, among others.[250]

Persons with Disabilities

Disability refers to any condition or impairment of the body or mind that limits a person's ability to do certain activities or restricts a person's participation in normal life activities, such as school, work, or recreation.[266] The term "disability" covers a wide variety and range of functional limitations related to expressive and receptive communication (hearing and speech), vision, cognition, and mobility. These factors, if not anticipated and accommodated before, during, and after extreme events, can result in illness and death.[267] The extent of disability, or its severity, is reflected in the affected person's need for environmental accessibility and accommodations for their impairment(s).[268]

Disability can occur at any age and is not uniformly distributed across populations. Disability varies by gender, race, ethnicity, and geographic location.[269] Approximately 18.7% of the U.S. population has a disability.[234] In 2010, the percent of American adults with a disability was approximately 16.6% for those aged 18–64 and 49.8% for persons 65 and older.[234] In 2014, working-age adults with disabilities were substantially less likely to participate in the labor force (30.2%) than people without disabilities (76.2%), and experience more than twice the rate of unemployment (13.9% and 6.0%, respectively).[270]

People with disabilities experience disproportionately higher rates of social risk factors, such as poverty and lower educational attainment, that contribute to poorer health outcomes during extreme events or climate-related emergencies. These factors compound the risks posed by functional impairments and disrupt planning and emergency response. Of the climate-related health risks experienced by people with disabilities, perhaps the most fundamental is their "invisibility" to decision-makers and planners.[271] There has been relatively limited empirical research documenting how people with disabilities fare during or after an extreme event.[272]

An increase in extreme weather can be expected to disproportionately affect populations with disabilities unless emergency planners make provisions to address their functional needs in preparing emergency response plans. In 2005, Hurricane Katrina had a significant and disproportionate impact on people with disabilities. Of the 986 deaths in Louisiana directly attributable to the storm, 103 occurred among individuals in nursing homes, presumably with a disability.[273] Strong social capital and societal connectedness to other people, especially through faith-based organizations, family networks, and work connections, were considered to be key enabling factors that helped people with disabilities to cope before, during, and

> *People with disabilities experience disproportionately higher rates of social risk factors, such as poverty and lower educational attainment, that contribute to poorer health outcomes during extreme events or climate-related emergencies.*

after the storm.[274] In the aftermath of Hurricane Sandy, the City of New York lost a lawsuit filed by the Brooklyn Center for Independence of the Disabled (*Brooklyn Center for Independence of the Disabled et al. v. Bloomberg et al.*, Case 1.11-cv-06690-JMF 2013), with the finding that the city had not adequately prepared to accommodate the social and medical support needs of New York residents with disabilities.

Risk communication is not always designed or delivered in an accessible format or media for individuals who are deaf or have hearing loss, who are blind or have low vision, or those with diminished cognitive skills.[275, 276] Emergency communication and other important notifications (such as a warning to boil contaminated water) simply may not reach persons with disabilities. In addition, persons with disabilities often rely on medical equipment (such as portable oxygen) that requires an uninterrupted source of electricity. Portable oxygen supplies must be evacuated with the patient.[277]

Persons with Chronic Medical Conditions

Preexisting medical conditions present risk factors for increased illness and death associated with climate-related stressors, especially exposure to extreme heat. In some cases, risks are mediated by the physiology of specific medical conditions that may impair responses to heat exposure. In other cases, the risks are related to unintended side effects of med-

Persons with disabilities often rely on medical equipment (such as portable oxygen) that requires an uninterrupted source of electricity.

ical treatment that may impair body temperature, fluid, or electrolyte balance and thereby increase risks. Trends in the prevalence of chronic medical conditions are summarized in Table 1 in Chapter 1: Introduction. In general, the prevalence of common chronic medical conditions, including cardiovascular disease, respiratory disease, diabetes, asthma, and obesity, is anticipated to increase over the coming decades (see Table 1 in Ch. 1: Introduction), resulting in larger populations at risk of medical complications from climate change related exposures.

Excess heat exposure has been shown to increase the risk of disease exacerbation or death for people with various medical conditions. Hospital admissions and emergency room visits increase during heat waves for people with diabetes, cardiovascular diseases, respiratory diseases, and psychiatric illnesses.[40, 58, 195, 278, 279, 280, 281, 282] Medical conditions like Alzheimer's disease or mental illnesses can impair judgment and behavioral responses in crisis situations, which can place people with those conditions at greater risk.[228]

Medications used to treat chronic medical conditions are associated with increased risk of hospitalization, emergency room admission, and in some cases, death from extreme heat. These medicines include drugs used to treat neurologic or psychiatric conditions, such as anti-psychotic drugs, anti-cholinergic agents, anxiolytics (anti-anxiety medicines), and some antidepressants (such as selective serotonin reuptake inhibitors or SSRIs; see also Ch. 8: Mental Health).[233, 283, 284] In addition, drugs used to treat cardiovascular diseases, such as diuretics and beta-blockers, may impair resilience to heat stress.[283, 285]

People with chronic medical conditions also can be more vulnerable to interruption in treatment. For example, interrupting treatment for patients with addiction to drugs or alcohol may lead to withdrawal syndromes.[286, 287, 288] Treatment for chronic medical conditions represents a significant proportion of post-disaster medical demands.[289] Communities that are both medically underserved and have a high prevalence of chronic medical conditions can be especially at risk.[290] While most studies have assessed adults, and especially the elderly, with chronic medical conditions, children with medical conditions such as allergic and respiratory diseases are also at greater risk of symptom exacerbation and hospital admission during heat waves.[144]

9.4 Measures of Vulnerability and Mapping

Vulnerability associated with exposures to climate-related hazards is closely tied to place. While an understanding of the individual-level factors associated with vulnerability is essential to assessing population risks and considering possible protective measures, understanding how potential exposures overlap with the geographic location of populations of concern is critical for designing and implementing appropri-

ate adaptations. Analytic capabilities provided by mapping tools allow public health and emergency response workers to consider multiple types of vulnerability and how they interact with place. The development of indices that combine different elements of vulnerability and allow visualization of areas and populations experiencing the highest risks is related to improved geographic information systems (GIS) capabilities.[291]

There are multiple approaches for developing vulnerability indices to identify populations of concern across large areas, such as state or multistate regions, or small areas, such as households within a county or several counties within a state.[293] The Social Vulnerability Index (SVI) developed by the CDC aggregates U.S. census data to estimate the social vulnerability of census tracts (which are generally subsets of counties; Figure 4). The SVI provides a measure of overall social vulnerability in addition to measures of elements that comprise social vulnerability (including socioeconomic status, household composition, race or ethnicity, native language, and infrastructure conditions). Each census tract receives a separate ranking for overall vulnerability and for each of the four elements, which are available at the census-tract level for the entire United States. A similar methodology has been

Mapping Social Vulnerability

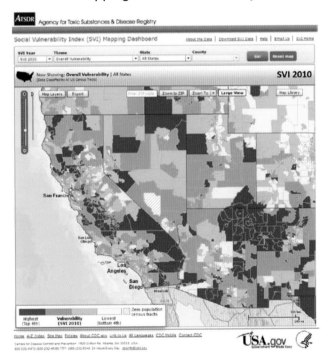

Figure 4: CDC Social Vulnerability Index (SVI): This interactive web map shows the overall social vulnerability of the U.S. Southwest in 2010. The SVI provides a measure of four social vulnerability elements: socioeconomic status; household composition; race, ethnicity, and language; and housing/transportation. Each census tract receives a separate ranking for overall vulnerability at the census-tract level. Dark blue indicates the highest overall vulnerability (the top quartile) with the lowest quartile in pale yellow. (Figure source: ATSDR 2015)[292]

Mapping Heat Vulnerability in Georgia

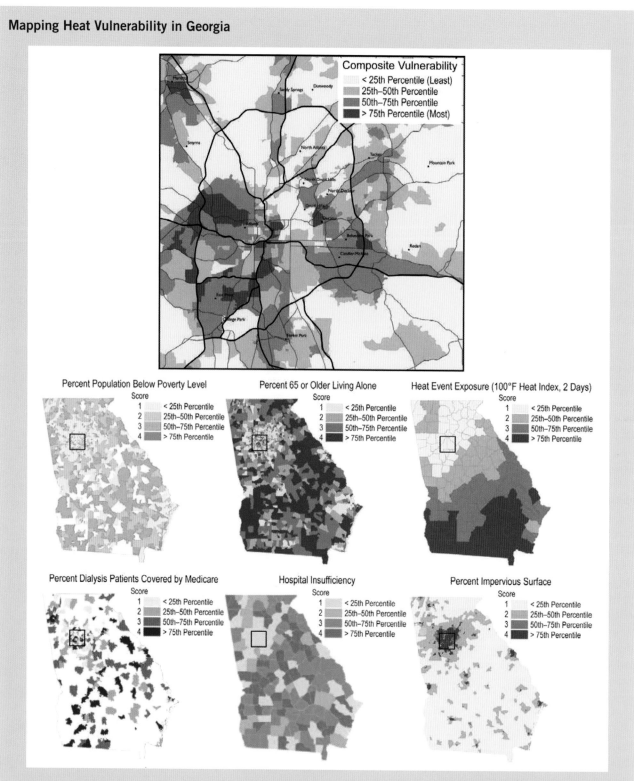

Figure 5: Vulnerability to heat-related illness in Georgia extends beyond urban zones. The map on top shows a composite measure of social vulnerability for the Atlanta, Georgia Metropolitan Area (darkest colors indicate the most vulnerable areas). The six state-wide maps below show the following six vulnerability factors: 1) percent population below the poverty level, 2) percent aged 65 and older living alone, 3) heat event exposure with Heat Index over 100°F for two consecutive days, 4) percent dialysis patients on Medicare, 5) hospital insufficiency based upon accessibility of hospital infrastructure, and 6) percent impervious surface. Areas located in rural southern Georgia experienced more hazardous heat events, had less access to health care, and had a higher percentage of people living alone. (Figure source: adapted from Manangan et al. 2014)[294]

Mapping Heat Vulnerability in Georgia, continued

The CDC conducted a case study of heat-related vulnerability in Georgia using data from 2002 to 2008. This climate and health vulnerability assessment, which identifies people and places that are most susceptible to hazardous exposures from climate change, uses GIS to overlay six maps depicting population-level sensitivity (poverty levels, elderly people living alone, preexisting health conditions, and people living in urban areas), adaptive capacity (a measure of access to healthcare), and exposure (a measure of heat events). The study found that vulnerability to heat-related illness in Georgia extends beyond urban zones. In fact, areas located in the southern portion of Georgia, which is more rural, experienced more hazardous heat events, had less access to health care, and had a higher percentage of people living alone. These types of studies allow researchers to use GIS to identify vulnerable communities, which can aid in the development of public health interventions and other adaptation strategies[294] (Figure 5).

used to develop a vulnerability index for climate-sensitive health outcomes which, in addition to socioeconomic data, incorporates data on climate-related exposures and adaptive capacity.[294]

Application of Vulnerability Indices

GIS—data management systems used to capture, store, manage, retrieve, analyze, and display geographic information—can be used to quantify and visualize factors that contribute to climate-related health risks. By linking together census data, data on the determinants of health (social, environmental, pre-existing health conditions), measures of adaptive capacity (such as health care access), and climate data, GIS mapping helps identify and position resources for at-risk populations.[4, 10, 294, 295, 296, 297] For instance, heat-related illnesses have been associated with social isolation in older adults, which can be mapped by combining data for persons living alone (determinants of health data), distribution of people aged 65 and older (census data), and frequency and severity of heat waves (climate data).

Vulnerability mapping can also enhance emergency and disaster risk management.[298, 299] Vulnerability mapping conducted at finer spatial resolution (for example, census tracts or census blocks) allows public health departments to target vulnerable communities for emergency preparedness, response, recovery, and mitigation.[300] Geographic characteristics of vulnerability can be used to determine where to position emergency medical and social response resources that are most needed before, during, and after climate change related events.[298, 299, 300]

Emergency response agencies can apply lessons learned by mapping prior events. For example, vulnerability mapping has been used to assess how social disparities affected the geography of recovery in New Orleans following Hurricane Katrina.[9] Maps displaying the intersection of social vulnerability (low, medium, high scores) and flood inundation (none, low, medium, high levels) showed that while the physical manifestation of the disaster had few race or class distinctions, the social vulnerability of communities influenced both pre-impact responses, such as evacuation, and post-event recovery.[9]

As climate change increases the probability of more frequent or more severe extreme weather events, vulnerability mapping is an important tool for preparing for and responding to health threats.

9.5 Research Needs

A number of research needs related to populations of concern have been identified. There are some limitations with current public health surveillance and monitoring of risk factors that impede the development of projections of vulnerability to climate change impacts. Obtaining detailed data on social, economic, and health factors that contribute to vulnerability is challenging, especially at the small spatial scales required for analyzing climate change impacts. Privacy concerns often limit the collection and use of personal health and socioeconomic data. Ultimately, data limitations determine the feasibility of developing alternative vulnerability indicators using existing data sources. The science requires comprehensive and standardized measures of vulnerability that combine data identification and collection with the development of appropriate vulnerability indices.

More comprehensive and robust projections of factors that contribute to population vulnerability would also enhance the value of predictive models. At present, there are only limited projections of health status of the U.S. population, and the U.S. Census no longer provides population projections at the state level. Projecting population vulnerability into the future, as well as the development of consensus storylines that characterize alternative socioeconomic scenarios, will facilitate more robust and useful assessments of future health impacts of climate change.

Future assessments can benefit from research activities that:

- improve understanding of the relative contributions and causal mechanisms of vulnerability factors (for example, genetic, physiological, social, behavioral) to risks of specific health impacts of climate change;

- investigate how available sources of data on population characteristics can be used to create valid indicators and help map vulnerability to the health impacts of climate change;

- understand how vulnerability to both medical and psychological health impacts of climate change affect cumulative stress and health status; and

- evaluate the efficacy of measures designed to enhance resilience and reduce the health impacts from climate change at the individual, institutional, and community levels.

Supporting Evidence

The chapter was developed through technical discussions of relevant evidence and expert deliberation by the report authors at several workshops, teleconferences, and email exchanges. The authors considered inputs and comments submitted by the public, the National Academies of Sciences, and Federal agencies. For additional information on the overall report process, see Appendices 2 and 3.

The author team identified a number of populations affected by climate change health impacts, including communities of color and low-income, immigrant, and limited English proficiency groups; Indigenous populations; children and pregnant women; older adults; certain occupational groups; persons with disabilities; and persons with chronic medical conditions. This list of populations was identified to reflect current understandings related to how the health of particular groups of people or particular places are affected by climate change in the United States. While not exhaustive, these populations of concern are those most commonly identified and discussed in reviews of climate change health impacts on vulnerable populations. In this chapter, the order of these populations is not prioritized. While there are other populations that may be threatened disproportionately by climate change, the authors focused the sections of this chapter on populations for which there is substantive literature. In addition to this chapter's summary of vulnerable populations, each of the health outcome chapters in the report includes discussion of populations of concern. Some populations may be covered more extensively in these other chapters; for instance, homeless populations are discussed in Chapter 8: Mental Health, as the literature on this population focuses primarily on mental health.

KEY FINDING TRACEABLE ACCOUNTS

Vulnerability Varies Over Time and Is Place-Specific

Key Finding 1: Across the United States, people and communities differ in their exposures, their inherent sensitivity, and their adaptive capacity to respond to and cope with climate change related health threats *[Very High Confidence]*. Vulnerability to climate change varies across time and location, across communities, and among individuals within communities *[Very High Confidence]*.

Description of evidence base

There is strong evidence from multiple current epidemiological studies on climate-sensitive health outcomes in the United States that health impacts will differ by location, pathways of exposure, underlying susceptibility, and adaptive capacity. The literature consistently finds that these disparities in health impacts will largely result from differences in the distribution of individual attributes in a population that confers vulnerability (such as age, socioeconomic status, and race), attributes of place that reduce or amplify exposure (such as floodplain, coastal zone, and urban heat island), and the resilience of public health infrastructure.

Across multiple studies, the following factors are consistently identified that contribute to exposure: occupation,[11] time spent in risk-prone locations,[12, 13, 14] displacement by weather extremes,[64] economic status,[15, 16] condition of infrastructure,[17, 18] and compromised mobility, cognitive function, and other mental or behavioral factors.[19]

There is consensus within the scientific literature that biologic sensitivity and adaptive capacity are tied to many of the same factors that contribute to exposures, and that all of these factors can change across time and life stage.[12, 20, 21] There is also strong evidence from multiple studies that social and economic factors affect disparities in the prevalence of chronic medical conditions that aggravate biological sensitivity.[22, 23]

Major uncertainties

Understanding how exposure, sensitivity, and adaptive capacity change over time and location for specific populations of concern is challenging, particularly when attempting to project impacts of climate change on health across long time frames (such as in the year 2100, a year for which climate projections often estimate impacts) or vast geographic areas. Uncertainties remain with respect to the underlying social determinants of health, public health interventions or outreach, adaptation options, and climate impacts at fine local scales.

Assessment of confidence and likelihood based on evidence

Based on the evidence presented in the peer-reviewed literature, there is **very high confidence** that climate change impacts on health will vary across place and time, as demonstrated by the complex factors driving vulnerability. Many qualitative and quantitative studies have been published with consistent findings and strong consensus that the impacts of climate change on human health will vary according to differential exposure, sensitivity, and adaptive capacity, which change over time and across places. These conclusions are well-documented and supported by high-quality evidence from multiple sources.

Health Impacts Vary with Age and Life Stage

Key Finding 2: People experience different inherent sensitivities to the impacts of climate change at different ages and life stages *[High Confidence]*. For example, the very young and the very old are particularly sensitive to climate-related health impacts.

Description of evidence base

There is strong, consistent evidence from multiple studies that children have inherent sensitivities to climate-related health impacts. There are multiple, high-quality studies concerning the impact of changes in ground-level ozone, particulate matter, and aeroallergens on increases in childhood asthma episodes and other adverse respiratory effects in children.[151, 152, 153, 156] In addition, the literature supports a finding that children are vulnerable to waterborne pathogens in drinking water and through exposures while swimming. There is a positive and statistically significant association between heavy rain and emergency department visits for children with gastrointestinal illness, though evidence comes from regional studies and is not at the national scale.[159, 160] The science also supports a finding that children's mental health is affected by exposures to traumatic weather events, which can undermine cognitive development and contribute to psychiatric disorders.[149, 150]

There is also strong, consistent evidence from multiple studies that older adults have inherent sensitivities to climate-related health impacts. In particular, exposure to extreme ambient temperature is an important determinant of health in older adults[24, 203] and has been associated with increased hospital admissions for cardiovascular, respiratory, and metabolic disorders.[26, 33] In addition, older adults are particularly affected by extreme weather events that compromise the availability and safety of food and water supplies; interrupt communications, utilities, and emergency services; and destroy or damage homes and the built environment.[209, 210, 211, 226] Some functional and mobility impairments make older adults less able to evacuate when necessary.[231, 301]

Major uncertainties

There is less information with which to quantify climate-related impacts on children and older adults at a national level given limited data availability. Some studies of age-related vulnerability have limited geographic scope or focus on single events in particular locations. Nevertheless, multiple factors, all with some degree of uncertainty, converge to determine climate-related vulnerability across age groups.

Assessment of confidence and likelihood based on evidence

Based on the evidence presented in the peer-reviewed literature, there is **high confidence** that a wide range of health effects exacerbated by climate change will be experienced by vulnerable age groups, especially young children and older adults. Both qualitative and quantitative studies have been published about the effects of age or life stage on vulnerability to health impacts, and that evidence is consistent and of good quality.

Social Determinants of Health Interact with Climate Factors to Affect Health Risks

Key Finding 3: Climate change threatens the health of people and communities by affecting exposure, sensitivity, and adaptive capacity *[High Confidence]*. Social determinants of health, such as those related to socioeconomic factors and health disparities, may amplify, moderate, or otherwise influence climate-related health effects, particularly when these factors occur simultaneously or close in time or space *[High Confidence]*.

Description of evidence base

The literature is consistent and the results are compelling that social determinants of health, such as those related to socioeconomic factors and health disparities, will contribute to the nature and extent of vulnerability and health effects due to climate change. The following factors illustrate the depth of the literature supporting the conclusions above regarding the relationship between climate change health threats, vulnerability (comprised of exposure, sensitivity, and adaptive capacity), and social determinants of health:

- *Occupation:* where workers are at risk due to their place of employment or the nature of their duties.[11]

- *Time spent in risk-prone locations*: There is an extensive literature base and broad consensus to support a finding that locations that experience greater risks include urban heat islands where exposed populations are likely to have limited adaptive capacity due to poor housing conditions, and inability to use or to afford air conditioning.[15, 24, 26, 34, 35, 38, 55, 56, 57].

- *Economic status*: In the literature, a significant relationship has been observed that links people living in poverty with being less likely to have adequate resources to prepare for or respond to extreme events or to access and afford necessary health or supportive services to cope with climate-related health impacts.[39, 42, 43, 44, 45, 46]

- *Condition of infrastructure*: Deteriorating infrastructure exposes people to increased health risks. The literature is consistent and of good quality to support a finding that persons who evacuate may be hampered by damage to transportation, utilities, and medical or communication facilities and by a lack of safe food or drinking water supplies.[12, 139, 226, 229]

- *Disparities in health conditions:* Health disparities contribute to the sensitivity of people to climate change. Numerous studies indicate increased sensitivity and health risk for people with chronic or preexisting medical or psychological illnesses, people of certain age or stage of life; and people with compromised mobility or cognitive functioning.[143, 289, 290] Social determinants of health contributing to disparities in rates of these conditions increase sensitivity of affected populations.[32, 206, 289, 290]

Health risks and vulnerability may increase in locations or instances where combinations of social determinants of health that amplify health threats occur simultaneously or close in time or space.[6, 7] For example, people with limited economic resources living in areas with deteriorating infrastructure are more likely to experience disproportionate impacts and are less able to recover following extreme events,[8, 9] increasing their vulnerability to climate-related health effects.

Major uncertainties

A wide range of non-climate factors are expected to interact with climate change health impacts to determine population vulnerability, all with some degree of uncertainty. The extent to which social determinants of health individually and collectively affect the different components of vulnerability is, in many cases, not well understood and not readily amenable to measurement or quantification. Assessing the extent and nature of non-climate impacts as compared to impacts related to climate change is limited by data availability. Many studies of climate change vulnerability have limited geographic scope or focus on single events in particular locations, which makes drawing national-level conclusions more challenging.

Assessment of confidence and likelihood based on evidence

Based on the evidence presented in the peer-reviewed literature, there is **high confidence** that climate change threatens the health of people and communities by affecting exposure, sensitivity, and adaptive capacity. This conclusion takes into account the consistent evidence presented in multiple studies regarding the causes of vulnerability to climate-related health effects and the role of social determinants of health. There is **high confidence** based on many peer-reviewed studies that social determinants of health, such as those related to socioeconomic factors and health disparities, may amplify, moderate, or otherwise influence climate-related health effects across populations of concern, and the evidence presented is of good quality, consistent, and compelling.

Mapping Tools and Vulnerability Indices Identify Climate Health Risks

Key Finding 4: The use of geographic data and tools allows for more sophisticated mapping of risk factors and social vulnerabilities to identify and protect specific locations and groups of people *[High Confidence]*.

Description of evidence base

Over the past decade, the literature on the use of GIS in a public health and vulnerability context has been steadily growing. Multiple studies provide strong, consistent evidence that spatial-analytic tools help facilitate analyses that link together spatially resolved representations of census data, data on the determinants of health (social, environmental, preexisting health conditions), measures of adaptive capacity (such as health care accessibility), and environmental data for the identification of at-risk populations.[4, 10, 294, 295, 296, 297] Similarly, the more recent additions to the literature indicate that demographic and environmental data can be integrated

to create an index that allows for analysis of the factors contributing to social vulnerability in a given geographic area.[292, 294] Multiple studies conclude that spatial mapping that identifies factors associated with relative vulnerability is an important step in developing prevention strategies or determining where to focus or position health or emergency response resources.[298, 299, 300] Fewer studies explicitly focus on vulnerability mapping in a climate change context, with the notable exception of the case study of heat-related vulnerability in Georgia.[294]

Major uncertainties

Multiple factors, all with some degree of uncertainty, determine geographic vulnerability to the health impacts of climate change. Although the literature indicates that mapping tools and vulnerability indices are useful in characterizing geographically based exposures, geocoded health data (particularly those data relevant to an analysis of climate change vulnerability) are not always available in some locations of interest. In addition, the extent of uncertainty increases at smaller spatial scales, which is typically the scale most relevant for targeting vulnerable communities. For instance, mental health outcome data are particularly challenging to obtain and geocode, partly because the majority of cases are underdiagnosed or underreported (see Ch. 8: Mental Health).

Assessment of confidence and likelihood based on evidence

Based on the evidence presented in the peer-reviewed literature, there is **high confidence** that geographic data used in mapping tools and vulnerability indices can help to identify where and for whom climate health risks are greatest. A number of published studies provide consistent and good quality evidence to support a finding regarding the utility of mapping tools and vulnerability indices in a public health context, but methods are still emerging to support the application of these tools in the context of climate change. Overall, evidence is strong that mapping tools and vulnerability indices can help to identify at-risk locations and populations for whom climate health risks are greatest. As the state of the science continues to evolve, substantial improvements in mapping and spatial analytic tools and methodologies are expected that will allow researchers to predict, for a certain geographic area, the probability that human health impacts will occur across time.

DOCUMENTING UNCERTAINTY

This assessment relies on two metrics to communicate the degree of certainty in Key Findings. See Appendix 4: Documenting Uncertainty for more on assessments of likelihood and confidence.

Confidence Level	
Very High	
Strong evidence (established theory, multiple sources, consistent results, well documented and accepted methods, etc.), high consensus	
High	
Moderate evidence (several sources, some consistency, methods vary and/or documentation limited, etc.), medium consensus	
Medium	
Suggestive evidence (a few sources, limited consistency, models incomplete, methods emerging, etc.), competing schools of thought	
Low	
Inconclusive evidence (limited sources, extrapolations, inconsistent findings, poor documentation and/or methods not tested, etc.), disagreement or lack of opinions among experts	

Likelihood	
Very Likely	
≥ 9 in 10	
Likely	
≥ 2 in 3	
As Likely As Not	
≈ 1 in 2	
Unlikely	
≤ 1 in 3	
Very Unlikely	
≤ 1 in 10	

PHOTO CREDITS

References

1. Luber, G., K. Knowlton, J. Balbus, H. Frumkin, M. Hayden, J. Hess, M. McGeehin, N. Sheats, L. Backer, C.B. Beard, K.L. Ebi, E. Maibach, R.S. Ostfeld, C. Wiedinmyer, E. Zielinski-Gutiérrez, and L. Ziska, 2014: Ch. 9: Human health. *Climate Change Impacts in the United States: The Third National Climate Assessment.* Melillo, J.M., T.C. Richmond, and G.W. Yohe, Eds. U.S. Global Change Research Program, Washington, D.C., 220-256. http://dx.doi.org/10.7930/J0PN93H5

2. IPCC, 2014: Climate Change 2014: Impacts, Adaptation, and Vulnerability. Part A: Global and Sectoral Aspects. Contribution of Working Group II to the Fifth Assessment Report of the Intergovernmental Panel on Climate Change. Field, C.B., V.R. Barros, D.J. Dokken, K.J. Mach, M.D. Mastrandrea, T.E. Bilir, M. Chatterjee, K.L. Ebi, Y.O. Estrada, R.C. Genova, B. Girma, E.S. Kissel, A.N. Levy, S. MacCracken, P.R. Mastrandrea, and L.L. White (Eds.), 1132 pp. Cambridge University Press, Cambridge, UK and New York, NY. http://www.ipcc.ch/report/ar5/wg2/

3. NRC, 2012: *Disaster Resilience: A National Imperative.* National Academies Press, Washington, D.C., 244 pp.

4. Turner, B.L., R.E. Kasperson, P.A. Matson, J.J. McCarthy, R.W. Corell, L. Christensen, N. Eckley, J.X. Kasperson, A. Luers, M.L. Martello, C. Polsky, A. Pulsipher, and A. Schiller, 2003: A framework for vulnerability analysis in sustainability science. *Proceedings of the National Academy of Sciences,* **100,** 8074-8079. http://dx.doi.org/10.1073/pnas.1231335100

5. Braveman, P., S. Egerter, and D.R. Williams, 2011: The social determinants of health: coming of age. *Annual Review of Public Health,* **32,** 381-398. http://dx.doi.org/10.1146/annurev-publhealth-031210-101218

6. Harlan, S.L., J.H. Declet-Barreto, W.L. Stefanov, and D.B. Petitti, 2013: Neighborhood effects on heat deaths: Social and environmental predictors of vulnerability in Maricopa County, Arizona. *Environmental Health Perspectives,* **121,** 197-204. http://dx.doi.org/10.1289/ehp.1104625

7. Reid, C.E., J.K. Mann, R. Alfasso, P.B. English, G.C. King, R.A. Lincoln, H.G. Margolis, D.J. Rubado, J.E. Sabato, N.L. West, B. Woods, K.M. Navarro, and J.R. Balmes, 2012: Evaluation of a heat vulnerability index on abnormally hot days: An environmental public health tracking study. *Environmental Health Perspectives,* **120,** 715-720. http://dx.doi.org/10.1289/ehp.1103766

8. Fothergill, A. and L.A. Peek, 2004: Poverty and disasters in the United States: A review of recent sociological findings. *Natural Hazards,* **32,** 89-110. http://dx.doi.org/10.1023/B:NHAZ.0000026792.76181.d9

9. Finch, C., C.T. Emrich, and S.L. Cutter, 2010: Disaster disparities and differential recovery in New Orleans. *Population and Environment,* **31,** 179-202. http://dx.doi.org/10.1007/s11111-009-0099-8

10. CSDH, 2008: Closing the Gap in a Generation: Health Equity through Action on the Social Determinants of Health. Final Report of the Commission on Social Determinants of Health. 247 pp. World Health Organization, Geneva. http://www.who.int/social_determinants/final_report/csdh_final-report_2008.pdf

11. Schulte, P.A. and H. Chun, 2009: Climate change and occupational safety and health: Establishing a preliminary framework. *Journal of Occupational and Environmental Hygiene,* **6,** 542-554. http://dx.doi.org/10.1080/15459620903066008

12. Balbus, J.M. and C. Malina, 2009: Identifying vulnerable subpopulations for climate change health effects in the United States. *Journal of Occupational and Environmental Medicine,* **51,** 33-37. http://dx.doi.org/10.1097/JOM.0b013e318193e12e

13. Uejio, C.K., O.V. Wilhelmi, J.S. Golden, D.M. Mills, S.P. Gulino, and J.P. Samenow, 2011: Intra-urban societal vulnerability to extreme heat: The role of heat exposure and the built environment, socioeconomics, and neighborhood stability. *Health & Place,* **17,** 498-507. http://dx.doi.org/10.1016/j.healthplace.2010.12.005

14. O'Neill, M.S., P.L. Kinney, and A.J. Cohen, 2008: Environmental equity in air quality management: Local and international implications for human health and climate change. *Journal of Toxicology and Environmental Health, Part A: Current Issues,* **71,** 570-577. http://dx.doi.org/10.1080/15287390801997625

15. Harlan, S.L., A.J. Brazel, L. Prashad, W.L. Stefanov, and L. Larsen, 2006: Neighborhood microclimates and vulnerability to heat stress. *Social Science & Medicine,* **63,** 2847-2863. http://dx.doi.org/10.1016/j.socscimed.2006.07.030

16. Woodruff, T.J., J.D. Parker, A.D. Kyle, and K.C. Schoendorf, 2003: Disparities in exposure to air pollution during pregnancy. *Environmental Health Perspectives,* **111,** 942-946. http://dx.doi.org/10.1289/ehp.5317

17. Pastor, M., R.D. Bullard, J.K. Boyce, A. Fothergill, R. Morello-Frosch, and B. Wright, 2006: *In the Wake of the Storm: Environment, Disaster, and Race After Katrina.* Russell Sage Foundation, New York. http://www.dscej.org/images/pdfs/In The Wake of the Storm.pdf

18. Bullard, R. and B. Wright, 2009: Introduction. *Race, Place, and Environmental Justice After Hurricane Katrina, Struggles to Reclaim Rebuild, and Revitalize New Orleans and the Gulf Coast.* Bullard, R. and B. Wright, Eds. Westview Press, Boulder, CO, 1-15.

19. Gamble, J.L., B.J. Hurley, P.A. Schultz, W.S. Jaglom, N. Krishnan, and M. Harris, 2013: Climate change and older Americans: State of the science. *Environmental Health Perspectives,* **121,** 15-22. http://dx.doi.org/10.1289/ehp.1205223

20. Shannon, M.W., D. Best, H.J. Binns, J.A. Forman, C.L. Johnson, C.J. Karr, J.J. Kim, L.J. Mazur, J.R. Roberts, and K.M. Shea, 2007: Global climate change and children's health. *Pediatrics,* **120,** 1149-1152. http://dx.doi.org/10.1542/peds.2007-2645

21. Sheffield, P.E. and P.J. Landrigan, 2011: Global climate change and children's health: Threats and strategies for prevention. *Environmental Health Perspectives,* **119,** 291-298. http://dx.doi.org/10.1289/ehp.1002233

22. Frumkin, H., J. Hess, G. Luber, J. Malilay, and M. McGeehin, 2008: Climate change: The public health response. *American Journal of Public Health,* **98,** 435-445. http://dx.doi.org/10.2105/AJPH.2007.119362

23. Keppel, K.G., 2007: Ten largest racial and ethnic health disparities in the United States based on Healthy People 2010 objectives. *American Journal of Epidemiology,* **166,** 97-103. http://dx.doi.org/10.1093/aje/kwm044

24. Luber, G. and M. McGeehin, 2008: Climate change and extreme heat events. *American Journal of Preventive Medicine,* **35,** 429-435. http://dx.doi.org/10.1016/j.amepre.2008.08.021

25. Frey, W., 2011: The New Metro Minority Map: Regional Shifts in Hispanics, Asians, and Blacks from Census 2010. Brookings Institution, Washington, D.C. http://www.brookings.edu/papers/2011/0831_census_race_frey.aspx

26. CDC, 2011: Health Disparities and Inequalities Report-United States, 2011. *MMWR. Morbidity and Mortality Weekly Report,* **60(Suppl),** 1-116. http://www.cdc.gov/mmwr/pdf/other/su6001.pdf

27. Miranda, M.L., D.A. Hastings, J.E. Aldy, and W.H. Schlesinger, 2011: The environmental justice dimensions of climate change. *Environmental Justice,* **4,** 17-25. http://dx.doi.org/10.1089/env.2009.0046

28. CDC, 2012: Table 2-1: Lifetime Asthma Prevalence Percents by Age, United States: National Health Interview Survey, 2012. Centers for Disease Control and Prevention, Atlanta, GA. http://www.cdc.gov/asthma/nhis/2012/table2-1.htm

29. CDC, 2013: Health Disparities and Inequalities Report--United States 2013. *MMWR. Morbidity and Mortality Weekly Report,* **62(Supp.3),** 1-187. http://www.cdc.gov/mmwr/pdf/other/su6203.pdf

30. Blackwell, D.L., J.W. Lucas, and T.C. Clarke, 2014: Summary Health Statistics for U.S. Adults: National Health Interview Survey, 2012. Vital and Health Statistics 10(260), 161 pp. National Center for Health Statistics, Hyattsville, MD. http://www.cdc.gov/nchs/data/series/sr_10/sr10_260.pdf

31. McGeehin, M.A. and M. Mirabelli, 2001: The potential impacts of climate variability and change on temperature-related morbidity and mortality in the United States. *Environmental Health Perspectives,* **109,** 185-189. http://dx.doi.org/10.2307/3435008

32. Basu, R. and B.D. Ostro, 2008: A multicounty analysis identifying the populations vulnerable to mortality associated with high ambient temperature in California. *American Journal of Epidemiology,* **168,** 632-637. http://dx.doi.org/10.1093/aje/kwn170

33. Lin, S., M. Luo, R.J. Walker, X. Liu, S.-A. Hwang, and R. Chinery, 2009: Extreme high temperatures and hospital admissions for respiratory and cardiovascular diseases. *Epidemiology,* **20,** 738-746. http://dx.doi.org/10.1097/EDE.0b013e3181ad5522

34. Semenza, J.C., J.E. McCullough, W.D. Flanders, M.A. McGeehin, and J.R. Lumpkin, 1999: Excess hospital admissions during the July 1995 heat wave in Chicago. *American Journal of Preventive Medicine,* **16,** 269-277. http://dx.doi.org/10.1016/s0749-3797(99)00025-2

35. Younger, M., H.R. Morrow-Almeida, S.M. Vindigni, and A.L. Dannenberg, 2008: The built environment, climate change, and health: Opportunities for co-benefits. *American Journal of Preventive Medicine,* **35,** 517-526. http://dx.doi.org/10.1016/j.amepre.2008.08.017

36. Larson, N.I., M.T. Story, and M.C. Nelson, 2009: Neighborhood environments: Disparities in access to healthy foods. *American Journal of Preventive Medicine,* **36,** 74-81. e10. http://dx.doi.org/10.1016/j.amepre.2008.09.025

37. Satia, J.A., 2009: Diet-related disparities: Understanding the problem and accelerating solutions. *Journal of the American Dietetic Association,* **109,** 610-615. http://dx.doi.org/10.1016/j.jada.2008.12.019

38. DHHS, 2014: National Healthcare Disparities Report 2013. AHRQ Publication No. 14-0006. U.S. Department of Health and Human Services, Agency for Healthcare Research and Quality, Rockville, MD. http://www.ahrq.gov/research/findings/nhqrdr/nhdr13/index.html

39. Maldonado, C.Z., R.M. Rodriguez, J.R. Torres, Y.S. Flores, and L.M. Lovato, 2013: Fear of discovery among Latino immigrants presenting to the emergency department. *Academic Emergency Medicine,* **20,** 155-161. http://dx.doi.org/10.1111/acem.12079

40. Knowlton, K., M. Rotkin-Ellman, G. King, H.G. Margolis, D. Smith, G. Solomon, R. Trent, and P. English, 2009: The 2006 California heat wave: Impacts on hospitalizations and emergency department visits. *Environmental Health Perspectives,* **117,** 61-67. http://dx.doi.org/10.1289/ehp.11594

41. McMichael, A.J., 2013: Globalization, climate change, and human health. *The New England Journal of Medicine,* **368,** 1335-1343. http://dx.doi.org/10.1056/NEJMra1109341

42. Ortega, A.N., H. Fang, V.H. Perez, J.A. Rizzo, O. Carter-Pokras, S.P. Wallace, and L. Gelberg, 2007: Health care access, use of services, and experiences among undocumented Mexicans and other Latinos. *Archives of Internal Medicine,* **167,** 2354-2360. http://dx.doi.org/10.1001/archinte.167.21.2354

43. Fuentes-Afflick, E. and N.A. Hessol, 2009: Immigration status and use of health services among Latina women in the San Francisco Bay Area. *Journal of Women's Health,* **18,** 1275-1280. http://dx.doi.org/10.1089/jwh.2008.1241

44. Vargas Bustamante, A., H. Fang, J. Garza, O. Carter-Pokras, S.P. Wallace, J.A. Rizzo, and A.N. Ortega, 2010: Variations in healthcare access and utilization among Mexican immigrants: The role of documentation status. *Journal of Immigrant and Minority Health,* **14,** 146-155. http://dx.doi.org/10.1007/s10903-010-9406-9

45. Eneriz-Wiemer, M., L.M. Sanders, D.A. Barr, and F.S. Mendoza, 2014: Parental limited English proficiency and health outcomes for children with special health care needs: A systematic review. *Academic Pediatrics,* **14,** 128-136. http://dx.doi.org/10.1016/j.acap.2013.10.003

46. Riera, A., A. Navas-Nazario, V. Shabanova, and F.E. Vaca, 2014: The impact of limited English proficiency on asthma action plan use. *Journal of Asthma,* **51,** 178-184. http://dx.doi.org/10.3109/02770903.2013.858266

47. Humes, K.R., N.A. Jones, and R.R. Ramirez, 2011: Overview of Race and Hispanic Origin: 2010. 2010 Census Briefs C2010BR-02, 23 pp. U.S. Census Bureau. http://www.census.gov/prod/cen2010/briefs/c2010br-02.pdf

48. U.S. Census Bureau, 2014: Annual Estimates of the Resident Population by Sex, Race, and Hispanic Origin for the United States, States, and Counties: April 1, 2010 to July 1, 2013. U.S. Census Bureau American Fact Finder. http://factfinder.census.gov/faces/tableservices/jsf/pages/productview.xhtml?src=bkmk

49. U.S. Census Bureau, 2008: An Older and More Diverse Nation by Midcentury. August 14. https://www.census.gov/newsroom/releases/archives/population/cb08-123.html

50. Whatley, M. and J. Batalova, 2013: Limited English Proficient Population of the United States. Migration Policy Institute, Washington, D.C. http://www.migrationpolicy.org/article/limited-english-proficient-population-united-states/

51. Passel, J.S. and D. Cohn, 2011: Unauthorized Immigrant Population: National and State Trends, 2010. 31 pp. Pew Research Center, Washington, D.C. http://www.pewhispanic.org/files/reports/133.pdf

52. Braveman, P.A., C. Cubbin, S. Egerter, S. Chideya, K.S. Marchi, M. Metzler, and S. Posner, 2005: Socioeconomic status in health research: One size does not fit all. *JAMA - Journal of the American Medical Association,* **294,** 2829-2888. http://dx.doi.org/10.1001/jama.294.22.2879

53. Kington, R.S. and J.P. Smith, 1997: Socioeconomic status and racial and ethnic differences in functional status associated with chronic diseases. *American Journal of Public Health,* **87,** 805-810. http://dx.doi.org/10.2105/AJPH.87.5.805

54. Wong, M.D., M.F. Shapiro, W.J. Boscardin, and S.L. Ettner, 2002: Contribution of major diseases to disparities in mortality. *New England Journal of Medicine,* **347,** 1585-1592. http://dx.doi.org/10.1056/NEJMsa012979

55. Ramin, B. and T. Svoboda, 2009: Health of the homeless and climate change. *Journal of Urban Health,* **86,** 654-664. http://dx.doi.org/10.1007/s11524-009-9354-7

56. Shonkoff, S.B., R. Morello-Frosch, M. Pastor, and J. Sadd, 2011: The climate gap: Environmental health and equity implications of climate change and mitigation policies in California—a review of the literature. *Climatic Change,* **109,** 485-503. http://dx.doi.org/10.1007/s10584-011-0310-7

57. Jesdale, B.M., R. Morello-Frosch, and L. Cushing, 2013: The racial/ethnic distribution of heat risk–related land cover in relation to residential segregation. *Environmental Health Perspectives,* **121,** 811-817. http://dx.doi.org/10.1289/ehp.1205919

58. Basu, R., 2009: High ambient temperature and mortality: A review of epidemiologic studies from 2001 to 2008. *Environmental Health,* **8,** 40. http://dx.doi.org/10.1186/1476-069x-8-40

59. Frumkin, H., 2002: Urban sprawl and public health. *Public Health Reports,* **117,** 201-217. http://www.ncbi.nlm.nih.gov/pmc/articles/PMC1497432/pdf/12432132.pdf

60. O'Neill, M.S., A. Zanobetti, and J. Schwartz, 2005: Disparities by race in heat-related mortality in four US cities: The role of air conditioning prevalence. *Journal of Urban Health,* **82,** 191-197. http://dx.doi.org/10.1093/jurban/jti043

61. Lane, K., K. Charles-Guzman, K. Wheeler, Z. Abid, N. Graber, and T. Matte, 2013: Health effects of coastal storms and flooding in urban areas: A review and vulnerability assessment. *Journal of Environmental and Public Health,* **2013,** Article ID 913064. http://dx.doi.org/10.1155/2013/913064

62. Joseph, N.T., K.A. Matthews, and H.F. Myers, 2014: Conceptualizing health consequences of Hurricane Katrina from the perspective of socioeconomic status decline. *Health Psychology,* **33,** 139-146. http://dx.doi.org/10.1037/a0031661

63. Arrieta, M.I., R.D. Foreman, E.D. Crook, and M.L. Icenogle, 2009: Providing continuity of care for chronic diseases in the aftermath of Katrina: From field experience to policy recommendations. *Disaster Medicine and Public Health Preparedness,* **3,** 174-182. http://dx.doi.org/10.1097/DMP.0b013e3181b66ae4

64. Donner, W. and H. Rodríguez, 2008: Population composition, migration and inequality: The influence of demographic changes on disaster risk and vulnerability. *Social Forces,* **87,** 1089-1114. http://dx.doi.org/10.1353/sof.0.0141

65. Eisenman, D.P., K.M. Cordasco, S. Asch, J.F. Golden, and D. Glik, 2007: Disaster planning and risk communication with vulnerable communities: Lessons from Hurricane Katrina. *American Journal of Public Health,* **97,** S109-S115. http://dx.doi.org/10.2105/ajph.2005.084335

66. Zoraster, R.M., 2010: Vulnerable populations: Hurricane Katrina as a case study. *Prehospital and Disaster Medicine,* **25,** 74-78. http://dx.doi.org/10.1017/s1049023x00007718

67. Collins, T.W., A.M. Jimenez, and S.E. Grineski, 2013: Hispanic health disparities after a flood disaster: Results of a population-based survey of individuals experiencing home site damage in El Paso (Texas, USA). *Journal of Immigrant and Minority Health,* **15,** 415-426. http://dx.doi.org/10.1007/s10903-012-9626-2

68. EPA, 2010: Our Nation's Air: Status and Trends through 2008. EPA-454/R-09-002, 54 pp. U.S. Environmental Protection Agency, Office of Air Quality Planning and Standards, Research Triangle Park, NC. http://www.epa.gov/airtrends/2010/

69. NRC, 2010: *Adapting to the Impacts of Climate Change. America's Climate Choices: Report of the Panel on Adapting to the Impacts of Climate Change.* National Research Council. The National Academies Press, Washington, D.C., 292 pp. http://dx.doi.org/10.17226/12783

70. Pope, C.A., III and D.W. Dockery, 1999: Epidemiology of particle effects. *Air Pollution and Health.* Holgate, S.T., J.M. Samet, S.K. Hillel, and R.L. Maynard, Eds. Academic Press, London, 673-706.

71. Donaldson, K., M.I. Gilmour, and W. MacNee, 2000: Asthma and PM10. *Respiratory Research,* **1,** 12-15. http://dx.doi.org/10.1186/rr5

72. Leikauf, G.D., 2002: Hazardous air pollutants and asthma. *Environmental Health Perspectives,* **110 Suppl 4,** 505-526. PMC1241200

73. Peden, D.B., 2002: Pollutants and asthma: Role of air toxics. *Environmental Health Perspectives,* **110 Suppl 4,** 565-568. PMC1241207

74. Pope, C.A., III, R.T. Burnett, M.J. Thun, E.E. Calle, D. Krewski, K. Ito, and G.D. Thurston, 2002: Lung cancer, cardiopulmonary mortality, and long-term exposure to fine particulate air pollution. *JAMA - Journal of the American Medical Association,* **287,** 1132-1141. http://dx.doi.org/10.1001/jama.287.9.1132

75. Gwynn, R.C. and G.D. Thurston, 2001: The burden of air pollution: Impacts among racial minorities. *Environmental Health Perspectives,* **109,** 501-506. PMC1240572

76. Finkelstein, M.M., M. Jerrett, P. DeLuca, N. Finkelstein, D.K. Verma, K. Chapman, and M.R. Sears, 2003: Relation between income, air pollution and mortality: A cohort study. *Canadian Medical Association Journal,* **169,** 397-402. PMC183288

77. Reynolds, P., J. Von Behren, R.B. Gunier, D.E. Goldberg, A. Hertz, and D.F. Smith, 2003: Childhood cancer incidence rates and hazardous air pollutants in California: An exploratory analysis. *Environmental Health Perspectives,* **111,** 663-668. http://dx.doi.org/10.1289/ehp.5986

78. Lin, S., X. Liu, L.H. Le, and S.-A. Hwang, 2008: Chronic exposure to ambient ozone and asthma hospital admissions among children. *Environmental Health Perspectives,* **116,** 1725-1730. http://dx.doi.org/10.1289/ehp.11184

79. Alderman, K., L.R. Turner, and S. Tong, 2012: Floods and human health: A systematic review. *Environment International,* **47,** 37-47. http://dx.doi.org/10.1016/j.envint.2012.06.003

80. HUD, 2007: Measuring Overcrowding in Housing. 38 pp. U.S. Department of Housing and Urban Development, Office of Policy Development and Research. http://www.huduser.org/portal//Publications/pdf/Measuring_Overcrowding_in_Hsg.pdf

81. Abara, W., S.M. Wilson, and K. Burwell, 2012: Environmental justice and infectious disease: Gaps, issues, and research needs. *Environmental Justice,* **5,** 8-20. http://dx.doi.org/10.1089/env.2010.0043

82. Hennessy, T.W., T. Ritter, R.C. Holman, D.L. Bruden, K.L. Yorita, L. Bulkow, J.E. Cheek, R.J. Singleton, and J. Smith, 2008: The relationship between in-home water service and the risk of respiratory tract, skin, and gastrointestinal tract infections among rural Alaska natives. *American Journal of Public Health,* **98,** 2072-2078. http://dx.doi.org/10.2105/ajph.2007.115618

83. DeGroote, J.P. and R. Sugumaran, 2012: National and regional associations between human West Nile virus incidence and demographic, landscape, and land use conditions in the coterminous United States. *Vector-Borne and Zoonotic Diseases,* **12,** 657-665. http://dx.doi.org/10.1089/vbz.2011.0786

84. Bowden, S.E., K. Magori, and J.M. Drake, 2011: Regional differences in the association between land cover and West Nile virus disease incidence in humans in the United States. *American Journal of Tropical Medicine and Hygiene,* **84,** 234-238. http://dx.doi.org/10.4269/ajtmh.2011.10-0134

85. Powell, L.M., S. Slater, D. Mirtcheva, Y. Bao, and F.J. Chaloupka, 2007: Food store availability and neighborhood characteristics in the United States. *Preventive Medicine,* **44,** 189-195. http://dx.doi.org/10.1016/j.ypmed.2006.08.008

86. Coleman-Jensen, A., M. Nord, M. Andrews, and S. Carlson, 2012: Household Food Security in the United States in 2011. Economic Research Report No. ERR-141, 29 pp. U.S. Department of Agriculture, Economic Research Service. http://www.ers.usda.gov/media/884525/err141.pdf

87. DHHS, 2001: Mental Health: Culture, Race, and Ethnicity—A Supplement to Mental Health: A Report of the Surgeon General. U.S. Department of Health and Human Services, Substance Abuse and Mental Health Services Administration, Center for Mental Health Services, Rockville, MD. http://www.ncbi.nlm.nih.gov/books/NBK44243/

88. Weisler, R.H., J.G. Barbee, and M.H. Townsend, 2006: Mental health and recovery in the Gulf Coast after Hurricanes Katrina and Rita. *JAMA,* **296,** 585-588. http://dx.doi.org/10.1001/jama.296.5.585

89. Espey, D.K., M.A. Jim, N. Cobb, M. Bartholomew, T. Becker, D. Haverkamp, and M. Plescia, 2014: Leading causes of death and all-cause mortality in American Indians and Alaska Natives. *American Journal of Public Health,* **104,** S303-S311. http://dx.doi.org/10.2105/ajph.2013.301798

90. Wong, C.A., F.C. Gachupin, R.C. Holman, M.F. MacDorman, J.E. Cheek, S. Holve, and R.J. Singleton, 2014: American Indian and Alaska Native infant and pediatric mortality, United States, 1999–2009. *American Journal of Public Health,* **104,** S320-S328. http://dx.doi.org/10.2105/ajph.2013.301598

91. Hoover, E., K. Cook, R. Plain, K. Sanchez, V. Waghiyi, P. Miller, R. Dufault, C. Sislin, and D.O. Carpenter, 2012: Indigenous peoples of North America: Environmental exposures and reproductive justice. *Environmental Health Perspectives,* **120,** 1645-1649. http://dx.doi.org/10.1289/ehp.1205422

92. Evans-Campbell, T., 2008: Historical trauma in American Indian/Native Alaska communities: A multilevel framework for exploring impacts on individuals, families, and communities. *Journal of Interpersonal Violence,* **23,** 316-338. http://dx.doi.org/10.1177/0886260507312290

93. Sarche, M. and P. Spicer, 2008: Poverty and health disparities for American Indian and Alaska Native children. *Annals of the New York Academy of Sciences,* **1136,** 126-136. http://dx.doi.org/10.1196/annals.1425.017

94. Milburn, M.P., 2004: Indigenous nutrition: Using traditional food knowledge to solve contemporary health problems. *The American Indian Quarterly,* **28,** 411-434. http://dx.doi.org/10.1353/aiq.2004.0104

95. Maldonado, J.K., R.E. Pandya, and B.J. Colombi (Eds.), 2013: Introduction: climate change and indigenous peoples of the USA. *Climatic Change,* **120,** 509-682. http://dx.doi.org/10.1007/978-3-319-05266-3

96. Nakashima, D.J., K. Galloway McLean, H.D. Thulstrup, A. Ramos Castillo, and J.T. Rubis, 2012: Weathering Uncertainty: Traditional Knowledge for Climate Change Assessment and Adaptation. 120 pp. UNESCO, Paris and UNU, Darwin. http://unesdoc.unesco.org/images/0021/002166/216613E.pdf

97. Ford, J.D., A.C. Willox, S. Chatwood, C. Furgal, S. Harper, I. Mauro, and T. Pearce, 2014: Adapting to the effects of climate change on Inuit health. *American Journal of Public Health,* **104 Suppl 3,** e9-e17. http://dx.doi.org/10.2105/ajph.2013.301724

98. Hinzman, L.D., N.D. Bettez, W.R. Bolton, F.S. Chapin, M.B. Dyurgerov, C.L. Fastie, B. Griffith, R.D. Hollister, A. Hope, H.P. Huntington, A.M. Jensen, G.J. Jia, T. Jorgenson, D.L. Kane, D.R. Klein, G. Kofinas, A.H. Lynch, A.H. Lloyd, A.D. McGuire, F.E. Nelson, W.C. Oechel, T.E. Osterkamp, C.H. Racine, V.E. Romanovsky, R.S. Stone, D.A. Stow, M. Sturm, C.E. Tweedie, G.L. Vourlitis, M.D. Walker, D.A. Walker, P.J. Webber, J.M. Welker, K.S. Winker, and K. Yoshikawa, 2005: Evidence and implications of recent climate change in northern Alaska and other Arctic regions. *Climatic Change,* **72,** 251-298. http://dx.doi.org/10.1007/s10584-005-5352-2

99. Laidler, G.J., J.D. Ford, W.A. Gough, T. Ikummaq, A.S. Gagnon, S. Kowal, K. Qrunnut, and C. Irngaut, 2009: Travelling and hunting in a changing Arctic: Assessing Inuit vulnerability to sea ice change in Igloolik, Nunavut. *Climatic Change,* **94,** 363-397. http://dx.doi.org/10.1007/s10584-008-9512-z

100. Wang, M. and J.E. Overland, 2012: A sea ice free summer Arctic within 30 years: An update from CMIP5 models. *Geophysical Research Letters,* **39,** L18501. http://dx.doi.org/10.1029/2012GL052868

101. Crozier, M.J., 2010: Deciphering the effect of climate change on landslide activity: A review. *Geomorphology,* **124,** 260-267. http://dx.doi.org/10.1016/j.geomorph.2010.04.009

102. Babaluk, J.A., J.D. Reist, J.D. Johnson, and L. Johnson, 2000: First records of sockeye (*Oncorhynchus nerka*) and pink salmon (*O. gorbuscha*) from Banks Island and other records of Pacific salmon in Northwest Territories, Canada. *Arctic,* **53,** 161-164. http://dx.doi.org/10.14430/arctic846

103. Ford, J.D. and B. Smit, 2004: A framework for assessing the vulnerability of communities in the Canadian arctic to risks associated with climate change. *Arctic,* **57,** 389-400. http://dx.doi.org/10.14430/arctic516

104. Weller, G., P. Anderson, and B. Wang, Eds., 1999: *Preparing for a Changing Climate: The Potential Consequences of Climate Variability and Change in Alaska. A Report of the Alaska Regional Assessment Group for the U.S. Global Change Research Program.* Center for Global Change and Arctic System Research, University of Alaska Fairbanks, Fairbanks, AK, 42 pp. http://www.besis.uaf.edu/regional-report/Preface-Ex-Sum.pdf

105. Ruckelshaus, M., S.C. Doney, H.M. Galindo, J.P. Barry, F. Chan, J.E. Duffy, C.A. English, S.D. Gaines, J.M. Grebmeier, A.B. Hollowed, N. Knowlton, J. Polovina, N.N. Rabalais, W.J. Sydeman, and L.D. Talley, 2013: Securing ocean benefits for society in the face of climate change. *Marine Policy,* **40,** 154-159. http://dx.doi.org/10.1016/j.marpol.2013.01.009

106. Kellogg, J., J. Wang, C. Flint, D. Ribnicky, P. Kuhn, E.G.l. De Mejia, I. Raskin, and M.A. Lila, 2010: Alaskan wild berry resources and human health under the cloud of climate change. *Journal of Agricultural and Food Chemistry,* **58,** 3884-3900. http://dx.doi.org/10.1021/jf902693r

107. Dybas, C.L., 2009: Minnesota's moose: Ghosts of the northern forest? *Bioscience,* **59,** 824-828. http://dx.doi.org/10.1525/bio.2009.59.10.3

108. Lenarz, M.S., M.E. Nelson, M.W. Schrage, and A.J. Edwards, 2009: Temperature mediated moose survival in northeastern Minnesota. *Journal of Wildlife Management,* **73,** 503-510. http://dx.doi.org/10.2193/2008-265

109. Redsteer, M.H., K.B. Kelley, H. Francis, and D. Block, 2011: Disaster Risk Assessment Case Study: Recent Drought on the Navajo Nation, Southwestern United States. Contributing Paper for the Global Assessment Report on Disaster Risk Reduction. 19 pp. United Nations Office for Disaster Risk Reduction and U.S. Geological Survey, Reston, VA. http://www.preventionweb.net/english/hyogo/gar/2011/en/bgdocs/Redsteer_Kelley_Francis_&_Block_2010.pdf

110. Cheruvelil, J.J. and B. Barton, 2013: Adapting to the Effects of Climate Change on Wild Rice. 12 pp. Great Lakes Lifeways Institute.

111. Donatuto, J.L., T.A. Satterfield, and R. Gregory, 2011: Poisoning the body to nourish the soul: Prioritising health risks and impacts in a Native American community. *Health, Risk & Society,* **13,** 103-127. http://dx.doi.org/10.1080/13698575.2011.556186

112. Donatuto, J., E.E. Grossman, J. Konovsky, S. Grossman, and L.W. Campbell, 2014: Indigenous community health and climate change: Integrating biophysical and social science indicators. *Coastal Management,* **42,** 355-373. http://dx.doi.org/10.1080/08920753.2014.923140

113. Egeland, G. and G.G. Harrison, 2013: Health disparities: Promoting Indigenous peoples' health through traditional food systems and self-determination. *Indigenous Peoples' Food Systems & Well-Being: Interventions & Policies for Healthy Communities.* Kuhnlein, H., B. Erasmus, D. Spigelski, and B. Burlingame, Eds. Food and Agriculture Organization of the United Nations, Rome, 9-22. http://www.fao.org/docrep/018/i3144e/I3144e02.pdf

114. Lynn, K., J. Daigle, J. Hoffman, F. Lake, N. Michelle, D. Ranco, C. Viles, G. Voggesser, and P. Williams, 2013: The impacts of climate change on tribal traditional foods. *Climatic Change,* **120,** 545-556. http://dx.doi.org/10.1007/s10584-013-0736-1

115. Garibaldi, A. and N. Turner, 2004: Cultural keystone species: Implications for ecological conservation and restoration. *Ecology and Society,* **9,** 1. https://dlc.dlib.indiana.edu/dlc/handle/10535/3108

116. Turner, N.J. and H. Clifton, 2009: "It's so different today": Climate change and Indigenous lifeways in British Columbia, Canada. *Global Environmental Change,* **19,** 180-190. http://dx.doi.org/10.1016/j.gloenvcha.2009.01.005

117. Turner, N.J., R. Gregory, C. Brooks, L. Failing, and T. Satterfield, 2008: From invisibility to transparency: Identifying the implications. *Ecology and Society,* **13,** 7. http://hdl.handle.net/10535/2984

118. Kuhnlein, H.V., O. Receveur, R. Soueida, and G.M. Egeland, 2004: Arctic Indigenous peoples experience the nutrition transition with changing dietary patterns and obesity. *The Journal of Nutrition,* **134,** 1447-1453. http://jn.nutrition.org/content/134/6/1447.full.pdf

119. Moerlein, K.J. and C. Carothers, 2012: Total environment of change: Impacts of climate change and social transitions on subsistence fisheries in northwest Alaska. *Ecology and Society,* **17,** 10. http://dx.doi.org/10.5751/es-04543-170110

120. Melillo, J.M., T.C. Richmond, and G.W. Yohe, Eds., 2014: *Climate Change Impacts in the United States: The Third National Climate Assessment.* U.S. Global Change Research Program, Washington, D.C., 842 pp. http://dx.doi.org/10.7930/J0Z31WJ2

121. Dijkstra, J.A., K.L. Buckman, D. Ward, D.W. Evans, M. Dionne, and C.Y. Chen, 2013: Experimental and natural warming elevates mercury concentrations in estuarine fish. *PLoS ONE,* **8,** e58401. http://dx.doi.org/10.1371/journal.pone.0058401

122. Cozzetto, K., K. Chief, K. Dittmer, M. Brubaker, R. Gough, K. Souza, F. Ettawageshik, S. Wotkyns, S. Opitz-Stapleton, S. Duren, and P. Chavan, 2013: Climate change impacts on the water resources of American Indians and Alaska Natives in the U.S. *Climatic Change,* **120,** 569-584. http://dx.doi.org/10.1007/s10584-013-0852-y

123. Booth, S. and D. Zeller, 2005: Mercury, food webs, and marine mammals: Implications of diet and climate change for human health. *Environmental Health Perspectives,* **113,** 521-526. http://dx.doi.org/10.1289/ehp.7603

124. Lewitus, A.J., R.A. Horner, D.A. Caron, E. Garcia-Mendoza, B.M. Hickey, M. Hunter, D.D. Huppert, R.M. Kudela, G.W. Langlois, J.L. Largier, E.J. Lessard, R. RaLonde, J.E.J. Rensel, P.G. Strutton, V.L. Trainer, and J.F. Tweddle, 2012: Harmful algal blooms along the North American west coast region: History, trends, causes, and impacts. *Harmful Algae,* **19,** 133-159. http://dx.doi.org/10.1016/j.hal.2012.06.009

125. Carter, L.M., J.W. Jones, L. Berry, V. Burkett, J.F. Murley, J. Obeysekera, P.J. Schramm, and D. Wear, 2014: Ch. 17: Southeast and the Caribbean. *Climate Change Impacts in the United States: The Third National Climate Assessment.* Melillo, J.M., T.C. Richmond, and G.W. Yohe, Eds. U.S. Global Change Research Program, Washington, D.C., 396-417. http://dx.doi.org/10.7930/J0NP22CB

126. EPA, Indian Health Service, Department of Agriculture, and Department of Housing and Urban Development, 2008: Meeting the Access Goal: Strategies for Increasing Access to Safe Drinking Water and Wastewater Treatment to American Indian and Alaska Native Homes. Infrastructure Task Force Access Subgroup, 2008. 34 pp. U.S. Environmental Protection Agency. http://www.epa.gov/tp/pdf/infra-tribal-access-plan.pdf

127. TWWG, 2012: Water in Indian Country: Challenges and Opportunities. 25 pp. Tribal Water Working Group. http://uttoncenter.unm.edu/pdfs/2012White_Paper.pdf

128. Sack, R.B., M. Santosham, R. Reid, R. Black, J. Croll, R. Yolken, L. Aurelian, M. Wolff, E. Chan, S. Garrett, and J. Froehlich, 1995: Diarrhoeal diseases in the White Mountain Apaches: Clinical studies. *Journal of Diarrhoeal Diseases Research,* **13,** 12-17.

129. Singleton, R.J., R.C. Holman, K.L. Yorita, S. Holve, E.L. Paisano, C.A. Steiner, R.I. Glass, and J.E. Cheek, 2007: Diarrhea-associated hospitalizations and outpatient visits among American Indian and Alaska Native children younger than five years of age, 2000-2004. *The Pediatric Infectious Disease Journal,* **26,** 1006-1013. http://dx.doi.org/10.1097/INF.0b013e3181256595

130. Doyle, J.T., M.H. Redsteer, and M.J. Eggers, 2013: Exploring effects of climate change on Northern Plains American Indian health. *Climatic Change,* **120,** 643-655. http://dx.doi.org/10.1007/s10584-013-0799-z

131. Hodge, D.R., G.E. Limb, and T.L. Cross, 2009: Moving from colonization toward balance and harmony: A Native American perspective on wellness. *Social Work,* **54,** 211-219. http://dx.doi.org/10.1093/sw/54.3.211

132. Hodge, D.R. and G.E. Limb, 2010: A Native American perspective on spiritual assessment: The strengths and limitations of a complementary set of assessment tools. *Health & Social Work,* **35,** 121-131. http://dx.doi.org/10.1093/hsw/35.2.121

133. Cunsolo Willox, A., 2012: Climate change as the work of mourning. *Ethics & the Environment,* **17,** 137-164. http://dx.doi.org/10.2979/ethicsenviro.17.2.137

134. Cunsolo Willox, A., S.L. Harper, V.L. Edge, K. Landman, K. Houle, J.D. Ford, and Rigolet Inuit Community Government, 2013: The land enriches the soul: On climatic and environmental change, affect, and emotional health and well-being in Rigolet, Nunatsiavut, Canada. *Emotion, Space and Society,* **6,** 14-24. http://dx.doi.org/10.1016/j.emospa.2011.08.005

135. Cunsolo Willox, A., E. Stephenson, J. Allen, F. Bourque, A. Drossos, S. Elgarøy, M.J. Kral, I. Mauro, J. Moses, T. Pearce, J.P. MacDonald, and L. Wexler, 2015: Examining relationships between climate change and mental health in the Circumpolar North. *Regional Environmental Change,* **15,** 169-182. http://dx.doi.org/10.1007/s10113-014-0630-z

136. Voggesser, G., K. Lynn, J. Daigle, F.K. Lake, and D. Ranco, 2013: Cultural impacts to tribes from climate change influences on forests. *Climatic Change,* **120,** 615-626. http://dx.doi.org/10.1007/s10584-013-0733-4

137. Alessa, L., A. Kliskey, P. Williams, and M. Barton, 2008: Perception of change in freshwater in remote resource-dependent Arctic communities. *Global Environmental Change,* **18,** 153-164. http://dx.doi.org/10.1016/j.gloenvcha.2007.05.007

138. ACIA, 2004: Impacts of a Warming Arctic: Arctic Climate Impact Assessment. 140 pp. Cambridge University Press, Cambridge, UK. http://www.amap.no/documents/doc/impacts-of-a-warming-arctic-2004/786

139. Redsteer, M.H., R.C. Bogle, and J.M. Vogel, 2011: Monitoring and Analysis of Sand Dune Movement and Growth on the Navajo Nation, Southwestern United States. U.S. Geological Survey Fact Sheet 2011-3085, 2 pp. U.S. Geological Survey, Reston, VA. http://pubs.usgs.gov/fs/2011/3085/

140. Draut, A.E., M. Hiza Redsteer, and L. Amoroso, 2012: Recent seasonal variations in arid landscape cover and aeolian sand mobility, Navajo Nation, southwestern United States. *Climates, Landscapes, and Civilizations.* Giosan, L., D.Q. Fuller, K. Nicoll, R.K. Flad, and P.D. Clift, Eds. American Geophysical Union, Washington, D.C., 51-60. http://dx.doi.org/10.1029/2012GM001214

141. Roberts, J.W., L.A. Wallace, D.E. Camann, P. Dickey, S.G. Gilbert, R.G. Lewis, and T.K. Takaro, 2009: Monitoring and reducing exposure of infants to pollutants in house dust. *Reviews of Environmental Contamination and Toxicology,* **201,** 1-39. http://dx.doi.org/10.1007/978-1-4419-0032-6_1

142. Bernstein, A.S. and S.S. Myers, 2011: Climate change and children's health. *Current Opinion in Pediatrics,* **23,** 221-226. http://dx.doi.org/10.1097/MOP.0b013e3283444c89

143. Xu, Z., P.E. Sheffield, W. Hu, H. Su, W. Yu, X. Qi, and S. Tong, 2012: Climate change and children's health—A call for research on what works to protect children. *International Journal of Environmental Research and Pulic Health,* **9,** 3298-3316. http://dx.doi.org/10.3390/ijerph9093298

144. Xu, Z., R.A. Etzel, H. Su, C. Huang, Y. Guo, and S. Tong, 2012: Impact of ambient temperature on children's health: A systematic review. *Environmental Research,* **117,** 120-131. http://dx.doi.org/10.1016/j.envres.2012.07.002

145. Gilchrist, J., T. Haileyesus, M. Murphy, R.D. Comstock, C. Collins, N. McIlvain, and E. Yard, 2010: Heat illness among high school athletes --- United States, 2005-2009. *MMWR. Morbidity and mortality weekly report,* **59,** 1009-1013. http://www.cdc.gov/mmwr/preview/mmwrhtml/mm5932a1.htm

146. Kerr, Z.Y., D.J. Casa, S.W. Marshall, and R.D. Comstock, 2013: Epidemiology of exertional heat illness among U.S. high school athletes. *American Journal of Preventive Medicine,* **44,** 8-14. http://dx.doi.org/10.1016/j.amepre.2012.09.058

147. Nelson, N.G., C.L. Collins, R.D. Comstock, and L.B. McKenzie, 2011: Exertional heat-related injuries treated in emergency departments in the U.S., 1997–2006. *American Journal of Preventive Medicine,* **40,** 54-60. http://dx.doi.org/10.1016/j.amepre.2010.09.031

148. Gottschalk, A.W. and J.T. Andrish, 2011: Epidemiology of sports injury in pediatric athletes. *Sports Medicine and Arthroscopy Review,* **19,** 2-6. http://dx.doi.org/10.1097/JSA.0b013e31820b95fc

149. Becker-Blease, K.A., H.A. Turner, and D. Finkelhor, 2010: Disasters, victimization, and children's mental health. *Child Development,* **81,** 1040-1052. http://dx.doi.org/10.1111/j.1467-8624.2010.01453.x

150. Fairbank, J.A., F.W. Putnam, and W.W. Harris, 2014: Child traumatic stress: Prevalence, trends, risk, and impact. *Handbook of PTSD: Science and Practice,* 2nd ed. Friedman, M.J., T.M. Keane, and P.A. Resick, Eds. Guilford Press, New York, 121-145.

151. Strickland, M.J., L.A. Darrow, M. Klein, W.D. Flanders, J.A. Sarnat, L.A. Waller, S.E. Sarnat, J.A. Mulholland, and P.E. Tolbert, 2010: Short-term associations between ambient air pollutants and pediatric asthma emergency department visits. *American Journal of Respiratory and Critical Care Medicine,* **182,** 307-316. http://dx.doi.org/10.1164/rccm.200908-1201OC

152. Ostro, B., L. Roth, B. Malig, and M. Marty, 2009: The effects of fine particle components on respiratory hospital admissions in children. *Environmental Health Perspectives,* **117,** 475-480. http://dx.doi.org/10.1289/ehp.11848

153. Parker, J.D., L.J. Akinbami, and T.J. Woodruff, 2009: Air pollution and childhood respiratory allergies in the United States. *Environmental Health Perspectives,* **117,** 140-147. http://dx.doi.org/10.1289/ehp.11497

154. CDC, 2014: Health, United States, 2013--At a Glance. Centers for Disease Control and Prevention, Atlanta, GA. http://www.cdc.gov/nchs/hus/at_a_glance.htm

155. Eder, W., M.J. Ege, and E. von Mutius, 2006: The asthma epidemic. *The New England Journal of Medicine,* **355,** 2226-2235. http://dx.doi.org/10.1056/NEJMra054308

156. Gauderman, W.J., E. Avol, F. Gilliland, H. Vora, D. Thomas, K. Berhane, R. McConnell, N. Kuenzli, F. Lurmann, E. Rappaport, H. Margolis, D. Bates, and J. Peters, 2004: The effect of air pollution on lung development from 10 to 18 years of age. *The New England Journal of Medicine,* **351,** 1057-1067. http://dx.doi.org/10.1056/NEJMoa040610

157. McCormack, M.C., P.N. Breysse, E.C. Matsui, N.N. Hansel, D.A. Williams, J. Curtin-Brosnan, P. Eggleston, and G.B. Diette, 2009: In-home particle concentrations and childhood asthma morbidity. *Environmental Health Perspectives,* **117,** 294-298. http://dx.doi.org/10.1289/ehp.11770

158. Kistin, E.J., J. Fogarty, R.S. Pokrasso, M. McCally, and P.G. McCormick, 2010: Climate change, water resources and child health. *Archives of Disease in Childhood,* **95,** 545-549. http://dx.doi.org/10.1136/adc.2009.175307

159. Drayna, P., S.L. McLellan, P. Simpson, S.-H. Li, and M.H. Gorelick, 2010: Association between rainfall and pediatric emergency department visits for acute gastrointestinal illness. *Environmental Health Perspectives,* **118,** 1439-1443. http://dx.doi.org/10.1289/ehp.0901671

160. McBride, G.B., R. Stott, W. Miller, D. Bambic, and S. Wuertz, 2013: Discharge-based QMRA for estimation of public health risks from exposure to stormwater-borne pathogens in recreational waters in the United States. *Water Research,* **47,** 5282-5297. http://dx.doi.org/10.1016/j.watres.2013.06.001

161. Dechet, A.M., P.A. Yu, N. Koram, and J. Painter, 2008: Nonfoodborne *Vibrio* infections: An important cause of morbidity and mortality in the United States, 1997–2006. *Clinical Infectious Diseases,* **46,** 970-976. http://dx.doi.org/10.1086/529148

162. Hilborn, E.D., V.A. Roberts, L. Backer, E. DeConno, J.S. Egan, J.B. Hyde, D.C. Nicholas, E.J. Wiegert, L.M. Billing, M. DiOrio, M.C. Mohr, F.J. Hardy, T.J. Wade, J.S. Yoder, and M.C. Hlavsa, 2014: Algal bloom–associated disease outbreaks among users of freshwater lakes — United States, 2009–2010. *MMWR. Morbidity and Mortality Weekly Report,* **63,** 11-15. http://www.cdc.gov/mmwr/preview/mmwrhtml/mm6301a3.htm

163. Janakiraman, V., 2008: Listeriosis in pregnancy: Diagnosis, treatment, and prevention. *Reviews in Obstetrics and Gynecology,* **1,** 179-185. PMC2621056

164. Chitra, T.V. and S. Panicker, 2011: Maternal and fetal outcome of dengue fever during pregnancy. *Journal of Vector Borne Diseases,* **48,** 210-213. http://www.mrcindia.org/journal/issues/484210.pdf

165. Rasmussen, S.A., D.J. Jamieson, and T.M. Uyeki, 2012: Effects of influenza on pregnant women and infants. *American Journal of Obstetrics and Gynecology,* **207,** S3-S8. http://dx.doi.org/10.1016/j.ajog.2012.06.068

166. Erwin, P.C., T.F. Jones, R.R. Gerhardt, S.K. Halford, A.B. Smith, L.E.R. Patterson, K.L. Gottfried, K.L. Burkhalter, R.S. Nasci, and W. Schaffner, 2002: La Crosse encephalitis in eastern Tennessee: Clinical, environmental, and entomological characteristics from a blinded cohort study. *American Journal of Epidemiology,* **155,** 1060-1065. http://dx.doi.org/10.1093/aje/155.11.1060

167. CCSP, 2008: Analyses of the Effects of Global Change on Human Health and Welfare and Human Systems. A Report by the U.S. Climate Change Science Program and the Subcommittee on Global Change Research. 205 pp. Gamble, J. L., (Ed.), Ebi, K.L., F.G. Sussman, T.J. Wilbanks, (Authors). U.S. Environmental Protection Agency, Washington, D.C. http://downloads.globalchange.gov/sap/sap4-6/sap4-6-final-report-all.pdf

168. CDC, 2015: Confirmed Lyme Disease Cases by Age and Sex--United States, 2001-2010. Centers for Disease Control and Prevention, Atlanta, GA. http://www.cdc.gov/lyme/stats/chartstables/incidencebyagesex.html

169. Nelson, G.C., M.W. Rosegrant, J. Koo, R. Robertson, T. Sulser, T. Zhu, C. Ringler, S. Msangi, A. Palazzo, M. Magalhaes, R. Valmonte-Santos, M. Ewing, and D. Lee, 2009: Climate Change: Impact on Agriculture and Costs of Adaptation. 30 pp. International Food Policy Research Institute, Washington, D.C. http://www.ifpri.org/sites/default/files/publications/pr21.pdf

170. Loladze, I., 2014: Hidden shift of the ionome of plants exposed to elevated CO2 depletes minerals at the base of human nutrition. *eLife,* **3,** e02245. http://dx.doi.org/10.7554/eLife.02245

171. Myers, S.S., A. Zanobetti, I. Kloog, P. Huybers, A.D.B. Leakey, A.J. Bloom, E. Carlisle, L.H. Dietterich, G. Fitzgerald, T. Hasegawa, N.M. Holbrook, R.L. Nelson, M.J. Ottman, V. Raboy, H. Sakai, K.A. Sartor, J. Schwartz, S. Seneweera, M. Tausz, and Y. Usui, 2014: Increasing CO2 threatens human nutrition. *Nature,* **510,** 139-142. http://dx.doi.org/10.1038/nature13179

172. Brown, M.E., J.M. Antle, P. Backlund, E.R. Carr, W.E. Easterling, M.K. Walsh, C. Ammann, W. Attavanich, C.B. Barrett, M.F. Bellemare, V. Dancheck, C. Funk, K. Grace, J.S.I. Ingram, H. Jiang, H. Maletta, T. Mata, A. Murray, M. Ngugi, D. Ojima, B. O'Neill, and C. Tebaldi, 2015: Climate Change, Global Food Security and the U.S. Food System. 146 pp. U.S. Global Change Research Program. http://www.usda.gov/oce/climate_change/FoodSecurity2015Assessment/FullAssessment.pdf

173. Walthall, C., P. Backlund, J. Hatfield, L. Lengnick, E. Marshall, M. Walsh, S. Adkins, M. Aillery, E.A. Ainsworth, C. Amman, C.J. Anderson, I. Bartomeus, L.H. Baumgard, F. Booker, B. Bradley, D.M. Blumenthal, J. Bunce, K. Burkey, S.M. Dabney, J.A. Delgado, J. Dukes, A. Funk, K. Garrett, M. Glenn, D.A. Grantz, D. Goodrich, S. Hu, R.C. Izaurralde, R.A.C. Jones, S.-H. Kim, A.D.B. Leaky, K. Lewers, T.L. Mader, A. McClung, J. Morgan, D.J. Muth, M. Nearing, D.M. Oosterhuis, D. Ort, C. Parmesan, W.T. Pettigrew, W. Polley, R. Rader, C. Rice, M. Rivington, E. Rosskopf, W.A. Salas, L.E. Sollenberger, R. Srygley, C. Stöckle, E.S. Takle, D. Timlin, J.W. White, R. Winfree, L. Wright-Morton, and L.H. Ziska, 2012: Climate Change and Agriculture in the United States: Effects and Adaptation. USDA Technical Bulletin 1935, 186 pp. U.S. Department of Agriculture, Washington, D.C. http://www.usda.gov/oce/climate_change/effects_2012/CC and Agriculture Report (02-04-2013)b.pdf

174. USDA, 2009: Access to Affordable and Nutritious Food: Measuring and Understanding Food Deserts and Their Consequences. Report to Congress. U.S. Department of Agriculture Economic Research Service, Washington, DC. http://www.ers.usda.gov/media/242675/ap036_1_.pdf

175. Crim, S.M., M. Iwamoto, J.Y. Huang, P.M. Griffin, D. Gilliss, A.B. Cronquist, M. Cartter, M. Tobin-D'Angelo, D. Blythe, K. Smith, S. Lathrop, S. Zansky, P.R. Cieslak, J. Dunn, K.G. Holt, S. Lance, R. Tauxe, and O.L. Henao, 2014: Incidence and trends of infections with pathogens transmitted commonly through food--Foodborne Diseases Active Surveillance Network, 10 U.S. sites, 2006-2013. *MMWR. Morbidity and Mortality Weekly Report,* **63,** 328-332. http://www.cdc.gov/mmwr/preview/mmwrhtml/mm6315a3.htm

176. Rylander, C., J.O. Odland, and T.M. Sandanger, 2013: Climate change and the potential effects on maternal and pregnancy outcomes: An assessment of the most vulnerable--the mother, fetus, and newborn child. *Global Health Action,* **6,** 19538. http://dx.doi.org/10.3402/gha.v6i0.19538

177. CDC, 2015: Reproductive Health: Preterm Birth. Centers for Disease Control and Prevention, Atlanta, GA. http://www.cdc.gov/reproductivehealth/maternalinfanthealth/pretermbirth.htm

178. March of Dimes, PMNCH, Save the Children, and WHO, 2012: Born Too Soon: The Global Action Report on Preterm Birth. Howson, C.P., M.V. Kinney, and J.E. Lawn (Eds.), 112 pp. World Health Organization, Geneva, Switzerland. http://www.who.int/pmnch/media/news/2012/201204_borntoosoon-report.pdf

179. Basu, R., B. Malig, and B. Ostro, 2010: High ambient temperature and the risk of preterm delivery. *American Journal of Epidemiology,* **172,** 1108-1117. http://dx.doi.org/10.1093/aje/kwq170

180. Kent, S.T., L.A. McClure, B.F. Zaitchik, T.T. Smith, and J.M. Gohlke, 2014: Heat waves and health outcomes in Alabama (USA): The importance of heat wave definition. *Environmental Health Perspectives,* **122,** 151–158. http://dx.doi.org/10.1289/ehp.1307262

181. Strand, L.B., A.G. Barnett, and S. Tong, 2011: Maternal exposure to ambient temperature and the risks of preterm birth and stillbirth in Brisbane, Australia. *American Journal of Epidemiology,* **175,** 99-107. http://dx.doi.org/10.1093/aje/kwr404

182. Van Zutphen, A.R., S. Lin, B.A. Fletcher, and S.-A. Hwang, 2012: A population-based case-control study of extreme summer temperature and birth defects. *Environmental Health Perspectives,* **120,** 1443-1449. http://dx.doi.org/10.1289/ehp.1104671

183. Polin, R.A. and S.H. Abman, 2011: Thermoregulation. *Fetal and Neonatal Physiology,* 4th ed. Elsevier, Philadelphia, PA, 615-670.

184. Choi, H., V. Rauh, R. Garfinkel, Y. Tu, and F.P. Perera, 2008: Prenatal exposure to airborne polycyclic aromatic hydrocarbons and risk of intrauterine growth restriction. *Environmental Health Perspectives,* **116,** 658-665. http://dx.doi.org/10.1289/ehp.10958

185. Makri, A. and N.I. Stilianakis, 2008: Vulnerability to air pollution health effects. *International Journal of Hygiene and Environmental Health,* **211,** 326-336. http://dx.doi.org/10.1016/j.ijheh.2007.06.005

186. Jayachandran, S., 2009: Air quality and early-life mortality: Evidence from Indonesia's wildfires. *Journal of Human Resources,* **44,** 916-954. http://dx.doi.org/10.3368/jhr.44.4.916

187. Ritz, B., M. Wilhelm, K.J. Hoggatt, and J.K.C. Ghosh, 2007: Ambient air pollution and preterm birth in the Environment and Pregnancy Outcomes Study at the University of California, Los Angeles. *American Journal of Epidemiology,* **166,** 1045-1052. http://dx.doi.org/10.1093/aje/kwm181

188. Dejmek, J., S.G. Selevan, I. Benes, I. Solanský, and R.J. Srám, 1999: Fetal growth and maternal exposure to particulate matter during pregnancy. *Environmental Health Perspectives,* **107,** 475-480. PMC1566587

189. Kim, J.J., M.W. Shannon, D. Best, H.J. Binns, C.L. Johnson, L.J. Mazur, D.W. Reynolds, J.R. Roberts, W.B.J. Weil, S.J. Balk, M. Miller, and K.M. Shea, 2004: Ambient air pollution: Health hazards to children. *Pediatrics,* **114,** 1699-1707. http://dx.doi.org/10.1542/peds.2004-2166

190. Callaghan, W.M., S.A. Rasmussen, D.J. Jamieson, S.J. Ventura, S.L. Farr, P.D. Sutton, T.J. Mathews, B.E. Hamilton, K.R. Shealy, D. Brantley, and S.F. Posner, 2007: Health concerns of women and infants in times of natural disasters: Lessons learned from Hurricane Katrina. *Maternal and Child Health Journal,* **11,** 307-311. http://dx.doi.org/10.1007/s10995-007-0177-4

191. Tees, M.T., E.W. Harville, X. Xiong, P. Buekens, G. Pridjian, and K. Elkind-Hirsch, 2010: Hurricane Katrina-related maternal stress, maternal mental health, and early infant temperament. *Maternal and Child Health Journal,* **14,** 511-518. http://dx.doi.org/10.1007/s10995-009-0486-x

192. Hamilton, B.E., P.D. Sutton, T.J. Mathews, J.A. Martin, and S.J. Ventura, 2009: The effect of Hurricane Katrina: Births in the U.S. Gulf Coast region, before and after the storm. *National Vital Statistics Reports,* **58,** 1-28, 32. http://www.cdc.gov/nchs/data/nvsr/nvsr58/nvsr58_02.pdf

193. Harville, E.W., X. Xiong, and P. Buekens, 2009: Hurricane Katrina and perinatal health. *Birth,* **36,** 325-331. http://dx.doi.org/10.1111/j.1523-536X.2009.00360.x

194. Yip, F.Y., W.D. Flanders, A. Wolkin, D. Engelthaler, W. Humble, A. Neri, L. Lewis, L. Backer, and C. Rubin, 2008: The impact of excess heat events in Maricopa County, Arizona: 2000–2005. *International Journal of Biometeorology,* **52,** 765-772. http://dx.doi.org/10.1007/s00484-008-0169-0

195. Green, R.S., R. Basu, B. Malig, R. Broadwin, J.J. Kim, and B. Ostro, 2010: The effect of temperature on hospital admissions in nine California counties. *International Journal of Public Health,* **55,** 113-121. http://dx.doi.org/10.1007/s00038-009-0076-0

196. Imhoff, B., D. Morse, B. Shiferaw, M. Hawkins, D. Vugia, S. Lance-Parker, J. Hadler, C. Medus, M. Kennedy, Matthew R. Moore, and T. Van Gilder, 2004: Burden of self-reported acute diarrheal illness in FoodNet surveillance areas, 1998–1999. *Clinical Infectious Diseases,* **38,** S219-S226. http://dx.doi.org/10.1086/381590

197. Filiberto, D., E. Wethington, K. Pillemer, N. Wells, M. Wysocki, and J.T. Parise, 2010: Older people and climate change: Vulnerability and health effects. *Generations,* **33,** 19-25. http://www.asaging.org/blog/older-people-and-climate-change-vulnerability-and-health-effects

198. IOM, 2010: *Providing Healthy and Safe Foods As We Age: Workshop Summary.* Institute of Medicine. The National Academies Press, Washington, D.C., 192 pp. http://www. nap.edu/catalog/12967/providing-healthy-and-safe-foods-as-we-age-workshop-summary

199. Wang, L., F.H.Y. Green, S.M. Smiley-Jewell, and K.E. Pinkerton, 2010: Susceptibility of the aging lung to environmental injury. *Seminars in Respiratory and Critical Care Medicine,* **31,** 539-553. http://dx.doi.org/10.1055/s-0030-1265895

200. Vincent, G.K. and V.A. Velkof, 2010: The Next Four Decades: The Older Population in the United States: 2010 to 2050. Current Population Reports #P25-1138, 16 pp. U.S. Department of Commerce, Economics and Statistics Administration, U.S. Census Bureau, Washington, D.C. http://www.census.gov/prod/2010pubs/p25-1138.pdf

201. U.S. Census Bureau, 2014: 2014 National Population Projections: Summary Tables. Table 9. Projections of the Population by Sex and Age for the United States: 2015 to 2060 U.S. Department of Commerce, Washington, D.C. http://www.census.gov/population/projections/data/national/2014/summarytables.html

202. Luber, G. and N. Prudent, 2009: Climate change and human health. *Transactions of the American Clinical and Climatological Association,* **120,** 113-117. PMC2744549

203. O'Neill, M.S. and K.L. Ebi, 2009: Temperature extremes and health: Impacts of climate variability and change in the United States. *Journal of Occupational and Environmental Medicine,* **51,** 13-25. http://dx.doi.org/10.1097/JOM.0b013e318173e122

204. O'Neill, M.S., R. Carter, J.K. Kish, C.J. Gronlund, J.L. White-Newsome, X. Manarolla, A. Zanobetti, and J.D. Schwartz, 2009: Preventing heat-related morbidity and mortality: New approaches in a changing climate. *Maturitas,* **64,** 98-103. http://dx.doi.org/10.1016/j.maturitas.2009.08.005

205. Thacker, M.T.F., R. Lee, R.I. Sabogal, and A. Henderson, 2008: Overview of deaths associated with natural events, United States, 1979–2004. *Disasters,* **32,** 303-315. http://dx.doi.org/10.1111/j.1467-7717.2008.01041.x

206. Zanobetti, A., M.S. O'Neill, C.J. Gronlund, and J.D. Schwartz, 2012: Summer temperature variability and long-term survival among elderly people with chronic disease. *Proceedings of the National Academy of Sciences,* **109,** 6608-6613. http://dx.doi.org/10.1073/pnas.1113070109

207. CDC, 2014: Diabetes Public Health Resource: Rate per 100 of Civilian, Noninstitutionalized Population with Diagnosed Diabetes, by Age, United States, 1980-2011. Centers for Disease Control and Prevention, Atlanta, GA. http://www.cdc.gov/diabetes/statistics/prev/national/figbyage.htm

208. Cherry, K.E., S. Galea, L.J. Su, D.A. Welsh, S.M. Jazwinski, J.L. Silva, and M.J. Erwin, 2010: Cognitive and psychosocial consequences of Hurricanes Katrina and Rita among middle-aged, older, and oldest-old adults in the Louisiana Healthy Aging Study (LHAS). *Journal of Applied Social Psychology,* **40,** 2463-2487. http://dx.doi.org/10.1111/j.1559-1816.2010.00666.x

209. Sanders, S., S.L. Bowie, and Y.D. Bowie, 2004: Chapter 2 Lessons learned on forced relocation of older adults: The impact of Hurricane Andrew on health, mental health, and social support of public housing residents. *Journal of Gerontological Social Work,* **40,** 23-35. http://dx.doi.org/10.1300/J083v40n04_03

210. Laditka, S.B., J.N. Laditka, S. Xirasagar, C.B. Cornman, C.B. Davis, and J.V.E. Richter, 2008: Providing shelter to nursing home evacuees in disasters: Lessons from Hurricane Katrina. *American Journal of Public Health,* **98,** 1288-1293. http://dx.doi.org/10.2105/ajph.2006.107748

211. Little, B., J. Gill, J. Schulte, S. Young, J. Horton, L. Harris, D. Batts-Osborne, C. Sanchez, J. Malilay, and T. Bayleyegn, 2004: Rapid assessment of the needs and health status of older adults after Hurricane Charley--Charlotte, DeSoto, and Hardee Counties, Florida, August 27-31, 2004. *MMWR. Morbidity and Mortality Weekly Report,* **53,** 837-840. http://www.cdc.gov/mmwr/preview/mmwrhtml/mm5336a2.htm

212. Laumbach, R.J., 2010: Outdoor air pollutants and patient health. *American Family Physician,* **81,** 175-180. http://www.aafp.org/afp/2010/0115/p175.pdf

213. Reid, C.E. and J.L. Gamble, 2009: Aeroallergens, allergic disease, and climate change: Impacts and adaptation. *EcoHealth,* **6,** 458-470. http://dx.doi.org/10.1007/s10393-009-0261-x

214. Baja, E.S., J.D. Schwartz, G.A. Wellenius, B.A. Coull, A. Zanobetti, P.S. Vokonas, and H.H. Suh, 2010: Traffic-related air pollution and QT interval: Modification by diabetes, obesity, and oxidative stress gene polymorphisms in the normative aging study. *Environmental Health Perspectives,* **118,** 840-846. http://dx.doi.org/10.1289/ehp.0901396

215. Patz, J.A. and W.K. Reisen, 2001: Immunology, climate change and vector-borne diseases. *Trends in Immunology,* **22,** 171-172. http://dx.doi.org/10.1016/S1471-4906(01)01867-1

216. Lindsey, N.P., J.E. Staples, J.A. Lehman, and M. Fischer, 2010: Surveillance for human West Nile virus disease--United States, 1999-2008. *MMWR Surveillance Summaries,* **59(SS02),** 1-17. http://www.cdc.gov/mmwr/preview/mmwrhtml/ss5902a1.htm

217. Bush, K.F., C.L. Fossani, S. Li, B. Mukherjee, C.J. Gronlund, and M.S. O'Neill, 2014: Extreme precipitation and beach closures in the Great Lakes region: Evaluating risk among the elderly. *International Journal of Environmental Research and Public Health,* **11,** 2014-2032. http://dx.doi.org/10.3390/ijerph110202014

218. Jagai, J.S., J.K. Griffiths, P.K. Kirshen, P. Webb, and E.N. Naumova, 2012: Seasonal patterns of gastrointestinal illness and streamflow along the Ohio River. *International Journal of Environmental Research and Public Health,* **9,** 1771-1790. http://dx.doi.org/10.3390/ijerph9051771

219. Naumova, E.N., A.I. Egorov, R.D. Morris, and J.K. Griffiths, 2003: The elderly and waterborne *Cryptosporidium* infection: Gatroenteritis hospitalizations before and during the 1993 Milwaukee outbreak. *Emerging Infectious Diseases,* **9,** 418-425. http://dx.doi.org/10.3201/eid0904.020260

220. Schwartz, J., R. Levin, and R. Goldstein, 2000: Drinking water turbidity and gastrointestinal illness in the elderly of Philadelphia. *Journal of Epidemiology and Community Health,* **54,** 45-51. http://dx.doi.org/10.1136/jech.54.1.45

221. Zimmerman, R., C.E. Restrepo, B. Nagorsky, and A.M. Culpen, 2007: Vulnerability of the elderly during natural hazard events. *Proceedings of the Hazards and Disasters Researchers Meeting,* July 11-12, Boulder, CO, pp. 38-40. http://create.usc.edu/sites/default/files/publications//vulnerabilityoftheelderlyduringnaturalhazardevents.pdf

222. Werner, C.A., 2011: The Older Population: 2010. 2010 Census Briefs C2010BR-09, 19 pp. U.S. Census Bureau. https://www.census.gov/prod/cen2010/briefs/c2010br-09.pdf

223. DHHS, 2009: A Profile of Older Americans: 2009. 17 pp. U.S. Department of Health and Human Services, Administration on Aging, Washington, D.C. http://www.aoa.gov/Aging_Statistics/Profile/2009/docs/2009profile_508.pdf

224. Browning, C.R., D. Wallace, S.L. Feinberg, and K.A. Cagney, 2006: Neighborhood social processes, physical conditions, and disaster-related mortality: The case of the 1995 Chicago heat wave. *American Sociological Review,* **71,** 661-678. http://dx.doi.org/10.1177/000312240607100407

225. Stone, B., J.J. Hess, and H. Frumkin, 2010: Urban form and extreme heat events: Are sprawling cities more vulnerable to climate change than compact cities? *Environmental Health Perspectives,* **118,** 1425-1428. http://dx.doi.org/10.1289/ehp.0901879

226. Haq, G., J. Whiteleg, and M. Kohler, 2008: Growing Old in a Changing Climate: Meeting the Challenges of an Ageing Population and Climate Change. 28 pp. Stockholm Environment Institute, Stockholm, Sweden. http://www.sei-international.org/mediamanager/documents/Publications/Future/climate_change_growing_old.pdf

227. Hansen, A., P. Bi, M. Nitschke, D. Pisaniello, J. Newbury, and A. Kitson, 2011: Perceptions of heat-susceptibility in older persons: Barriers to adaptation. *International Journal of Environmental Research and Public Health,* **8,** 4714-4728. http://dx.doi.org/10.3390/ijerph8124714

228. Hansen, A., P. Bi, M. Nitschke, D. Pisaniello, J. Newbury, and A. Kitson, 2011: Older persons and heat-susceptibility: The role of health promotion in a changing climate. *Health Promotion Journal of Australia,* **22,** 17-20. http://dx.doi.org/10.1071/HE11417

229. Sheridan, S.C., A.J. Kalkstein, and L.S. Kalkstein, 2009: Trends in heat-related mortality in the United States, 1975–2004. *Natural Hazards,* **50,** 145-160. http://dx.doi.org/10.1007/s11069-008-9327-2

230. Ren, C., M.S. O'Neill, S.K. Park, D. Sparrow, P. Vokonas, and J. Schwartz, 2011: Ambient temperature, air pollution, and heart rate variability in an aging population. *American Journal of Epidemiology,* **173,** 1013-1021. http://dx.doi.org/10.1093/aje/kwq477

231. Kovats, R.S. and S. Hajat, 2008: Heat stress and public health: A critical review. *Annual Review of Public Health,* **29,** 41-55. http://dx.doi.org/10.1146/annurev.publhealth.29.020907.090843

232. Conti, S., M. Masocco, P. Meli, G. Minelli, E. Palummeri, R. Solimini, V. Toccaceli, and M. Vichi, 2007: General and specific mortality among the elderly during the 2003 heat wave in Genoa (Italy). *Environmental Research,* **103,** 267-274. http://dx.doi.org/10.1016/j.envres.2006.06.003

233. Martin-Latry, K., M.P. Goumy, P. Latry, C. Gabinski, B. Bégaud, I. Faure, and H. Verdoux, 2007: Psychotropic drugs use and risk of heat-related hospitalisation. *European Psychiatry,* **22,** 335-338. http://dx.doi.org/10.1016/j.eurpsy.2007.03.007

234. Brault, M.W., 2012: Americans With Disabilities: 2010. Current Population Reports #P70-131, 23 pp. U.S. Census Bureau, Washington, D.C. http://www.census.gov/prod/2012pubs/p70-131.pdf

235. Plassman, B.L., K.M. Langa, G.G. Fisher, S.G. Heeringa, D.R. Weir, M.B. Ofstedal, J.R. Burke, M.D. Hurd, G.G. Potter, W.L. Rodgers, D.C. Steffens, R.J. Willis, and R.B. Wallace, 2007: Prevalence of dementia in the United States: The aging, demographics, and memory study. *Neuroepidemiology*, **29**, 125-132. http://dx.doi.org/10.1159/000109998

236. Lundgren, K., K. Kuklane, C. Gao, and I. Holmer, 2013: Effects of heat stress on working populations when facing climate change. *Industrial Health*, **51**, 3-15. http://dx.doi.org/10.2486/indhealth.2012-0089 <Go to ISI>://WOS:000314383700002

237. Spector, J.T. and P.E. Sheffield, 2014: Re-evaluating occupational heat stress in a changing climate. *Annals of Occupational Hygiene*, **58**, 936-942. http://dx.doi.org/10.1093/annhyg/meu073

238. Xiang, J., P. Bi, D. Pisaniello, and A. Hansen, 2014: Health impacts of workplace heat exposure: An epidemiological review. *Industrial Health*, **52**, 91-101. http://dx.doi.org/10.2486/indhealth.2012-0145

239. Kjellstrom, T., R.S. Kovats, S.J. Lloyd, T. Holt, and R.S.J. Tol, 2009: The direct impact of climate change on regional labor productivity. *Archives of Environmental & Occupational Health*, **64**, 217-227. http://dx.doi.org/10.1080/19338240903352776

240. Nilsson, M. and T. Kjellstrom, 2010: Invited Editorial: Climate change impacts on working people: How to develop prevention policies. *Global Health Action*, **3**. http://dx.doi.org/10.3402/gha.v3i0.5774

241. Gubernot, D.M., G.B. Anderson, and K.L. Hunting, 2014: The epidemiology of occupational heat exposure in the United States: A review of the literature and assessment of research needs in a changing climate. *International Journal of Biometeorology*, **58**, 1779-1788. http://dx.doi.org/10.1007/s00484-013-0752-x

242. Roelofs, C. and D. Wegman, 2014: Workers: The climate canaries. *American Journal of Public Health*, **104**, 1799-1801. http://dx.doi.org/10.2105/AJPH.2014.302145

243. Noyes, P.D., M.K. McElwee, H.D. Miller, B.W. Clark, L.A. Van Tiem, K.C. Walcott, K.N. Erwin, and E.D. Levin, 2009: The toxicology of climate change: Environmental contaminants in a warming world. *Environment International*, **35**, 971-986. http://dx.doi.org/10.1016/j.envint.2009.02.006

244. USFA, 2013: Firefighter Fatalities in the United States in 2012. 63 pp. U.S. Department of Homeland Security, Federal Emergency Management Agency, U.S. Fire Administration. https://www.usfa.fema.gov/downloads/pdf/publications/ff_fat12.pdf

245. Hardy, K. and L.K. Comfort, 2015: Dynamic decision processes in complex, high-risk operations: The Yarnell Hill Fire, June 30, 2013. *Safety Science*, **71(Part A)**, 39-47. http://dx.doi.org/10.1016/j.ssci.2014.04.019

246. NIOSH, 2012: Fact Sheet: Wildland Fire Fighting. Hot Tips to Stay Safe and Healthy. DHHS (NIOSH) Publication No. 2013-158, 2 pp. U.S. Department of Health and Human Services, Centers for Disease Control and Prevention, National Institute for Occupational Safety and Health, Washington, D.C. http://www.cdc.gov/niosh/docs/2013-158/pdfs/2013-158v2.pdf

247. Bartra, J., J. Mullol, A. del Cuvillo, I. Davila, M. Ferrer, I. Jauregui, J. Montoro, J. Sastre, and A. Valero, 2007: Air pollution and allergens. *Journal of Investigational Allergology and Clinical Immunology*, **17 Suppl 2**, 3-8. http://www.jiaci.org/issues/vol17s2/vol17s2-2.htm

248. Ziska, L.H., R.C. Sicher, K. George, and J.E. Mohan, 2007: Rising atmospheric carbon dioxide and potential impacts on the growth and toxicity of poison ivy (*Toxicodendron radicans*). *Weed Science*, **55**, 288-292. http://dx.doi.org/10.1614/ws-06-190

249. Estrada-Peña, A., 2002: Increasing habitat suitability in the United States for the tick that transmits Lyme disease: A remote sensing approach. *Environmental Health Perspectives*, **110**, 635-640. PMC1240908

250. DOD, 2014: Quadrennial Defense Review. 64 pp. U.S. Department of Defense. http://www.defense.gov/pubs/2014_Quadrennial_Defense_Review.pdf

251. DOD, 2012: Department of Defense FY 2012 Climate Change Adaptation Roadmap. Appendix to DOD's Strategic Sustainability Performance Plan 2012. U.S. Department of Defense, Alexandria, VA. http://www.acq.osd.mil/ie/download/green_energy/dod_sustainability/2012/Appendix A - DoD Climate Change Adaption Roadmap_20120918.pdf

252. DOD, 2010: Quadrennial Defense Review Report. 105 pp. U.S. Department of Defense. http://www.defense.gov/qdr/qdr as of 26jan10 0700.pdf

253. AFHSC, 2015: Update: Heat injuries, active component, U.S. Armed Forces. *MSMR: Medical Surveillance Monthly Report*, **22(3)**, 17-20. https://www.afhsc.mil/documents/pubs/msmrs/2015/v22_n03.pdf

254. Hess, J.J., S. Saha, and G. Luber, 2014: Summertime acute heat illness in U.S. emergency departments from 2006 through 2010: Analysis of a nationally representative sample. *Environmental Health Perspectives,* **122,** 1209-1215. http://dx.doi.org/10.1289/ehp.1306796

255. Hurt, L. and K.A. Dorsey, 2014: The geographic distribution of incident Lyme disease among active component service members stationed in the continental United States, 2004-2013. *MSMR: Medical Surveillance Monthly Report,* **21(5),** 13-15. http://www.afhsc.mil/documents/pubs/msmrs/2014/v21_n05.pdf - Page=13

256. AFHSC, 2014: The geographic distribution of incident coccidioidomycosis among active component service members, 2000-2013. *MSMR: Medical Surveillance Monthly Report,* **21(6),** 12-14. http://www.afhsc.mil/documents/pubs/msmrs/2014/v21_n06.pdf

257. Crum-Cianflone, N.F., 2007: Coccidioidomycosis in the U.S. militray. *Annals of the New York Academy of Sciences,* **1111,** 112-121. http://dx.doi.org/10.1196/annals.1406.001

258. AFHSC, 2015: Chikungunya in the Americas Surveillance Summary. 8 July 2015. Armed Forces Health Surveillance Center. http://www.afhsc.mil/Store/Document/DIBSurveillanceSummary/AFHSC Caribbean Chikungunya Surveillance Summary 8 JUL 2015 Unclassified.pdf

259. Reeves, W.K., N.M. Rowe, R.K. Kugblenu, and C.L. Magnuson, 2015: Case Series: Chikungunya and dengue at a forward operating location. *MSMR: Medical Surveillance Monthly Report,* **22(5),** 9-10. http://www.afhsc.mil/documents/pubs/msmrs/2015/v22_n05.pdf - Page=9

260. Gibbons, R.V., M. Streitz, T. Babina, and J.R. Fried, 2012: Dengue and US military operations from the Spanish-American War through today. *Emerging Infectious Diseases,* **18,** 623-630. http://dx.doi.org/10.3201/eid1804.110134

261. AFHSC, 2007: Leishmaniasis in relation to service in Iraq/Afghanistan, U.S. Armed Forces, 2001-2006. *MSMR: Medical Surveillance Monthly Report,* **14(1),** 2-5. http://www.afhsc.mil/documents/pubs/msmrs/2007/v14_n01.pdf

262. AFHSC, 2015: Update: Malaria, U.S. Armed Forces, 2014. *MSMR: Medical Surveillance Monthly Report,* **22(1),** 2-6. http://www.afhsc.mil/documents/pubs/msmrs/2015/v22_n01.pdf

263. Klein, T.A., L.A. Pacha, H.-C.S. Lee, H.-C. Kim, W.-J. Lee, J.-K. Lee, G.-G. Jeung, W.J. Sames, and J.C. Gaydos, 2009: *Plasmodium vivax* malaria among U.S. Forces Korea in the Republic of Korea, 1993-2007. *Military Medicine,* **174,** 412-418. http://dx.doi.org/10.7205/MILMED-D-01-4608

264. Kotwal, R.S., R.B. Wenzel, R.A. Sterling, W.D. Porter, N.N. Jordan, and B.P. Petruccelli, 2005: An outbreak of malaria in US Army Rangers returning from Afghanistan. *JAMA - Journal of the American Medical Association,* **293,** 212-216. http://dx.doi.org/10.1001/jama.293.2.212

265. Whitman, T.J., P.E. Coyne, A.J. Magill, D.L. Blazes, M.D. Green, W.K. Milhous, T.H. Burgess, D. Freilich, S.A. Tasker, R.G. Azar, T.P. Endy, C.D. Clagett, G.A. Deye, G.D. Shanks, and G.J. Martin, 2010: An outbreak of *Plasmodium falciparum* malaria in U.S. Marines deployed to Liberia. *American Journal of Tropical Medicine and Hygiene,* **83,** 258-265. http://dx.doi.org/10.4269/ajtmh.2010.09-0774

266. CDC, 2015: Disability and Health: Disability Overview. Centers for Disease Control and Prevention, Atlanta, GA. http://www.cdc.gov/ncbddd/disabilityandhealth/disability.html

267. Kailes, J.I. and A. Enders, 2007: Moving beyond "special needs": A function-based framework for emergency management and planning. *Journal of Disability Policy Studies,* **17,** 230-237. http://dx.doi.org/10.1177/10442073070170040601

268. WHO, 2001: International Classification of Functioning, Disability and Health. 303 pp. World Health Organization, Geneva.

269. NCHS, 2014: Table 49. Disability measures among adults aged 18 and over, by selected characteristics: United States, selected years 1997–2012. *Health, United States, 2013: With Special Feature on Prescription Drugs.* National Center for Health Statistics, Hyattsville, MD, 172-173. http://www.cdc.gov/nchs/data/hus/hus13.pdf

270. BLS, 2015: Economic News Release: Table A. Employment Status of the Civilian Noninstitutional Population by Disability Status and Age, 2012 and 2013 Annual Averages. U.S. Department of Labor, Bureau of Labor Statistics. http://www.bls.gov/news.release/disabl.a.htm

271. Wolbring, G. and V. Leopatra, 2012: Climate change, water, sanitation and energy insecurity: Invisibility of people with disabilities. *Canadian Journal of Disability Studies,* **1,** 66-90. http://dx.doi.org/10.15353/cjds.v1i3.58

272. Lemyre, L., S. Gibson, J. Zlepnig, R. Meyer-Macleod, and P. Boutette, 2009: Emergency preparedness for higher risk populations: Psychosocial considerations. *Radiation Protection Dosimetry,* **134,** 207-214. http://dx.doi.org/10.1093/rpd/ncp084

273. Brunkard, J., G. Namulanda, and R. Ratard, 2008: Hurricane Katrina deaths, Louisiana, 2005. *Disaster Medicine and Public Health Preparedness,* **2,** 215-223. http://dx.doi.org/10.1097/DMP.0b013e31818aaf55

274. Fox, M.H., G.W. White, C. Rooney, and A. Cahill, 2010: The psychosocial impact of Hurricane Katrina on persons with disabilities and independent living center staff living on the American Gulf Coast. *Rehabilitation Psychology,* **55,** 231-240. http://dx.doi.org/10.1037/a0020321

275. Lazrus, H., B.H. Morrow, R.E. Morss, and J.K. Lazo, 2012: Vulnerability beyond stereotypes: Context and agency in hurricane risk communication. *Weather, Climate, and Society,* **4,** 103-109. http://dx.doi.org/10.1175/wcas-d-12-00015.1

276. Nick, G.A., E. Savoia, L. Elqura, M.S. Crowther, B. Cohen, M. Leary, T. Wright, J. Auerbach, and H.K. Koh, 2009: Emergency preparedness for vulnerable populations: People with special health-care needs. *Public Health Reports,* **124,** 338-343. 2646456 http://www.ncbi.nlm.nih.gov/pmc/articles/PMC2646456/pdf/phr124000338.pdf

277. Jan, S. and N. Lurie, 2012: Disaster resilience and people with functional needs. *The New England Journal of Medicine,* **367,** 2272-2273. http://dx.doi.org/10.1056/NEJMp1213492

278. Schifano, P., G. Cappai, M. De Sario, P. Michelozzi, C. Marino, A. Bargagli, and C.A. Perucci, 2009: Susceptibility to heat wave-related mortality: A follow-up study of a cohort of elderly in Rome. *Environmental Health,* **8,** Article 50. http://dx.doi.org/10.1186/1476-069x-8-50

279. Hansen, A.L., P. Bi, P. Ryan, M. Nitschke, D. Pisaniello, and G. Tucker, 2008: The effect of heat waves on hospital admissions for renal disease in a temperate city of Australia. *International Journal of Epidemiology,* **37,** 1359-1365. http://dx.doi.org/10.1093/ije/dyn165

280. Ostro, B., S. Rauch, R. Green, B. Malig, and R. Basu, 2010: The effects of temperature and use of air conditioning on hospitalizations. *American Journal of Epidemiology,* **172,** 1053-1061. http://dx.doi.org/10.1093/aje/kwq231

281. Åström, D.O., F. Bertil, and R. Joacim, 2011: Heat wave impact on morbidity and mortality in the elderly population: A review of recent studies. *Maturitas,* **69,** 99-105. http://dx.doi.org/10.1016/j.maturitas.2011.03.008

282. Kravchenko, J., A.P. Abernethy, M. Fawzy, and H.K. Lyerly, 2013: Minimization of heatwave morbidity and mortality. *American Journal of Preventive Medicine,* **44,** 274-282. http://dx.doi.org/10.1016/j.amepre.2012.11.015

283. Stöllberger, C., W. Lutz, and J. Finsterer, 2009: Heat-related side-effects of neurological and non-neurological medication may increase heatwave fatalities. *European Journal of Neurology,* **16,** 879-882. http://dx.doi.org/10.1111/j.1468-1331.2009.02581.x

284. Nordon, C., K. Martin-Latry, L.d. Roquefeuil, P. Latry, B. Bégaud, B. Falissard, F. Rouillon, and H. Verdoux, 2009: Risk of death related to psychotropic drug use in older people during the European 2003 heatwave: A population-based case–control study. *The American Journal of Geriatric Psychiatry,* **17,** 1059-1067. http://dx.doi.org/10.1097/JGP.0b013e-3181b7ef6e

285. Hausfater, P., B. Megarbane, S. Dautheville, A. Patzak, M. Andronikof, A. Santin, S. André, L. Korchia, N. Terbaoui, G. Kierzek, B. Doumenc, C. Leroy, and B. Riou, 2009: Prognostic factors in non-exertional heatstroke. *Intensive Care Medicine,* **36,** 272-280. http://dx.doi.org/10.1007/s00134-009-1694-y

286. Miller, A.C. and B. Arquilla, 2008: Chronic diseases and natural hazards: Impact of disasters on diabetic, renal, and cardiac patients. *Prehospital and Disaster Medicine,* **23,** 185-194. http://dx.doi.org/10.1017/S1049023X00005835

287. Tofighi, B., E. Grossman, A.R. Williams, R. Biary, J. Rotrosen, and J.D. Lee, 2014: Outcomes among buprenorphine-naloxone primary care patients after Hurricane Sandy. *Addiction Science & Clinical Practice,* **9,** 3. http://dx.doi.org/10.1186/1940-0640-9-3

288. Carlisle Maxwell, J., D. Podus, and D. Walsh, 2009: Lessons learned from the deadly sisters: Drug and alcohol treatment disruption, and consequences from Hurricanes Katrina and Rita. *Substance Use & Misuse,* **44,** 1681-1694. http://dx.doi.org/10.3109/10826080902962011

289. Jhung, M.A., N. Shehab, C. Rohr-Allegrini, D.A. Pollock, R. Sanchez, F. Guerra, and D.B. Jernigan, 2007: Chronic disease and disasters: Medication demands of Hurricane Katrina evacuees. *American Journal of Preventive Medicine,* **33,** 207-210. http://dx.doi.org/10.1016/j.amepre.2007.04.030

290. Davis, J.R., S. Wilson, A. Brock-Martin, S. Glover, and E.R. Svendsen, 2010: The impact of disasters on populations with health and health care disparities. *Disaster Medicine and Public Health Preparedness,* **4,** 30-38. http://dx.doi.org/10.1017/S1935789300002391

291. Preston, B.L., E.J. Yuen, and R.M. Westaway, 2011: Putting vulnerability to climate change on the map: A review of approaches, benefits, and risks. *Sustainability Science,* **6,** 177-202. http://dx.doi.org/10.1007/s11625-011-0129-1

292. ATSDR, 2015: Social Vulnerability Index (SVI) Mapping Dashboard. Agency for Toxic Substances & Disease Registry. http://svi.cdc.gov/map.aspx

293. Ebi, K., P. Berry, D. Campbell-Lendrum, C. Corvalan, and J. Guillemot, 2013: Protecting Health from Climate Change: Vulnerability and Adaptation Assessment. 62 pp. World Health Organization, Geneva. http://www.who.int/globalchange/publications/vulnerability-adaptation/en/

294. Manangan, A.P., C.K. Uejio, S. Saha, P.J. Schramm, G.D. Marinucci, C.L. Brown, J.J. Hess, and G. Luber, 2014: Assessing Health Vulnerability to Climate Change: A Guide for Health Departments. 24 pp. Climate and Health Technical Report Series, Centers for Disease Control and Prevention, Atlanta, GA. http://www.cdc.gov/climateandhealth/pubs/AssessingHealthVulnerabilitytoClimateChange.pdf

295. Friel, S., M. Marmot, A.J. McMichael, T. Kjellstrom, and D. Vågerö, 2008: Global health equity and climate stabilisation: A common agenda. *The Lancet,* **372,** 1677-1683. http://dx.doi.org/10.1016/s0140-6736(08)61692-x

296. Gubler, D.J., P. Reiter, K.L. Ebi, W. Yap, R. Nasci, and J.A. Patz, 2001: Climate variability and change in the United States: Potential impacts on vector- and rodent-borne diseases. *Environmental Health Perspectives,* **109,** 223-233. http://dx.doi.org/10.2307/3435012

297. Smit, B., O. Pilifosova, I. Burton, B. Challenger, S. Huq, R.J.T. Klein, and G. Yohe, 2001: Adaptation to climate change in the context of sustainable development and equity. *Climate Change 2001: Impacts, Adaptation and Vulnerability. Contribution of Working Group II to the Third Assessment Report of the Intergovernmental Panel on Climate Change.* McCarthy, J.J., O.F. Canziani, N.A. Leary, D.J. Dokken, and K.S. White, Eds. Cambridge University Press, Cambridge, UK, 877-912. https://www.ipcc.ch/ipccreports/tar/wg2/pdf/wg2TARchap18.pdf

298. Blaikie, P., T. Cannon, I. Davis, and B. Wisner, 1994: *At Risk: Natural Hazards, People's Vulnerability, and Disasters,* 2nd ed. Routledge, New York, 284 pp.

299. Hutton, D., 2010: Vulnerability of children: More than a question of age. *Radiation Protection Dosimetry,* **142,** 54-57. http://dx.doi.org/10.1093/rpd/ncq200

300. Keim, M.E., 2008: Building human resilience: The role of public health preparedness and response as an adaptation to climate change. *American Journal of Preventive Medicine,* **35,** 508-516. http://dx.doi.org/10.1016/j.amepre.2008.08.022

301. Agarwal, S., J.C. Driscoll, X. Gabaix, and D. Laibson, 2009: The age of reason: Financial decisions over the life cycle and implications for regulation. *Brookings Papers on Economic Activity,* **2009,** 51-117. http://dx.doi.org/10.1353/eca.0.0067

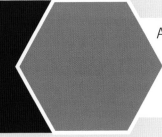

APPENDIX

1 TECHNICAL SUPPORT DOCUMENT: MODELING FUTURE CLIMATE IMPACTS ON HUMAN HEALTH

Lead Authors

Allison Crimmins*
U.S. Environmental Protection Agency

John Balbus
National Institutes of Health

Janet L. Gamble
U.S. Environmental Protection Agency

Contributing Authors

David R. Easterling
National Oceanic and Atmospheric Administration

Kristie L. Ebi
University of Washington

Jeremy Hess
University of Washington

Kenneth E. Kunkel
Cooperative Institute for Climate and Satellites–North Carolina

David M. Mills
Abt Associates

Marcus C. Sarofim
U.S. Environmental Protection Agency

Acknowledgements: Jennifer Parker, Centers for Disease Control and Prevention

Recommended Citation: Crimmins, A., J. Balbus, J.L. Gamble, D.R. Easterling, K.L. Ebi, J. Hess, K.E. Kunkel, D.M. Mills, and M.C. Sarofim, 2016: Appendix 1: Technical Support Document: Modeling Future Climate Impacts on Human Health. *The Impacts of Climate Change on Human Health in the United States: A Scientific Assessment.* U.S. Global Change Research Program, Washington, DC, 287–300. http://dx.doi.org/10.7930/J0KH0K83

*Chapter Coordinator

1 TECHNICAL SUPPORT DOCUMENT: MODELING FUTURE CLIMATE IMPACTS ON HUMAN HEALTH

Models are an important component of climate change impact projections. In general, quantitative evaluations of health impacts require projections of 1) physical climate changes, 2) future socioeconomic characteristics, and 3) the relationships between these factors and the health outcome of interest. Uncertainties exist in each of these areas, and aligning the spatial and temporal parameters used in climate models with epidemiological data to assess health outcomes can be challenging. Despite these challenges, health impact modeling continues to improve, increasing our understanding of the quantitative impacts associated with climate change (for example, Melillo et al. 2014; Tamerius et al. 2007; Post et al. 2012).[1, 2, 3]

A1.1 Quantitative Evaluations of Health Impacts

Projecting Climate Change Impacts

Climate models are used to analyze past changes in the long-term averages and variations in temperature, precipitation, and other climate indicators and to make projections of how these trends may change in the future. Since there is no universally accepted set of metrics to identify the "best" climate models, it is standard practice to use an ensemble (a collection of simulations from different models) in order to present a range of results and provide a measure of the certainty in the results. In addition, because climate model results can depend on initial conditions (the state of the atmosphere and ocean at the moment the modeling run begins), even for a single model, multiple model simulations can be used to similarly present a range of results and improve understanding of variability. Climate model outputs may require additional processing, such as the use of downscaling methods when higher resolutions are needed, or coupling to an atmospheric chemistry model in order to examine and incorporate changes in local air quality.

Projections of climate changes are usually based on scenarios (or sets of assumptions) regarding how future emissions may change as a result of population, energy, technology, and economics. Over the past decade, climate change simulations were based primarily on emissions scenarios developed in the Intergovernmental Panel on Climate Change (IPCC) Special Report on Emissions Scenarios (SRES).[4] These scenarios were used as inputs to climate models in order to develop projections used in the Coupled Model Intercomparison Project Phase 3 (CMIP3). The global climate

model (GCM) simulations included in CMIP use a standard experimental protocol so that their outputs can be compared. The IPCC Fifth Assessment Report[5] drew upon model simulations from the Coupled Model Intercomparison Project Phase 5 (CMIP5), which collected simulation data from more recent models, used Representative Concentration Pathways (RCPs) in place of SRES scenarios, and incorporated updated historical forcing trends and other exogenous model inputs.

CMIP5 contains approximately 60 climate representations from 28 different modeling centers.[6] The spatial resolution of most model grid cells is about 1° to 2° of latitude and longitude, or about 60 to 130 square miles. CMIP5 experiments simulate both

a. the 20th century climate using the best available estimates of the temporal variations in external forcing factors (such as greenhouse gas concentrations, solar output, and volcanic aerosol concentrations); and

b. the 21st century climate based on future greenhouse gas concentration pathways resulting from various emissions scenarios.

Four RCP emissions pathways were used for the CMIP5 simulations: RCP2.6, RCP4.5, RCP6.0, and RCP8.5. These pathways are named according to the increase in radiative forcing (a measure of the total change in Earth's energy balance) projected for that pathway in the year 2100 relative to preindustrial levels, measured in Watts per square meter (Wm^{-2}). For example, RCP6.0 projects that the end-of-century radiative forcing increase will be 6.0 Wm^{-2} above preindustrial levels. The range of simulated global average surface temperature changes under both SRES and RCPs is shown in Figure 1.

Projecting Socioeconomic Development

Along with the RCPs, used to provide a range of possible future greenhouse gas concentrations for climate models, the modeling of climate change impacts can be improved by acknowledging scenarios that describe future societal characteristics. For the IPCC's Fifth Assessment Report,[5] impact modelers discussed the use of new scenarios constructed from three building blocks:

Scenarios of Future Temperature Rise

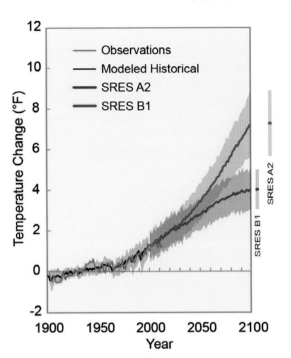
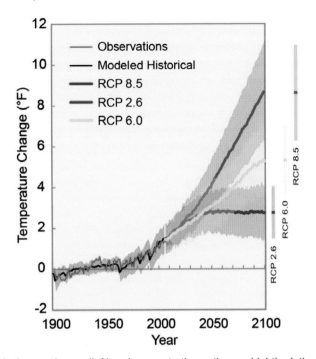

Figure 1: Projected global average temperature rise for specific emissions pathways (left) and concentration pathways (right) relative to the 1901–1960 average. Shading indicates the range (5th to 95th percentile) of results from a suite of climate models. Projections in 2099 are indicated by the bars to the right of each panel. In all cases, temperatures are expected to rise, although the difference between lower and higher pathways is substantial.

The left panel shows the two main CMIP3 scenarios (SRES) used in this assessment: A2 assumes continued increases in emissions throughout this century, and B1 assumes significant emissions reductions beginning around 2050. The right panel shows the newer CMIP5 scenarios using Representative Concentration Pathways (RCPs). CMIP5 includes both lower and higher pathways than CMIP3. The lowest concentration pathway shown here, RCP2.6, assumes immediate and rapid reductions in emissions and would result in about 2.5°F of warming in this century. The highest pathway, RCP8.5, roughly similar to a continuation of the current path of global emissions increases, is projected to lead to more than 8°F warming by 2100, with a high-end possibility of more than 11°F. (Data from CMIP3, CMIP5, and NOAA NCEI). (Figure source: adapted from Melillo et al. 2014)[1]

- Representative Concentration Pathways (RCPs)

- Shared Socioeconomic Pathways (SSPs)

- Shared Climate Policy Assumptions (SPAs)

Shared Socioeconomic Pathways define plausible alternative states of global human and natural societies at a macro scale, including qualitative and quantitative factors such as demographic, political, social, cultural, institutional, lifestyle, economic, and technological variables and trends. Also included are the human impacts on ecosystems and ecosystem services, such as air and water quality.[7, 8, 9]

As with the IPCC Fifth Assessment Report, SSPs are not explicitly used in the analyses highlighted in this assessment. However, because these scenarios are likely to be used by the impacts modeling community over the next few years, placing the approach taken in this assessment in context is a valuable exercise.

Five reference SSPs, referred to as SSP1 through SSP5,[9] describe challenges to adaptation (efforts to adapt to climate change) and mitigation (efforts to reduce the amount of climate change) that change over time irrespective of climate change.[7, 8, 9] Although the SSPs describe broad-scale global trends across multiple sectors, these trends are relevant to projections of health impacts in the United States; trends within each SSP represent different challenges for maintaining and improving the health of Americans. For example, future vulnerability to changing concentrations of air pollutants, particularly ozone, will in part depend on demographics, urbanization, policies to control air pollutants, and hemispheric transport of emissions from areas outside the region.

The combination of RCP6.0 (used by most of the analyses highlighted in the Temperature-Related Death and Illness, Air Quality Impacts, Vector-Borne Diseases, and Water-Related Illness chapters—see Section A1.2) and the population parameters for the SRES B2 emissions pathway (used in the extreme heat and ozone analyses highlighted in Ch. 2: Temperature-Related Death and Illness and Ch. 3: Air Quality Impacts) can be partially mapped to the SSP2 storyline.[9, 10] SSP2 depicts a world where global health improves at an intermediate pace. Under SSP2, multiple factors contribute to some countries making slower progress in reducing health burdens, including, in some low-income countries, high burdens of climate-related diseases combined with moderate to high population growth. In the United States, challenges to public health infrastructure and health care under this socioeconomic pathway could include inadequate resources and international commitment for 1) integrated monitoring and surveillance systems, 2) research on and modeling of the health risks of climate change, 3) iterative management approaches, 4) training and education of health care and public health professionals and practitioners, and 5) technology development and deployment.[7]

The SSPs do not include any explicit climate policy assumptions. This role is reserved for the Shared Climate Policy Assumptions (SPAs) which capture key policy attributes such as the goals, instruments, and obstacles of mitigation and adaptation measures up to the global and century scale.[11] In this way, the SPAs provide the link between RCPs and SSPs by allowing for a variety of alternative socioeconomic evolutionary paths to be coupled with a library of climate model simulations created using the RCPs. SPAs are also not used in the analyses highlighted in this assessment.

Projecting Health Outcomes

Public health officials often require information on health risks at relatively local geographic scales. Climate models, on the other hand, are better at projecting changes on national to global scales and over timescales of decades to centuries. Figure 2 shows two illustrative resolutions for eastern North American topography. The top figure has a grid cell resolution of 68 miles by 68 miles, which is comparable to high resolution global models with projections at a 1° latitude by 1° longitude resolution. The lower figure shows how the same topography would look using smaller grid cells with a resolution of 19 miles by 19 miles. The finer detail at the higher resolution (note, for example, the better representation of the elevation changes of the Appalachian Mountains) would potentially improve a model's ability to provide local information, as temperature, winds, and other features of the model simulation are all influenced by topography. On the other hand, models with higher resolution are not necessarily better at capturing large-scale climate changes and weather patterns.

In addition to higher spatial resolutions, public health officials are also generally most interested in short-term projections of future conditions (for example, one to five years). This is in part due to the fact that these officials work in resource-constrained environments where relative priorities and associated funding decisions can shift, often quickly. In addition, they provide services to populations with characteristics that are likely to change in response to changing economic conditions, immigration patterns, or impacts of extreme weather events. In this short timeframe, public health officials typically focus on information regarding the timing and magnitude of specific events or combinations of events that

Example of Increasing Spatial Resolution of Climate Models

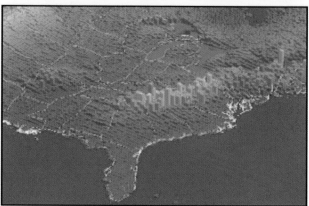

Figure 2: Top: Illustration of eastern North American topography in a resolution of 68 miles x 68 miles (110 x 110 km). Bottom: Illustration of eastern North America at a resolution of 19 miles x 19 miles (30 x 30 km). Global climate models are constantly being enhanced as scientific understanding of climate improves and as computational power increases. For example, in 1990, the average model divided up the world into grid cells measuring more than 300 miles per side. Today, most models divide the world up into grid cells of about 60 to 100 miles per side, and some of the most recent models are able to run short simulations with grid cells of only 15 miles per side. Supercomputer capabilities are the primary limitation on grid cell size. Newer models also incorporate more of the physical processes and components that make up the Earth's climate system. (Figure source: Melillo et al. 2014)[1]

would stress existing programs and systems (for example, heat waves, tropical storms, wildfires, and air quality events). The one- to five-year information requirements of public health providers can contrast with the information climate modelers can develop, which project future conditions for timescales of decades to centuries and often derive impacts in 2050 or 2100. Climate models provide less guidance in terms of changes in near-term impacts because short-term variability from natural sources such as ocean circulation can obscure the long-term climate trends produced by increasing greenhouse gas concentrations. As such, climate projections over longer time periods typically serve more as a guide to emerging issues and as an input to longer-range planning.

A1.2   Modeling Highlighted in the Assessment

The four chapters that highlight modeling studies conducted for this assessment (Temperature-Related Death and Illness, Air Quality Impacts, Vector-Borne Diseases, and Water-Related Illness) analyzed a subset of the full CMIP5 dataset (see Table 1). The air quality analyses required the most intensive processing of the CMIP5 model output; calculating air quality changes at the appropriate geographic scale requires modelers to use a technique known as dynamical downscaling to generate climate data at the desired small-scale resolution, and then run an atmospheric chemistry model, both of which are computationally intensive processes. Thus the ozone analysis was limited to two model–scenario examples (see Table 1). By contrast, the water-related illness analyses examined results from 21 of the CMIP5 models, though only for one particular scenario.

In general, the authors of the studies highlighted in this assessment used historical data in order to calibrate their historical results and to improve geographic resolution. These downscaling approaches determine the climate signal by taking the difference between the modeled future and the modeled historical period at the grid cell resolution (often averaged over 30 years). This climate signal can then be added to observed historical data at a resolution potentially much finer than the model grid cell scale. For example, any given weather station might be, on average, cooler in the summer than the grid cell average because it is located next to a lake. By adding the modeled climate signal to the historical data from the weather station, the projected future temperatures can more effectively account for microclimate effects, from lakes or hills for example, that are consistent with historical variation at a spatial resolution smaller than the modeled grid scale. More sophisticated calibrations can also adjust model

variability to match historical variability by using a technique known as quantile mapping.[12]

The modeling studies highlighted in this assessment use several approaches. The three different historical reference periods used in the highlighted studies (1985–2000, 1992–2007, and 1976–2006) are slightly warmer than the 1971–2000 period used in the 2014 National Climate Assessment (NCA), by 0.3°F to 0.8°F. In addition, different sets of climate models were used.

A sensitivity analysis was conducted to test for two potential impacts of using different modeling approaches: the use of different historical reference periods and the use of different sets of CMIP5 models. Figure 3 shows the change in temperature from the 2014 NCA reference period (1971–2000) for three historical reference periods used in the studies highlighted (first column). The differences among these three historical reference periods are small compared with the warming projected for the middle of this century by the different sets of models used (second column). For the sets of 21, 11, and 5 models used in the studies of *Vibrio/Alexandrium* species, *Gambierdiscus* species, and Lyme disease, respectively, the multi-model mean warming for the middle of the 21st century are within 0.5°F of each other, although the set of 11 models does not include a few of the cooler models and the set of 5 models spans a narrower range. The two models used in the extreme temperature study are slightly warmer than the mean of the entire set of models, while the single model used in the air quality (ozone) study is slightly cooler. However, these differences in mean warming among the five approaches shown in the second column are small compared to projected warming.

Sensitivity Analysis of Differences in Modeling Approaches

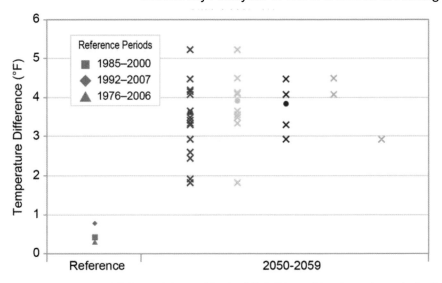

Figure 3: A sensitivity analysis was conducted to test for potential impacts of differences in the modeling approaches (use of different historical reference periods and use of different sets of CMIP5 models) in the research studies highlighted in this assessment (see Research Highlights in Chapters 2, 3, 5, and 6). The values in the first column are temperature changes for three different reference periods used in this assessment, relative to the 1971–2000 reference period used in the 2014 NCA. The sets of values in the second column show future temperature changes for individual climate models for 2050–2059, relative to 1971–2000, for those studies that used the RCP6.0 scenario.

From left to right, the vertical sets of values represent (a) 21 models used in the *Vibrio/Alexandrium* bacteria study (red), (b) 11 models used in the *Gambierdiscus* study (green), (c) the 5 models used in the Lyme disease study (purple), (d) the 2 models used in the extreme temperature study (blue), and (e) the single model used in the air quality study (orange). Each "x" represents a single model. The filled-in circle is the mean temperature change for all models in the column. (Figure source: NOAA NCEI / CICS-NC)

Each modeling approach requires different input from the climate models. For example, the temperature mortality analysis required only temperature data, while the analysis in the Water-Related Illness chapter used sea surface temperature data. However, the ambient air quality modeling required temperature, precipitation, ventilation, and other data in order to provide boundary conditions for the dynamical downscaling approach. Besides climate data, modeling teams also used other inputs. The main sources of additional data were the Integrated Climate and Land Use Scenarios (ICLUS) model for population projections and the Environmental Benefits Mapping and Analysis Program (BenMAP) model for baseline mortality data, which were used for the extreme temperature and air quality modeling efforts.[13, 14] The analysis in the Water-Related Illness chapter required salinity, light, and other oceanographic data not provided by the CMIP5 models.

Table 1: See Research Highlights in Ch. 2: Temperature-Related Death and Illness; Ch. 3: Air Quality Impacts; Ch. 6: Water-Related Illness; Ch. 5: Vector-Borne Diseases.

Chapter	Modeled Endpoint	Time-frame	Temporal Resolution	Scenarios/Pathways	Models	Bias Correction and/or Downscaling	Geographic Scope	Climate Variables	Additional Data Inputs
Temperature-Related Death and Illness	Mortality[15]	2030, 2050, 2100	30 years	RCP6.0	GFDL–CM3, MIROC5	Statistical downscaling, then delta approach	209 U.S. cities	Temperature (0–5 day lags)	BenMAP baseline mortality data
Air Quality	Mortality/Morbidity from changes in ozone[16]	2030	3 years within 11 year span	RCP6.0	GISS-E2	Dynamic downscaling	National	Temperature, precipitation, ventilation, others	ICLUS population data, BenMAP health model, SES, air condition prevalence, baseline health status data
		2030	11 year average	RCP8.5	CESM	Dynamic downscaling	National	Temperature, precipitation, ventilation, others	
	Changes in air exchange that drive indoor air quality[17]	2040–70	30 years	SRES A2	CCSM, CGM3, GFDL, HadCM3	Dynamic downscaling	9 U.S. cities	Temperature, wind speed at 3-hour resolution	NA
Water-Related Illness	*Vibrio* bacteria seasonality[18]	2030, 2050, 2095	Decadal average of monthly data	RCP6.0	21 CMIP5 models	Statistical downscaling; mean and variance bias correction	Chesapeake Bay	SST (driven by surface air temperature)	NA
	Vibrio bacteria geographic range[18]	2030, 2050, 2090	Decadal average for August	RCP6.0	4 CMIP5 models	Statistical downscaling; mean and variance bias correction	Alaskan coast	SST (driven by surface air temperature)	NA
	Alexandrium bloom seasonality[18]	2030, 2050, 2095	Decadal average of monthly data	RCP6.0	21 CMIP5 models	Statistical downscaling; mean and variance bias correction	Puget Sound	SST (driven by surface air temperature)	NA
	Growth rates of 3 *Gambierdiscus* algae species[19]	2000–2099	Annual	RCP6.0	11 CMIP5 models	Mean and variance bias correction, then temporal disaggregation	Gulf of Mexico and Caribbean	SST	Salinity, light, and other growth variables
Vector-Borne Disease	Lyme disease onset week[20]	2025–2040 and 2065–2080	16 year periods	RCP2.6, RCP4.5, RCP6.0, RCP8.5	CESM1 (CAM5), GFDL–CM3, GISS–E2–R, HadGEM2-ES, MIROC5	Statistical downscaling, then delta approach	12 U.S. states where Lyme is prevalent	Temp (growing degree days) precipitation, and saturation deficit (assume constant relative humidity)	Distance to coast in decimal degrees

The modeling approaches also included different geographic scales. The Water-Related Illness analyses examined individual bodies of water such as the Chesapeake Bay, Puget Sound, and the Gulf of Mexico. The vector-borne disease projections of Lyme disease concentrated on the 12 U.S. states where Lyme disease is most prevalent. The temperature mortality analysis examined 209 U.S. cities that had sufficient data for a historical epidemiology analysis. The ozone analysis was able to address the entire contiguous United States.

A1.3   Sources of Uncertainty

The use of the term "uncertainty" in climate assessments refers to a range of possible futures. Uncertainty about the future climate arises from the complexity of the climate system and the ability of models to represent timing, magnitude, and location of changes, as well as the difficulties in predicting the decisions that society will make. There is also uncertainty about how climate change, in combination with other stressors, will affect people and natural systems.[1]

Though quantitative evaluations of climate change impacts on human health are continually improving, there is always some degree of uncertainty when using models to gain insight into future conditions (see Figure 4). The presence of uncertainty, or the fact that there is a range in potential outcomes, does not negate the knowledge we have, nor does it mean that actions cannot be taken. Everyone makes decisions, in all aspects of their life, based on limited knowledge or certainty about the future. Decisions like where to go to college or what job to take, what neighborhood to live in or which restaurant to eat in, whom to befriend or marry, and so on are all made in light of uncertainty, which can sometimes be considerable. Recent years have seen considerable progress in the development of improved methods to describe and deal with uncertainty in modeling climate change impacts on human health (for example, Melillo et al. 2014; Tamerius et al. 2007; Post et al. 2012).[1,2,3]

Uncertainty in Projecting Climate Change

Two of the key uncertainties in projecting future global temperatures are 1) uncertainty about future concentrations of greenhouse gases, and 2) uncertainty about how much warming will occur for a given increase in greenhouse gas concentrations. Future concentrations depend on both future emissions and how long these emissions remain in the atmosphere (which can vary depending on how natural systems process those emissions). Because of uncertainty in future greenhouse gas concentrations, climate modelers analyze multiple future emissions pathways in order to determine the range of varying impacts of lower emissions compared to higher emissions. In terms of how much warming will occur for a given increase in greenhouse gas concentrations, the most recent assessment by the IPCC found the most likely response of the climate system to a doubling of carbon dioxide (CO_2) concentrations, referred to as the sensitivity in climate models, lies between a 1.5°C and 4.5°C (2.7°F to 8.1°F) increase in global average temperature (see Figure 1).[5]

Climate scientists have greater confidence in predicting the average temperature of the whole planet than what the temperature will be in any given region or locale. Global average temperatures may not, however, be particularly informative for determining health impacts at a local scale. An increase in global temperatures will, at local scales, result in different warming rates in different locations, different seasonal warming rates, different warming rates during the day compared to the night, and different changes in day-to-day or year-to-year variability. Despite these possible differences, it is highly likely that warming will occur almost everywhere.[21] In addition to temperature, changes in precipitation, humidity, and weather systems are all important drivers of local impacts. However, future changes in these variables are less certain than changes in temperature.

Uncertainty in Public Health Surveillance and Monitoring

Improvements in understanding future health impacts can result from better understanding current health impacts. Obtaining this understanding is complicated by the fact that, in the United States, there is no single source of health data and surveillance often involves acquiring, analyzing, and interpreting data from several sources collected using potentially different techniques and systems.[22, 23] This is further complicated by a number of additional limitations, including the fact that data are often incomplete, may not include a representative sample of all members of society, and rely on reporting of disease status. Estimates of disease patterns or trends may also vary across geographic locations.[23] Understanding the surveillance and monitoring limitations regarding population health data and spatial variability can enable more accurate estimations of the confidence in the links between health impacts and climate drivers, and this can be used to estimate uncertainty in future projections of health impacts.

Having complete socioeconomic, geographic, demographic, and health data at an individual level for everyone would improve our understanding of connections between these attributes and deaths and illnesses. However, such complete data are not available for both practical and confidentiality reasons. Mandatory reporting, disease records, and administrative sources, including data from medical records or vital records, can be used to estimate counts of given health impacts and these counts can be divided by population estimates to produce health impact rates. Uncertainty in the data can differ depending on the type of population health estimate and the existing surveillance data source used (such as using registries versus surveys).

In addition to uncertainty regarding the quality and usefulness of data, confidence in estimates of health impacts depends on the extent of useable data. In general, the larger the data set (larger

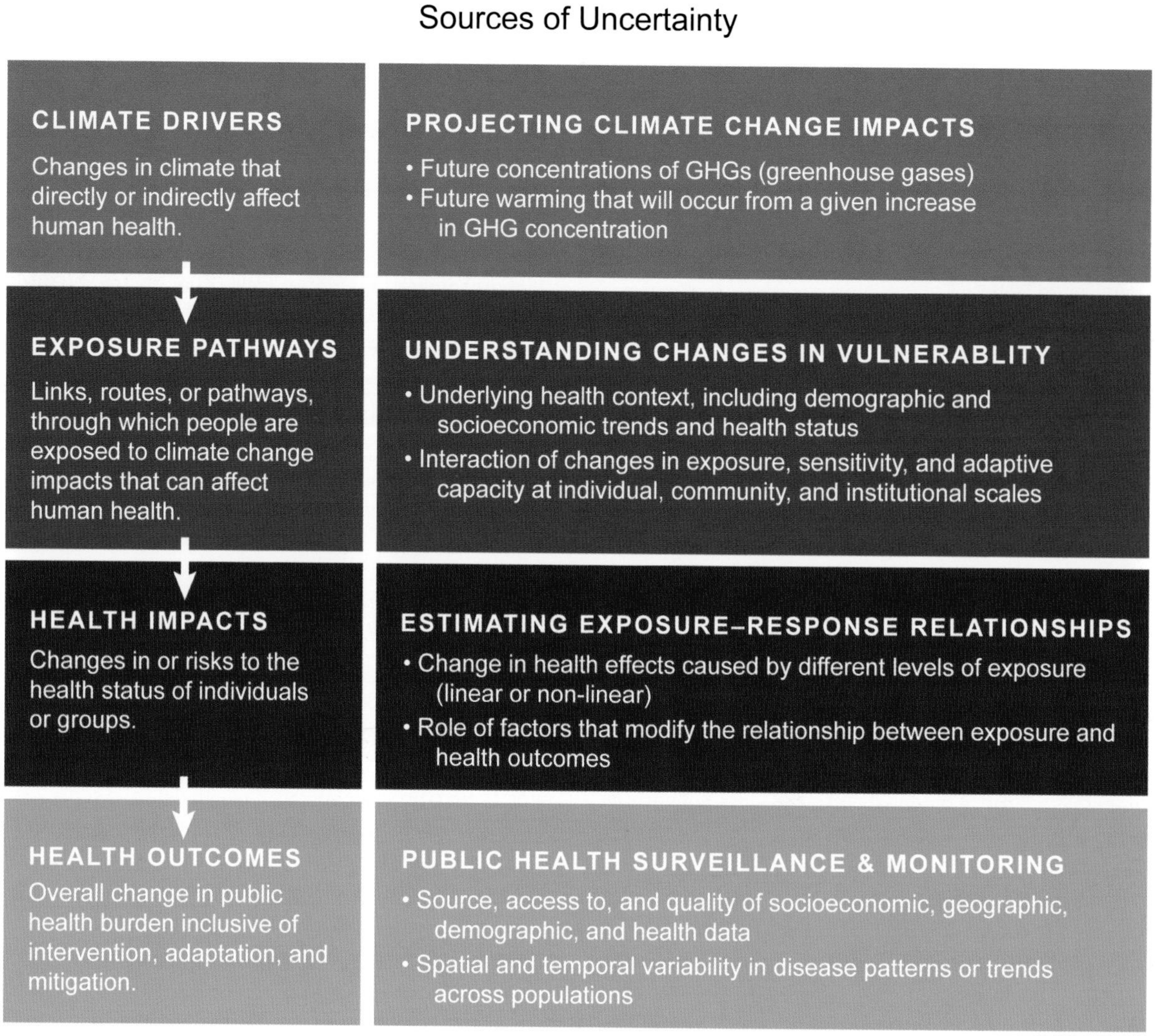

Sources of Uncertainty

CLIMATE DRIVERS

Changes in climate that directly or indirectly affect human health.

PROJECTING CLIMATE CHANGE IMPACTS

• Future concentrations of GHGs (greenhouse gases)
• Future warming that will occur from a given increase in GHG concentration

EXPOSURE PATHWAYS

Links, routes, or pathways, through which people are exposed to climate change impacts that can affect human health.

UNDERSTANDING CHANGES IN VULNERABLITY

• Underlying health context, including demographic and socioeconomic trends and health status
• Interaction of changes in exposure, sensitivity, and adaptive capacity at individual, community, and institutional scales

HEALTH IMPACTS

Changes in or risks to the health status of individuals or groups.

ESTIMATING EXPOSURE–RESPONSE RELATIONSHIPS

• Change in health effects caused by different levels of exposure (linear or non-linear)
• Role of factors that modify the relationship between exposure and health outcomes

HEALTH OUTCOMES

Overall change in public health burden inclusive of intervention, adaptation, and mitigation.

PUBLIC HEALTH SURVEILLANCE & MONITORING

• Source, access to, and quality of socioeconomic, geographic, demographic, and health data
• Spatial and temporal variability in disease patterns or trends across populations

Figure 4: Examples of sources of uncertainty in projecting impacts of climate change on human health. The left column illustrates the exposure pathway through which climate change can affect human health. The right column lists examples of key sources of uncertainty surrounding effects of climate change at each stage along the exposure pathway.

populations or longer time periods), and the more common the health condition, the more confidence there is in estimated rates, and changes in those rates, across time periods, demographic groups, or other attributes.[22]

Uncertainty in Estimating Stressor-Response Relationships

Exposure–response or stressor–response relationships describe the change in the health status associated with different levels of exposure to a stressor or concentration of a stressor (also see Ch. 1: Introduction, Section 1.4). Some environmental exposures, such as air quality and ambient temperature, have a relatively direct effect on deaths and illness, which is captured in stressor-response relationships in epidemiological studies. For example, increases in temperature can affect a range of chronic illnesses and infectious diseases. In other situations, climate change will have health effects through intermediaries such as changes in ecological conditions like pollen distribution (causing allergies)

and the distribution of infectious disease pathogens and vectors (causing vector-borne, foodborne, and waterborne infectious diseases). Modeling exposure–response relationships can be particularly challenging for outcomes involving multiple intermediary stressors along an exposure pathway, each of which may be influenced by climate change. Even for relatively direct impacts, the same exposure can produce different responses for different health outcomes. Moreover, responses for a given exposure can vary by location (for example, different impacts of extreme heat in dry areas versus humid areas) and across sub-populations (different socioeconomic and demographic groups). For each pairing of exposure and health response, the exposure–response relationship may be represented as a quantitative estimate (such as the increase in number of deaths for a 1°F increase in maximum temperature) or in a qualitative manner (such as a determination that increases in extreme precipitation events may increase exposure to indoor molds).

In recent decades, progress has been made in modeling exposure–response relationships for a wide range of climate-sensitive environmental exposures and health responses. For example, we have gained a better understanding in recent years of the relationships between exposure to varying temperatures, concentrations of ozone and fine particulate matter, and the health response in terms of a range of illnesses and premature death (for example, Samoili et al. 2005. Bobb et al. 2014; see also Ch. 2: Temperature-Related Death and Illness and Ch. 3: Air Quality Impacts).[24, 25] Quantitative exposure–response functions are often used in understanding how health risks from these exposures vary across locations; these are also used in modeling efforts to project the health impacts of climate change in specific locations. However, it is important to carefully consider uncertainty when developing and using exposure–response functions, as the environmental processes affecting human health are complex.

Exposure–response functions may not remain constant over time or space. One source of uncertainty arises from the potential that high levels of exposure could be associated with proportionately larger effects compared to low levels of exposure (non-linearity, see for example Gasparrini 2014 and Burnett et al 2014).[26, 27] Further, as the nature of the exposure and the potential for changes in human behavior and adaptive capacity change over time, so can the response function change. Representing health response for a singular point estimate of exposure instead of a range of exposure values could lead to imprecise assessment of the health risk. The large amounts of data required for reliable and accurate estimation of exposure–response functions may not be available at suitable resolutions for all locations (for example, Hubbell et al 2009).[28] In some cases, estimating health outcomes by using exposure–response functions from other locations in the absence of reliable locally specific exposure–response relationships introduces uncertainty (for example, Wardekker et al. 2012).[29] The exposure–response estimates may also vary within sub-populations in a location, being relatively high for particularly vulnerable communities (for example, the elderly population will have a higher exposure–response relationship from extreme heat compared to the rest of the population).

Another challenge in characterizing the relationship between exposure and health impacts is determining when a relationship is correlative, as opposed to causative. For example, statistical analyses would adjust for other factors that could be influencing health outcomes, such as age, race, year, day of the week, insurance status, and the concentrations of other air pollutants. Evaluating and integrating evidence across epidemiological, toxicological, and controlled human exposure studies allows researchers to conclude whether there is a causal relationship between human exposure to air pollution and a given health outcome. As evidence mounts, as is the case for associations between ozone concentration and adverse health impacts,[30, 31, 32, 33, 34, 35] the hypothesis of a causal relationship is strengthened, and observed exposure–response associations can be used with greater confidence.

Users of exposure–response relationships in risk assessments or disease burden projection need to carefully consider the context in which the estimates were derived prior to their use. Carefully designed meta-analyses, leveraging the information obtained from multiple studies, can provide summary estimates of relationships and ensure consistency in application (for example, Normand 1999).[36]

Approach to Reporting Uncertainty in Key Findings

Despite the sources of uncertainty described above, the current state of the science allows an examination of the likely direction of and trends in the health impacts of climate change. Over the past ten years, the models used for climate and health assessments have become more useful and more accurate (for example, Melillo et al. 2014; Tamerius et al. 2007; Post et al. 2012).[1, 2, 3] This assessment builds on that improved capability. A more detailed discussion of the approaches to addressing uncertainty from the various sources can be found in the Guide to the Report (Front Matter) and Appendix 4: Documenting Uncertainty: Confidence and Likelihood.

Two kinds of language are used when describing the uncertainty associated with specific statements in this report: confidence language and likelihood language. Confidence in the validity of a finding is based on the type, amount, quality, strength, and consistency of evidence and the degree of expert agreement on the finding. Confidence is expressed qualitatively and ranges from low confidence (inconclusive evidence or disagreement among experts) to very high confidence (strong evidence and high consensus).

Likelihood language describes the likelihood of occurrence based on measures of uncertainty expressed probabilistically (in other words, based on statistical analysis of observations or model results or on expert judgment). Likelihood, or the probability of an impact, is a term that allows a quantitative estimate of uncertainty to be associated with projections. Thus likelihood statements have a specific probability associated with them, ranging from very unlikely (less than or equal to a 1 in 10 chance of the outcome occurring) to very likely (greater than or equal to a 9 in 10 chance). The likelihood rating does not consider severity of the health risk or outcome, particularly as it relates to health risk factors not associated with climate change, unless otherwise stated in the Key Finding.

Each Key Finding includes confidence levels; where possible, separate confidence levels are reported for 1) the impact of climate change, 2) the resulting change in exposure or risk, and 3) the resulting change in health outcomes. Where projections can be quantified, both a confidence and likelihood level are reported. Determination of confidence and likelihood language involves the expert assessment and consensus of the chapter author teams. The author teams determine the appropriate level of confidence or likelihood by assessing the available literature, determining the quality and quantity of available evidence, and evaluating the level of agreement across different studies. Often, the underlying studies will provide their own estimates of uncertainty and confidence intervals. When available, these confidence intervals are used by the chapter authors in making their own expert judgments.

DOCUMENTING UNCERTAINTY

This assessment relies on two metrics to communicate the degree of certainty in Key Findings. See Appendix 4: Documenting Uncertainty for more on assessments of likelihood and confidence.

Confidence Level

Very High	
Strong evidence (established theory, multiple sources, consistent results, well documented and accepted methods, etc.), high consensus	
High	
Moderate evidence (several sources, some consistency, methods vary and/or documentation limited, etc.), medium consensus	
Medium	
Suggestive evidence (a few sources, limited consistency, models incomplete, methods emerging, etc.), competing schools of thought	
Low	
Inconclusive evidence (limited sources, extrapolations, inconsistent findings, poor documentation and/or methods not tested, etc.), disagreement or lack of opinions among experts	

Likelihood

Very Likely	
≥ 9 in 10	
Likely	
≥ 2 in 3	
As Likely As Not	
≈ 1 in 2	
Unlikely	
≤ 1 in 3	
Very Unlikely	
≤ 1 in 10	

References

1. Melillo, J.M., T.C. Richmond, and G.W. Yohe, Eds., 2014: *Climate Change Impacts in the United States: The Third National Climate Assessment*. U.S. Global Change Research Program, Washington, D.C., 842 pp. http://dx.doi.org/10.7930/J0Z31WJ2

2. Post, E.S., A. Grambsch, C. Weaver, P. Morefield, J. Huang, L.-Y. Leung, C.G. Nolte, P. Adams, X.-Z. Liang, J.-H. Zhu, and H. Mahone, 2012: Variation in estimated ozone-related health impacts of climate change due to modeling choices and assumptions. *Environmental Health Perspectives*, **120**, 1559-1564. http://dx.doi.org/10.1289/ehp.1104271

3. Tamerius, J.D., E.K. Wise, C.K. Uejio, A.L. McCoy, and A.C. Comrie, 2007: Climate and human health: Synthesizing environmental complexity and uncertainty. *Stochastic Environmental Research and Risk Assessment*, **21**, 601-613. http://dx.doi.org/10.1007/s00477-007-0142-1

4. IPCC, 2000: Special Report on Emissions Scenarios. A Special Report of Working Group III of the Intergovernmental Panel on Climate Change. Nakicenovic, N. and R. Swart (Eds.), 570 pp. Cambridge University Press, Cambridge, UK. http://www.ipcc.ch/ipccreports/sres/emission/index.php?idp=0

5. IPCC, 2013: Climate Change 2013: The Physical Science Basis. Contribution of Working Group I to the Fifth Assessment Report of the Intergovernmental Panel on Climate Change. Stocker, T.F., D. Qin, G.-K. Plattner, M. Tignor, S.K. Allen, J. Boschung, A. Nauels, Y. Xia, V. Bex, and P.M. Midgley (Eds.), 1535 pp. Cambridge University Press, Cambridge, UK and New York, NY. http://www.climatechange2013.org

6. Meehl, G.A., L. Goddard, J. Murphy, R.J. Stouffer, G. Boer, G. Danabasoglu, K. Dixon, M.A. Giorgetta, A.M. Greene, E. Hawkins, G. Hegerl, D. Karoly, N. Keenlyside, M. Kimoto, B. Kirtman, A. Navarra, R. Pulwarty, D. Smith, D. Stammer, and T. Stockdale, 2009: Decadal Prediction. *Bulletin of the American Meteorological Society*, **90**, 1467-1485. http://dx.doi.org/10.1175/2009bams2778.1

7. Ebi, K.L., 2014: Health in the new scenarios for climate change research. *International Journal of Environmental Research and Public Health*, **11**, 30-46. http://dx.doi.org/10.3390/ijerph110100030

8. O'Neill, B.C., E. Kriegler, K. Riahi, K.L. Ebi, S. Hallegatte, T.R. Carter, R. Mathur, and D.P. van Vuuren, 2014: A new scenario framework for climate change research: The concept of shared socioeconomic pathways. *Climatic Change*, **122**, 387-400. http://dx.doi.org/10.1007/s10584-013-0905-2

9. O'Neill, B.C., E. Kriegler, K.L. Ebi, E. Kemp-Benedict, K. Riahi, D.S. Rothman, B.J. van Ruijven, D.P. van Vuuren, J. Birkmann, K. Kok, M. Levy, and W. Solecki, 2015: The roads ahead: Narratives for shared socioeconomic pathways describing world futures in the 21st century. *Global Environmental Change*, **In press.** http://dx.doi.org/10.1016/j.gloenvcha.2015.01.004

10. van Vuuren, D.P., E. Kriegler, B.C. O'Neill, K.L. Ebi, K. Riahi, T.R. Carter, J. Edmonds, S. Hallegatte, T. Kram, R. Mathur, and H. Winkler, 2014: A new scenario framework for climate change research: Scenario matrix architecture. *Climatic Change*, **122**, 373-386. http://dx.doi.org/10.1007/s10584-013-0906-1

11. Kriegler, E., J. Edmonds, S. Hallegatte, K.L. Ebi, T. Kram, K. Riahi, H. Winkler, and D.P. van Vuuren, 2014: A new scenario framework for climate change research: The concept of shared climate policy assumptions. *Climatic Change*, **122**, 401-414. http://dx.doi.org/10.1007/s10584-013-0971-5

12. Wood, A.W., L.R. Leung, V. Sridhar, and D.P. Lettenmaier, 2004: Hydrologic implications of dynamical and statistical approaches to downscaling climate model outputs. *Climatic Change*, **62**, 189-216. http://dx.doi.org/10.1023/B:CLIM.0000013685.99609.9e

13. EPA, 2009: Land-Use Scenarios: National-Scale Housing-Density Scenarios Consistent with Climate Change Storylines. EPA/600/R-08/076F, 137 pp. U.S. Environmental Protection Agency, National Center for Environmental Assessment, Global Change Research Program, Washington D.C. http://cfpub.epa.gov/ncea/cfm/recordisplay.cfm?deid=203458

14. EPA, 2014: Environmental Benefits Mapping and Analysis Program--Community Edition (BenMAP-CE). U.S. Environmental Protection Agency. http://www2.epa.gov/benmap

15. Schwartz, J.D., M. Lee, P.L. Kinney, S. Yang, D. Mills, M. Sarofim, R. Jones, R. Streeter, A. St. Juliana, J. Peers, and R.M. Horton, 2015: Projections of temperature-attributable premature deaths in 209 U.S. cities using a cluster-based Poisson approach. *Environmental Health*, **14**. http://dx.doi.org/10.1186/s12940-015-0071-2

16. Fann, N., C.G. Nolte, P. Dolwick, T.L. Spero, A. Curry Brown, S. Phillips, and S. Anenberg, 2015: The geographic distribution and economic value of climate change-related ozone health impacts in the United States in 2030. *Journal of the Air & Waste Management Association*, **65**, 570-580. http://dx.doi.org/10.1080/10962247.2014.996270

17. Ilacqua, V., J. Dawson, M. Breen, S. Singer, and A. Berg, 2015: Effects of climate change on residential infiltration and air pollution exposure. *Journal of Exposure Science and Environmental Epidemiology*, Published online 27 May 2015. http://dx.doi.org/10.1038/jes.2015.38

18. Jacobs, J., S.K. Moore, K.E. Kunkel, and L. Sun, 2015: A framework for examining climate-driven changes to the seasonality and geographical range of coastal pathogens and harmful algae. *Climate Risk Management*, **8**, 16-27. http://dx.doi.org/10.1016/j.crm.2015.03.002

19. Kibler, S.R., P.A. Tester, K.E. Kunkel, S.K. Moore, and R.W. Litaker, 2015: Effects of ocean warming on growth and distribution of dinoflagellates associated with ciguatera fish poisoning in the Caribbean. *Ecological Modelling*, **316**, 194-210. http://dx.doi.org/10.1016/j.ecolmodel.2015.08.020

20. Monaghan, A.J., S.M. Moore, K.M. Sampson, C.B. Beard, and R.J. Eisen, 2015: Climate change influences on the annual onset of Lyme disease in the United States. *Ticks and Tick-Borne Diseases*, **6**, 615-622. http://dx.doi.org/10.1016/j.ttbdis.2015.05.005

21. Walsh, J., D. Wuebbles, K. Hayhoe, J. Kossin, K. Kunkel, G. Stephens, P. Thorne, R. Vose, M. Wehner, J. Willis, D. Anderson, S. Doney, R. Feely, P. Hennon, V. Kharin, T. Knutson, F. Landerer, T. Lenton, J. Kennedy, and R. Somerville, 2014: Ch. 2: Our changing climate. *Climate Change Impacts in the United States: The Third National Climate Assessment*. Melillo, J.M., T.C. Richmond, and G.W. Yohe, Eds. U.S. Global Change Research Program, Washington, D.C., 19-67. http://dx.doi.org/10.7930/J0KW5CXT

22. Klein, R.J., S.E. Proctor, M.A. Boudreault, and K.M. Turczyn, 2002: Healthy People 2010 Criteria for Data Suppression. Statistical Notes No. 24, 12 pp. National Center for Health Statistics, Hyattsville, MD. http://198.246.124.22/nchs/data/statnt/statnt24.pdf

23. NCHS, 2015: Health, United States, 2014: With Special Feature on Adults Aged 55-64. 473 pp. National Center for Health Statistics, Centers for Disease Control and Prevention, Hyattsville, MD. http://www.cdc.gov/nchs/data/hus/hus14.pdf

24. Bobb, J.F., R.D. Peng, M.L. Bell, and F. Dominici, 2014: Heat-related mortality and adaptation to heat in the United States. *Environmental Health Perspectives*, **122**, 811-816. http://dx.doi.org/10.1289/ehp.1307392

25. Samoli, E., A. Analitis, G. Touloumi, J. Schwartz, H.R. Anderson, J. Sunyer, L. Bisanti, D. Zmirous, J.M. Vonk, J. Pekkanen, P. Goodman, A. Paldy, C. Schindler, and K. Katsouyanni, 2005: Estimating the exposure-response relationships betwen particulate matter and mortality within the APHEA multicity project. *Environmental Health Perspectives*, **113**, 88-95. http://dx.doi.org/10.1289/ehp.7387

26. Gasparrini, A., 2014: Modeling exposure-lag-response associations with distributed lag non-linear models. *Statistics in Medicine*, **33**, 881-889. http://dx.doi.org/10.1002/sim.5963

27. Burnett, R.T., C.A. Pope, III, M. Ezzati, C. Olives, S.S. Lim, S. Mehta, H.H. Shin, G. Singh, B. Hubbell, M. Brauer, H.R. Anderson, K.R. Smith, J.R. Balmes, N.G. Bruce, H. Kan, F. Laden, A. Pruss-Ustun, M.C. Turner, S.M. Gapstur, W.R. Diver, and A. Cohen, 2014: An integrated risk function for estimating the global burden of disease attributable to ambient fine particulate matter exposure. *Environmental Health Perspectives*, **122**, 397-403. http://dx.doi.org/10.1289/ehp.1307049

28. Hubbell, B., N. Fann, and J. Levy, 2009: Methodological considerations in developing local-scale health impact assessments: Balancing national, regional, and local data. *Air Quality, Atmosphere & Health*, **2**, 99-110. http://dx.doi.org/10.1007/s11869-009-0037-z

29. Wardekker, J.A., A. de Jong, L. van Bree, W.C. Turkenburg, and J.P. van der Sluijs, 2012: Health risks of climate change: An assessment of uncertainties and its implications for adaptation policies. *Environmental Health*, **11**, Article 67. http://dx.doi.org/10.1186/1476-069x-11-67

30. Bell, M.L., A. McDermott, S.L. Zeger, J.M. Samet, and F. Dominici, 2004: Ozone and short-term mortality in 95 US urban communities, 1987-2000. *JAMA: The Journal of the American Medical Association*, **292**, 2372-8. http://dx.doi.org/10.1001/jama.292.19.2372

31. Jerrett, M., R.T. Burnett, C.A. Pope, K. Ito, G. Thurston, D. Krewski, Y. Shi, E. Calle, and M. Thun, 2009: Long-term ozone exposure and mortality. *The New England Journal of Medicine*, **360**, 1085-1095. http://dx.doi.org/10.1056/NEJMoa0803894

32. Ji, M., D.S. Cohan, and M.L. Bell, 2011: Meta-analysis of the association between short-term exposure to ambient ozone and respiratory hospital admissions. *Environmental Research Letters*, **6**, 024006. http://dx.doi.org/10.1088/1748-9326/6/2/024006

33. Fann, N., A.D. Lamson, S.C. Anenberg, K. Wesson, D. Risley, and B.J. Hubbell, 2012: Estimating the national public health burden associated with exposure to ambient PM2.5 and ozone. *Risk Analysis,* **32,** 81-95. http://dx.doi.org/10.1111/j.1539-6924.2011.01630.x

34. Vinikoor-Imler, L.C., E.O. Owens, J.L. Nichols, M. Ross, J.S. Brown, and J.D. Sacks, 2014: Evaluating potential response-modifying factors for associations between ozone and health outcomes: A weight-of-evidence approach. *Environmental Health Perspectives,* **122,** 1166-1176. http://dx.doi.org/10.1289/ehp.1307541

35. EPA, 2013: Integrated Science Assessment for Ozone and Related Photochemical Oxidants. EPA 600/R-10/076F, 1251 pp. U.S. Environmental Protection Agency, National Center for Environmental Assessment, Office of Research and Development, Research Triangle Park, NC. http://cfpub.epa.gov/ncea/isa/recordisplay.cfm?deid=247492

36. Normand, S.-L.T., 1999: Meta-analysis: Formulating, evaluating, combining, and reporting. *Statistics in Medicine,* **18,** 321-359. http://dx.doi.org/10.1002/(SICI)1097-0258(19990215)18:3%3C321::AID-SIM28%3E3.0

2 PROCESS FOR LITERATURE REVIEW

The systematic literature review included a comprehensive search of the literature, collection and incorporation of information submitted by the public, screening and assessment of the eligibility of the collected literature, and synthesis of the collected literature. Authors were provided with detailed guidance, including Information Quality Act (IQA) procedures and the following process for the literature review.

A2.1 Identification of Literature Sources

The sources of literature and information assessed for this report were derived from a comprehensive literature search conducted by the National Institute of Environmental Health Sciences (NIEHS), literature submitted for consideration during public engagement opportunities, references included in the Third National Climate Assessment (2014 NCA),[1] and additional sources of information or data identified by the chapter authors.

NIEHS, coordinating closely with the Interagency Crosscutting Group on Climate Change and Human Health (CCHHG), developed an updated (2012–2014) Health Sector Literature Review and Bibliography as part of the larger literature review for the 2014 NCA. The NIEHS search covered multiple electronic databases (such as PubMed, Scopus, and Web of Science) as well as web search engines such as Google Scholar. Overall, searches were limited to publication dates of 2007 or later and to English-language citations. NIEHS conducted an eligibility screening of the information retrieved from the citation databases.[2]

A Federal Register Notice (FRN) published by the U.S. Environmental Protection Agency (EPA) on behalf of the U.S. Global Change Research Program (USGCRP) on February 7, 2014, called for submissions of relevant, peer-reviewed, scientific and/or technical research studies on observed and/or projected climate change impacts on human health in the United States.[3] A second FRN was published on April 7, 2015, announcing a public comment period, in which many commenters suggested additional sources of literature for consideration.[4] Chapter authors were responsible for screening and assessing the eligibility of literature submitted by the public using the same process developed by NIEHS.

In the process of performing the review and evaluating the literature, authors identified additional relevant literature, not captured in the NIEHS literature search or public call for information. Chapter authors screened and assessed the eligibility of these sources using the same process developed by NIEHS.

A2.2 Screening for Eligibility

Throughout the process of drafting this assessment, guidance was provided to authors regarding the requirements of the IQA. In accordance with these requirements, chapter authors considered information quality when deciding whether or not to use source material in their chapter. The literature review guidance provided to authors required consideration of the following criteria for each source of information used in the assessment:

- **Utility:** Is the particular source important to the topic of your chapter?

- **Transparency and traceability:** Is the source material identifiable and publicly available?

- **Objectivity:** Why and how was the source material created? Is it accurate and unbiased?

- **Information integrity and security:** Will the source material remain reasonably protected and intact over time?

The Supporting Evidence sections of each chapter include "Traceable Accounts" for the Key Findings. The Traceable Accounts identify the key studies for explaining a particular issue or answering a particular question, and which form the basis of support for Key Findings. Key studies exhibit the general attributes defined below:

- **Focus:** the work not only addresses the area of inquiry under consideration but also contributes to its understanding;

- **Verify:** the work is credible within the context of the wider body of knowledge/literature or, if not, the new or varying information is documented within the work;

- **Integrity:** the work is structurally sound; in a piece of research, the design or research rationale is logical and appropriate;

- **Rigor:** the work is important, meaningful, and non-trivial relative to the field and exhibits sufficient depth of intellect rather than superficial or simplistic reasoning;

- **Utility:** the work is useful and professionally relevant; it makes a contribution to the field in terms of the practitioners' understanding or decision-making on the topic; and

- **Clarity:** it is written clearly and appropriately for the nature of the study.

Authors were responsible for certifying adherence to IQA requirements by applying the process outlined in the Author's Guidance documents (see Appendix 3: Report Requirements, Development Process, Review, and Approval).

Recommended Citation: USGCRP, 2016: Appendix 2: Process for Literature Review. *The Impacts of Climate Change on Human Health in the United States: A Scientific Assessment.* U.S. Global Change Research Program, Washington, DC, 301–302. http://dx.doi.org/10.7930/J0FQ9TJT

References

1. Melillo, J.M., T.C. Richmond, and G.W. Yohe, Eds., 2014: *Climate Change Impacts in the United States: The Third National Climate Assessment.* U.S. Global Change Research Program, Washington, D.C., 842 pp. http://dx.doi.org/10.7930/J0Z31WJ2

2. USGCRP, 2012: National Climate Assessment Health Sector Literature Review and Bibliography. Technical Input for the Interagency Climate Change and Human Health Group. National Institute of Environmental Health Sciences. http://www.globalchange.gov/what-we-do/assessment/nca-activities/available-technical-inputs

3. 40 CFR Part 82, 2014: Request for Public Engagement in the Interagency Special Report on the Impacts of Climate Change on Human Health in the United States. U.S. Environmental Protection Agency on behalf of the United States Global Change Research Program. http://www.gpo.gov/fdsys/pkg/FR-2014-02-07/pdf/2014-02304.pdf

4. 80 FR 18619, 2015: Notice of Availability of Draft Scientific Assessment for Public Comment. U.S. Environmental Protection Agency on behalf of the U.S. Global Change Research Program. https://www.federalregister.gov/articles/2015/04/07/2015-07629/notice-of-availability-of-draft-scientific-assessment-for-public-comment

APPENDIX

3 REPORT REQUIREMENTS, DEVELOPMENT PROCESS, REVIEW, AND APPROVAL

A3.1 Scoping the Report

In early 2013, the Interagency Crosscutting Group on Climate Change and Human Health (CCHHG), a working group of the U.S. Global Change Research Program (USGCRP), established that developing a climate and health assessment was a priority action and convened a Climate and Health Assessment Steering Committee in June 2013. The Steering Committee determined the scope of the Climate and Health Assessment with input from a scoping workshop, held November 21, 2013. The CCHHG participants in this workshop discussed the focus and breadth of the report outline, roles and responsibilities of authors and contributors, the process and timing for report development, and the goals of leveraging federal expertise and ongoing research/analyses across CCHHG agencies and synthesizing multiple efforts into a single robust product. A draft prospectus outlining the proposed focus areas and scope of the report was developed by the Steering Committee and published in a Federal Register Notice (FRN) on February 7, 2014.1 The prospectus proposed plans for scoping, drafting, reviewing, producing, and disseminating the report.

A3.2 Author Selection

A team of more than 100 experts was involved in writing this report. The selection of authors was limited to Federal employees and their contractors or affiliates. Each chapter had an author team consisting of Lead and Contributing Authors, who were responsible for a chapter or subsection of a chapter based on their expertise. Lead and Contributing Authors came from multiple agencies across the government, including the U.S. Department of Health and Human Services (HHS; National Institutes of Health [NIH], Centers for Disease Control and Prevention [CDC], National Institute for Occupational Safety and Health [NIOSH], the Assistant Secretary for Preparedness and Response [ASPR], U.S. Food and Drug Administration [FDA], and Substance Abuse and Mental Health Services Administration [SAMHSA]), National Oceanic and Atmospheric Administration (NOAA), U.S. Environmental Protection Agency (EPA), U.S. Department of Agriculture (USDA), National Aeronautics and Space Administration (NASA), U.S. Geological Survey (USGS), U.S. Department of Defense (DOD; the Uniformed Services University of the Health

Sciences [USUHS]), and the U.S. Department of Veterans Affairs (VA) (see author lists in the front matter and in each chapter for full affiliations). Lead Authors were nominated and selected by the CCHHG and include CCHHG members, attendees of the first scoping workshop, and other Federal experts and contractors/grantees with relevant expertise. Contributing Authors were nominated by the Lead Authors, CCHHG or other interagency members, and the general public. Public nominations were accepted through the FRN dated February 7, 2014, which provided an opportunity for external (non-Federal) subject matter experts to be hired under a Federal contract as Contributing Authors. These nominees were screened according to criteria established by the Steering Committee and selected through an independent process.

A3.3 Drafting the Report

The report was drafted between spring 2014 and spring 2015. Guidance and resources provided to authors included:

- **Literature Review Guidance.** Guidance was provided to authors on reviewing and assessing the literature, screening for eligibility and information quality, and documenting their process for inclusion in the assessment. Please see Appendix 2: Process for Literature Review for more information on the literature review and selection process.

- **Author Guidance.** Guidance was provided to authors on chapter development, including basic and technical guidance on scope, chapter preparation and outlines, and meeting information quality guidance. Guidelines were also provided for transparent reporting of likelihood, confidence, and uncertainty.

- **Modeling Guidance.** Guidance was provided to the authors for the four chapters within the assessment that highlight recent peer-reviewed modeling and/or quantitative analyses. These analyses were conducted by the chapter authors for the purpose of this assessment, in addition to their assessment of the broader body of literature. Please see Appendix 1: Technical Support Document, for more information on modeling approaches.

- **Style and Language Guidance.** The Steering Committee, in conjunction with USGCRP staff and the NOAA Technical Support Unit (TSU), developed a style guide to ensure consistent style, tone, formatting, use of graphics, and documentation of metadata across the report.

- **Author Resource Portal.** An online platform was developed by the NOAA TSU to provide author teams with a shared online workspace, help structure the drafting and revising process, and document metadata on report figures.

- **Drafting Workshop.** An all-authors workshop was held on September 10–11, 2014, to review guidelines and timelines and to discuss cross-cutting issues among and between author teams.

A3.4 Public Engagement

The Steering Committee provided a number of opportunities for public engagement in scoping, informing, and reviewing the report. On February 7, 2014, EPA released a FRN on behalf of USGCRP announcing a request for public engagement in a Public Forum (held March 13, 2014) and establishing a 30-day period to submit public comments on the draft prospectus, suggestions for scientific information to inform the assessment, and nominations for Contributing Authors. A second FRN, released by EPA on behalf of USGCRP on April 7, 2015,[2] announced a 60-day period to submit public comments on the draft assessment. Responses to each comment are posted on the USGCRP Climate and Health Assessment website (http://www.globalchange.gov/health-assessment). Finally, Steering Committee members and authors further engaged the community of experts and the general public about the report and public comment periods at scientific meetings, conferences, and symposia.

A3.5 Peer-Review and Clearance

The draft assessment was peer-reviewed by a committee convened by the National Academies of Sciences, Engineering, and Medicine. Based on comments from the public and the National Academies' report,[3] the authors extensively reviewed and revised the assessment. The assessment was reviewed and approved by the USGCRP agencies and the Federal Committee on Environment, Natural Resources, and Sustainability (CENRS). This report meets all Federal requirements associated with the Information Quality Act (see Appendix 2: Process for Literature Review), including those pertaining to public comment and transparency.

Recommended Citation: USGCRP, 2016: Appendix 3: Report Requirements, Development Process, Review, and Approval. *The Impacts of Climate Change on Human Health in the United States: A Scientific Assessment.* U.S. Global Change Research Program, Washington, DC, 303–304. http://dx.doi.org/10.7930/J09Z92TG

References:

1. 40 CFR Part 82, 2014: Request for Public Engagement in the Interagency Special Report on the Impacts of Climate Change on Human Health in the United States. U.S. Environmental Protection Agency on behalf of the United States Global Change Research Program. http://www.gpo.gov/fdsys/pkg/FR-2014-02-07/pdf/2014-02304.pdf

2. 80 FR 18619, 2015: Notice of Availability of Draft Scientific Assessment for Public Comment. U.S. Environmental Protection Agency on behalf of the U.S. Global Change Research Program. https://www.federalregister.gov/articles/2015/04/07/2015-07629/notice-of-availability-of-draft-scientific-assessment-for-public-comment

3. National Academies of Sciences Engineering and Medicine, 2015: *Review of the Draft Interagency Report on the Impacts of Climate Change on Human Health in the United States.* National Academies Press, Washington, D.C. http://www.nap.edu/catalog/21787/review-of-the-draft-interagency-report-on-the-impacts-of-climate-change-on-human-health-in-the-united-states

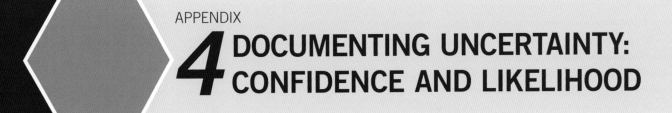

4 DOCUMENTING UNCERTAINTY: CONFIDENCE AND LIKELIHOOD

The authors have assessed a wide range of information in the scientific literature and various technical reports to arrive at their Key Findings. Similar to the 2014 NCA1 and the Intergovernmental Panel on Climate Change's Fifth Assessment Report,2 this assessment relies on two metrics to communicate the degree of certainty in Key Findings:

1. **Confidence** in the validity of a finding based on the type, amount, quality, strength, and consistency of evidence (such as mechanistic understanding, theory, data, models, and expert judgment); the skill, range, and consistency of model projections; and the degree of agreement within the body of literature.

2. **Likelihood,** or probability of an effect or impact occurring, is based on measures of uncertainty expressed probabilistically (in other words, based on statistical analysis of observations or model results or on the authors' expert judgment).

Key sources of information used to develop these characterizations are referenced in the Supporting Evidence section found at the end of each chapter. The Supporting Evidence sections include "Traceable Accounts" for each Key Finding that 1) document the process and rationale the authors used in reaching the conclusions in their Key Finding, 2) provide additional information to readers about the quality of the information used, 3) allow traceability to resources and data, and 4) describe the level of likelihood and confidence in the Key Finding. Thus, the Traceable Accounts represent a synthesis of the chapter author team's judgment of the validity of findings, as determined through evaluation of evidence and agreement in the scientific literature. The Traceable Accounts also identify areas where data are limited or emerging or where scientific uncertainty limits the authors' ability to estimate future climate change impacts. Each Traceable Account includes 1) a description of the evidence base, 2) major uncertainties, and 3) an assessment of confidence based on evidence.

A4.1 Evaluation of Confidence in the Validity of a Finding

Assessments of confidence in the Key Findings are based on the expert judgment of the chapter authors. Authors provide supporting evidence for each of the chapter's Key Findings in the Traceable Accounts. Confidence is expressed qualitatively and ranges from low confidence (inconclusive evidence or disagreement among experts) to very high confidence (strong evidence and high consensus) (see Figure 1). Confidence levels are reported even where confidence is low. Confidence should not be interpreted probabilistically, as it is distinct from statistical likelihood.

Figure 1: Likelihood and Confidence Evaluation

Confidence Level	Likelihood
Very High	**Very Likely**
Strong evidence (established theory, multiple sources, consistent results, well documented and accepted methods, etc.), high consensus	≥ 9 in 10
	Likely
High	≥ 2 in 3
Moderate evidence (several sources, some consistency, methods vary and/or documentation limited, etc.), medium consensus	**As Likely As Not**
	≈ 1 in 2
Medium	**Unlikely**
	≤ 1 in 3
Suggestive evidence (a few sources, limited consistency, models incomplete, methods emerging, etc.), competing schools of thought	**Very Unlikely**
	≤ 1 in 10
Low	
Inconclusive evidence (limited sources, extrapolations, inconsistent findings, poor documentation and/or methods not tested, etc.), disagreement or lack of opinions among experts	

A4.2 Evaluation of Likelihood of Risk

For the purposes of this assessment, likelihood is the chance of occurrence of an effect or impact based on measures of uncertainty expressed probabilistically (in other words, based on statistical analysis of observations or model results or on expert judgment). Authors came to a consensus using expert judgment, based on the synthesis of the literature assessed, to arrive at an estimation of the likelihood that a particular impact will occur within the range of possible outcomes. Where it is considered justified to report the likelihood of particular impacts within the range of possible outcomes, this report takes a plain-language approach to expressing the expert judgment of the chapter team, based on the best available evidence. For example, an outcome termed "likely" has at least a 66% chance of occurring; an outcome termed "very likely," at least a 90% chance (see Figure 1).

A4.3 Uncertainty Language in Key Findings

All Key Findings include a description of confidence. Where it is considered scientifically justified to report the likelihood of particular impacts within the range of possible outcomes, Key Findings also include a likelihood designation. Where possible, levels of confidence and likelihood are provided for different steps along the exposure pathway to enable separate reporting of levels of uncertainty in understanding climate impacts, changes in exposure, the role of moderating or exacerbating factors, and observed or projected health outcomes.

Confidence and likelihood levels are based on the expert assessment and consensus of the chapter author teams. These teams determined the appropriate level of confidence or likelihood by assessing the available literature, determining the quality and quantity of available evidence, and evaluating the level of agreement across different studies. Often, the underlying studies provided their own estimates of uncertainty and confidence intervals. When available, these confidence intervals were assessed by the chapter authors in making their own expert judgments. For specific descriptions of the process by which each chapter author team came to consensus on the Key Findings and the assessment of confidence and likelihood, see the Traceable Accounts in the Supporting Evidence section of each chapter.

Recommended Citation: USGCRP, 2016: Appendix 4: Documenting Uncertainty: Confidence and Likelihood. *The Impacts of Climate Change on Human Health in the United States: A Scientific Assessment.* U.S. Global Change Research Program, Washington, DC, 305–306. http://dx.doi.org/10.7930/J0668B3G

References

1. Melillo, J.M., T.C. Richmond, and G.W. Yohe, Eds., 2014: *Climate Change Impacts in the United States: The Third National Climate Assessment.* U.S. Global Change Research Program, Washington, D.C., 842 pp. http://dx.doi.org/10.7930/J0Z31WJ2

2. IPCC, 2014: Climate Change 2014: Impacts, Adaptation, and Vulnerability. Part A: Global and Sectoral Aspects. Contribution of Working Group II to the Fifth Assessment Report of the Intergovernmental Panel on Climate Change. Field, C.B., V.R. Barros, D.J. Dokken, K.J. Mach, M.D. Mastrandrea, T.E. Bilir, M. Chatterjee, K.L. Ebi, Y.O. Estrada, R.C. Genova, B. Girma, E.S. Kissel, A.N. Levy, S. MacCracken, P.R. Mastrandrea, and L.L. White (Eds.), 1132 pp. Cambridge University Press, Cambridge, UK and New York, NY. http://www.ipcc.ch/report/ar5/wg2/

Glossary

Acclimatization. Physiological and behavioral adjustments to a change of climatic environment.

Acute. Occurring over a short period of time (as opposed to chronic).

Adaptive capacity. The ability of communities, institutions, or people to adjust to potential hazards, to take advantage of opportunities, or to respond to consequences.

Aeroallergens. Various airborne substances, such as pollen or spores, which can cause an allergic response.

Aerosol (atmospheric). Aerosols are fine solid or liquid particles, caused by people or occurring naturally, that are suspended in the atmosphere. Aerosols can cause cooling by scattering incoming radiation or by affecting cloud cover. Aerosols can also cause warming by absorbing radiation. Related terms: *Aerosolize, aerosols.*

Algae. Photosynthetic organisms forming the base of the food chain in freshwater and marine ecosystems. Algae range in in size from single-celled microalgae to large macroalgae, like kelp. Related term: *Harmful algal blooms (HABs).*

Algal bloom. A sudden, rapid growth of algae in lakes and coastal oceans caused by a variety of factors including, for example, warmer surface waters or increased nutrient levels- Some algal blooms may be toxic or harmful to humans and ecosystems.

Allergy/allergic. Reactions of the immune system to substances that, in most people, do not cause symptoms. Allergenicity refers to a substance being able to cause an allergic response.

Anxiety. Feelings of worry, nervousness, distress or a sense of apprehension.

Asthma. A chronic respiratory disease or condition characterized by recurrent breathing problems.

Bacteria. Small single-celled organisms. Though common and vital to ecosystems, particular species or groups may cause illness in humans and other organisms. See *Cyanobacteria.*

Baseline. A starting point or reference used as the basis for comparison.

Carbon dioxide (CO_2). A colorless, odorless, greenhouse gas produced by combustion, respiration, and organic decomposition.

Carbon monoxide (CO). A colorless, odorless, poisonous gas produced by incomplete combustion. Related term: *Carbon monoxide poisoning.*

Cardiovascular. Referring to the heart and blood vessels. Cardiovascular disease (CVD) includes all diseases and conditions of the cardiovascular system.

Chronic. Occurring over a long period of time (as opposed to acute).

Chronic obstructive pulmonary disease (COPD). A group of diseases that cause airflow blockage and breathing-related problems.

Climate. The long-term statistical average of weather. Climate typically refers to the mean and variability of relevant weather variables, such as temperature, precipitation, and wind, over long time scales (30 years or more).

Climate change. Changes in average weather conditions that persist over multiple decades or longer. Climate change encompasses both increases and decreases in temperature, as well as shifts in precipitation, changing risk of certain types of severe weather events, and changes to other features of the climate system.

Climate variability. Natural changes in climate that fall within the observed range of extremes for a particular region, as measured by temperature, precipitation, and frequency of events. Drivers of climate variability include the El Niño Southern Oscillation and other phenomena. Related terms: *Natural variability.*

Cognitive. Referring to intellectual activity like thinking, reasoning, remembering, imagining, or learning.

Cold wave. A period of abnormally cold weather lasting days to weeks.

Contaminant. A contaminant is any physical, chemical, biological, or radiological substance or matter found in any media where it does not belong, particularly at concentrations that may pose a threat to human health or the environment.

Cryptosporidium. A one-celled (protozoan) parasite that infects the intestines of people and animals. Cryptosporidiosis is an infection caused by *Cryptosporidium.*

Cumulative (health effects). The combination of successive or concurrent impacts on health.

Cyanobacteria. A photosynthetic group of bacteria that are functionally similar to algae.

Demographic. Related to the characteristics of a population such as age, gender, ethnicity, and race.

Dengue fever. A viral disease spread by mosquitoes.

Depression. A common, but serious, illness that interferes with daily life and is characterized by a sustained sad mood or inability to experience pleasure.

Diabetes. A group of diseases that affect the ability of the pancreas to produce insulin and thus affect how the body uses blood sugar (glucose).

Disability. A physical or mental condition that limits a person from doing one or more major life activities, including walking, talking, hearing, seeing, breathing, learning, performing manual tasks, and caring for oneself. A functional disability is any long-term limitation in activity resulting from a condition or health problem. Related term: *Functional limitations.*

Downscaling. Methods that use models to estimate future climate at local scales (for example, county, state, region).

Drought. A period of abnormally dry weather marked by little or no rain that lasts long enough to cause water shortage for people and natural systems.

Ecosystem. All the living things in a particular area as well as components of the physical environment with which they interact, such as air, soil, water, and sunlight.

Ecosystem services. The benefits produced by ecosystems on which people depend, including, for example, fisheries, drinking water, fertile soils for growing crops, climate regulation, and aesthetic and cultural value.

Electrolyte imbalance. Minerals (such as sodium, calcium, and potassium) in the body that have an electric charge. Electrolyte imbalance is when levels of these minerals are too high or too low.

Emissions. The release of climate-altering gases and particles into the atmosphere from human and natural sources.

Emissions scenarios. Quantitative illustrations of how the release of different amounts of climate-altering gases and particles into the atmosphere from human and natural sources will produce different future climate conditions. Scenarios are developed using a wide range of assumptions about population growth, economic and technological development, and other factors. Related term: *emissions scenario, emission scenario.* See *Scenario.*

Endemic. The constant or usual presence of a disease or infectious agent within a geographic area or population.

Enteric. Relating to the intestines of humans and animals. See *Gastrointestinal.*

Environmental justice. The fair treatment and meaningful involvement of all people regardless of race, color, national origin, or income with respect to the development, implementation, and enforcement of environmental laws, regulations, and policies.

Epidemiology. The study of the distribution and determinants of health conditions, states, or events in specified populations. Related term: *Epidemiological*

Exposure. Contact between a person and one or more biological, psychosocial, chemical, or physical stressors, including stressors affected by climate change.

Extreme events. A weather event that is rare at a particular place and time of year, including, for example, heat waves, cold waves, heavy rains, periods of drought and flooding, and severe storms. Related terms: *Extreme weather, Extreme weather event.*

Food security. When all people at all times have both physical and economic access to sufficient food to meet their dietary needs for a productive and healthy life.

Foodborne illness. Illness or disease caused by foods or drinks contaminated with biological or chemical toxins or pathogens, including disease-causing microbes or toxic chemicals. Related terms: *Foodborne disease, Foodborne infection.*

Forcing. Factors that affect the Earth's climate. For example, natural factors such as volcanoes and human factors such as the emission of heat-trapping gases and particles through fossil fuel combustion.

Gastrointestinal. Gastrointestinal refers to the stomach and intestinal tract. Gastroenteritis is inflammation of the stomach and intestines. Related term: *Enteric.*

Global Climate Models (GCM). Mathematical models that simulate the physics, chemistry, and biology that influence the climate system. Related term: *General Circulation Model.*

Greenhouse gases. Gases that absorb heat in the atmosphere near the Earth's surface, preventing it from escaping into space. If the atmospheric concentrations of these gases rise, the average temperature of the lower atmosphere will gradually increase, a phenomenon known as the greenhouse effect. Greenhouse gases include, for example, carbon dioxide, water vapor, and methane.

Health. A state of physical, mental and social well-being, and not just the absence of disease.

Heat wave. A period of abnormally hot weather lasting days to weeks.

Heatstroke. A serious health condition that occurs when the body's heat regulating mechanisms—such as sweating and respiration—fail.

Hypertension. Abnormally high arterial blood pressure.

Hyperthermia. Unusually high body temperature.

Hypothermia. Unusually low body temperature that causes a rapid, progressive mental and physical collapse.

Incidence. A measure of the frequency with which an event, such as a new case of illness, occurs in a population over a period of time.

Indicator. An observation or calculation that allows scientists, analysts, decision makers, and others to track environmental trends, understand key factors that influence the environment, and identify effects on ecosystems and society.

Infectious. A characterization of a disease indicating it can be transmitted between organisms.

Infrastructure. The physical structures, services, and institutions (for example, roads, electric utilities, legal systems) needed by a community, organization or country.

Land cover. The physical characteristics of the land surface, such as crops, trees, or concrete. See *Land use.*

Land use. Activities taking place on land, such as growing food, cutting trees, or building cities. Related term: *Land-use patterns.* See *Land cover.*

Lyme disease. A bacterial disease caused by microorganism *Borrelia burgdorferi* and transmitted by *Ixodes* ticks, commonly known as deer ticks.

Mental illnesses. Conditions that affect a person's thinking, feeling, mood, or behavior.

Metabolic rate. The rate at which a person or animal uses calories over time, especially as estimated by food consumption, energy released as heat, or oxygen used in processes of the body.

Meteorological. Referring to the atmosphere and its phenomena, particularly weather and weather forecasting.

Microbial. Referring to microbes, also known as microorganisms, including disease-causing bacteria, viruses or parasites.

Mitigation. Measures to reduce the amount and speed of future climate change by reducing emissions of heat-trapping gases or removing carbon dioxide from the atmosphere. Related terms: *Mitigate.*

Morbidity. A disease or condition that reduces health and quality of life.

Mortality. Death as a health outcome. The mortality rate is the number of deaths in a defined population during a specified time period.

Neurologic/neurological. Referring to the nervous system (including the brain, spinal cord, and nerves), particularly its structure, functions, and diseases.

Nutrients. Chemicals (such as nitrogen and phosphorus) that plants and animals need to live and grow. At high concentrations, particularly in water, nutrients can become pollutants.

Obesity. Having greater body fat relative to lean body mass than what is considered healthy. Related Term: *Obese.*

Ozone (O₃). A colorless gas consisting of three atoms of oxygen, readily reacting with many other substances. Ozone in the upper atmosphere protects the Earth from harmful levels of ultraviolet radiation from the Sun. In the lower atmosphere ozone is an air pollutant with harmful effects on human health.

Parasite. An organism that lives inside or on a host organism, while causing harm to the host organism.

Particulate matter. Tiny airborne pieces of solid or liquid matter such as soot, dust, fumes, mists, aerosols, haze, and smoke.

Pathogen. Microorganisms (such as bacteria or viruses) that cause disease.

Permafrost. Ground that remains at or below freezing for at least two consecutive years.

Populations of concern. Vulnerable groups of people. Related Terms: *Vulnerable populations, Populations at risk.*

Postpartum. The time period after a woman gives birth.

Post-traumatic stress disorder (PTSD). A mental health problem that can occur after war, assault, accident, natural disaster, or other trauma.

Premature (early) death. Death that occurs earlier than a specified age, often the average life expectancy at birth.

Preparedness. Actions taken to build, apply, and sustain the capabilities necessary to prevent, protect against, and ameliorate negative effects.

Prevalence (in health context). A measure of the number or proportion of people with a specific disease or condition at a specific point in time.

Protozoa. A kind of single-celled microorganism that can be free-living or parasitic. See *Parasite.*

Psychiatric. Referring to mental illnesses and treatment. Psychiatric illnesses are mental health conditions affecting a person's thinking, feeling, mood, or behavior. See *Mental illness, Psychological.*

Psychological. Of, affecting, or arising in the mind; refers to the mental and emotional state of a person.

Renal. Renal refers to the kidneys and surrounding region. Related terms: *Kidney disease/disorder, Kidney/renal failure.*

Representative Concentration Pathways (RCPs). Greenhouse gas concentration trajectories from the Intergovernmental Panel on Climate Change's (IPCC's) Fifth Assessment Report (2014) that reflect possible increases in radiative forcing associated with emissions over time. See *Forcing.*

Resilience. A capability to anticipate, prepare for, respond to, and recover from significant multi-hazard threats with minimum damage to social well-being, the economy, and the environment.

Respiratory. Related to the system of organs and tissue the body uses for breathing, including the airways, the lungs and linked blood vessels, and the muscles that enable breathing.

Risk. Risks are threats to life, health and safety, the environment, economic well-being, and other things of value. Risks are often evaluated in terms of how likely they are to occur (probability) and the damages that would result if they did happen (consequences).

Risk assessment. Studies that estimate the likelihood of specific sets of events occurring and their potential positive or negative consequences.

Risk perception. The psychological and emotional factors that affect people's behavior and beliefs about potential negative hazards or consequences.

Salmonellosis. An infection with the *Salmonella enterica* bacteria that causes diarrhea, fever, and abdominal cramps.

Scenario. Sets of assumptions used to help understand potential future conditions such as population growth, land use, and sea level rise. Scenarios are neither predictions nor forecasts. Scenarios are commonly used for planning purposes. See *Emissions scenarios.*

Sensitivity. The degree to which people or communities are affected, either adversely or beneficially, by climate variability and change.

Social determinants of health. The conditions in which people are born, grow, live, work, and age as shaped by the distribution of money, power, and resources.

Socioeconomic. Referring to a combination of social and economic factors, such as the education, income, and work status of individuals or communities.

Special Report on Emissions Scenarios (SRES). A set of emission scenarios from the IPCC Special Report on Emission Scenarios released in 2000 that describe a wide range of potential future socioeconomic conditions and resulting emissions. See *Emissions scenario, Scenario.*

Storm surge. The sea height during storms such as hurricanes that is above the normal level expected at that time and place based on the tides alone.

Stratification. The layering of water by temperature and density that can occur in lakes or other bodies of water, often seasonally.

Stressor. Something that has an effect on people and on natural, managed, and socioeconomic systems. Multiple stressors can have compounded effects, such as when economic or market stress combines with drought to negatively impact farmers.

Surveillance. The collection, analysis, interpretation, and dissemination of health data.

Thermoregulation. The process of maintaining the core internal temperature of the body. Normally, a person's core temperature remains relatively constant at 98.6°F (37°C).

Toxin. Biological, chemical, or physical agents (such as radiation) that can cause harmful effects on people. Related term: *Toxic.*

Trauma. An adverse physical or psychological state caused by physical injury or mental stress. Related terms: *Traumatic injury, Psychological trauma.*

Uncertainty (climate change). An expression of the degree to which future climate is unknown. Uncertainty about the future climate arises from the complexity of the climate system and the ability of models to represent it, as well as the inability to predict the decisions that society will make. There is also uncertainty about how climate change, in combination with other stressors, will affect people and natural systems.

Urban heat island effect. The tendency for higher air temperatures to persist in urban areas as a result of heat absorbed and emitted by buildings and asphalt, tending to make cities warmer than the surrounding countryside.

Vector (disease). An organism, such as an insect or a tick, which transmits disease-causing microorganisms such as viruses, bacteria, or protozoa. Vector-Borne diseases include, for example, malaria, dengue fever, West Nile virus, and Lyme disease. Related terms: *Vector-Borne disease.*

Virus. A microorganism that can cause disease by infecting and then growing and multiplying in cells. Related terms: *Enterovirus, Rotavirus, Norovirus, Hantavirus.*

Vulnerability. The tendency or predisposition to be adversely affected by stressors or impacts, including climate-related health effects.

Waterborne illness. Diseases contracted through contact with water that is infected with pathogens such as *Vibrio cholerae, Campylobacter, Salmonella, Shigella,* and the diarrhea-causing *Escherichia coli.*

Watershed. An area of land that drains water to a particular stream, river, lake, bay, or ocean.

Weather. The day-to-day variations in temperature, precipitation, and other aspects of the atmosphere around us.

West Nile virus. A virus carried by birds and most often transmitted to people by infected mosquitos.

Wildfire. An unplanned fire that occurs in forest, shrubland, or grassland.

Zoonotic disease. A disease that can spread to people from other vertebrate animals. Examples of zoonotic diseases include dengue fever, avian flu, West Nile virus, and bubonic plague. Related term: *Zoonoses.*

Abbreviations and Acronyms

BenMAP – Benefits Mapping and Analysis Program

CCHHG – Interagency Crosscutting Group on Climate Change and Human Health

CDC – Centers for Disease Control and Prevention

CENRS – Committee on Environment, Natural Resources, and Sustainability

CICS-NC – Cooperative Institute for Climate and Satellites–North Carolina

CMIP – Coupled Model Intercomparison Project

CO – carbon monoxide

CO$_2$ – carbon dioxide

COPD – chronic obstructive pulmonary disease

CVD – cardiovascular disease

DoD – U.S. Department of Defense

DOE – U.S. Department of Energy

EPA – U.S. Environmental Protection Agency

FDA – U.S. Food and Drug Administration

GCM – global climate model, also referred to as general circulation model

GHG – greenhouse gas

GIS – geographic information systems

HAB – harmful algal bloom

HHS – U.S. Department of Health and Human Services

ICLUS – Integrated Climate and Land Use Scenarios

ICS – inhaled corticosteroids

IPCC – Intergovernmental Panel on Climate Change

IQA – Information Quality Act

LEP – limited English proficiency

NASA – National Aeronautics and Space Administration

NCA – National Climate Assessment

NCEI – National Centers for Environmental Information, formerly the National Climatic Data Center

NIEHS – National Institute of Environmental Health Sciences

NIH – National Institutes of Health

NIOSH – National Institute for Occupational Safety and Health

NOAA – National Oceanic and Atmospheric Administration

NO$_x$ – nitrogen oxides

NRC – National Research Council

O$_3$ – ozone

PM – particulate matter

PTSD – post-traumatic stress disorder

RCP – Representative Concentration Pathway

SES – socioeconomic status

SO$_2$ – sulfur dioxide

SRES – Special Report on Emissions Scenarios

USDA – U.S. Department of Agriculture

USGCRP – U.S. Global Change Research Program

USGS – U.S. Geological Survey

VA – U.S. Department of Veterans Affairs

WNV – West Nile virus

Recommended Citation: USGCRP, 2016: Appendix 5: Glossary and Acronyms. *The Impacts of Climate Change on Human Health in the United States: A Scientific Assessment.* U.S. Global Change Research Program, Washington, DC, 307–312. http://dx.doi.org/10.7930/J02F7KCR

Climate change is affecting the health of Americans. As
the climate continues to change, the risks to human health
will grow, exacerbating existing health threats and creating
new public health challenges. This assessment significantly
advances what we know about the impacts of climate change
on public health, and the confidence with which we know it.

U.S. Global Change
Research Program